SUBSURFACE HYDROLOGY

SUBSURFACE HYDROLOGY

GEORGE F. PINDER
University of Vermont

MICHAEL A. CELIA
Princeton University

WILEY-INTERSCIENCE

A JOHN WILEY & SONS, INC. PUBLICATION

Published by John Wiley & Sons, Inc., Hoboken, New Jersey.
Published simultaneously in Canada.

For general information on our other products and services or for technical support, please contact our Customer Care Department within the United States at (800) 762-2974, outside the United States at (317) 572-3993 or fax (317) 572-4002.

Wiley also publishes its books in a variety of electronic formats. Some content that appears in print may not be available in electronic formats. For more information about Wiley products, visit our web site at www.wiley.com.

Library of Congress Cataloging-in-Publication Data:

Pinder, George Francis, 1942–
    Subsurface hydrology / George F. Pinder.
        p. cm.
    Includes index.
    ISBN-13: 978-0-471-74243-2 (cloth)
    ISBN-10: 0-471-74243-0 (cloth)
    1. Groundwater. 2. Groundwater—Pollution. I. Celia, Michael Anthony. II. Title.
    GB1003.2.P53 2006
    551.49—dc22

                                                        2006001185

Printed in the United States of America.

10 9 8 7 6 5 4 3 2 1

To Phyllis and Lin

CONTENTS

7 NUMERICAL SOLUTIONS OF THE GROUNDWATER FLOW EQUATION   263

8 CONTAMINATION OF SUBSURFACE WATER   320

PREFACE

Subsurface Hydrology was written as a textbook for use in teaching subsurface hydrology at the advanced undergraduate to graduate level. It has been classroom tested for many years and benefits from comments provided by students over this period.

A review of the table of contents will reveal that the book addresses subsurface hydrology from many perspectives and at different levels of mathematical sophistication. Thus, if used as a textbook, the teacher can select those topics that are of primary interest and those chapters that are presented at a suitable mathematical level for the students.

The book can also be used as a professional reference. The groundwater professional will find most topics of current interest considered in the book, albeit at different levels of detail.

As with any book, there are a number of persons who deserve to be recognized for their contributions. Since neither of the authors of this book are experts in water chemistry, we especially wish to thank Claudia Spita, a chemical engineer currently employed by the SAIC Corporation, for reviewing the section on chemical concepts presented in Chapter 10. We also wish to thank Professor Donna Rizzo of the University of Vermont for reviewing the section on spatial statistics.

GEORGE F. PINDER
MICHAEL A. CELIA

University of Vermont
Princeton University
March 2006

CHAPTER 1

WATER AND THE SUBSURFACE ENVIRONMENT

1.1 GROUNDWATER HYDROLOGY

Groundwater hydrology involves the study of the subsurface and of the overall science of water movement therein. In this book, under the general topic of groundwater hydrology, we focus on water movement as well as the movement of various kinds of pollutants in the subsurface. For some problems we will also consider movement of fluids other than water, for example, air in the shallow subsurface and oil, gasoline, and other "non aqueous liquids," which may serve as sources of water contamination. The study of groundwater hydrology involves concepts from a number of traditional sciences, including geology, soil science, physics, chemistry, and biology. Together these form the foundation of the science of groundwater hydrology.

While we will consider a variety of subsurface fluids, the focus of groundwater hydrology is water, a fascinating and unique substance. Of all the resources that are critical to life, none is more important than water. Humans are composed of about 60% water (by weight). Approximately 5% of that water needs to be replaced daily to support vital bodily functions. Therefore water supply to humans is a critical part of our engineered environment. Plants depend on water for photosynthesis and for delivery of virtually all nutrients into their biomass via water uptake through their root systems. And water plays a central role in heat exchange and regulation of the earth's temperature. Water has guided the historical development of civilization, has provided a backdrop for some of the largest engineering feats, and continues to provide political challenges in many parts of the United States and the world.

Its unique physical and chemical properties allow water to be a key regulator of many natural systems and cycles on earth. For example, its molecular structure makes it an almost universal solvent, able to dissolve a very wide variety of compounds. Water has a very high specific heat, which provides temperature regulation to the entire earth

Subsurface Hydrology   By George F. Pinder and Michael A. Celia
Copyright © 2006 John Wiley & Sons, Inc.

system as well as to individual organisms, including humans, whose composition, as already noted, is dominated by water. Water also has very high latent heats associated with phase change, which greatly affects the energy balance of the earth. For example, approximately 25% of the incoming solar energy is used to evaporate water. Compare this to 0.02% used in all of photosynthesis. Clearly water is a unique and singularly important substance on planet Earth.

The temperatures and pressures on the earth's surface allow water to exist, under natural conditions, in all three of its states: liquid, vapor, and solid. Liquid water is plentiful and supports virtually all life on the planet. At the same time, a significant amount of water is circulating on the planet in the vapor state, and large amounts of water are stored on the earth's surface in the solid state (ice). Earth is the only planet in our solar system that allows water to exist in all three of its states under natural conditions; as such, it is not surprising that life arose on Earth and not on any other planets (at least as far as we can identify and define life). Water cycling on the planet is sufficiently important that the water cycle is considered one of the critical biogeochemical cycles in nature. It is instructive to examine this cycle to see how groundwater fits into the overall water cycle.

1.2 GROUNDWATER AND THE HYDROLOGIC CYCLE

Approximately 1,386,000,000 km^3 of water exists on earth. Approximately 1,338,000,000 km^3, or 96.5%, is found in the oceans. Of the remaining 3.5%, 1.70% is below ground in the form of groundwater (about 45% of this is fresh water, with the remainder being saline), and 1.74% exists in the solid state as ice in glaciers and permanent snow cover. The remaining 0.06% accounts for water in all other forms and containments, including water in lakes, rivers, biomass, and the atmosphere. Clearly oceans dominate the water system in terms of volume stored, and groundwater dominates the volume of liquid fresh water. The large amount of fresh water stored below the ground makes groundwater an important natural resource. Other parts of the hydrologic cycle, including atmospheric water vapor and water flowing in rivers, also play important roles in the water cycle.

Water in these different "compartments" is continually moving within a given compartment (e.g., ocean circulation or groundwater movement) and is also exchanging between compartments (e.g., evaporation from the oceans to the atmosphere). This constant movement and exchange of water is referred to as the *hydrologic cycle* or *water cycle*. We can provide a simplified representation of this cycle diagrammatically by representing the compartments as storage reservoirs, each of which contains a total volume of water (often referred to as a *stock*). Between these reservoirs are ongoing exchanges of water, which are usually referred to as *fluxes* or *flows*. Figure 1.1 provides such a picture of the hydrologic cycle. The numbers presented in the figure were taken from Dingman [1]. Other sources [2] give somewhat different numbers. This is to be expected, given the difficulty in their estimation. Note that in the groundwater compartment, only the freshwater part (10,500,000 km^3) is included in the figure.

Based on the numbers presented in this figure, we can compute *residence times* for the various compartments. The residence time is defined as the ratio of volume to flux rate through the reservoir. This provides a kind of average measure of how long it takes

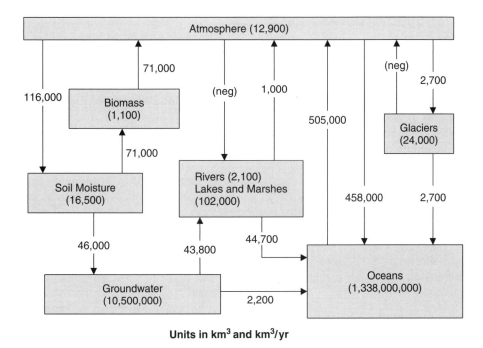

FIGURE 1.1. The hydrologic cycle (from [1]). The numbers identified with the arrows are the flux rates and the quantities in parentheses are the volumes.

to move through that particular reservoir. First, consider the residence time for rivers. If we assume that much of the flow into and out of the box in Figure 1.1 that includes rivers and lakes goes through the rivers, then we find a residence time of $(2100 \ \text{km}^3) \div (44,700 \ \text{km}^3/\text{yr}) = 0.047$ years $= 17$ days. So the average residence time in a river is just about 2 weeks. Conversely, if we consider the groundwater component, we find an average residence time of $(10,500,000 \ \text{km}^3) \div (46,000 \ \text{km}^3/\text{yr}) = 229$ years. Therefore, once water enters the groundwater system (typically through infiltration associated with rainfall events), it remains there for a long time.

These dramatically different residence times have a variety of implications. For example, if we think about contamination and water pollution, while pollutants will travel quickly (and therefore potentially spread over large distances in relatively short times) in river systems, they can also flush out of such systems rapidly. In the case of groundwater, very slow velocities associated with the long residence times mean that contamination takes a long time to move from one place to another. While this means slow spreading of pollutants, it also means that much more time is needed to clean up underground pollution. Also, notice that while surface water and groundwater have very different properties (such as flow rates and residence times), they are linked together in important ways. For example, the slow release of underground water to surface water bodies such as rivers allows rivers to continue to flow even when no rain has fallen for a number of days or even weeks. In rivers, this is referred to as *baseflow*, and it accounts for a significant portion of the flow in rivers. Similar relationships between the different components of the water cycle may be seen in Figure 1.1.

1.3   GROUNDWATER AS A RESOURCE

In 1990 groundwater supplied 51% of the population of the United States. In rural areas, groundwater supplied 95% of the population [3]. The percentage of the population using groundwater tabulated by state is presented in Figure 1.2. How the groundwater is used is shown schematically in Figure 1.3. Worldwide groundwater supplies approximately two-thirds of the world's population.

Water is typically extracted from the subsurface via wells. In its simplest form, a *well* is simply a narrow vertical hole in the ground that penetrates deep enough to

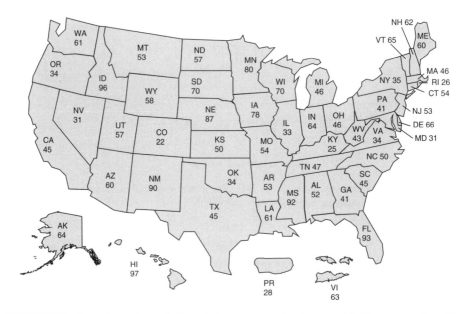

FIGURE 1.2. Percentage of population relying on groundwater as a drinking source (modified from [4].

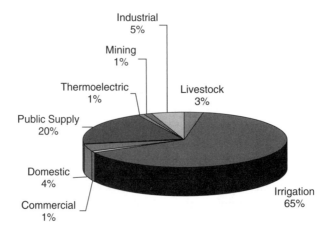

FIGURE 1.3. Categories of groundwater use (from [4]).

reach water that can be extracted in useful quantities. Modern wells are drilled using mechanical drilling devices, usually mounted on drilling rigs. Pumps are usually used to deliver the water from the subsurface to the land surface, and piping installations are used to distribute the water. The use of wells can be traced back in history for more than 3000 years [5, 6], and detailed specifications regarding current drilling and completion techniques may be found in, for example, Driscoll [7].

From a utilization perspective, *irrigation* is by far the largest category. Especially in the arid western United States, irrigation wells provide water for financially attractive crops that would not otherwise grow under the existent climatic conditions. Air photographs of these irrigated areas typically show a circular crop pattern with the irrigation well supplying groundwater to the irrigation distribution system located at the center of the irrigated circle.

Municipal wells may serve a large municipality or a relatively small development. Well water may also be combined with surface water. These large-capacity wells tend to be designed to provide large volumes to water distribution networks. Well discharge rates of hundreds of gallons per minute are typical. *Industrial wells* are similar to municipal wells in terms of their design and their discharge rate. In some instances, water quality demands in manufacturing favor the use of groundwater over the use of alternative surface-water supplies.

Domestic supplies tend to be from wells yielding relatively small discharges, often on the order of tens of gallons per minute. However, such wells are important in providing domestic supplies because there are many homes that are distant from the nearest water distribution network. The cost to provide the piping network required to provide municipal supplies to these homes may be prohibitive, especially in remote areas.

Given the importance of groundwater as an irrigation, industrial, municipal, and domestic supply, the contamination of this supply, either by natural processes or as a result of humankind's activities, is a matter of considerable practical as well as philosophical concern.

1.4 GROUNDWATER AND THE SUBSURFACE

In order to extract subsurface water for human use, water must flow into an extraction well at relatively large flow rates, so that the volume produced satisfies the specific demand. Geological formations that provide water in these quantities are referred to as *aquifers* (from the Latin meaning water bearing). The ability to transmit water in large quantities requires a relatively low resistance to flow, or a high *permeability* to flow. While aquifers are clearly important geological formations, there are other types of geological formations that we also need to analyze. A formation that can transmit water but only in small quantities, and therefore cannot usually be used for water supply, is called an *aquitard*, while a formation that contains water but cannot transmit it is called an *aquiclude*.

Many geological formations occur in a vertically layered structure, with some of the layers being aquifers and others being aquitards or perhaps aquicludes. An aquifer that is bounded above and below by aquicludes is referred to as a *confined aquifer*, while an aquifer bounded by aquitards is referred to as a *leaky aquifer*. If an aquifer is not bounded above by either an aquitard or an aquiclude, but instead corresponds to a formation that has the land surface as its upper boundary, then that aquifer is referred to as an *unconfined aquifer* or a *phreatic aquifer*. Sometimes a confined aquifer contains water at sufficiently high pressure that when a well is drilled into the aquifer, the pressure pushes water all the way to the land surface. In that case the aquifer is referred to as an *artesian aquifer*.

Layered subsurface systems are often characterized by layers that have very large areal coverage, especially when compared to their thickness. For example, Figure 1.4 shows a vertical cross section of materials that form the coastal plane in New Jersey. Notice that the ratio of horizontal to vertical extent for each of these layers is very large. This is not obvious by a simple look at the figure, until we realize that the vertical length scale has been exaggerated by a factor of 160:1. That is, the vertical lengths have been blown up so that we can see the vertical dimensions more easily. If this graph were redrawn so that the horizontal and vertical scales were the same, the layers would be very thin lines on the page.

There are two important messages in this figure. The first is that geological formations tend to be much more extensive in the areal plane than in the vertical direction. The second is that vertical cross sections are almost always drawn with significant vertical exaggeration of the axes. This latter point must be kept in mind when viewing these kinds of figures.

One important consequence of the use of vertical exaggeration in Figure 1.4 is the appearance of a significant slope in the layered structure. The magnitude of this slope is greatly exaggerated by the expanded vertical scale. If this figure were drawn with consistent scales on the two axes, the angle of the slope would be very small indeed. Coupled with our earlier observation that on consistent scales the thickness is very small, we are left with the observation that many of these kinds of layered systems have areally extensive layers that are quite thin, and the layers are very close to the horizontal in their orientation. Later in the book we will refer back to these observations when we consider whether it makes sense to treat these layers as effectively two-dimensional structures, and whether or not we can assume they are essentially horizontal in their orientation.

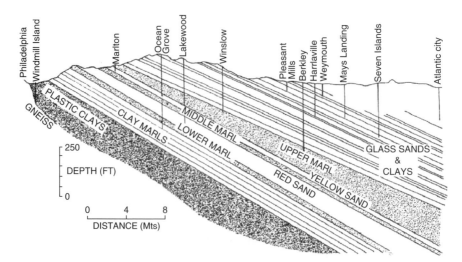

FIGURE 1.4. Hydrogeologic cross section from Philadelphia, Pennsylvania to Atlantic City, New Jersey (from [8]). Vertical exaggeration is approximately 160:1.

1.5   THE NEAR-SURFACE ENVIRONMENT

The *near-surface environment*, as used herein, refers to that portion of the earth's crust that contains, or has the potential to contain, mobile fresh water. In this section, we consider the physical properties of the subsurface environment, including the rock and soil matrix as well as the fluids contained therein. We focus on the near-surface environment because our concern is mainly with freshwater resources, and these occur in shallow regions, with deeper regions containing saline waters and brines.

1.5.1   Soil

Soil is an assemblage of particles containing a mixture of organic and inorganic materials that is capable of sustaining vegetation. It is generally derived from the weathering of rock, although it may also have purely organic origins. Generally, the first stage in soil formation is the accumulation of unconsolidated rock fragments. As we shall see in Section 3.1.1, there are several mechanisms available to transport these rock fragments. Alternatively, soil may form essentially 'in place' through the chemical and physical weathering of rock. Once soil formation begins it forms layers in a process known as *horizon differentiation*. How the various horizons form and their final topology depend on which of several chemical and biological processes dominate. The transformation of organic matter contained in the soil, the deposition or solution of soluble minerals, and the formation of new compounds such as clay minerals all compete to define the final soil horizons and their composition.

In groundwater hydrology, we are interested in not only the soils found at the earth's surface but also those in the near-surface environment. In Figure 1.5 we provide a diagrammatic representation of a soil profile. While soils capable of sustaining crops and other forms of vegetation, such as horizons O, A, B, and C in Figure 1.5, are of great interest from an agronomical point of view, we are interested in them primarily as a reservoir that can store and transmit fluids. As we will see, the groundwater reservoir of interest to us will lie in zones A, B, C, and, in some instances, R.

The soil affects the nature of the vegetation resident in it and the vegetation affects the host soil. *Soil water* provides the resident plant with a supply of both water and key

Soil Profile	
Organic debris lodged on the soil	Organic horizons
The Solum Genetic horizons formed by soil-forming processes	A. Horizons of maximum biological activity and where there is removal of materials dissolved or suspended in soil or both
	B. Horizons accumulation of suspended material, residual concentration of sesquioxides or silicate clays, or soil-forming structure accompanied by alteration of material from its original condition
The weathered mineral material or other underlying material	C. Weathered mineral material, little affected by pedogenic processes
	R. Unweathered rock

FIGURE 1.5. Soil profiles (modified from [9]).

nutrients. On the other hand, the plant acts as a biological pump, drawing water from the soil and, through *evaporation* and *transpiration*, transferring water to the atmosphere. In addition, when plants die they leave channels where their roots were located. These channels can and often do form effective secondary pathways for the movement of subsurface water. However, at this point, we will focus on some of the basic properties of soils.

Pore Structure and Packing   The storage and transmission of fluids, in particular, water and air, are affected primarily, although not exclusively, by the physical properties of the soil pores. One can readily imagine that all continuous void spaces do not have the same properties from the perspective of fluid flow, mass transport, or energy transport. Some pathways may be relatively straight while others may be sinuous. The size and shape of the pore may vary both in terms of its overall geometry and its cross section. In a porous medium composed of relatively spherical grains, the pores will have a geometry quite different from that found in media characterized by "plate-like" particles such as clay. For example, in Figure 1.6 the pore occupying the center of this cubically arranged set of eight spherical grains has approximately the same geometry when viewed from any of the six faces. However, the pore illustrated in Figure 1.7 reveals a different pore geometry depending on whether you view it from the top, bottom, or side.

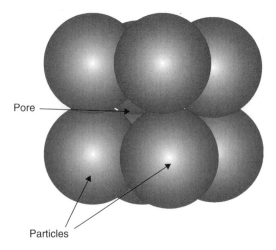

FIGURE 1.6. Cubic packing of spheres generates a porosity of 0.48.

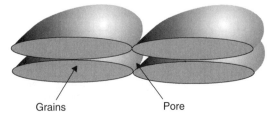

FIGURE 1.7. Two "plate-like" clay particles form a pore with a geometry that mimics the particles.

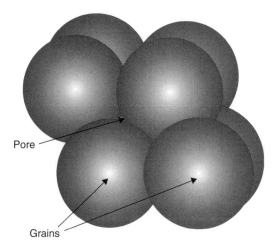

FIGURE 1.8. Rhombohedral packing of spheres of equal size.

The *pore geometry* can also be influenced by the arrangement of the grains. In Figure 1.8 the equally sized spheres are arranged in a *rhombohedral packing*. It differs from the *cubic packing* of Figure 1.6 through a lateral shift of the centers of the spheres occupying the top layer. Through this translation, the amount of void space in the assembled spheres changes from 48% to 26%. Thus the packing of the grains in a porous medium, such as a soil, significantly impacts the percentage of void space in a porous medium.

The percentage of void space is also a function of the variability of the grain sizes present in the sample. Consider, for example, the porous medium sample presented in Figure 1.9. The inclusion of a second grain size reduces the amount of void space in the sample. The small grains tend to occupy part of the void space generated by the packing arrangement of the larger spheres.

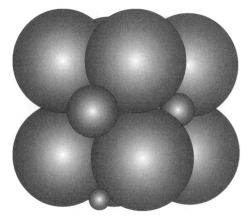

FIGURE 1.9. In this packing arrangement involving two grain sizes, the small grains tend to occupy spaces left between larger grains to yield a smaller percentage of void space in the sample.

We will define the measure of the volume of the pore space as the *volume of voids* and give it the symbol V_v. The ratio of the volume of voids to the total volume V we call the *porosity* and give it the symbol ε, that is,[1]

$$\varepsilon \equiv \frac{V_v}{V}. \tag{1.1}$$

To provide a physical picture of what is meant by porosity, we will conduct a simple experiment. First we pour 500 mL of water into a 1000 mL graduated cylinder (see Figure 1.10). A 500 mL volume of sand of uniform grain size (the soil of choice in our experiment) located in a second container is now poured into the graduated cylinder that contains the water. The difference between the water levels in the 1000 mL graduated cylinder before and after the sand is added to it is used to calculate the volume of sand. It is found to be 400 mL. Since the air plus the solid grains originally occupied 500 mL and the volume of the grains is found to be 400 mL, then the volume of pore space must be 100 mL. This assumes that the packing of the sand in the water-filled cylinder is the same as it was in its original container.

We can now calculate the porosity from our definition, provided in Eq. (1.1), that is,

$$\varepsilon \equiv \frac{V_v}{V} = \frac{100}{500} = 0.20. \tag{1.2}$$

While we can measure the total volume of pore space, *we are unable, in general, to measure the volume of individual pores*. In other words, we do not have a measure of individual pore space size or its geometry. Numerous concepts exist for attempting to quantify pore geometry and shape. Even the more elementary concepts of pore diameter or pore size have no unique definition [10]. Several possible definitions are presented

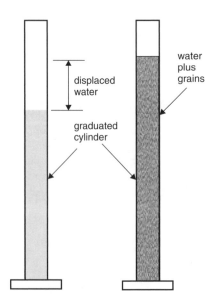

FIGURE 1.10. Experimental apparatus for determining bulk porosity of sand.

[1]The symbol $\equiv$ we use to denote *defined as*.

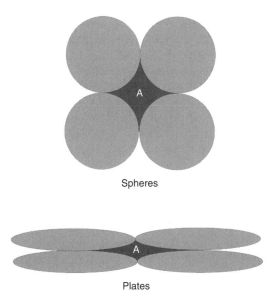

Spheres

Plates

FIGURE 1.11. Cross section of pores associated with cubic packing of spheres and packing plate-like particles. The letter A denotes the pore location.

in Dullien [10], but all have serious limitations. Consider, for example, the difference in pore geometry identified with Figures 1.6 and 1.7, when viewed in cross section as shown in Figure 1.11.

Figures 1.6 and 1.9 show that variability in grain size creates variability in pore size. If one accepts the concept of the pore volume being defined vis-à-vis the *pore throat*, one can use the mercury porosimetry method to obtain pore size distributions. Comparison of distributions created via *mercury porosimetry* with those obtained using photomicrographic methods shows the method does have merit. Evidence indicates that the mercury porosimetry approach does not recognize the large pores and allocates that volume to pore sizes corresponding to the pore-entry necks.

Comparison of distributions based on mercury porosimetry with those based on detailed studies using photomicrographs shows that the porosimetry method can produce reasonable estimates of pore size distributions for the pore throats in the material [10]. Note that the resistance to fluid flow is generally governed by the pore throats, as they are the narrowest parts of the flow paths, and therefore the information derived from mercury porosimetry has practical application to important aspects of flow dynamics.

Because of the difficulty associated with the direct measurement of pore size and geometry, it would be helpful to find a surrogate measure related, albeit indirectly, to pore size. *Grain size* is often used for this purpose. As we have seen, different sizes of uniformly packed grains give the same porosities. However, other properties, like resistance to flow in the pores, do depend on the size of the particles. While factors other than grain size, including grain shape and packing arrangements, affect the resulting pore structure, we still find that grain size can be a reasonable indicator of porous-medium properties. Therefore the *grain size distribution* is used as a basic characterization of soils and other unconsolidated materials. In an effort to quantify grain size distributions, standard engineering tests have been developed and are routinely applied.

Grain Size Distribution   Grain size is easily measured through the use of *sieves*. A standard sieve is a metal cylinder approximately 5 cm in height and approximately 20 cm in diameter. It is open at one end and contains a metal screen at the other. Sieves are normally stacked with the sieve with the smallest screen size opening, or *mesh size*, at the base of the stack. To facilitate stacking, the top of one sieve is designed to receive the bottom edge of the one above. A pan is placed below the last sieve to collect those grains smaller than the smallest grain size (see Figure 1.12).

To sieve a sample of soil, a known weight of the soil is placed in the uppermost sieve of the stack of sieves and the top sieve is covered. A shaking apparatus is used to vibrate the column of sieves while they remain approximately vertical. The grains smaller than the opening in the top sieve eventually pass to the next lower sieve. This sieve, in turn, retains those grains with a diameter larger than its mesh size and smaller than the mesh size of the upper sieve. The filtering process continues in successive sieves until the grains retained in the container at the bottom of the column are smaller than the diameter of the sieve with the smallest mesh. The soil fraction retained on each sieve is removed and weighed and the results are plotted. We examine such plots on page 15.

Sieve sizes are designated in a number of different ways. Some sieves provide the sieve diameter in inches or millimeters. Others designate the sieve by a number that has no obvious relationship to the mesh size. Typical sieve sizes are shown in the following table.

U.S. Standard Test Sieves (ASTM)[2]

Sieve Designation		Nominal Sieve Opening	
Standard	Alternative	Inches	Millimeters
25.0 mm	1 in.	1	25.7
11.2 mm	$\frac{7}{16}$ in.	0.438	11.2
4.75 mm	No. 4	0.187	4.76
1.70 mm	No. 12	0.0661	1.68
0.075 mm	No. 200	0.0029	0.063

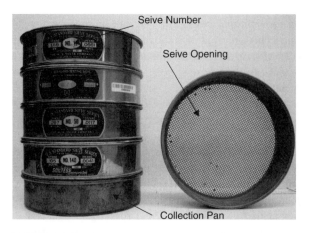

FIGURE 1.12. Stack of sieves used for grain size analysis.

[2]Information taken from [11].

Normally, material smaller than that captured by the No. 200 mesh screen is very difficult to screen further and is therefore analyzed via a "wet" method called the *hydrometer method*. This approach is used to separate silt from clay sized particles (terms that will be defined formally on page 15) and is based on the use of Stokes's law and a knowledge of the density of the water–soil suspension. Stokes's law is needed because it relates the velocity of a spherical particle falling through a fluid to its diameter and specific gravity. Why this is needed becomes evident in the following description of the procedure.

Imagine we have already weighed the portion of a sieved sample that has been collected in the pan underlying the No. 200 mesh screen. We are now confronted with the task of determining the size distribution of this sample.

The first step is to place the smaller-than-200-mesh-screen sample in a graduated cylinder and add water until the resulting suspension is 1000 mL. Next, we add a defloc-culating agent so that particle dissociation is achieved. The resulting suspension is then agitated by covering the open end of the cylinder with one hand and inverting the cylinder several times.

For reasons that will become evident shortly, we next place a hydrometer in the solution (see Figure 1.13) and measure the density of the suspension that is found above a selected, but arbitrary, depth below its surface. We make this measurement at 0.5, 1.0, 2.0, 4.0, 8.0, 15.0, 30.0, 60.0, 120.0, . . . , 5760.0 minutes after the suspension is made[3]. Now we examine how we use this information to achieve our goal of determining the weight of the fractions of the sample composed of the various grain sizes smaller than the No. 200 mesh screen.

FIGURE 1.13. Hydrometer placed in 1000 mL graduated cylinder.

[3]Note that this selection of measurement times is not universally accepted. Different references in the literature recommend different measurement times.

From Stokes's law[4] we can calculate the size (in the sense of diameter) of the grain particle that is passing by our arbitrary plane at each of the above times, that is, 0.5, 1.0, ... minutes. Stokes's law is given by

$$v = \frac{2}{9} \frac{\gamma_s - \gamma_f}{\eta} \left(\frac{D}{2}\right)^2,\qquad(1.3)$$

where v is the velocity of fall of the sphere in the suspension (cm/s), γ_s is the *specific weight* of the sphere (N/cm^3), γ_f is the specific weight of fluid (N/cm^3), η is the *dynamic viscosity* of the fluid $(\text{dyn} \cdot \text{s/cm}^2 \cdot (\text{g/[cm} \cdot \text{s]}))$, and D is the diameter of the sphere (cm). Solving for D we obtain

$$D = \sqrt{\frac{18\eta v}{\gamma_s - \gamma_w'}},\qquad(1.4)$$

where γ_w' is the *specific weight of water*. The reported range of validity of this equation is $0.0002 \text{ mm} \le D \le 0.2 \text{ mm}$ [12]. The use of Eq. (1.4) requires that we define the distance L in Figure 1.14. This is achieved using the formula

$$L = L_1 + \frac{1}{2}\left(L_2 - \frac{V_b}{A}\right),\qquad(1.5)$$

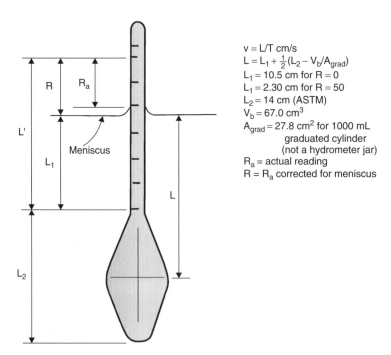

$v = L/T$ cm/s
$L = L_1 + \frac{1}{2}(L_2 - V_b/A_{grad})$
$L_1 = 10.5$ cm for $R = 0$
$L_1 = 2.30$ cm for $R = 50$
$L_2 = 14$ cm (ASTM)
$V_b = 67.0$ cm^3
$A_{grad} = 27.8$ cm^2 for 1000 mL
graduated cylinder
(not a hydrometer jar)
R_a = actual reading
$R = R_a$ corrected for meniscus

FIGURE 1.14. Definition sketch for hydrometer measurements (from [12]).

[4]An important assumption that is made in using Stokes's law is that the grains are spherical. While this may be appropriate for sand-sized particles, clay particles tend to be plate-like and some calibration of the procedure may be necessary.

where V_b is the volume of the hydrometer bulb and A is the cross-sectional area of a graduated cylinder used to contain the suspension. Using the velocity v from Stokes's law computed using the distance L and the time t, that is, $v = L/t$ for any selected t, we can determine the size of the grains passing by the horizontal plane defined at the distance L from the fluid surface at the time t.

Since we know the size of the grains passing the plane of interest at the measurement times, the outstanding question is what weight of particles of a specified size has passed the arbitrary plane at each of these times. The answer lies in the fact that we know, from our earlier measurement, the density of the solution at the elevation of the specified plane at the measurement times. Thus we know the mass of soil particles in suspension on that plane at these times. The remainder of the soil particles must have passed by earlier and therefore must be larger than the size calculated to have been passing the plane at the selected times. If we keep track of the weight of particles in suspension at each time along the plane, we can determine the weight of particles along the plane smaller than the particle size calculated using Stokes's law at each time. From this we can determine the weight of particles in the entire solution that are smaller; from this, we can calculate, using our knowledge of the weight of the original sample, the percentage by weight of the soil smaller than the calculated particle diameter.

Table 1.1 provides the results of such an experiment conducted using a soil sample with a smaller than 200 mesh screen size fraction of 150 g.

It is because the density measurements are usually obtained using a hydrometer that the method is called the hydrometer method. Worked examples of the hydrometer method can be found in both Bowles [12] and McArthur and Roberts [13].

The information gained from a sieve analysis reveals more than just the range of grain sizes. It can also help to classify the soil as to its type, that is, sand, silt, silty sand, and soon. In addition, it reveals the degree of sorting of the soil. Finally, the shape of the resulting soil size distribution curves can also reveal information regarding the history of the soil[5].

In Figure 1.15 the *grain size distribution curves* for two soil samples are plotted. The grain size is plotted along the horizontal axis. On the vertical axis is plotted the percent weight finer than the indicated grain size. For example, the percent by weight of grains smaller than 0.01 mm in the clayey sandy silt sample is approximately 40%. Saying the same thing slightly differently, 40% of the grains, by weight, in the sample are smaller

TABLE 1.1. Experimental Results from a Hydrometer Method Experiment for Determining Fine Grain Size Distributions

Grain Size (mm)	Weight Smaller (g)	Percentage Smaller
0.040	147.0	98
0.010	127.5	85
0.005	91.5	61
0.002	42.2	28
0.001	22.5	15

[5]We will consider this in more detail in Chapter 2.

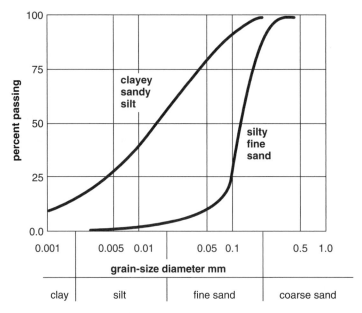

FIGURE 1.15. The grain size distribution indicates the soil classification of a sample and its degree of gradation.

than 0.01 mm. Similarly, in the case of the silty fine sand sample, approximately 25% of the grains are smaller than 0.1 mm.

It is evident that the clayey sandy silt sample is finer grained than the silty fine sand sample. In fact, by referring to the soil classification designations found beneath the distribution curve, it is evident how these samples acquired their classification.

However, there is more information that can be obtained from these curves. It is clear that the slope of the silty fine sand curve is much steeper than is the clayey sandy silt curve. This tells us that the silty fine sand has a more uniform size distribution. In other words, the range of grain sizes in this sample is small relative to the clayey sandy silt sample. The silty-fine sand is considered to be better sorted or more poorly graded than the clayey sandy silt. Figures 1.16 and 1.17 illustrate the concept of gradation in soil.

A measure has been developed to describe the range in grain sizes of a soil sample. It is called the *uniformity coefficient* and is defined as

$$C_u = \frac{D_{60}}{D_{10}}, \tag{1.6}$$

where D_{60} refers to the grain size corresponding to the percent passing of 60%. In other words, 60% of the grains by weight are smaller than D_{60}. The denominator designated as D_{10} is also known as the *effective grain size*. The larger the value of the uniformity coefficient, the better graded (more poorly sorted) is the sample. In our example, the uniformity coefficient of the clayey sandy silt sample is

$$C_u = \frac{0.02}{0.001} = 20, \tag{1.7}$$

FIGURE 1.16. Well-graded soil (from [14]).

FIGURE 1.17. Poorly graded soil (from [14]).

and that of the silty fine sand is

$$C_u = \frac{0.15}{0.05} = 3, \tag{1.8}$$

which confirms our earlier statement that the clayey sandy silt is a better graded soil[6].

Soil Classification In Figure 1.15 we introduced the concept of grain size distribution to classify soils. In addition to this specification format, there are other alternatives. One possibility is shown in Figure 1.18. On the vertices of this triangle are located the points representing 100% clay, 100% silt, and 100% sand. The definition of these terms is

[6]See the *Earth Manual* [15] for more information regarding grain size distributions and soil classification.

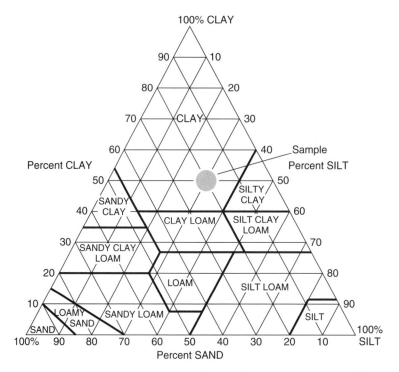

FIGURE 1.18. U.S. Department of Agriculture soil classification chart.

TABLE 1.2. U.S. Department of Agriculture Size Classification

Texture	Grain Size
Clay	Less than 0.002 mm
Silt	0.002 to 0.05 mm
Sand	0.05 to 2.0

shown in Table 1.2. To plot a point, first determine the percentage by weight of each of the soil classes found in Table 1.2. The percentage of clay is plotted by moving along the left-hand side of the triangle, beginning at the corner marked 100% sand. The line parallel to the bottom of the triangle and corresponding to the desired percentage of clay is noted. Now the same procedure is repeated for the silt fraction, this time beginning at the corner identified as 100% clay and moving down the right-hand side of the triangle until the desired fraction is located. Note the line that is parallel to the left-hand side of the triangle and of the correct silt percentage. The intersection of these two lines will define the soil, since the sand fraction is determined by the requirement that the percentages sum to 100. The sample shown in Figure 1.18 represents 30% silt, 20% sand, and 50% clay. The soil sample, according to this classification system, would be designated as a clay.

TABLE 1.3. Soil Types and Particle Sizes

Principal Soil Type	Descriptive Term	Size	Familiar Example
Coarse-grained soils	Cobble	76 mm or larger	Grapefruit or orange
	Coarse gravel	76 to 19 mm	Walnut or grape
	Fine gravel	19 to 5 mm	Pea
	Coarse sand	5 to 2 mm	Rock salt
	Medium sand	2 to 0.4 mm	Window screen opening
Fine-grained soils	Fine sand	0.4 to 0.074 mm	Table salt or sugar
	Silt or clay	Microscopic and submicroscopic	
Organic	Peat or muck		Decaying vegetable matter

Source: Adapted from [14].

The classification system shown in Figure 1.18 is suitable for soils composed of sand, silt, and clay. In order to address both fine-grained and coarse-grained soils, the classification system employed by the U.S. Army Corps of Engineers (USACE) is provided as Table 1.3.

1.6 POROSITY

1.6.1 Primary Porosity

As noted earlier, by definition, porosity is the volume of void space V_v in a given volume V of soil and is designated by ε (see Eq. (1.1)). When the soil is unconsolidated, that is, it has not been subject to compression and deformation and the individual grains retain their identity, the pore space is called *primary porosity*. In essence, it is porosity that is attributable to and can be identified with the original sediment. When sediments are consolidated through burial or the application of tectonic stress, the pores may change their shape, but even after the deformation occurs, the rock exhibits primary porosity. The term *porous medium* is usually identified with material that exhibits primary porosity.

1.6.2 Secondary Porosity

Secondary porosity is porosity attributable to geological processes that occur after the formation of a layer of sediments. *Planes of dislocation*, or *fractures* such as shown in Figures 1.19 and 1.20, are one form of secondary porosity. In unconsolidated soils, such planes are normally attributable to movement of soil blocks along fracture planes. Such planes may be open or filled with coarse-grained secondary deposits that are more permeable than the intervening soil blocks and therefore form preferential pathways to water. Planes of dislocation can also be filled with sediment less permeable than the intervening blocks, such that the planes of dislocation act as barriers.

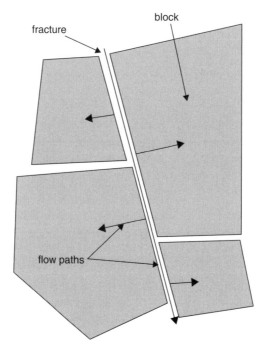

FIGURE 1.19. Diagrammatic representation of water movement from fractures to blocks under conditions where the pressure in the fracture exceeds that in the block.

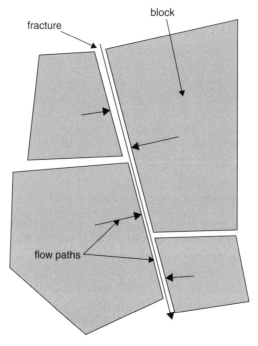

FIGURE 1.20. Diagrammatic representation of a fractured porous medium where water movement is from blocks into fractures due to a lower pressure in the fractures than in the blocks.

Secondary porosity can occur on various scales. At Love Canal, a well-known super-fund site location in New York State, waste was placed in an open pit cut into clay that was characterized by near-vertical fractures. The planes of dislocation in the clay tended to form a connected polygonal pattern and were found to occur at intervals on the order of a meter. In this case the fractures were believed to have been caused by drying of the clay surface due to exposure to the atmosphere subsequent to deposition. These fractures were the primary conduits for the contaminants that migrated from the canal into the surrounding subsurface environment.

In contrast, in the San Fernando Valley in California, the planes of dislocation appear to be filled with lower permeability material and their surface traces are on the order of a kilometer in length and separated one from another by hundreds of feet. Because of the low-permeability nature of these fracture planes, they act as impediments to groundwater flow such that the water level elevation on the upstream side of the fracture planes can be considerably higher than on the downstream side.

Heretofore we focused on granular aquifer material. However, there are many geological materials that are not porous media in the classical sense. For example, the rock called granite (which is often used for tombstones and countertops because of its hardness) is essentially a solid mass of crystals. It is observed, however, that planes of dislocation, or fractures, are found periodically in granite when large sampling areas are examined[7]. Thus granite is not a solid mass as it initially appears. Rather, granite rock and many other rocks as well are composed of solid massive blocks separated by fracture plains.

Because large segments of North America, and indeed the world, are underlain by massive rocks rather than granular porous media, it is important to consider the attributes of this kind of groundwater reservoir.

As we have seen, fractured rocks are characterized by having blocks of soil or rock separated from one another by fractures. For crystalline rocks the blocks are generally of very low porosity since they tend to contain relatively little void space. Moreover, in many instances, the void space that is available tends to be disconnected and to occur as isolated voids that are relatively unimportant from a flow and transport perspective. On the other hand, blocks of clay separated by fractures may have relatively high porosity, as is characteristic of clay soils in general. In this instance, it is the fact that the blocks are of low permeability, and therefore do not readily transmit fluids at a significant rate, that is of importance.

While, as noted in the Love Canal example, the blocks of clay do not transmit water easily, they nevertheless play an important role in the fractured groundwater reservoir. In essence, the clay blocks and, to a more limited degree, rock blocks act as water reservoirs. In other words, they hold water in storage. When the pressure in the fractures is high relative to the blocks, water moves into the block and remains resident there until the pressure in the fractures decreases below that in the blocks (see Figure 1.19). At this point, the water begins to move out of the block and the block water pressure begins to decrease (see Figure 1.20).

Blocks can also play an important role in contaminant hydrology. When contaminants move along fractures, the contaminated water can move into blocks and reside there until the pressure in the fractures decreases. At that point the contaminants begin to drain, but the drainage rate may be very slow. As a result, the blocks may act as sources

---

[7]When we speak of large with respect to the occurrence of fractures in massive rock, we are thinking in terms of tens of feet between significant fracture planes.

FIGURE 1.21. Cementation of fractures reveals their geometry in this hydrothermally altered rock.

of contamination long after the contaminants in the fractures would otherwise have been removed.

An example of a fractured porous medium is found in Figure 1.21. The sample is approximately 13 cm long, so these are fractures on a relatively small scale.

In contrast, we see in Figure 1.22 the Champlain thrust fault. While there is some uncertainty as to the length of this fault, it is quite possible that it extends for as much as 199 miles from Rosenberg, Canada to the Catskill Plateau in east-central New York State. At the location of this photograph at Lone Rock Point near Burlington, Vermont, the fault cuts through 2275 feet of rock. The actual rock displacement along the fault has been estimated to be as high as 62 miles [16].

It is evident that a plane of dislocation of this size, along with its associated zone of crushed and deformed rock, will have a significant impact on groundwater flow. Major faults can act as both conduits and barriers, depending on the composition of the crush

FIGURE 1.22. Champlain thrust as viewed at Lone Rock Point in Vermont.

rock in the neighborhood of the plane of dislocation. Fine fracture filling tends to impede flow as noted earlier in the case of the San Fernando Valley.

It is interesting that the Champlain fault is a *thrust fault* in which the rock on the top of the fault (Lower Cambrian[8] dolostone) is older than the rock beneath (Middle Ordovician shale).

Diffusion (a phenomenon we discuss in greater detail in Section 2.6.1), normally secondary in importance when compared to convection and dispersion, may be a significant player in the case of fractured porous media. The contaminants in the fractures may move into the block perimeter, whereupon diffusion may carry the contaminants deep inside the blocks. The process, of course, will also work in reverse when the fractures are free of contamination. When the concentration in the fractures, and thus in the perimeter of the blocks, is decreased, diffusion will move the contaminants out of the blocks. However, since the process of diffusion is slow, long periods of time may be required to reduce the level of contamination in the blocks to a point where the blocks are not a significant contaminant source. Moreover, one also observes that the time required to move contaminants into a low-permeability formation is smaller than the time required to move the same mass of contaminant back out.

The fractures tend to be the contaminant highways. Although the contribution of the fracture openings to the porosity of the overall fractured rock may be very small, perhaps on the order of a few percent, their permeability is often very high. Thus groundwater and dissolved substances, should they exist, tend to move rapidly along fractures. Contaminant transport along fractures may extend over distances measured in kilometers. The positive side of this rapid transport of contaminants is that wells placed judiciously can reverse the direction of flow and retrieve the contaminants quickly. Delayed drainage of contaminants from blocks, however, may extend the cleanup period (as described above).

While it is somewhat unusual to think about fractured rock unless you are a geotechnical engineer perhaps building a dam, one often does encounter examples of the closely related *cavernous rock* situation, which constitutes another form of secondary porosity. Caves are the visible representation of cavernous rock. They occur because groundwater moving along permeable fractures in soluble rock, such as limestone, over long periods of time, can dissolve away the rock adjacent to the fractures to form caverns. For various geological and climatological reasons, the groundwater subsequently drained from these caverns and they are now accessible to humans. There are, however, many more caverns that are not visible because they are located at greater depths below the land surface and are often flooded with groundwater.

Cavernous rocks are similar to fractured rocks inasmuch as they tend to be the groundwater superhighways. However, unlike fractures that tend to move small amounts of water quickly, caverns, because of their size, can conduct enormous quantities of groundwater rapidly over long distances.

Void Ratio   In the engineering discipline of soil mechanics, the porosity is expressed in the context of the *void ratio*. The void ratio, e, is defined as

$$e = \frac{V_v}{V_s}, \tag{1.9}$$

---

[8]See Section 3.5 for a discussion of geologic time and the nomenclature associated with it.

where V_v is the volume of void space and V_s is the volume of solid grains. The reason this measure has been adopted in soil mechanics is that the denominator V in the definition of porosity (see Eq. (1.1)) changes as the soil compacts or consolidates. Because it is advantageous in some engineering calculations to have the denominator remain constant during consolidation, in such instances the void ratio is a more convenient parameter than the porosity. The void ratio is related to the porosity as follows:

$$\varepsilon = \frac{V_v}{V} = \frac{V_v/V_s}{V/V_s} = \frac{e}{1+e}.$$

Multiple Fluid Phases   A porous medium often consists of soil grains and multiple fluids. The coexistence of two fluid phases, for example, water and air, defines a multi-phase- fluid system, which in groundwater hydrology is usually referred to simply as a "multiphase" system, where the "multi" is interpreted to mean "multifluid." One phase is separated within individual pores from a second phase by a sharp interface. The most commonly encountered multiphase system is that associated with the vadose zone, sometimes referred to as the unsaturated or partially saturated zone. The vadose zone exists immediately beneath the soil surface and extends down to the point where all soil pores contain only water. The amount of water present, relative to air, is described by either the *water content* or the *degree of saturation*. The *gravimetric water content*, $\theta_g(\%)$, is defined as

$$\theta_g(\%) = \frac{W_w}{W_s} \times 100,$$

where W_w is the weight of the water and W_s is the weight of the soil grains, taken as the weight of the oven-dry soil matter. The *volumetric water content*, θ_v, on the other hand, is defined as

$$\theta_v = \frac{V_w}{V}.$$

The *degree of saturation* of water in the soil, $S_w(\%)$, is defined as

$$S_w(\%) = \frac{V_w}{V_v} \times 100,$$

where V_w is the volume of water in the soil. The saturation of air, $S_a(\%)$, is given by

$$S_a(\%) = \frac{V_a}{V_v} \times 100,$$

and the total saturation $S_w + S_a = 1$.

The degree of saturation, which we will define as *saturation*[9], S_w, when it is taken as a ratio rather than a percentage, is related to the volumetric water content according to the relationship

$$\theta_v = \frac{V_w}{V_v} \times \frac{V_v}{V} = S_w \times \varepsilon.$$

[9]The saturation when taken as a fraction is also known as the *saturation ratio*.

Connectivity and Tortuosity To this point, porosity has been viewed primarily as the ratio of pore volume to total soil volume, irrespective of the pore geometry. Pore geometry plays an important role in subsurface flow and transport. The ability of a fluid to flow in soil depends not only on the existence of pores but also on their connectivity. Consider, for example, the volcanic rock in Figure 1.23. The rock, according to our definition of porosity, is very porous. Many voids exist in the rock because of gas that was trapped during the formation of the lava. However, for the most part, the pores

Isolated pore

FIGURE 1.23. Volcanic rock specimen showing porosity by very little connectivity.

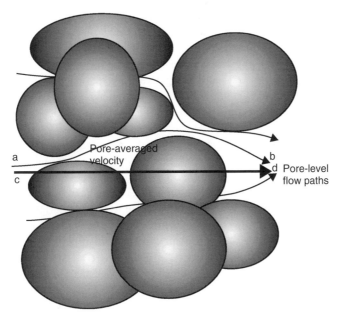

Pore-averaged velocity

a

b

c

d Pore-level flow paths

FIGURE 1.24. Diagrammatic representation of the movement of particles through a porous medium.

are disconnected. From the point of view of fluid flow, the disconnected cavities in this sample are ineffective.

To be of interest from a fluid flow and transport perspective, the void space in a porous medium must be connected. It must form channels through which the water will flow. Dead-end pores are of interest, but only to the degree that they are hydrodynamically connected to other connected pores through which fluid may flow.

Tortuosity addresses the distance that a particle must travel in a porous medium between two points relative to the straight-line distance between the two points. Consider the paths illustrated in Figure 1.24. The particle traveling from position a to position b takes a tortuous path of distance $a - b$ through the pores. The distance traveled by a particle moving at the mean flow velocity obtained by averaging over the local sample of porous medium is given by the distance between c and d, that is, distance $c - d$. Assuming that the two particles started at a common location $a = c$ and ended at a common location $b = d$, the measure of the tortuosity of the particle would be given as length $a - b$ divided by length $c - d$. The phenomenon of tortuosity is an important one in terms of understanding the physics of flow through porous media, especially as it pertains to solute transport.

1.7   SOIL WATER

Water that exists in soil can be defined as *mobile, adsorbed, capillary,* and *pendular*. Mobile water moves freely between the soil grains primarily under the influence of hydrodynamic rather than chemical forces. In general, this form of water is of primary interest to the groundwater professional. Adsorbed water, in contrast, exhibits behavior that is governed largely by forces of attraction associated with the bipolar structure of the water molecules and the solid mineral surface. A few tens of molecules adjacent to the grain surface are influenced by these forces, which are very strong but decrease dramatically with distance from the soil surface. Water, under these conditions, is characterized by physical properties that differ very much from mobile water.

Capillary water exhibits physical properties dictated by the fact that it exists in the *capillary fringe*, wherein water is under negative pressure, or suction. As discussed in Section 2.4, capillary water occurs because of the phenomenon of capillarity, which causes water to rise above the zero-pressure surface a distance dependent on, in part, the grain size of the soil.

Pendular water is essentially immobile. It is found as residual water around grain-to-grain contact points and is largely disconnected in a hydrodynamic sense from other water. It is essentially isolated and is continuous only in the sense that it may connect to more mobile water via a very thin film around each grain. In this state water may not have the capacity to transmit pressure from one point to another and therefore is referred to as *hydrodynamically disconnected*. Water that occurs in *dead-end pores*, that is, those pores that have but one opening to the flowing fluid and are effectively hydrodynamically isolated, is also considered to be immobile.

The soil profile in Figure 1.25 illustrates the concept of the *vadose zone*. On the left side is the soil profile; on the right side is a plot of the degree of saturation as a function of depth. Below the water table the soil is 100% saturated with water as is indicated by the fact that the saturation curve shows a value of 100 at this elevation. Above the water table, the saturation continues to remain at nearly 100%. However, above the water table

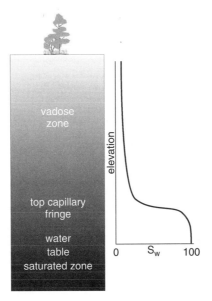

FIGURE 1.25. Diagrammatic representation of water in the shallow subsurface. It is assumed that the system is in dynamic equilibrium—that is, in some sense, a steady state.

the water is held under tension, as will be discussed in greater detail in Section 2.4. At the elevation of the top of the capillary fringe, the saturation begins to decrease significantly and at higher elevations continues to do so. The soil at elevations above the water table are denoted as the *vadose zone,* and the area below the water table is denoted as the *saturated zone*.

When the water table fluctuates due to either precipitation, drainage to surface water bodies, or anthropogenic activities, water either enters into or is released from storage in the soil pores. Lowering the water table results in water draining from the soil that was originally saturated. Similarly, when the water table rises, soil that once contained air becomes water saturated. One can think of this phenomenon as replenishing or depleting the groundwater reservoir.

The term applied to the ability of a groundwater reservoir to store or release water in this way is specific yield. Formally, the *specific yield,* θ_y, is defined as the volume of water drained from a soil column of unit horizontal cross-sectional area per unit decline in the elevation of the water table. The volume of water that remains in the volume defined by the original and terminal locations of the water table is called the *specific retention* and denoted as θ_r.

1.8 GROUNDWATER CONTAMINATION

Compounds found either dissolved in groundwater or as a separate liquid phase often endanger this resource for domestic, municipal, industrial, or agricultural use. When such endangerment occurs, the compounds are denoted as contaminants. Figure 1.26 presents the results of U.S. Environmental Protection Agency (EPA) survey wherein 37 states provided information related to contaminant sources. The information is presented

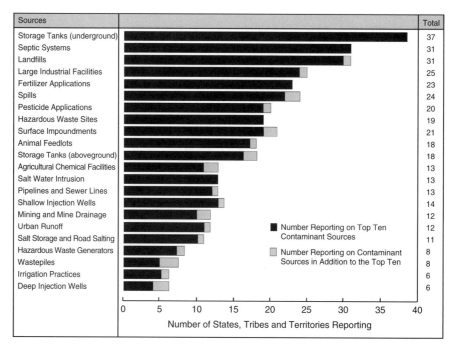

FIGURE 1.26. Major sources of contamination as reported by political divisions (from [3]).

by source rather than contaminant. *Storage tanks, septic systems*, and *landfills* are of major concern. *Large industrial facilities* are significant. *Fertilizer applications* and *spills* round out the top six contamination sources according to this survey. In our ensuing discussion we will subdivide contaminants, somewhat arbitrarily, into *naturally occurring contaminants* and *anthropogenic* or *human-derived contaminants*.

1.8.1   Naturally Occurring Groundwater Contaminants

While there are many naturally occurring groundwater contaminants, the three that are the most notorious are salt, arsenic, and radon. These naturally occurring compounds are widely encountered and in some instances are of considerable economic, environmental, and public health concern. We will deal briefly with each of these.

Salt Water   The intrusion of salt water into otherwise potable-water wells is of enormous importance, especially in coastal areas. Under normal conditions groundwater derived from precipitation moves, at great depth, toward the sea. Figure 1.27 illustrates the very interesting hydrodynamic tug-of-war that goes on in a coastal aquifer. Seawater, having a density of approximately 1.025 g/cm^3, versus that of water at 1.000 g/cm^3 is driven by gravity landward in an attempt to totally occupy the aquifer in contact with the sea bottom. However, precipitation entering the aquifer inland raises the water table and sets up an energy gradient that drives the fresh water seaward. As the fresh water moves toward its ultimate discharge location along the sea coast, it encounters the inward migrating, heavier, seawater. The salt water is being pushed landward and the fresh water seaward. The result is a zone where the two come into contact and mixing occurs. The

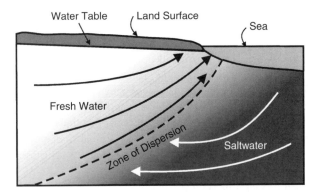

FIGURE 1.27. Groundwater hydrodynamics in a coastal aquifer (adapted from [17]).

mixing zone is indicated in Figure 1.27 as the *zone of dispersion*. In this mixing zone, salt water enters the fresh water and then is carried seaward to be discharged into the ocean bottom, often as very large springs. The salt water moves in to replace that being carried seaward by the fresh water and a dynamic equilibrium results. As long as there are no major changes in the hydrodynamics of this system, there will be a salt-water interface established near the coast. Inward of this interface, the water is relatively fresh and potable; seaward, it is salty and of little use for any practical purpose.

When humans attempt to utilize the groundwater residing in the coastal aquifer, the dynamic equilibrium along the salt-water interface is disturbed. Fluid dynamic forces associated with groundwater withdrawal cause the salt-water interface to rise. An idealized representation of this concept is provided in Figure 1.28. In this conceptual model of the system, the salt-water interface is nearly horizontal, as is often the case. The well

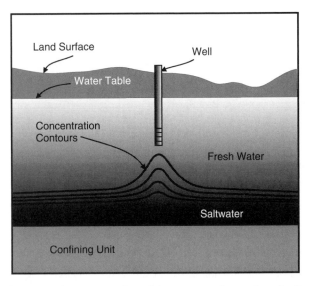

FIGURE 1.28. Diagrammatic representation of the concept of upconing of salt water in response to groundwater withdrawal (adapted from [18]).

is withdrawing fresh water but, in the process, is inducing the salt water to rise and enter the well. The rising salt concentration contours foretell an upcoming disaster as the salt water is drawn ever closer to the well intakes. Once the well is contaminated, it is likely to be of little use as a water supply. We call this contamination process salt-water intrusion. If the well is withdrawn from use, the salt-water interface will eventually return to its original location, but this may take a very long time. While in this example the interface is moving vertically upward, it may also move laterally.

In Figure 1.29 we observe the impact of secondary permeability on the movement of saline water into the production well. The near-vertical fault provides a preferential conduit for the salt water from the lower saline formations to those containing fresh water used for a water supply.

Salt-water intrusion is an important problem in coastal areas where large quantities of fresh water are needed for agriculture. Where salt-water intrusion has been observed or is anticipated, various injection well strategies, in which fresh (salt-free) water is introduced inland from the salt-water interface, are used to keep the salt-water interface seaward of water-extraction wells.

A related challenge involves modification of natural groundwater discharge to coastal ecosystems. Modification of nutrient budgets and salinity may adversely affect the biology of coastal waterways and wetlands.

An illustration of the concern exhibited by coastal communities regarding salt-water intrusion is provided by the following statement found on an EPA *non-point-source* pollution website *NPS* [21]:

> Saltwater intrusion is a major problem for the Southwest Florida Water Management District, where it has affected numerous wells along the southwestern coast of Florida. Projections

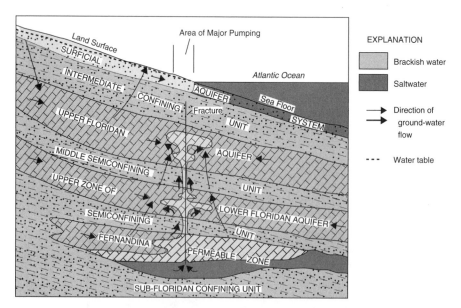

FIGURE 1.29. Idealized cross section of the coastal aquifer of Florida showing the impact salt water may have on production wells. Note that the near-vertical fracture provides a preferred pathway from the deeper formations to the production well (from [19], original reference [20]).

by the District forecast continuing saltwater movement inland at a rate of several inches per day. At this rate, hundreds of wells in coastal Hillsborough, Manatee, and Sarasota Counties are at risk...

Arsenic   Naturally occurring *arsenic* in groundwater supplies is an important public health issue. A naturally occurring element in rocks and the water in contact with them, it has recently received increased scrutiny. While recognized as a toxic element for centuries, it is now recognized also as a *carcinogen*; that is, it is a contributing factor in the initiation of cancer.

Drinking water standards have been developed that appear to take a relatively conservative position. Several regulatory guidelines exist. The 1974 federal Safe Drinking Water Act requires that EPA establish safe levels of hazardous chemicals in drinking water. The criteria used in establishing these levels are based solely on health risks and exposure and are called the maximum contaminant level goals (MCLGs). The MCLG for chromium has been set at 0.1 ppm, which is also the MCLG for Cr(VI).

EPA established, based on the MCLG, an enforceable standard called the *maximum contaminant level* (MCL). The MCL takes into account not only potential health problems but also the ability of the public water systems to detect and remove contaminants using currently available technology. The MCL has been established at 0.1 ppm because this level is believed to be the lowest level to which water systems can reasonably be expected to remove the contaminant. The stated standards and associated regulations are called the *National Primary Drinking Water Regulations,* and all public water supplies are required to abide by them [22].

The National Research Council has recently recommended that the maximum contaminant level (MCL) be reduced from the current 50 µg/L due to a concern that this level may not protect against bladder and other cancers [23].

The U.S. Geological Survey collected and analyzed for arsenic water quality samples drawn from irrigation, industrial, water supply, and research wells. Over the past twenty years, 18,850 wells located across the country were sampled. That investigation revealed that large areas, especially in the Southwest, Midwest, and Northeast of the country, are characterized by arsenic groundwater concentrations at or above 10 µg/L, the concentration level that is currently the World Heath Organization's provisional guideline for maximum allowable concentration and considered as a possible new MCL in the United States. While currently just over 1% of the water supply systems have arsenic exceeding 50 µg/L, considerably more could exceed the new MCL once announced. In such instances, the provider of the water will either need to treat it or obtain an alternative source that complies with the new MCL.

Radon   All rocks contain uranium, usually between one and three parts per million (ppm). Soil associated with a rock will have a uranium content similar to the rock from which it was derived. The formation of radon from uranium is described as follows [24]:

Just as uranium is present in all rocks and soils, so are radon and radium because they are daughter products formed by the radioactive decay of uranium.

Each atom of radium decays by ejecting from its nucleus an alpha particle composed of two neutrons and two protons. As the alpha particle is ejected, the newly formed radon atom recoils in the opposite direction, just as a high-powered rifle recoils when a bullet is fired. Alpha recoil is the most important factor affecting the release of radon from mineral grains.

The location of the radium atom in the mineral grain (how close it is to the surface of the grain) and the direction of the recoil of the radon atom (whether it is toward the surface or the interior of the grain) determine whether or not the newly formed radon atom enters the pore space between mineral grains. If a radium atom is deep within a big grain, then regardless of the direction of recoil, it will not free the radon from the grain, and the radon atom will remain embedded in the mineral. Even when a radium atom is near the surface of a grain, the recoil will send the radon atom deeper into the mineral if the direction of recoil is toward the grain's core. However, the recoil of some radon atoms near the surface of a grain is directed toward the grain's surface. When this happens, the newly formed radon leaves the mineral and enters the pore space between the grains or the fractures in the rocks.

The recoil of the radon atom is quite strong. Often newly formed radon atoms enter the pore space, cross all the way through the pore space, and become embedded in nearby mineral grains. If water is present in the pore space, however, the moving radon atom slows very quickly and is more likely to stay in the pore space.

Because radon moves more slowly in water than in air and decays rapidly, most of it decays within an inch of transport in an aquifer system. However, that which remains in solution in water can enter homes through their groundwater supply systems. Especially in closed water supply systems with short transit times, radon-containing water can enter a home before it has an opportunity to decay. Once in the home, the radon "outgasses" or escapes from the water through showers and other household activities involving water [25].

1.8.2 Anthropogenic Contaminants

Anthropogenic contaminants are those that can be attributed directly or indirectly to humans and their activities. Figure 1.30 is a schematic that illustrates many of the sources of contamination presented in Figure 1.26. Also shown are the various water supply facilities that draw from the potentially contaminated subsurface reservoir.

Chromium   Chromium is a naturally occurring compound that can occur in nine different oxidation states but is normally found as *trivalent chromium* Cr(III) and *hexavalent chromium* Cr(VI). The metallic form of chromium, Cr(0), is a steel-gray solid with a high melting point. It does not occur naturally and is produced from chrome ore [26]. Cr(III) occurs naturally in rocks, soil, plants, and animals and in emanations from active volcanoes. A divalent form of chromium, Cr(II), forms compounds that are often very soluble but rapidly oxidize to Cr(III). Under certain circumstances, Cr(III) can oxidize to Cr(VI), an observation that is important as we will see momentarily.

Chromium is widely used in numerous industrial processes such as the manufacture of dyes and pigments, the manufacture of metals and alloys, as a corrosion inhibitor, and in leather tanning and wood preserving. It enters the environment by both natural processes and human activities. While there are many mechanisms for discharge to the air and into waterways, increased chromium concentrations in soil are primarily the result of disposal of commercial products containing chromium, chromium waste from industry, and coal ash from electric utilities [26]. Cr(VI) is generated as a by-product of heating Cr(III) in the presence of mineral bases and oxygen. According to ASTDR [26], "the concentration of chromium in air (both Cr(III) and Cr(VI)) generally ranges between[10] 0.01 and

---

[10]One microgram (μg) equals 1/1,000,000 of a gram.

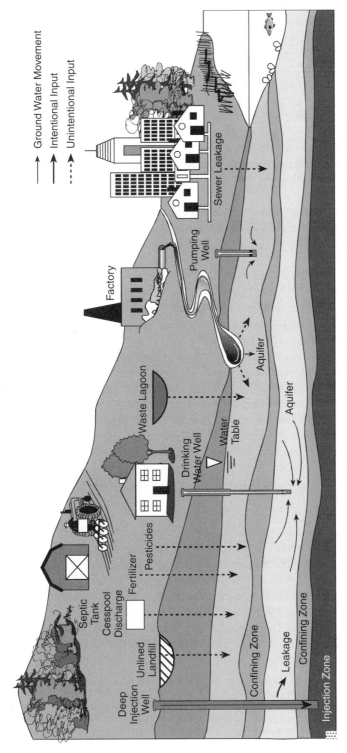

FIGURE 1.30. Sources of human-generated pollution impacting groundwater supplies (from [25]).

0.03 µg per cubic meter of air ($\mu g/m^3$). Chromium concentrations in drinking water (mostly as Cr(III)) are generally very low, less than 2 parts of chromium in a billion parts of water (2ppb)[11]. Contaminated well water may contain Cr(VI)." As a point of reference, the background concentration in soils in a comprehensive survey conducted in Canada showed an average total chromium concentration of 10–100 µg/g dry weight [27][12]. The maximum concentration observed was at a wood-preserving plant where a concentration of 5280 µg/g dry soil was found.

In the environment, one normally finds Cr(III) in relatively inert solid phases. It is readily adsorbed onto solid surfaces. Consequently, Cr(III) tends to accumulate and remain in soils for long periods of time, but its ability for plant uptake may be limited [27]. Cr(VI), on the other hand, is quite soluble and not readily adsorbed onto particulate matter, although adsorption to certain forms of clay has been reported. It can persist in bioavailable form in soil water but may be reduced to Cr(III) under anaerobic conditions [27].

Cr(III) is the form of chromium found almost exclusively in unpolluted soils. It is released through the natural weathering process of rocks and soils. If Cr(VI) is found in concentrations in excess of a few ppb, a man-made source of contamination should be considered [27].

All forms of chromium are not equally important from a public health viewpoint. Cr(III) is an essential ingredient for normal body function. It is required to allow for the effective use of insulin in the body's utilization of sugar, protein, and fat. In fact, Cr(III) compounds have been taken as dietary supplements and are considered beneficial when taken at recommended doses [26].

Cr(VI), however, is a potentially nocuous contaminant. The evidence is strong that Cr(VI) in the form of calcium chromate, chromium trioxide, lead chromate, strontium chromate, and zinc chromate is a human carcinogen when inhaled. The evidence is both epidemiological and derived from rat experiments. However, according to the U.S. Department of Health and Human Services, some forms of Cr(VI) have been found to be noncarcinogenic; "they are the monochromates and dichromates of hydrogen, lithium, sodium, potassium, rubidium, cesium, ammonium, and chromium(VI) oxide" [28]. It has been noted, however, that the perceived lower toxicity of Cr(III) relative to Cr(VI) may be due "as much to the lower solubility of Cr(III) solids in aquifer environments compared to Cr(VI) solids as to the inherent toxicity of the two valence states of chromium" [29].

The carcinogenic effects of Cr(VI) when ingested in water or food have not been unequivocally demonstrated. It is known that ingested Cr(VI) is rapidly converted to relatively innocuous Cr(III) in the stomach. Others argue that too little is known about how Cr(VI) may enter cells of the body and that additional research may reveal as yet unknown pathways that lead to organ damage. The MCLG for chromium has been set at 0.1 ppm, which is also the MCLG for Cr(VI).

In February 1999, the California Environmental Protection Agency (Office of Environmental Health Hazard Assessment, OEHHA) established a *Public Health Goal* of

[11]There is some variability in the estimate of the concentrations found in tap water. According to ATSDR [26] concentrations of chromium range from 400 to 8000 ppb with a mean value of 1800 ppb.

[12]Note that the units of measure used for the soil are different from those used for water. In the case of soil, the concentration is the weight of chromium per unit weight of dry soil, whereas in the case of water, the concentration is normally expressed as weight of solute per unit weight of solution or weight of solute per unit volume of solution.

0.0025 ppm of chromium as their standard. On November 9, 2001 the agency withdrew this goal because the agency determined that the study on which it had relied was flawed. It announced that it would develop a *preliminary remediation goal* (PRG) for Cr(VI) by the spring of 2003.

The two approaches normally used to address chromium contamination problems involve redox reactions. Subsurface permeable reactive walls, a relatively new concept, and natural attenuation have been identified as a methodology to transform Cr(VI) into less nocuous Cr(III).

Contaminant plumes of Cr(VI) exist and are of concern. As an example, we consider the plume generated by the Liberty Aircraft Corporation in South Farmingdale, Nassau County, Long Island. The following description is taken from Pinder [30]:

> During a routine sanitary survey, 0.1 mg/L of chromium[13] was detected in a private well located relatively close to a pit used for disposal of liquid industrial wastes from the plant. The chief use of the chromium in Nassau County is anodizing aluminum and aluminum alloys to protect them from corrosion and to prepare their surfaces to take paints. With the advancements in the military aircraft industry in the 1940s the use of chromic acid on aircraft parts became common. In 1945, when trained personnel again became available after World War II, a series of shallow test wells were installed in an area several hundred feet south of the aircraft plant. The samples collected from these wells indicated chromium concentrations from zero to a trace, and subsurface contamination was discounted as a threat to New York City's auxiliary groundwater system at Massapequa several miles to the south. The shallow test wells and an additional shallow domestic well about 450 meters south of the disposal pond were sampled in 1948, and 1.0–3.5 mg/L of hexavalent chromium was reported along with cadmium, copper and aluminum.
>
> Recognizing the potential danger to public water supplies, the Nassau County departments of health and public works made a joint investigation of the contaminated area in 1949 and 1950. This program, which included the drilling and sampling of about 40 test wells, concentrated on defining the extent of contamination of toxic hexavalent chromium. By this time, about 9 years after the start of disposal of the plating wastes, the contaminant plume had assumed a cigar shape in plan view, was about 1,200 meters long, and had a maximum width of about 260 meters.
>
> In 1953, 22 new sampling wells were constructed, and analyses were made for cadmium as well as chromium. Additional wells were drilled and samples collected in 1958 in the northern part of the plume; by this time the leading edge probably was a short distance west of Massapequa Creek.
>
> The latest and most detailed examination of the extent, chemical composition, and pattern of movement of the contaminated water was initiated in 1962 by the U.S. Geological Survey in cooperation with the Nassau County departments of health and public works. About 100 test wells were installed and sampled between the plating waste disposal basins and Massapequa Creek [Figure 1.31]. To determine the vertical distribution of the chromium, samples were collected at 5-foot intervals until there was no detectable contamination. This investigation showed that the plume was about 2,300 meters long and had a maximum width of about 300 meters. The leading edge of the plume converged as it approached and moved beneath Massapequa Creek, where some of the contaminated water was discharged. The peak concentration had decreased from an observed high of 40 mg/L in 1940 to 14 mg/L in 1962. The maximum concentration at this time was located 900 meters down the direction of flow from the disposal ponds.

---

[13] A measurement of 0.1 mg/L is approximately equivalent to 100,000 ppb.

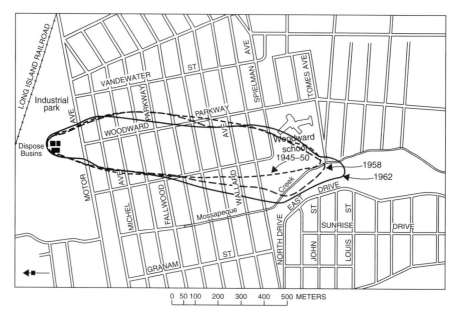

FIGURE 1.31. Chromium plume distribution at three stages of its evolution on Long Island, New York (from [30]).

The above is one of the earliest and most carefully documented cases of Cr(VI) contamination and illustrates the scale at which such environmental insults can occur.

Underground Storage Tanks Underground storage tanks (USTs) were reported as a potential threat to groundwater quality by 35 out of the 37 states whose data are integrated into Figure 1.26. The various compounds reported are shown in Figure 1.32. Many liquids emanating from USTs are only slightly soluble and form a separate liquid phase when in contact with water. If the soil is only partially saturated—in other words, there is air present in the soil along with the water and slightly soluble nonaqueous liquid—the system has three fluid phases: air, water, and nonaqueous liquid.

Because such slightly soluble contaminants form a separate liquid phase, they have been called *nonaqueous phase liquids*, or NAPLs for short[14]. Petroleum compounds are

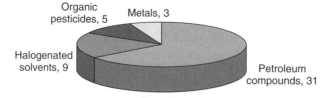

FIGURE 1.32. Number of states reporting a contaminant associated with leaking USTs [3].

[14]The first time the term NAPL is reported is during the investigation of the S Area in Niagara Falls, New York. where a separate liquid-contaminant phase was observed during subsurface investigations. The S Area is adjacent to the notorious Love Canal site.

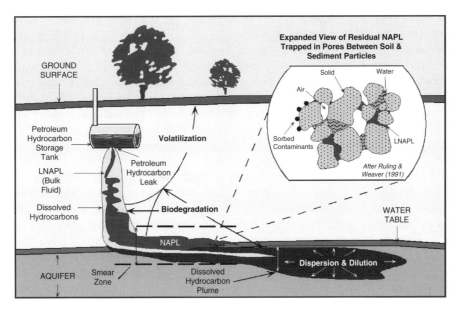

FIGURE 1.33. Groundwater contamination as a result of petroleum spillage. Petroleum migrates to the water table, where it begins to spread and move in the direction of the slope of the water table. Being slightly soluble, the components of the petroleum dissolve in the groundwater as shown. The dissolved component moves in the direction of groundwater flow (from [31]).

a frequently encountered form of NAPL. A diagrammatic representation of a leaking tank containing petroleum is provided in Figure 1.33. Moving under the force of gravity, such compounds migrate nearly vertically from their source locations, which are usually at the land surface, to the water table. During this downward migration, small droplets, or blobs as they are sometimes called, are left behind, trapped in the pores. In the inset of Figure 1.33 a diagrammatic representation of the physical–chemical state is provided. Note that all three phases—water, air, and LNAPL—are represented. Contaminants that have moved from the fluid phases to the solid surface of the grains are also represented as *sorbed contaminants*.

Once the NAPL has reached the water table, it begins to form a pancake-like layer on the water table. Because it is lighter than water it floats, and so such NAPLs have come to be known as *lighter-than water nonaqueous phase liquids* or LNAPLs. Once the LNAPL has reached the water table, it moves by gravitational forces along the water-table surface as shown in Figure 1.33.

While in transit, the LNAPL dissolves into the groundwater around it. The resulting plume of contaminated groundwater moves in a direction dictated by water-table slope, as is the case with the LNAPL, but in this case there is usually a vertical downward component to the dissolved contaminant plume. The downward component causes the dissolved contaminant plume to move deeper into the aquifer.

When the water table rises and falls due primarily to changes in net infiltration caused by rainfall, the water phase may rise and surround the blobs of petroleum that have been trapped on their travels down to the water table. A complex mixture of oil, water, and air therefore results in the zone through which the water table has moved. The zone characterized by either water and oil or water, air, and oil is sometimes denoted as a *smear zone*. The larger the vertical movement up and down that the water table experiences, the larger is the smear zone.

Some components of LNAPLs such as gasoline, may be *volatile*, meaning that they can evaporate into the gaseous (air) phase. As the volatile LNAPL moves vertically downward through the unsaturated zone, some compounds in the LNAPL may change from the liquid to the gaseous phase. The resulting gas phase moves as a fluid through the unsaturated zone and may migrate in directions quite different from those of either the LNAPL or its associated aqueous contaminant plume. When such migrating gases are explosive and enter enclosed areas, such as basements, the consequences can be serious.

When the NAPL entering the subsurface is heavier than water, it is called a *dense non-aqueous phase liquid* or DNAPL. Figure 1.34 illustrates the complex physical–chemical behavior of this type of contaminant. Like LNAPL, DNAPL moves vertically downward through the vadose zone under the influence of gravity and may volatilize in transit. Unlike LNAPL, DNAPL does not accumulate at the water table because it has a density greater than that of water. Rather, it will continue its vertical movement until it encounters a geologic formation that exhibits the physical–chemical characteristics to impede its flow. Because a DNAPL is generally less attracted to soil grains than water, the DNAPL must overcome the resistance of the water, which prefers to occupy the pores, in order to displace it. The tenacity with which water attempts to thwart the efforts of the DNAPL depends on its physical–chemical properties, a topic we will consider in Chapter 11.

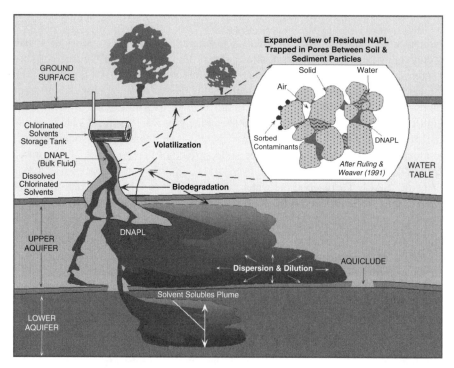

FIGURE 1.34. A slightly soluble liquid with a density greater than that of water enters the subsurface and moves vertically downward through the water table to contaminate both an upper and a lower aquifer (from [31]).

In Figure 1.34 the DNAPL has passed though the water table, through an upper aquifer and has entered a lower aquifer via a breach in the *aquiclude* or relatively impermeable layer. It has now entered the lower aquifer and will continue to migrate in a near-vertical trajectory until encountering another inhibiting layer. Notice that the aquiclude has caused the DNAPL to pool and then to move horizontally.

As in the case of a LNAPL, a DNAPL forms blobs as it migrates. Thus the further it travels, the less there remains of the mobile DNAPL body. Eventually, it will consist only of blobs and will be immobile. Even while immobile it may volatilize to form a vapor phase or dissolve to generate a contaminant plume. Such plumes can migrate large distances.

The primary causes of UST contamination are faulty construction and corrosion of tanks and pipelines. The U.S. EPA reports that [3]:

> As of March 1996, more than 300,000 releases from USTs had been confirmed. EPA esti-
> mates that nationally 60% of these leaks have impacted groundwater quality and, in some
> states, the percentage is as high as 90%.

While there has been a justifiable focus on commercial establishments as the primary source of groundwater contamination from USTs, and many of the most newsworthy cases have been from this UST category, other sources are also of concern. EPA presents the following perspective [3]:

> The "registered USTs" and "facilities" described above represent tanks used for commercial
> and industrial purposes. Hundreds of thousands of household fuel oil USTs are not included
> in the numbers presented above. Many of these household USTs, installed 20-to-30 years
> ago as suburban communities were developed across the country, have reached or surpassed
> their normal service lifespans. Some of these tanks are undoubtedly leaking and threatening
> groundwater supplies. Because household tanks are not regulated as commercial facilities
> are, however, it is not possible to determine the extent to which groundwater quality is
> threatened by them. In addition, since the cost of replacing leaking USTs would be borne
> by the homeowner, there is little incentive for the homeowner to investigate the soundness
> of his/her home oil tank.

Landfills   Landfills were cited as the third most significant source of contaminants in the 1996 EPA survey. Municipal waste landfills are typically used to dispose of relatively inert substances. Thus the groundwater contamination from such landfills tends to be excessive total dissolved solids and high chemical and biochemical oxygen demand. Volatile organic compounds are also found, usually in relatively minor amounts.

Industrial landfills tend to be more specific in the materials they receive. According to EPA, common materials that may be disposed of in industrial landfills include [3] "plastics, metals, fly ash, sludges, coke, tailings, waste pigment particles, low-level radioactive wastes, polypropylene, wood, brick, cellulose, ceramics, and synthetics." Associated with such landfills, EPA reports contamination in the form of "heavy metals, high sulfates, and volatile organic compounds." Specifically reported in the 1996 EPA survey were "metals, halogenated solvents, and petroleum compounds. To a lesser extent, organic and inorganic pesticides were also cited as a contaminant of concern."

The *unlined landfill* is of particular concern because of its typical location. Abandoned gravel pits, sand pits, strip mines, marshlands, and sinkholes[15] were often used as unlined

---

[15]A sinkhole is a depression in the ground's surface attributable to the collapse of the supporting roof of a solution cavern at depth.

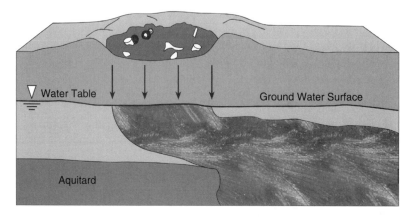

FIGURE 1.35. Groundwater contamination as a result of unlined landfill disposal (from [3]).

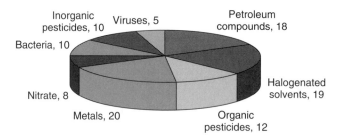

FIGURE 1.36. Number of states reporting a contaminant grouping associated with landfills (from [3]).

landfills. Often the water table was at or very near the surface of these locations and thus transport to the groundwater was facilitated Figure 1.35. The suite of contaminants reported in the 1996 EPA survey is shown in Figure 1.36. Of these, the halogenated solvents tend to have captured the most attention given their potential as carcinogens. Of the metals, hexavalent chromium attracted the most interest, again because of its public health risk potential.

Septic Systems   Examination of Figure 1.26 reveals that the second most cited source of groundwater contamination in the EPA survey was septic systems. Septic systems consist of buried holding tanks and fluid distribution systems that are designed to convey fluids to permeable leaching beds (see Figure 1.37). From there the fluids move into the shallow soil, where the dissolved waste compounds are attacked by microorganisms in the soil or are degraded by other natural processes. When improperly constructed, poorly maintained, or located in areas where leaching is ineffective due to soil characteristics, septic systems may lead to groundwater contamination. The reasons that septic system contamination is considered an important problem are their sheer number and the fact that they are often located near water supplies.

The contaminants of concern discharged to domestic septic systems include bacteria, viruses, nitrates, phosphates from detergents, and various chemicals associated primarily

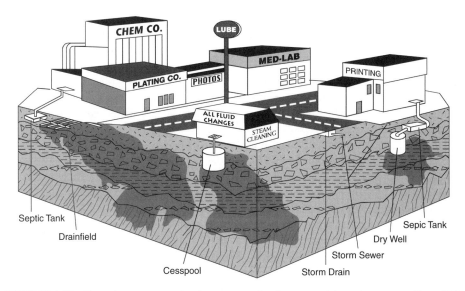

FIGURE 1.37. Groundwater contamination as a result of commercial septic systems (from [3]).

with household cleaners. EPA reports that "American households dispose of an estimated 3.5 billion gallons of liquid waste into septic systems each day" [3].

While most septic systems are domestic, similar systems are also used by small businesses. The chemicals discharged into such systems can often find their way into groundwater. Of particular concern are automotive repair and service businesses that may dispose of automobile engine fluids, fuels, and cleaning solvents [3]. In the case of commercial septic systems, EPA reports that "as much as 4 million pounds of waste per year are disposed of by commercial sites into septic systems that have affected the drinking water of approximately 1.3 million Americans." The pie chart presented in Figure 1.38 is an indication of the type of contaminants resulting from septic systems.

1.8.3 Superfund

In response to widespread groundwater contamination legislation known as the Comprehensive Environmental Response, Compensation and Liability Act (CERCLA) was enacted at the federal level in 1980. This legislation authorized the realization of a

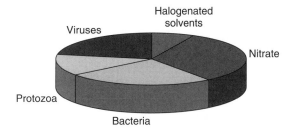

FIGURE 1.38. Number of states reporting a contaminant grouping in association with septic systems (from [3]).

concept generally know as "superfund." It is designed primarily to address abandoned contaminated toxic waste sites. It provides federal funds to address groundwater cleanup, although up until now it has been used primarily to address the wastes themselves. This statute also sets down very stringent monitoring and response standards.

In general, this statute addresses what EPA calls "priority sites." If a site is investigated and found to be of sufficient concern to be on the National Priorities List, which is normally abbreviated NPL, it is targeted for federal cleanup under *superfund*. In 1986 there were 888 superfund sites selected out of an estimated 16,000 sites that were believed to cause or have the potential to cause groundwater contamination.

Basically, the law works like this. EPA has the right to enter any property, or even any facility, and to determine whether there is a need for the agency to conduct a *response action*. If they determine that there is *"imminent and substantial endangerment"* and a response action is required, they can also enter the property to conduct it.

If a facility has good cause to refuse to comply with the EPA order, it cannot immediately obtain judicial relief to set aside the order. The law specifically says that the federal court has no jurisdiction to review the order until either EPA seeks to enforce it or EPA sues to recover the costs associated with the response action directed in the order.

As a result, once a company receives an "administrative order," it has essentially two choices: it can comply with the order or it can face judicial action by EPA. *EPA can, and often does, fund and implement the response action, but it will eventually sue the company to recover its costs*.

Because Congress, when it wrote the superfund legislation, was anxious to expedite environmental compliance, it provided in the legislation a special opportunity for a company that complies with a cleanup order applicable to a NPL site to recover the costs of compliance. The protocol requires filing a claim against the superfund. However, such a claim can be successful only if the company can show that it was not a liable party or that the response action ordered was arbitrary, capricious, or contrary to law.

If a company simply refuses or fails to comply, there are very serious penalties. The company may be forced to pay $25,000 per day of the violation. That can add up to quite a lot of money, even for a large company! What is even more frightening is the possibility of incurring punitive damages that may be up to three times the amount of the costs incurred as a result of the company's failure to comply. *In other words, the company may end up not only paying for the remedy but also penalties and punitive damages*[16].

In the case of a landfill that is leaking leachate and contaminating the groundwater, there would normally be many parties that placed toxic compounds at the site. How does the EPA figure out how much each party would be required to pay to remediate the site?

The answer to this question is one of the most controversial aspects of the superfund law. The EPA has successfully argued in court that the commingling of contaminants at a site makes it impossible to divide the responsibility for the contamination between the various responsible parties. As a result, the defendants in a multiparty action are each considered responsible for the entire cost of the site remediation. In other words, each party is *jointly and severally liable*, irrespective of its actual contributions. However, in subsequent legal proceedings, it is possible to bring up the matter of the amount and toxicity of hazardous substances contributed to the site in determining the final allocation of costs between the various parties.

---

[16]Punitive damages are those damages that are awarded in addition to normal compensation to a plaintiff in order to punish the defendant for a serious wrong.

In addition to these federal statutes, there are also state statutes. For example, in 1983 New Jersey passed a law that placed the responsibility for cleaning up a contaminated property on the shoulders of the owners. The law requires that the site be cleaned up before the owners can sell their property. The New Jersey Department of Environmental Protection can void any sale and can fine property owners that are not in compliance with the law.

Determination of whether a site is selected for the *National Priorities List* (NPL) is based on three mechanisms [32]:

- The first mechanism is EPA's *hazard ranking system* (HRS).
- The second mechanism allows states or territories to designate one top-priority site regardless of score.
- The third mechanism allows listing a site if it meets all three of the following requirements:

1. The *Agency for Toxic Substances and Disease Registry* (ATSDR) of the U.S. Public Health Service has issued a health advisory that recommends removing people from the site.
2. EPA determines the site poses a significant threat to public health.
3. EPA anticipates it will be more cost-effective to use its remedial authority (available only at NPL sites) than to use its emergency removal authority to respond to the site.

The HRS is defined as follows [32]:

[A] numerically based screening system that uses information from initial, limited investigations—the preliminary assessment and the site inspection—to assess the relative potential of sites to pose a threat to human health or the environment. The HRS uses a structured analysis approach to scoring sites. This approach assigns numerical values to factors that relate to risk based on conditions at the site. The factors are grouped into three categories:

- likelihood that a site has released or has the potential to release hazardous substances into the environment;
- characteristics of the waste (e.g., toxicity and waste quantity); and
- people or sensitive environments (targets) affected by the release.

Four pathways can be scored under the HRS:

- groundwater migration (drinking water);
- surface water migration (drinking water, human food chain, sensitive environments);
- soil exposure (resident population, nearby population, sensitive environments); and
- air migration (population, sensitive environments).

After scores are calculated for one or more pathways, they are combined using a root-mean-square equation to determine the overall site score.

The electronic scoring system, *PREscore (Preliminary Ranking Evaluation Score)*, can be used to do the scoring calculations. If all pathway scores are low, the site score is low. However, the site score can be relatively high even if only one pathway score is high. This is an important requirement for HRS scoring, because some extremely dangerous sites pose threats through only one pathway.

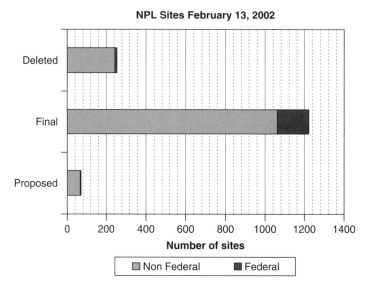

FIGURE 1.39. National Priorities List (NPL) Sites as of February 13, 2002. The sites are subdivided into nonfederal and federal sites (from [33]).

Proposed, final, and deleted NPL sites as of February 13, 2002 are presented in Figure 1.39[17]. The graph subdivides the sites into federal and nonfederal.

Organic chemicals detected and reported at NPL sites are plotted in Figure 1.40. Organic compounds are the most frequently reported contaminants. Chlorinated hydrocarbons tend to be the most frequently reported compounds. The chemical 1,1,1-trichloroethylene is a widely used degreasing compound and its occurrence in groundwater is ubiquitous. Some of the compounds found in this figure may be daughter products of this chemical. For example, 1,1,2-trichloroethylene may degrade to *trans*-1,2-dichloroethylene, *cis*-1,2-dichloroethene, and 1,1-dichloroethene. *trans*-1,2-Dichloroethylene and *cis*-1,2-dichloroethene compounds may, in turn, degrade to 1,2-dichloroethane and eventually vinyl chloride. *trans*-1,2-Dichloroethylene and 1,1-dichloroethene may degrade directly to vinyl chloride without the intermediate 1,2-dichloroethane step. A more thorough discussion of organic chemicals and their evolution is found in Chapter 10.

Inorganic compounds are also common NPL site contaminants. The information provided in Figure 1.41, taken from the 1998 EPA report [3], illustrates the frequency with which the most commonly encountered chemicals are reported.

1.9   QUANTITATIVE ANALYSIS OF GROUNDWATER PROBLEMS

A description of the science and engineering technology required to solve subsurface flow problems is an important element of the remainder of this book. Such a description must include a presentation of the physical, chemical, and biochemical attributes of the problems to be addressed and the modeling and measurement tools available to study

---

[17]As of May 26, 2006 the total number of deleted sites was 309, the number of proposed sites was 59, and the number of final sites was 1244 [35].

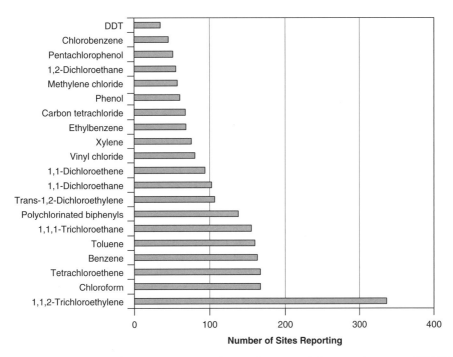

FIGURE 1.40. Organic contaminants most frequently reported in groundwater at CERCLA National Priorities List (NPL) sites (from [3]).

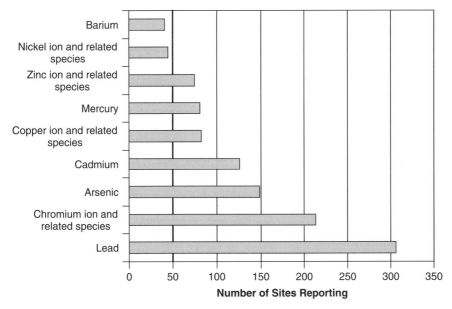

FIGURE 1.41. Inorganic compounds most frequently reported in groundwater at CERCLA National Priorities List (NPL) sites (from [3]).

them. The discussion will focus on (1) a mathematical–physical–chemical description of the groundwater system, (2) field measurements that can be used to determine the parameters that make the resulting equations tractable, and (3) an interpretation of the behavior of groundwater systems as defined by observation and mathematical analysis. Having gained these insights, we will use the tools we have developed to study groundwater hydrology, especially as it pertains to problems of water supply. Since water supply has both a quality of water as well as quantity of water dimension, this study will require consideration of the impact, remediation, and monitoring of groundwater contaminants.

1.9.1   Governing Equations

In this section we will consider only conceptually the mathematics behind the analysis of groundwater flow and groundwater mass transport and the use of these quantitative tools in the solution of problems of practical importance. Fundamental to the understanding and solution of groundwater problems are the balance equations that describe the *conservation of mass, energy*, and *momentum*. A fourth equation that comes into play in certain fairly abstract applications is the equation describing the *entropy inequality*. We will focus on the first three—mass, energy, and momentum.

The balance equations, as commonly presented, are based on the concept of *continuum mechanics* or *mixture theory* as it is applied to the flow of fluids through porous media such as encountered in groundwater flow and mass transport in unconsolidated sediments. The inclusion of the deformation of the porous medium itself can also be considered. This particular area of analysis is addressed in the study of engineering *soil mechanics*. Inclusion of reaction chemistry allows for the study of the chemical evolution of the soil matrix as well as the contained fluids. The energy-balance equation describes the behavior of temperature-dependent systems as diverse as those encountered in frozen soils and geothermal energy. Extensions of these concepts also allow for the mathematical description of the behavior of fluids moving through fractured media.

The balance equations alone do not fully describe a physical system. In fact, the balance equations are so general that they are applicable to virtually all physical systems of practical interest. To make these equations unique—that is, to make them descriptive of a specified physical system—requires a set of auxiliary conditions called constitutive relationships. Constitutive relationships are derived through experiment. The experiments must be conducted on the same scale as that at which the phenomena of interest are measured. For example, experiments applicable to the description of the behavior of fluids flowing through porous media must be conducted at the porous-medium scale, not the molecular scale. We will learn more about these relationships in the next chapter.

Built into the constitutive equations are parameters that are unique to the system of interest. In the case of a model of groundwater flow, a parameter of interest may be the porosity. In addition to a directly measurable property such as porosity, there are also properties that can be determined only indirectly from constitutive experiments. The hydraulic conductivity of a porous medium describes, in some sense, its ability to transmit fluid. This parameter, which will be discussed in the next chapter, can be determined only indirectly through experiment.

The combination of the balance equations and the constitutive relationships provides what are called the governing equations for a particular physical process, for example, the flow of fluids in a porous medium. Since a mathematical model of a system represents, by its nature, a subset of the universe, it is necessary to define precisely what subset one is considering. For example, a model of a contaminant site might represent a small portion of the aquifer in which the contamination occurs. To define the subarea of interest it is necessary to provide conditions descriptive of the behavior of the state variable, say, the concentration in the case of a contaminant transport problem, on the perimeter of the model. The conditions defining the behavior of the state variable on the boundary are called the boundary conditions. In the case of groundwater-contaminant transport, a typical boundary condition would be, for example, the contaminant concentration defined everywhere along the model perimeter.

If the problem of interest involves time, it is necessary to describe the conditions of the system at the point in time when the analysis begins. The state variable or variables of interest must be defined everywhere in the model at this initial time. Such conditions are called initial conditions. In our example of contaminant transport, the initial conditions would be the concentration everywhere in the model area at the beginning of the analysis.

1.9.2 Field Data

Ideally the boundary conditions on a model are readily discernible from a field investigation. Knowledge of the concentration of a contaminant at a source location, such as a leaking tank, could provide the information needed to define a concentration boundary condition at that point. A surface-water body such as a lake would provide information on the water level elevations at the contact between the lake and the aquifer, thereby providing boundary condition information useful in the simulation of groundwater flow.

A change in lithology from sand to clay would suggest a change in hydraulic conductivity. The change in material properties could be incorporated into the model as a spatially variable parameter. Such spatial variability is called heterogeneity. Thus knowledge of the geology of an aquifer, especially information on its lithology, can help to define the spatially dependent properties needed for model parameter specification.

Since indirectly measured aquifer parameters are best determined using experiments conducted at the same scale as that for which the model results are required, field experiments are, generally speaking, a very helpful element of model development. Experiments involve artificially stressing the aquifer, that is, deliberately changing the state of the system, and then observing the system response. For example, a well pumped at a specified rate for a specified time will cause the water level in the vicinity of the well to change. Careful measurement of these changes can provide state-variable information that can lead to the indirect determination of the hydraulic conductivity. We will consider such experiments in greater detail in Chapter 6.

Aquifer systems respond to applied stresses, such as changes in net infiltration or changes in barometric pressure. Withdrawal of water from wells can be treated as stresses. In some instances stresses can be recast as boundary conditions, depending on the nature of the stress and the dimensionality (number of space dimensions) of the model. Stresses applicable to a model must be either measured or estimated. If they change over time, the change in time may also require estimation depending on the time scale of the analysis. For example, if one is interested in the behavior of the aquifer over a period of many years, daily fluctuations in stresses may be of minor importance. On the other hand, if the period of interest is measured in terms of days, daily or even hourly stress variability may be of interest.

In summary, field data can provide information on boundary conditions, initial conditions, stresses, and model parameters.

1.9.3   Behavior of Groundwater Systems

The goal of a model is to mimic and sometimes predict the behavior of a system. If the prediction is initiated from an earlier starting point, the model may be used to describe past events. While prediction is essential in the management of groundwater systems, it may also be helpful in understanding the natural system. An attempt to model a system using various hypotheses regarding its boundary conditions, parameters, and so on can facilitate testing concepts regarding aquifer system dynamics. At a more applied level, models can be used to extract information about prior or future impacts of groundwater contamination, concepts that are of great interest in groundwater litigation. Efforts to determine the history of contaminant concentrations in the production wells in Woburn, Massachusetts was at the heart of the litigation described in the best-selling novel *A Civil Action* [34] and in the movie by the same name.

As noted earlier, in the next chapter we are going to focus on the basic physical concepts that govern the behavior of fluid flow in the subsurface. Fundamental to an understanding of groundwater flow, this chapter will provide the foundation for many of the ideas found in the remainder of this book.

1.10   SUMMARY

In this introductory chapter we introduce the main topic of this book, groundwater (within the context of hydrology), the hydrological cycle, and the role groundwater plays as a natural resource. Also considered is the groundwater environment, the properties of the subsurface that are important in its behavior, and the challenges faced in maintaining this resource while it is threatened by natural and human-induced contamination and aggressive exploitation as a water supply. Finally, we provide a preview into the physical processes and laws that dictate the behavior of groundwater and its chemical constituents.

For additional information, the reader may want to peruse [36–44].

1.11   PROBLEMS

1.1. Consider the fluxes and storage volumes associated with the hydrologic cycle, as given in Figure 1.1. Based on the numbers shown in that figure, calculate the residence time for water vapor in the atmosphere. Explain the meaning of the term *residence time*, and then comment on why the concept is useful.

1.2. Prove that the porosity associated with the simple cubic packing of equally sized spheres, shown in Figure 1.6, is 0.48. Does this value of porosity change if the grain size changes from a radius of 0.1 mm to a radius of 1.0 mm?

1.3. Consider the cubic packing shown in Figure 1.6. If each sphere has radius R, determine the largest sphere that could fit within the void space between the eight spheres shown in Figure 1.6. Would you consider this to be a reasonable measure of the size of the pore space associated with this packing?

1.4. Is the porosity associated with the rhombohedral packing larger than, smaller than, or the same as, the porosity associated with the cubic packing?

1.5. When does one use the hydrometer method to determine grain size distribution in soils?

1.6. Is a soil with a uniformity coefficient of 20 better sorted (more poorly graded) than one with a uniformity coefficient of 2? Explain.

1.7. A sieving experiment has provided the information contained in the following table, where "Pan" refers to the amount of material passing through the finest sieve. Using the plot provided in Figure 1.42, (a) plot the percent passing versus grainsize diameter on the graph and construct the associated curve; (b) classify the soil according to the U.S. Department of Agriculture soil classification system.

Sieve	Opening	Mass Retained
a	0.01 mm	35 grams
b	0.02 mm	25 grams
c	0.05 mm	30 grams
Pan		10 grams

1.8. A soil has porosity of 35%. Determine the void ratio of this soil. If 50% of the void space is filled with water, with the remaining 50% filled with air, determine the volumetric water content, θ_v, and the degree of saturation, S_w.

1.9. Explain the difference between naturally occurring groundwater contaminants and anthropogenic groundwater contaminants.

1.10. According to the Environmental Protection Agency, what is the most widely reported source of groundwater contamination? Give an example of such a source from your town or city.

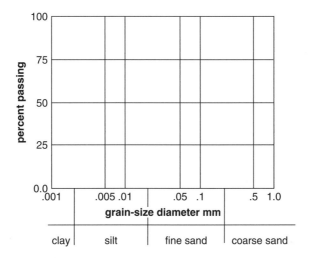

FIGURE 1.42. Reference figure for problem 1.7.

1.11. Explain the difference between LNAPLs and DNAPLs. Give an example of each of these types of contaminants.

1.12. Go to the website www.scorecard.org and investigate pollution sources and other interesting information about your town, city, or county. You may also consult the Enviromapper resource at maps.epa.gpv. Prepare a short (1 page) report on the environmental health of your local community.

BIBLIOGRAPHY

[1] S. L. Dingman, *Physical Hydrology,* (Appendix B—Water as a Substance), Macmillan, New York, 1994.

[2] R. Reeburgh, A web page on elemental cycles (http://ess1.ps.uci.edu/~reeburgh/figures. html)hydrologiccycle.

[3] United States Environmental Protection Agency, *National Water Quality Inventory, 1996 Report to Congress*, EPA-816-R-98-011, 1998.

[4] U.S. Geological Survey, *U.S. Geological Survey Report to Congress*, 1998.

[5] D. K. Todd, *Groundwater Hydrology*, 2nd ed., Wiley, Hoboken, NJ, 1980.

[6] A. K. Biswas, *History of Hydrology*, Elsevier, New York, 1970.

[7] F. G. Driscoll, *Groundwater and Wells*, U.S. Filter/Johnson Screens, St. Paul, MN, 1986.

[8] C. M. Epstein, Discovery of the aquifers of the New Jersey coastal plain in the nineteenth century, in E. R. Landa and S. Ince (eds.), *History of Geophysics*, Vol. 3, American Geophysical Union, Washington, D.C., 1987.

[9] F. L. Duley, Soil, in *McGraw-Hill Encyclopedia of Science and Technology*, Vol. 12, McGraw-Hill, New York, 1982, p. 517.

[10] F. A. L. Dullien, *Porous Media, Fluid Transport and Pore Structure*, Academic Press, New York, 1979.

[11] F. A. L. Dullien, *Porous Media: Fluid Transport and Pore Structure*, 2nd ed., Academic Press, San Diego, CA, 1992.

[12] J. E. Bowles, *Engineering Properties of Soils and Their Measurement*, 4th ed., McGraw-Hill, New York, 1992.

[13] T. McArthur and J. Roberts, *Understanding Soil Mechanics*, Delmar Publishers, Clifton Park, New York, 1995.

[14] U.S. Army Corps of Engineers, *Engineering and Design Soil Sampling*, www.usace. army.mil/inet/usace-docs/eng-manuals/em1110-1-1906, 2002.

[15] U.S. Department of the Interior, *Earth Manual*, U.S. Government Printing Office, Washington, DC, 1963.

[16] R. S. Stanley, The Champlain thrust fault, Lone Rock Point, Burlington, Vermont, *Geological Society of America Field Guide—Northeastern Section, The Geological Society of America Centennial Field Guide*, Vol. 5, Geological Society of America, 1987, p. 225.

[17] H. H. Cooper, *A Hypothesis Concerning the Dynamic Balance of Fresh Water and Salt Water in a Coastal Aquifer,* U.S. Geological Survey Water-Supply Paper 1613-C, 1964.

[18] T. E. Reilly, Analysis of ground-water systems in freshwater–saltwater environments, in W. M. Alley (ed.), *Regional Ground-water Quality*, Van Nostrand, New York, 1993, p. 634.

[19] U.S. Geological Survey, web site address, http://water.usgs.gov/ogw/gwrp/saltwater/fig5. html, update 2001.

[20] R. M. Spechler, *Saltwater Intrusion and Quality of Water in the Floridan Aquifer System, Northeastern Florida,* U.S. Geological Survey Water-Resources Investigations Report 92-4174, p. 76, 1994.

[21] U.S. EPA website, http://www.epa.gov/owow/NPS/sec6/salty.html, update, 2000.

[22] U.S. EPA, *Ground Water and Drinking Water, List of Contaminants,* http://www.epa.gov/OGWDW/dwh/c-ioc/chromium.html, 2001.

[23] National Research Council, *Arsenic in Drinking Water*, National Academy Press, Washington, DC, 1999, p. 273.

[24] U.S. Geological Survey, "The Geology of Radon," http://energy.cr.usgs.gov/radon/georadon/3.html, update 1995.

[25] United States Environmental Protection Agency, *Safe Drinking Water Act, Section 1429, Ground Water Report to Congress*, EPA 816-R-99-016, 1999.

[26] ATSDR, "Public Health Statement for Chromium," CAS# 7440-47-3, http://www.atsdr.cdc.gov/ToxProfiles/phs8810.html, 2000.

[27] The National Contaminated Sites Remediation Program, *Canadian Soil Quality Guidelines for Contaminated Sites, Human Health Effects: Chromium, Final Report,* Canadian Government Report, 1996.

[28] U.S. Department of Health and Human Services, *9th Report on Carcinogens, Revised January 2001,* Public Health Service, National Toxicology Program, http://ehp.niehs.nih.gov/roc/toc9.html, 2001.

[29] U.S. EPA, *Workshop on Monitoring Oxidation–Reduction Processes for Ground-water Restoration,* Workshop Summary, Dallas, TX, 2000.

[30] G. F. Pinder, A Galerkin–finite element simulation of groundwater contamination on Long Island, New York, *Water Resour. Res.* **9**(6): 1657, 1973.

[31] U.S. EPA, *Monitored Natural Attenuation of Petroleum Hydrocarbons*, EPA/600/F-98/021, 1999.

[32] U.S. EPA, "Introduction to the HRS," http://www.epa.gov/superfund/programs/npl_hrs/hrsint.htm, update 2001.

[33] U.S. EPA, "NPL Site Totals by Status and Milestone as of February 13, 2002," http://www.epa.gov/superfund/sites/query/queryhtm/npltotal.htm, update 2002.

[34] J. Harr, *A Civil Action*, Random House, New York, 1995.

[35] U.S. EPA, National Priorities List, http://www.epa.gov/superfund/sites/query/queryhtm/npltotal.htm.

[36] K. E. Anderson, *Ground Water Handbook*, National Groundwater Association, Dublin, OH, 1992, p. 40.

[37] H. L. Ritter, and L.C. Drake, Pore-size distribution in porous materials. Pressure porosimeter and determination of complete macropore-size distributions, *Ind. Eng. Chem. Anal. Ed.* **17**:782, 1945.

[38] L. D. Leet and S. Judson, *Physical Geology*, Prentice Hall, Englewood Cliffs, NJ, 1959.

[39] U.S. Geological Survey, *National Water-Quality Assessment, National Analysis of Trace Elements, Arsenic in Groundwater of the United States,* water.usgs.gov/nawqa/trace/arsenic/, 2002.

[40] U.S. Geological Survey, *Arsenic in Ground-water Resources of the United States,* USGS Fact Sheet 063-00, 2000.

[41] J. C. I. Dooge, On the study of water, *Hydrological Sci. J.* **28**(1):23, 1983.

[42] T. E. A. van Hylckama, Water, something peculiar, *Hydrological Sci. J.* **24**(4): 499, 1979.

[43] F. G. Driscoll, *Groundwater and Wells,* 2nd ed., Johnson Division Publication, St. Paul, MN, 1986.

CHAPTER 2

FLUID FLOW AND MASS TRANSPORT

In this chapter we introduce the fundamental concepts that govern fluid flow in the near-surface environment. We begin with a discussion of fluid pressure, fluid potential, and hydraulic head, each constituting a measure of energy. We then explore Darcy's law, a relationship that relates the rate of flow of water in a porous medium to the energy gradient. Having provided this introduction to the flow of fluids in porous media, we will examine some of the more subtle issues involving matters of scale, dimensionality, and directional preference. We begin with the concept of fluid pressure.

2.1 FLUID PRESSURE

Let us first consider the case of *soil water pressure* and along the way introduce the associated concept of *hydraulic head.* To set the stage, consider the following elementary experiment. Take a cylinder slightly more than 1.0 m long and closed at the bottom except for a device to allow for the controlled drainage of water from the base of the column if desired (see Figure 2.1). Beginning at the bottom of the column and moving upward, water-pressure measurement devices are installed at 10 cm intervals. Thus, we have eleven pressure measuring points along the length of the column. We now fill the column with sand up to 1m, so that sand fills all of the cylinder except a small distance (say, 1 cm) at the top, which remains unfilled. Next, we add water containing blue dye to the column until it appears to be saturated, with any additional water overflowing at the top of the column. This gives us a column of water that fills the pore space of the 100 cm of sand, as well as the extra centimeter of open column at the top.

If the water in this column is static, and the top of the column is open to the atmosphere, then the pressure measurement devices show a pressure profile like that shown in Figure 2.1. The pressure increases linearly as depth increases. The slope of

Subsurface Hydrology   By George F. Pinder and Michael A. Celia
Copyright © 2006 John Wiley & Sons, Inc.

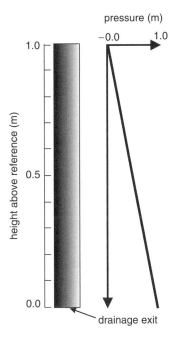

FIGURE 2.1. Pressure head representation in a saturated column.

the pressure-versus-depth function is equal to the density of water multiplied by the gravitational constant, that is, $\Delta p_w / \Delta z = \rho_w g$, where p_w denotes the water pressure $[M/LT^2]$, ρ_w is the water density $[M/L^3]$, and g is the gravitational acceleration constant $[L/T^2]$.

We can understand this by a simple consideration of the weight of water. For the column system, the pressure of the water at the top of the column is equal to atmospheric pressure, which is a measure of the weight of the atmospheric gases above the land surface, expressed as a force-per-area measure. As we move downward in the column, to some depth d, the added weight of the water above this depth adds to the force acting on a horizontal plane in the water. If we perform a simple force balance across this plane within the column at depth d, then we find that the water pressure pushing upward must balance the weight of the air and water pushing downward (if these forces did not balance then the fluids would not be static, they would move). Recall that pressure has units of force per area. So the force balance takes the form (see Figure 2.2)

$$p_w \times \left(\pi r^2 \right) = \rho_w g d \times \left(\pi r^2 \right) + p_a \times \left(\pi r^2 \right),$$

where r is the radius of the cylindrical volume and p_a denotes air (atmospheric) pressure.

We see directly that the pressure as a function of depth d is

$$p_w = \rho_w g d + p_a.$$

If we define the gauge pressure as the pressure relative to atmospheric pressure, so that the reference pressure (the pressure where we set $p = 0$) corresponds to atmospheric pressure, then the water pressure is simply the specific gravity of water, defined as the density times the gravitational constant, times the depth. If z is the vertical coordinate, defined

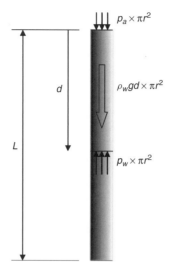

FIGURE 2.2. Diagrammatic sketch of the pressure profile.

to be positive upward, and if the reference elevation is set so that $z = 0$ corresponds to the bottom of the column, then the water pressure is

$$p_w\left(z\right) = \rho_w g\left(L - z\right), \tag{2.1}$$

where L is the length of the column.

Finally, we often use a measure of pressure that has the dimension of length. We call this normalized measure *pressure head*, denoted by the symbol ψ_w, defined by

$$\psi_w = \frac{p_w}{\rho_w g}. \tag{2.2}$$

Notice that the pressure head profile in our column of static water is given, via combination of Eq. (2.1) and (2.2), by

$$\psi_w\left(z\right) = L - z.$$

Next, consider a more interesting, and complex, experiment, in which we allow some of the water to drain out the bottom of the column. We allow the water to drain until it is observed, based on the color of the dyed water, to be saturated up to a little above the halfway point in the column, that is, up to a few centimeters above 50 cm. We then close the bottom to establish a static fluid system again. We then observe that water in the bottom half of the column, up to a few centimeters above 50 cm, shows a uniform, deep blue color. Further up the column, the color becomes progressively less visible, which corresponds to the fact that as water drains out of some of the pores, air replaces the water, and the air does not have a blue color. Examination of the pressure measurements for this partially drained static system again indicates a linear relationship with depth, as shown in Figure 2.3. However, now the location of $p_w = p_a$ is at a depth of 50 cm. Because the $p_w\left(z\right)$ function has a slope equal to $\rho_w g$, the region of the column where $z > 50$ cm shows pressures that are below atmospheric, meaning negative gauge pressures. Such negative water pressures, relative to atmospheric pressure, are characteristic of the soil

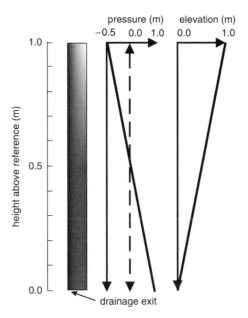

FIGURE 2.3. Relationship between fluid pressue head and elevation in a variably saturated column of sand.

zone where both water and air coexist. The question may arise as to how negative can a negative water pressure get. Atmospheric pressure is equivalent to the pressure exerted by 10.3 m (33.9 ft) of water. Once we are at a distance greater than 10.3 m above the *phreatic surface* (the elevation where the pressure is atmospheric), we have a problem.

If the water is continuous (in the sense of connected to the saturated zone boundary), and the water is stationary, the weight of the water would lead to a continuous decrease in pressure as elevation increases. But then the pressure in the water above 10.3 m would become less than absolute zero pressure. Since this is not possible, one of our assumptions must be wrong. The invalid assumption is that the water would be continuous at this elevation. Because the maximum negative pressure we could encounter would be -10.3 m of water, this is the maximum height above the phreatic surface at which continuous water can be found in a soil, assuming the fluid is static. Above this elevation the water will only exist in isolated packets such as *pendular rings* around the contact point of two soil grains, or as thin films that might coat the soil grains. The resulting information is shown in Figure 2.3.

Example of a Pressure Calculation To determine the pressure in cgs units of water at an elevation above a reference of 30 cm, we would employ the following logic. The amount of fluid above the point of measurement (30 cm) is $(100 - 30) = 70$ cm. Thus $\Delta h = 70$ cm. From Eq. (2.1) we have

$$p_0 = \rho g \, (70 \text{ cm}) = 1 \left(\frac{\text{g}}{\text{cm}^3} \right) \times 980 \frac{\text{cm}}{\text{s}^2} \times 70 \text{ cm}$$

$$= 68{,}600 \frac{\text{g}}{\text{s}^2 \cdot \text{cm}} = 68{,}600 \frac{\text{dynes}}{\text{cm}^2}.$$

This is the pressure that may be expressed equivalently as that beneath 70 cm of water, with the top surface of the container open to the atmosphere.

2.2 HYDRAULIC HEAD

To examine the pressure profile more closely, we plot not only the pressure but also the elevation along the column (see Figure 2.3). Note that in the case of no fluid movement, such as represented in Figure 2.3, these two lines slope in opposite directions. In other words, the pressure head, expressed in length units, increases in the fluid as the elevation of the measuring point decreases. This seems reasonable, at least for that portion of the column below 50 cm, inasmuch as the weight of the water at a point at lower elevation should be greater due to the increasing height of the overlying water column. The existence of negative pressure or pressure less than atmospheric above the 50 cm point is more curious.

Let us now take the matter one step further. Sum the values appearing on the two curves found in Figure 2.3 to produce a third curve such as found in Figure 2.4.

The sum of the pressure head, again expressed in length units, plus the elevation, both measured at any point along the column, is found to be a constant. We denote this sum as the *hydraulic head*. Sometimes this is called the *total head* because it is made up of two parts, the pressure head as illustrated in the first graph to the right of our experimental column and the elevation head illustrated in the next curve to the right. Thus, if we assume for the moment that water is incompressible, we see that we have the relationship

$$h(z) = h_p(z) + \zeta(z), \tag{2.3}$$

where $h(z)$ represents the total head (a function of the vertical coordinate z), $h_p(z)$ is the pressure head, and $\zeta(z)$ is the elevation head[1]. Note that to assure that this analysis makes sense, one must use a common datum, or reference point, as well as common units for the measurement of each of these quantities. In other words, the reference pressure is to be measured at the same physical point as the reference elevation. Also, recall that the pressure head for water corresponds to ψ_w, defined earlier in Eq. (2.2). Therefore we have

$$h_p = \psi_w = \frac{p(z)}{\rho g}. \tag{2.4}$$

We noted in our observation of the column containing the sand and water that the color changed as one moved upward from the 50 cm mark. The reason for this is that the amount of water contained in the pores is changing: the less the amount of water the less intense the color. The reason for this is that above the 50 cm line the water is held in the soil by surface tension, much as a sponge holds water. The ability to suspend water in a porous medium via surface tension is called capillarity.

---

[1] In this experiment the elevation head is directly proportional to the vertical distance along z, but this is not always the case since the coordinate z is not necessarily always vertical.

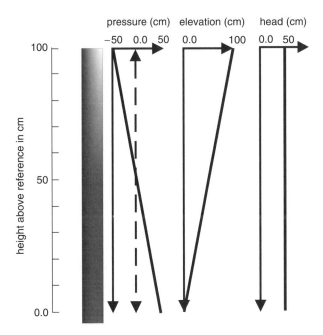

FIGURE 2.4. The head is found to be the sum of the pressure head plus the elevation head.

2.3 FLUID POTENTIAL

The fundamental theoretical concept governing flow of fluids in porous media is that *fluids move in response primarily to gradients in potential energy*. We are interested in this because rather than use velocity as a primary (or state) variable we would prefer to express velocity in terms of a more practical primary variable. A candidate is the *fluid potential*, which, as we will soon observe, is closely related to fluid pressure and the concept of head.

The total potential energy stored in a unit mass parcel of fluid at a specified state is the work required to transform that fluid from an arbitrary reference state to the specified state under consideration. For example, if we were to transform a unit mass parcel of liquid initially at the reference state—elevation $z = 0$, pressure $p = p_{atm}$, specific volume $V = V(p_{atm})^2$, and velocity $v = 0$—to another state—z, p, $V(p)$, and v—the *total potential per unit mass* with respect to the reference state, which we denote by ϕ, is given by [1]

$$\phi = gh = g \int_{z_0}^{z} dz + \int_{P_{atm}}^{p} \frac{dp}{\rho(p)} + \frac{v^2}{2}. \tag{2.5}$$

To understand the meaning of Eq. (2.5) it is helpful to recognize that ϕ is a measure of energy per unit mass. This is easily verified by examining the units of this expression. The first term on the right-hand side of Eq. (2.5) describes the *gravitational potential energy* acquired by the parcel as it is moved from the reference point $z = z_0$ to a point z^3, where g is once again the acceleration due to gravity. The second term is the *pressure potential*

---

[2]For constant unit mass, the fluid density is related to the specific volume as $\rho(p) = 1/V(p)$.

[3]Note that, in this example, the z-axis is oriented positive upward.

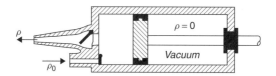

FIGURE 2.5. Pump for transforming liquid from standard to final state (from [1]).

at z, where the atmospheric pressure, p_{atm}, is taken as the datum. *For the pressure integral to be unique, the density of the fluid, ρ, can only be a function of fluid pressure.* The third term describes the component of the total potential attributable to kinetic energy.

From a physical perspective, one can describe the pressure potential of our reference parcel as the change in energy state per unit mass as the fluid parcel experiences a change in pressure from p_{atm} to p. The change may be due to moving the parcel through an overlying static water column or to an additional pressure increment due to fluid flow within the system. The change in pressure results in a change in volume of the parcel from $V(p_{atm})$ to $V(p)$, which, for constant unit mass, is related to the density as $\rho(p) = 1/V(p)$. Thus the *elasticity* of the fluid in the parcel results in an increase (or decrease) in stored energy. Finally, the third term is the *kinetic energy potential*. It describes the energy per unit mass required to accelerate the parcel from rest to a velocity v.

In his classic paper, "Theory of Groundwater Motion," M. King Hubbert [1] presented, through the vehicle of a conceptual experiment, an insightful explanation of the meaning of the fluid potential. We paraphrase Hubbert's explanation here.

The potential of a fluid at a specified point is the work required to transform a unit of mass of the fluid from an arbitrarily chosen standard state to the state at the point under consideration. For the standard state it is convenient to employ an elevation of zero, a pressure of 1 atmosphere, and a velocity of zero (relative to the earth's surface). Let the fluid in its final state at the point P be characterized by an elevation z, a pressure p, and a velocity v. Let V_0 be the volume per unit mass, or specific volume, of the fluid in its standard state, and let V be that for the final state. Let the corresponding densities be ρ_0 and ρ. We may also note that the density is the reciprocal of the specific volume: $V = 1/\rho$. We wish to find the work required to transform a unit mass of the fluid from the initial to the final state, and to do this we imagine a pump constructed along the lines indicated by Figure 2.5. This consists of a cylinder with frictionless piston, on the front of which is the fluid chamber and on the back a perfect vacuum. Inlet and outlet valves are provided. We then imagine the transformation to be effected by the following successive steps:

1. Under standard conditions, we slowly withdraw the piston and charge the cylinder with unit mass of the fluid. The work done by the piston on the fluid is then

$$w_1 = -p_0 V_0.$$

2. Next, we lift the pump with its fluid contents to the point P of elevation z. The work expended for this is

$$w_2 = gz + m_p gz, \tag{2.6}$$

where gz is the work required to lift the unit mass of the fluid, and $m_p gz$ is that required to lift the pump alone.

3. The contents of the cylinder are injected into the system at point P. The work required for this is

$$w_3 = \int_V^{V_0} p \cdot dV + pV. \tag{2.7}$$

The first term on the right-hand side of Eq. (2.7) is the work of compression of the fluid in order to raise its pressure from p_0 to p before it can be injected. The pV term is the work of injection against the pressure p.

4. The fluid is accelerated from a velocity of zero to that of v, requiring an amount of work

$$w_4 = \frac{v^2}{2}. \tag{2.8}$$

5. The cylinder is returned to its initial position at zero elevation, thus completing the cycle. This requires an amount of work

$$w_5 = -m_p gz. \tag{2.9}$$

The sum of these separate amounts of work is the potential ϕ of the fluid at the point P. Performing the addition and canceling out terms that repeat with opposite signs gives us

$$\phi = gz - p_0 V_0 + \int_V^{V_0} p dV + pV + \frac{v^2}{2}. \tag{2.10}$$

In this equation, the first and last terms on the right-hand side are the gravitational potential energy and the kinetic energy, respectively. The significance of the other three terms is best visualized by means of the "indicator diagram" of Figure 2.6, in which the pressure in the cylinder is plotted against piston displacement for both compressible and incompressible fluids. If the fluid is incompressible, a condition satisfied approximately by liquids under ordinary pressure ranges,

$$\int p \cdot dV = 0 \quad \text{and} \quad V = V_0. \tag{2.11}$$

In this case the pressure–volume work reduces to $(p - p_0) V$, and Eq. (2.10) simplifies to

$$\phi = gz + (p - p_0) V + \frac{v^2}{2}. \tag{2.12}$$

By a mathematical transformation we can convert Eq. (2.10) into another form whose physical significance may not be immediately apparent, but which will prove to be of great usefulness later. To effect this we make use of the fact that

$$d(pV) = p \cdot dV + V \cdot dp,$$

or

$$\int p \cdot dV = \int d(pV) - \int V \cdot dp. \tag{2.13}$$

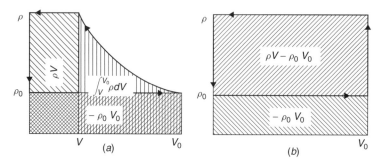

FIGURE 2.6. Pump indicator diagrams: (a) for gases and (b) for liquids (from [1]).

Then, as a definite integral, this becomes

$$\int_V^{V_0} p \cdot dV = \int_{pV}^{p_0 V_0} d(pV) - \int_p^{p_0} V \cdot dp$$

$$= p_0 V_0 - pV + \int_{p_0}^{p} V \cdot dp. \tag{2.14}$$

Substituting this value for $\int_V^{V_0} p\,dV$ into Eq. (2.10) transforms that equation into

$$\phi = gs + \int_{p_0}^{p} V \cdot dp + \frac{v^2}{2}, \tag{2.15}$$

which, when we substitute $1/\rho$ for V, becomes

$$\phi = gz + \int_{p_0}^{p} \frac{dp}{\rho} + \frac{v^2}{2}. \tag{2.16}$$

A graphical interpretation of Eq. (2.14) is readily afforded by noting that the area enclosed by the indicator diagram of Figure 2.6a is equal to the work performed by the pump per cycle and that this is given by

$$\int_{p_0}^{p} V \cdot dp \quad \text{or} \quad \text{by} \quad \int_{p_0}^{p} \frac{dp}{\rho}.$$

Note that Eq. (2.16) is essentially the same as our Eq. (2.5) given $\phi \equiv gh$ and the same reference conditions used by Hubbert.

Because the flow velocities in groundwater are small, the kinetic energy potential is normally neglected and will be neglected in our formulation. Thus there are two major components of our mechanical energy potential which contribute significantly to fluid flow in groundwater: the *gravitational potential* and the *pressure potential*. To expose the relationship between the fluid potential and the hydraulic head defined earlier, let us rewrite Eq. (2.5) in terms of the *potential per unit weight*, which is ϕ/g. This yields

$$h = \frac{\phi}{g} = \int_{z_0}^{z} dz + \int_{p_{\text{atm}}}^{p} \frac{dp}{g\rho(p)}, \tag{2.17}$$

where we have divided both sides by g, and $\rho(p)$ is defined by a fluid compressibility relationship. Note that the total potential per unit weight, h, has units of length. We also observe that this expression corresponds to our earlier definition of total head, h, as introduced in Eq. (2.3). For the special case where ρ is a constant and $z_0 = 0$, Eq. (2.17) becomes

$$h = z + \frac{p - p_{\text{atm}}}{\rho g}, \tag{2.18}$$

where

$$\frac{p - p_{\text{atm}}}{\rho g} = h_p = \psi_w$$

is called the pressure head and was presented earlier in Eq. (2.4).

The most general way to approach the concept of potential is in terms of potential per unit volume:

$$\rho g h = \rho g z + (p - p_{\text{atm}}). \tag{2.19}$$

This form of potential has units of pressure and puts no restrictions on the density function. The pressure potential of a fluid can be either positive or negative depending respectively, on whether the fluid saturates the pore space or shares the pore space with one or more fluids.

The term $(p - p_{\text{atm}})$ was considered earlier in our discussion of pressure less than atmospheric. In Eq. (2.19) we see that when water is at rest, the increase in elevation is offset by the decrease in pressure head until an elevation of 10.3 m above the water table (the surface where $p = p_{\text{atm}}$) is approached. At heights greater than 10.3 m one could maintain a constant potential only if the pressure p were below absolute zero, which is impossible. In such cases, additional forces enter into the definition of potential including short-range surface forces associated with thin films of water on the solid surfaces.

The head value can be measured at a point with a device called a *piezometer*. The elevation of the water level in the piezometer relative to a reference data base is a measure of the head at the point of measurement located at the tip of the piezometer and is called the *piezometric head*. The adjective piezometric has no other physical significance.

Example of a Head Calculation   Let us determine the head at two locations $z = 3$ cm and $z = 7$ cm along the length of a water column of height $l = 10$ cm wherein the water is at rest. For the location $z = 3$ cm, we have from Eq. (2.18)

$$h(3) = 3 + \frac{\rho g \times 7}{\rho g} = 10 \text{ cm,}$$

where we have used Eq. (2.1) in determining the value of the pressure difference $(p - p_{\text{atm}})$.

At the point $z = 7$ cm, we have

$$h(7) = 7 + \frac{\rho g \times 3}{\rho g} = 10 \text{ cm.}$$

The fact that the head was the same at both $z = 3$ cm and $z = 7$ cm in a column of water at rest tells us that the head is a constant in a fluid at rest, which is consistent with our observations in Section 2.2.

2.4 CONCEPT OF SATURATION

Before we pursue further the matter of pressure less than atmospheric, it is helpful to revisit the concept of saturation introduced in Section 1.6. Recall that the amount of water present in the soil is known as the *degree of saturation*. It is formally defined as the volume of water in the pores divided by the total volume of pore space. Therefore, if we designate the degree of saturation (or volumetric saturation) by the symbol S and express it as a percent, we have

$$S = \frac{V_w}{V_v} \times 100. \tag{2.20}$$

In Figure 2.7 we plot the saturation observed in our original water-and-sand-filled column. Notice that the saturation drops off very quickly above the 50 cm mark in the column. The shape of this curve when the saturation is less than 100% depends very much on the soil. A coarse sandy soil will show a very rapid decrease in saturation with height and a fine-grained soil will show a slower decrease. Thus, in general, fine-grained soils will hold water under tension more readily than will coarse-grained soils.

Examination of the curves appearing in Figure 2.7 reveals that there is a relationship between the water saturation and the pressure. We find that the saturation decreases as the gauge pressure ($p - p_{atm}$) becomes more negative. We will learn why this is the case shortly. The point where the pressure is zero we will call the water table and it will play an important role in the remainder of this book. Above the water table there is a zone that is still saturated, although the pressure is negative. We call this zone the capillary fringe. Above the capillary fringe the saturation decreases with increasing elevation.

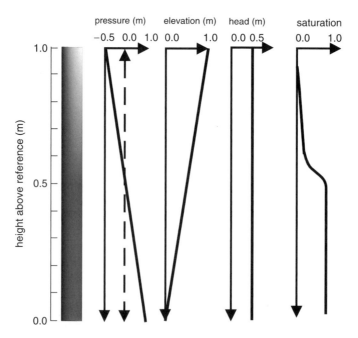

FIGURE 2.7. The saturation curve shows that the volume of water held in the pores by capillarity decreases with height above the 100% saturation point in the column.

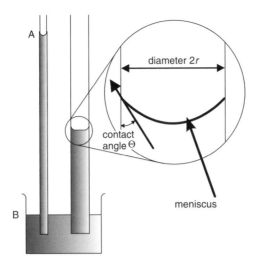

FIGURE 2.8. Capillary effects are due to the influence of molecular forces between the water and the glass. The height of water in the capillary tube is proportional to the inverse of the radius of the tube.

To understand why water rises to different elevations in different soils, we need yet another simple experiment. We place the ends of two vertical clean glass tubes (*capillary tubes*) of different radii in a beaker of water (see Figure 2.8). Several minutes later we examine the water level elevations in each tube. We observe that water has entered the tubes and risen above the level of the water in the beaker. Moreover, the water level in the smaller diameter tube has risen significantly higher than in the tube with the larger diameter.

What do we conclude from this experiment? We note primarily that, all other things being equal, the rise in the water level seems to be a function of the diameter of the tube.

To understand the significance of this observation, we utilize the fact that, at points below the surface of the water, the pressure on the inside of the tube at a specified elevation must be the same as on the outside. If this were not the case, we would expect the water at the higher pressure to flow toward the lower pressure, something that is not, in fact, happening. From this we can conclude that the pressure inside the tube at the elevation of the surface of the water in the beaker must be zero, assuming atmospheric pressure to be zero.

Then, what is the pressure of the water in the capillary tube at elevations above that of the water surface in the beaker? Consider, for example, the point A in Figure 2.8. If we were able to measure the pressure below point A in the capillary tube, we would find that it is less than atmospheric. Recall that we have observed pressure less than atmospheric before in the experiment illustrated in Figure 2.3. As we noted in Section 2.3, when the fluid is stationary, the total head is constant along the length of the tube and the pressure decreases in direct proportion to the height in the tube above the point B in Figure 2.8. This observation is important.

To determine why the water is rising in the tube, we must introduce the concept of *wettability*. Wettability is the tendency of a liquid to attach itself to a solid material surface, in our case the solid material is the porous-medium grains. In the air–water system that we are considering in the above experiments, water preferentially wets the surface of the tubes relative to air. Thus the water is drawn up the capillary tube because of its affinity for the sides of the tube.

One can also think of this from the solid grains point of view. Wettability could be thought of as the relative preference of the solid for one fluid relative to another.

The height of the *capillary rise* can be determined by a study of the balance of forces at the meniscus between the water and the air. From Figure 2.8 we see that the force pulling the water up the tube is the upward component of the *surface tension* γ_{aw}. Surface tension, you may recall, is the force acting on the surface of a liquid in contact with a gas (in this case air) that acts to minimize its surface area. If the angle between the meniscus and the tube wall is θ, then the upward component is $\gamma_{aw} \cos(\theta)$. Since it is acting along the line of contact between the air–water interface and the glass, the total force is $2\pi r \gamma_{aw} \cos(\theta)$. This is balanced by the weight of water pulling down. If the height of the water column above the surface of the liquid in the beaker is h_c, then the total downward force is $\pi r^2 h_c \rho g$. Balancing forces, we have

$$2\pi r \gamma_{aw} \cos(\theta) = \pi r^2 h_c \rho g, \qquad (2.21)$$

from which we obtain

$$h_c = \frac{2\gamma_{aw} \cos(\theta)}{r\rho g}. \qquad (2.22)$$

Defining $r^* = r/\cos(\theta)$ as the *radius of curvature* of the meniscus, we have

$$h_c = \frac{2\gamma_{aw}}{\rho g r^*}. \qquad (2.23)$$

We can now see why the water in the smaller capillary tube was drawn to a higher elevation. It had the smaller radius of curvature of the meniscus because the tube had the smaller diameter. This is also the reason why the water in the sand moved above the phreatic surface in the sand-packed column. The pathways between the grains act, in some sense, as capillary tubes. Thus the water is drawn into them, generating a pressure less than atmospheric in the neighborhood of the granular "capillary tube." Finer-grained material, which would normally have smaller diameter pores, causes water to move vertically upward from the water table further than water residing in the coarser-grained materials, which would normally have larger pores. An important corollary here is that, in a soil with a mixture of grain sizes, the *water is going to have a tendency to move higher in the finer-grained portions of the soil*.

Let us now discuss an example of capillary rise. Consider the case where the capillary rise in a capillary tube (or equivalent porous-medium channel) with a diameter of 1.0 mm is compared to that of a capillary tube of diameter 0.2 mm. Determine the relative height of the capillary rise above a common datum of one relative to the other. The rise of water in the 1 mm diameter tube is

$$h_{c_1} = \frac{2\gamma_{aw} \cos(\theta)}{r_1 \rho g} = \frac{2\gamma_{aw} \cos(\theta)}{(1\text{mm})\rho g}$$

and that for the capillary of 0.2 mm is

$$h_{c_2} = \frac{2\gamma_{aw} \cos(\theta)}{r_2 \rho g} = \frac{2\gamma_{aw} \cos(\theta)}{(0.2\text{mm})\rho g}.$$

The relative elevation of the water in the two tubes is

$$\frac{h_1}{h_2} = \frac{\dfrac{2\gamma_{aw}\cos(\theta)}{(1\text{mm})\rho g}}{\dfrac{2\gamma_{aw}\cos(\theta)}{(0.2\text{mm})\rho g}} = 0.2.$$

Thus the water level in the small diameter tube (0.2 mm) is five times greater than that in the 1 mm tube.

Let us now return to our earlier discussion of pressure. Since we know the relationship between the pressure head and the pressure—that is, Eq. (2.2)—we can determine from Eq. (2.23) the pressure due to capillarity. It is

$$p_c = h_c \rho g, \tag{2.24}$$

where we will define p_c as the capillary pressure. Capillary pressure is defined to remain nonnegative whenever there is more than one fluid present in the system. Because water is the wetting fluid, it will have a pressure less than the air pressure whenever both fluids are present, so we define $p_c \equiv (p_{\text{air}} - p_w)$. The air pressure p_{air} is often taken to equal atmospheric pressure, which is the reference pressure, so that $p_c = -p_w$ in this special case. The relationship found in Eq. (2.24) indicates that the higher the water rises in the capillary tube, the more negative the pressure at the top of the water column becomes. Do you think the pressure just above the point B in Figure 2.8 is influenced by the height of the water in the tube above this point? The answer lies in the fact that the total head must remain constant in a stationary fluid. Therefore, since the total head is the sum of the elevation head and the pressure head as given in Eq. (2.3), that is,

$$h(z) = h_p(z) + \zeta(z), \tag{2.25}$$

the pressure depends only on the height of the observation point above the water level in the beaker. In other words, if the left-hand side of Eq. (2.25) remains constant in time and space, as it must in a stationary fluid, the pressure head $h_p(z)$ (and therefore the pressure) must become more negative as the elevation head $\zeta(z)$ (the height above the water surface in the beaker) becomes more positive, that is, as the observation point moves up the tube. Thus the amount of water that exists above this point is immaterial. A point to ponder is whether this argument holds for points below the elevation of the surface of the water in the beaker.

Recall that we found from our earlier experiments that there appears to be a relationship between the saturation S and the negative pore-water pressure. This was illustrated in Figure 2.7. Thus, via Eq. (2.24), we also have a relationship between the capillary pressure and the saturation. If we were to conduct a series of experiments wherein we examined the saturation that is observed as the soil sample is subjected to various capillary pressures, we would be able to determine a relationship between the capillary pressure and the saturation. We might do this by very slowly raising the soil column relative to the water surface in the beaker and observing the change in pressure and saturation that occurs at a fixed point A on the soil column illustrated in Figure 2.9. Based on what we have just learned about the behavior of pressure in a capillary tube, we would expect that as the column is raised, the saturation should decrease at a given distance from the base of the column and the capillary pressure should increase.

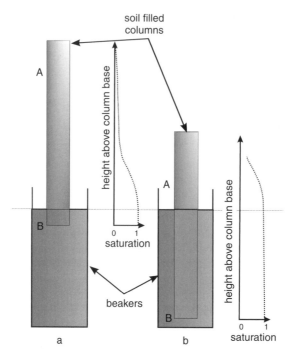

FIGURE 2.9. The experiment consists of raising the column of soil from the location pictured in (b) to the location pictured in (a). As a result of raising the column, the saturation at point A in the column is reduced due to drainage and the pressure becomes more negative.

In Figure 2.9a we have plotted the saturation curves observed when the column in panel Figure 2.9b is lifted. As expected, the saturation at the top of the column in (b) is higher than that at the top of the column in (a) since the capillary forces are capable of retaining more water at the shorter distance above the water surface. If you were to plot the head distribution, what would you expect to find? Since the water is not flowing, the head value must be the same anywhere in the system, so the head value will be equal to the elevation of the water level in the beaker. What would be the water pressure along the length of the column? The pressure head plus the elevation head must sum to the total head, which we have determined is a constant. Thus the pressure head will decrease linearly moving upward from the beaker surface since the elevation head is increasing linearly in that direction, assuming, as we learned earlier in Section 2.3, that the height of the observation point above the liquid surface is less than 10.3 m. Similarly, using the same logic, the water pressure below the beaker-water surface also will increase linearly. But unlike pressures less than atmospheric, there is no upper bound on this pressure that exceeds atmospheric.

The relationship between the capillary pressure and the saturation observed during the drainage of water out of the soil column *at any given point along its length* is called the *primary drainage curve.* The primary drainage curve is measured in a more formal setting using a device that operates on the principles illustrated in Figure 2.10. A saturated soil sample of known porosity is placed in a cell, which, in turn, is in contact with air at the top and water at the bottom. The sample is of sufficient size that it is statistically representative of the soil being examined. Such a sample has been given the

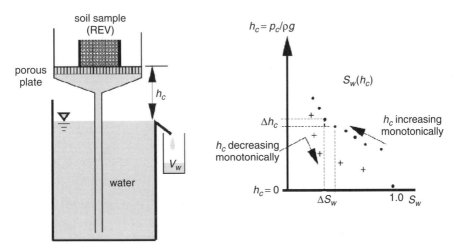

FIGURE 2.10. Experimental apparatus for the measurement of the saturation–capillary pressure curve. The plot on the right-hand panel shows both the drainage and imbibition (wetting) data points (from [2]).

name *representative elementary volume* or REV. The concept of a REV will be discussed further in Section 4.2.

The measurement process proceeds in a series of steps. The capillary head is increased in small increments, Δh_c, during each step. The increment Δh_c can be achieved by raising the sample relative to the water level in the beaker. Since the elevation head of the sample has now been increased, the pressure head must decrease, thus decreasing h_c. The key element of the process is the porous plate located between the bottom of the sample and the column of water. The porous plate is made of a material, usually ceramic, that has wetting properties and pore sizes such that water can flow through the plate, but air cannot. In other words, the pore openings are sufficiently small that the capillary pressure required for air to enter the pore (see Eq. (2.23)) is never reached in the experiment. If the air entry pressure of the plate—that is, the difference in pressure between the air at the plate surface and the water in the plate—is exceeded, air will enter the plate and eventually penetrate it. Once this occurs and air breakthrough is observed beneath the plate, the experiment is no longer viable.

After equilibrium in the system is reached (i.e., the water level in the measuring beaker no longer changes), the change in saturation, ΔS_w, is measured. It is easy to determine ΔS_w by observing the increase in volume of the water collected in the small beaker containing the volume V_w in Figure 2.10. Each data point appearing as a dot ($\bullet$) in the right-hand-side panel in Figure 2.10 represents a specified capillary pressure, and therefore capillary head, and an associated saturation. A plot of the best-fitting curve through these points would produce the primary drainage curve $S_w(h_c)$. The quantity $\Delta S_w / \Delta h_c$ (see Figure 2.10) is the slope of the primary drainage curve and is called the *specific water capacity*. The specific water capacity is a measure of the water held in storage in the unsaturated soil column.

We would expect the results of this experiment to yield a relationship such as shown in Figure 2.11. This curve, which is obtained using data such as shown in Figure 2.10, shows how the saturation and pressure change as the soil is drained. Let us begin at the point

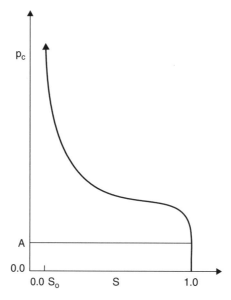

FIGURE 2.11. The saturation of water in an unsaturated soil is related to the capillary pressure (and therefore the total head), as shown diagrammatically.

where the saturation is 100% (the saturation value of 1.0 in Figure 2.11) and the pressure is 0.0. We now begin a series of experiments wherein the pressure is controlled. As the water pressure is reduced below atmospheric[4], the saturation at the point of measurement initially does not change. This phenomenon is indicated by the vertical portion of the curve on the right-hand side of Figure 2.11. The water pressure must continue to decrease (meaning capillary pressure p_c increases) beyond the value indicated by the letter A before the saturation will drop below 1.0. In other words, there is a minimum capillary pressure below which the water in the pores is held by capillarity and will not move. The point at which the water begins to drain is called the *threshold pressure*[5] (or entry pressure as used on page 67). It is at this point that the air begins to enter the largest pores and drainage begins. In fact, the curve shown in Figure 2.11 is called the *drainage curve*.

Interestingly, it is not possible to achieve a water saturation of 0.0 due only to drainage. Experiments show that there will always be some water isolated in the pores that cannot drain. We call this the *irreducible saturation* or *residual saturation* and denote it by the symbol S_0.[6] This curve is called the *primary drainage curve*, which, by definition, starts at $S = 1$ and terminates at $S = S_0$.

One can get to zero saturation if a phase change is invoked. Evaporation of water, for example, will eventually lower the water saturation below $S = S_0$ and eventually reduce it to zero.

We now continue our experiment. By raising the column as far as possible without lifting the lower end above the surface of the water in the beaker, we reduce the saturation as far as possible without resorting to a change of phase from water to gas (such as would

---

[4]A pressure below 0.0 is a positive suction pressure since suction is defined as a value below 0.0.

[5]The threshold pressure is also called the *critical pressure,* the *bubbling pressure,* and the *nonwetting fluid* (pressure) *entry* value.

[6]The residual saturation is also represented by the symbols S_r, S_{res}, and S_w^{res}.

occur in evaporation). Having achieved this goal, we now begin to lower the column back into the beaker (increase the pressure) and thereby allow water to enter the column from below. The experimental observations are recorded in Figure 2.12. The equivalent procedure has produced the plus signs (+) shown in Figure 2.10.

We observe that as water enters the porous medium, that is, as the saturation of water increases, the water pressure simultaneously increases (capillary pressure decreases). However, much to our surprise, the saturation observed for a given pressure during the drainage of the water is not the same as that obtained during the *imbibition* stage. The experiment shows that there is a different pressure–saturation relationship on drainage than there is on imbibition. Perhaps most remarkable is the observation that complete water saturation is never achieved, not even at a pressure of 0.0. This is due to the fact that some air will become isolated (completely surrounded by invading water) and thereby trapped in the sand column as water infiltrates. Thus complete saturation is not possible except through removal of the air through dissolution by the water. This irreducible gas phase saturation is called the *residual gas phase saturation*, S_{gr}, that is, $(1 - I)$ in Figure 2.12.

What is being observed here is the phenomenon of *hysteresis*. When a physical phenomenon is not exactly reversible, that is, the system responds differently depending on its history, it is exhibiting hysteresis. In our case, the saturation can have two different values for the same pressure, depending on whether one is observing drainage or imbibition. In fact, if we were to proceed to drain the column beginning at this residual phase saturation, we would obtain a drainage curve different from the initial curve (referred to as the primary drainage curve). This curve that begins at residual saturation is called a main drainage curve as it originates at $S < 1$ and terminates at $S = S_0$. With each subsequent complete drainage and imbibition cycle, we would find our experiment retraces the main curves. As such, the main curves are also called bounding curves. However, should

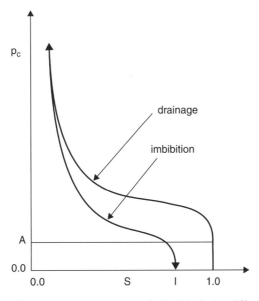

FIGURE 2.12. The capillary pressure–saturation relationship looks different on imbibition than on drainage. This is called hysteresis.

we begin drainage or imbibition from a point other than the endpoints just discussed, a new curve would form called a scanning curve. There is no practical limit to the number of scanning curves that can be generated. Therefore for any given pressure there are an infinite number of possible saturations, bounded by the main drainage and main imbibition values.

A set of experimentally determined scanning curves is presented in Figure 2.13. Let us examine the results of this experiment within the hysteresis context. The primary drainage curve is identified by the curve a. At initiation of drainage, the saturation is 1.0 at a pressure of atmospheric (i.e., 0.0). As the capillary pressure increases (water pressure decreases), the saturation reduces to a value of S_{lr}, the residual liquid water saturation. The capillary pressure is then increased with a concomitant increase in saturation as described by the primary imbibition curve, d. Imbibition continues until the saturation reaches a maximum of S_{gr}, the residual gas saturation. If the capillary pressure is increased, drainage is described by curve b, the main drainage curve. Finally, if the pressure is decreased beginning at a saturation of about 0.36, the imbibition is described by curve c. The cycle need not stop at this point but can continue with the subsequent generation of additional scanning curves.

Why the phenomenon of hysteresis occurs is an interesting topic—too intricate to be covered here but addressed in more detail in Chapter 11.

2.5 THE DARCY EXPERIMENT

In the preceding section we focused on mechanical energy and its role in defining the state of a fluid at rest and in equilibrium. In this section we extend these concepts to

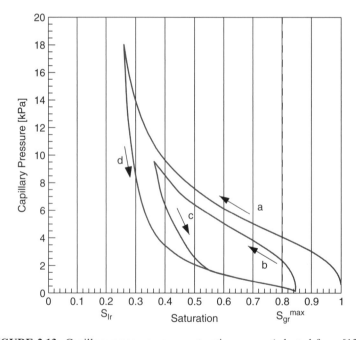

FIGURE 2.13. Capillary pressure versus saturation curves (adapted from [13]).

consider the physical laws that govern the movement of fluids in a porous medium, such as encountered in the subsurface. In 1856 Henry Darcy, Inspector General of Bridges and Roads in the town of Dijon, France, published an article entitled "Les Fontaines Publiques de la Ville de Dijon, Exposition et Application des Principes a Suivre et des Formules a Employer dan les Questions de Distribution d'Eau Ouvrage Terminé par un Appendice Relatif aux Fournitures d'Eau de Plusieurs Villes au Filtrage des Eau et la Fabrication des Tuyaux de Fonte, de Plomb, de Tole et de Bitume."[7] According to Hubbert [4] Henry Darcy had been "engaged in modernizing and enlarging the public water works of the town of Dijon" and the above referenced document was an engineering report on that effort. Of interest to us is the brief, four page appendix to this report entitled "Determination des Lois d'Ecoulement de l'eau à travers le sable" which translates [4] to "Determination of the Law of Flow of Water Through Sand." The report is in response to problems encountered by Darcy in his efforts to develop a suitable filter for the water supply of Dijon. His goal was to determine how large a filter would be required to provide filtration to a specified amount of water per day.

To understand the importance of Henry Darcy within the context of his time, the following description provided by Paul Darcy, a later relative of Henry Darcy, is of interest [5]:

> ...to understand his enthusiastic recognition by the people of Dijon, it must be recalled that the town had at its disposition only wells plus the waters, often inadequate, always in short supply and dirty, of the Ouche, that the well waters were not protected from contamination of all kinds; that the town was crossed by the ancient bed of a small stream, the Suzon, and that this ditch, uncovered over almost all its length with no part paved, served over a length of 1,300 meters as the main sewer for wastes of every kind, for dwellings and adjoining industrial establishments (such as the tripe-makers of the rue du Bourg); that it was never cleaned, that it was never flushed out, and that, above all during hot weather, the town was poisoned by pestilential odors [translated by J. Phillip [6]]

In addition to the important contribution that Darcy[8]: made to the quality of drinking water in Dijon through his introduction of the concept of filtration, he was also recognized for his unselfish devotion to the betterment of the city. For example, he forwent his professional fee of approximately $1,500,000 in payment for his work as designer and project manager of the water system, for a medal struck in his honor, upon which was inscribed [6,7]:

> *To H-P-G Darcy*
> *Chief Engineer*
> *of the Department of the Côte d'Or*
> *He conceived the project*
> *made all the studies*
> *pursued to the end the execution*
> *of the works to which Dijon*
> *owes the creation and abundance of its public waters*
> *He accepted neither pecuniary remuneration*

---

[7] A loose translation of the title is: Presentation and Application of the Guiding Principles and Formulas to Use Relating to Questions of Water Distribution. An appendix lists suppliers, for a number of cities, of water filters and fabricators of pipes used in water distribution made of cast iron, lead, sheet metal, and bitumen.

[8] For a very interesting tour through the life of Henry Darcy, the reader is referred to http://biosystems.okstate.edu/darcy/index.htm.

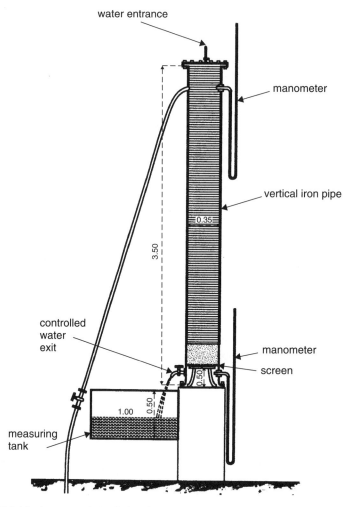

FIGURE 2.14. Apparatus intended to determine the law of the water flow through sand.

nor even payment of his personal expenses
The Municipal Council offers this testament
of public appreciation to H-P-G Darcy
doubly benefactor
of his native town
through his talent
through his désintéressement

Let us now examine the legendary experiment conducted by Darcy. The apparatus he used is reproduced as Figure 2.14.

The experiment, as described by Darcy, is as follows[9]:

[9]The following passage is a translation of the original paper and provided in Brown [7].

The apparatus employed pl. 24, fig.3. [our Figure 2.14] consists of a vertical column 2.50 m in height[10] formed from a portion of conduit 0.35 m interior diameter, and closed at each of its ends by a bolted plate.

In the interior and 0.20 m above the bottom, is a horizontal partition with an open screen, intended to support the sand, which divides the column into two chambers. This partition is formed by the superposition upwards on a iron grid with prismatic bars of 0.007 m, a cylindrical grizzly of 0.005 m, and finally a metal cloth with mesh of 0.002 m. The spacing of the bars of each grid is equal to their thickness, and the two grids are positioned so that their bars are perpendicular to one another.

The higher chamber of the column receives water by a pipe connected to the hospital water supply, and whose tap makes it possible to moderate the flow at will. The lower chamber opens by a top on a gauging basin, 1 meter on a side.

The pressure at the two ends of the column is indicated by mercury U-tube manometers[11]. Finally each of the chambers is provided with an air tap, which is essential for filling apparatus.

The experiments were made with siliceous sand of the Saone[12], composed as follows[13]:

Percent Larger	Grain Size
58	0.77 mm
13	1.10 mm
12	2.00 mm
17	Undifferentiated gravel shells, etc.

It has approximately 38/100 void[14].

The sand was placed and pack [sic] in the column, which beforehand had been filled with water, so that the sand filter voids contained no air, and the height of sand was measured at the end of each series of experiments, after that the passage of water had suitably pack [sic] it.

Each experiment consisted of establishing in the higher chamber of the column, by the operation of the supply tap, a given pressure. Then, when by two observations one had ensured oneself that the flow had become appreciably uniform, one noted the flow in the filter during a certain time and one concluded the medium flow per minute for it.

For weak heads, the almost complete lack of motion of the mercury in the manometer made it possible to measure to the millimeter, representative of 26.2 mm of water. When one operated under strong pressures, the supply tap was almost entirely opened, and then the manometer, in spite of the diaphragm that it was provided, had continuous oscillations. Nevertheless, the strong oscillation[s] were random, and one could appreciate, except for 5 mm, the average height of mercury, i.e. know the water pressure within 1.30 m.

All these manometer oscillation[s] were due to water hammer produced by the play of the many public faucets in the hospital, where the experimental apparatus was placed.

---

[10]The diagram of the apparatus shows a height of 3.5 m.

[11]A manometer is a device that measures pressure. It normally defines pressure as the height of a liquid, often mercury, that is required to counterbalance the measured pressure.

[12]The Saone is a river in eastern France, flowing south into the Rhone at Lyon and passing to the southeast of Dijon.

[13]The following table has been modified to clarify the intent of the original table.

[14]The fracton 38/100 indicates a porosity of 38%.

Experiment Number	Duration min	Mean Flow L/min	Mean Pressure m	Ratio of Volumes and Pressure	Observations
1st Series, with a thickness of sand of 0.58 m					
1	25	3.60	1.11	3.25	Sand was not washed
2	20	7.65	2.36	3.24	
3	15	12.00	4.00	3.00	
4	18	14.28	4.90	2.91	]
5	17	15.20	5.02	3.03	
6	17	21.80	7.63	2.86	]The manometer column
7	11	23.41	8.13	2.88	
8	15	24.50	8.58	2.85	]had weak movements
9	13	27.80	9.86	2.82	
10	10	29.40	10.89	2.70	]
					]
					]Very strong oscillations.
					]
					] Strong manometer] oscillations.
2nd Series, with a thickness of sand of 1.14 m					
1	30	2.66	2.60	1.01	Sand not washed.
2	21	4.28	4.70	0.91	
3	26	6.26	7.71	0.81	
4	18	8.60	10.34	0.83	
5	10	8.90	10.75	0.83	
6	24	10.40	12.34	0.84	]
					]Very strong oscillations.
					]
3rd Series, with a thickness of sand of 1.71 m					
1	31	2.13	2.57	0.83	Washed sand
2	20	3.90	5.09	0.77	
3	17	7.25	9.46	0.76	] Very strong oscillations.
4	20	8.55	12.35	0.69	
					]
4th Series, with a thickness of sand of 1.70 m					
1	20	5.25	6.98	0.75	Sand washed, with a grain size a little coarser than the proceeding.
2	20	7.00	9.95	0.70	
3	20	10.30	13.93	0.74	Low oscillations because of the partial blockage of the manometer opening

FIGURE 2.15. Table of the experiments made in Dijon on October 29 and 30, and November 2, 1855.

All pressures have be report [sic] relative to the level of lower face of the filter, and no account has been taken of friction in the higher part of the column, which is obviously negligible.

The table of the experiments [see our Figure 2.15] like their chart, show that the flow of each filter grows proportional with the head.

For the filters operated, the flow per square meter-second, (q) is related very roughly to the load, $(h_1 - h_0)$ by the following relations[15]:

1st series $q = 0.493 \ (h_1 - h_0)$ $\quad$ 3rd series $q = 0.126 \ (h_1 - h_0)$
2nd series $q = 0.145 \ (h_1{}_-h_0)$ $\quad$ 4th series $q = 0.123 \ (h_1 - h_0)$

By calling I, the load proportional per meter thickness of the filter, these formulas change into the following[16].

1st series $q = 0.286 \ I$ $\quad$ 3rd series $q = 0.216 \ I$
2nd series $q = 0.165 \ I$ $\quad$ 4th series $q = 0.332 \ I$

The differences between the values of coefficient q/I results from the sand employed not being constantly homogeneous. For the 2nd series, it had not been washed; for the 3rd, it was washed; and for the 4th, it was very well washed and a little larger in grain size.

It thus appears that for sand of comparable nature, one can conclude that output volume is proportional to the head and inversely related to the thickness of the layer traversed. . . .

In summary, Darcy constructed a saturated, sand-filled, vertically oriented column. In each experiment he specified a flow rate of water entering the top of the column and draining from the bottom. Using two mercury manometers he was able to measure the drop in head between the upper manometer and the lower. The length of the sand-filled portion of the column and the characteristics of the packed sand were variable from one experiment to another. He reported pressures in terms of the height of water (in meters) that one would observe in a water manometer, although his measurements were, in fact, made on mercury manometers.[17]

Using ideas drawn from our earlier discussion of pressure, head, and fluid potential in Section 2.1, we can gain further insight into the observations made by Darcy. Darcy measured the pressure below the sand using the lower manometer and the pressure above the sand using the upper manometer. Note that the sand thickness was, at the most, 1.71 m and the column length was 2.5 m,[18] so there was a water column above the top of the sand. Thus he was measuring the pressure, in terms of meters of water in the fluid volume above and below the sand rather than water pressure in the sand itself.

In the experiment discussed above, Darcy reported his pressures as water level elevations relative to the level of the lower face of the filter. By reporting the water levels relative to the bottom of the sand column, Darcy was actually reporting the head values. In other words, he is adding the elevation head, as determined from the manometer entrance to the column, to the pressure head, as determined by the gauge pressure in the manometer at the manometer entrance (see Figure 2.16).

Darcy presented the proportionality relationships later in his paper as

$$Q = \frac{KA}{l}(h_1 - h_0), \tag{2.26}$$

[15]The nomenclature in these and the equations to follow in this quote has been changed to be consistent with that used elsewhere in this book. The head h_1 is measured in the upper manometer and h_0 is the value measured in the lower.

[16]$I \equiv (h_1 - h_0)/e$ where e is the length of the sand-filled column.

[17]The pressure of 1 mm of mercury is equivalent to 13.6 mm of water.

[18]If one accepts the improbable dimensions appearing on the apparatus drawing as accurate, the column length was a meter longer.

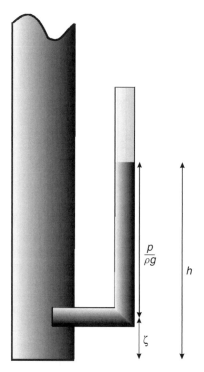

FIGURE 2.16. The strategy Darcy used to measure his pressure was to report pressure as water level elevations relative to the bottom of the column. Thus he was providing the quantity h in this figure. The quantity h is the head as we have defined it.

where Q is the total discharge from the bottom of the column, A is the cross sectional area of the sand column, K is the permeability coefficient[19], and l is the length of the sand column. Eq. (2.26) is the famous *Darcy's law*.

In Figure 2.17 we present a plot of the measurements reported in the table presented as Figure 2.15. In Figure 2.15, the first column records the number of the experiment within each series. Each experiment involves a different value of fluid flow. Each series, in turn, reports the results for a specified sand and specified sand column length. It is interesting to note that the shorter column allowed for the realization of higher flow rates for a given inlet pressure[20], so more experiments are reported in this series. The second column records the duration of the experiment, from which one can, knowing the volume of water collected in the gauging basin, determine the average discharge rate, which is reported in the third column in units of liters per minute. The fourth column is deceptive in that it purports to present the pressure, while in fact it is reporting the difference in pressure between the two manometers. Recall, in this experiment, that the difference in pressure is actually the difference in head because the water pressure is reported in terms of height of water above the base of the column of sand. The fifth column divides the mean flow reported in column three by the head loss reported in column four.

[19]Darcy did not call K the permeability coefficient, but rather "a coefficient dependent on the permeability of the layer."

[20]The maximum inlet pressure was limited by the pressure of the water supply system of the hospital.

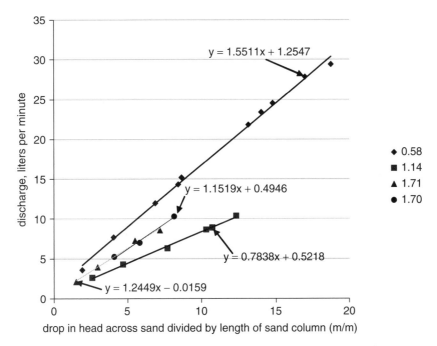

FIGURE 2.17. Results of a series of four experiments conducted by Darcy in 1855. The equations are those of the best-fitting linear regression line; the key indicates the thickness of the sand.

In Figure 2.17, for each experiment in each series, we have plotted on the ordinate the discharge Q and on the abscissa the head difference divided by the sand column length l. The equation of the best-fitting linear regression line for each series is presented along with the data. The key in the figure shows that the series are identified by the length of the sand column, which is the same approach used in the table.

The coefficient multiplying x in these equations represents the slope of the line. Thus, for series 0.58, for example, $\Delta Q/(\Delta h/e) = 1.5511$. The units of this ratio are $[L/T]/[L/L] = L/T$. From Eq. (2.26) we see by rearrangement that

$$K = \frac{Ql}{A\,(h_1 - h_0)},\tag{2.27}$$

from which we can obtain, using the information from Figure 2.17, the information in Table 2.1.

TABLE 2.1. Calculation of K Using the Experimental Data Provided by Darcy

Series	$Ql/\,(h_1 - h_0)$	$K = Ql/\,(h_1 - h_0)/A$ (L/min)	K (cm/s)
0.58	1.55	0.0161	0.0268
1.14	0.784	0.00816	0.0136
1.71	1.25	0.0130	0.0217
1.70	1.15	0.0119	0.0198

The parameter K appearing in Table 2.1 is generally called the *hydraulic conductivity* and the term *permeability coefficient* is used in another context to be discussed later. Thus we conclude that the proportionality factor that relates the gradient in head to the total discharge, that is, $(h_1 - h_0)/l$, is the hydraulic conductivity multiplied by the cross-sectional area of the sand column. Typical values of this parameter for a range of soil types are found in Figures 2.18 to 2.22.

Having considered the experiments by Darcy, a few questions regarding the results and interpretation come to mind.

1. The column was oriented, quite logically, such that its axis was vertical. Would the results have been different had the column been oriented otherwise, horizontally perhaps?

2. Water was introduced at the top of the column and drained from the bottom. What would one expect had the water entered the bottom and exited the top?

3. The pressure measurements were made above and beneath the sand column. If the measurements were made in the sand column itself, would the results be different?

4. The column was saturated with water; no air was present. Would unsaturated flow, that is, flow occurring in the presence of some air in the pores, be described by the same experimental relationship, namely, Eq. (2.26)?

Would the results of the experiment vary if the apparatus were oriented such that the long axis of the column were not vertical? The answer lies in an examination of the equation derived by Darcy, that is, Eq. (2.26). The driving force in this equation is the difference in the fluid potential, here expressed in terms of a head difference between the two manometers. The potential energy gradient defined by the difference in these head values is transferred to thermal energy by virtue of friction losses caused by the movement of the fluid around the grains. If the amount of potential energy loss in the two experiments is a constant, that is, the difference in head between the two manometers remains the same for each experiment, and the length of the column and therefore the energy dissipated by friction does not change, then the flow rate should not change. In conclusion, the orientation of the column does not make any difference. In fact, the water could have been injected into the bottom of the column and have exited through the top and the results of the experiment would be the same. The stated equivalence assumes that the velocity is below that required to cause the sand grains to move vertically upward against gravity. The first and second questions are therefore addressed.

The answer to question 3 draws once again on our knowledge of pressure and hydraulic head. The lower manometer is located at the base of the sand and would appear to represent to a reasonable degree the pressure and head at that point. The manometer at the top of the column, however, is located, and measures the pressure, above the sand (recall that the sand does not entirely fill the column). Does the measurement of the pressure in the fluid column above the sand represent conditions at the top of the sand?

Recall that in Section 2.1 we found that the head in a fluid column at rest is a constant. Any increase in elevation head is offset by a decrease in pressure head. Thus, if the fluid in the column above the sand were at rest—that is, motionless—the head would be the same anywhere along its length, whether at the top of the sand or the top of the column. But, you argue, the water in the column is not motionless; it must be moving because water is entering the top of the column and exiting the bottom. Correct, but the loss in energy due to the movement in the column of water above the sand is extremely small

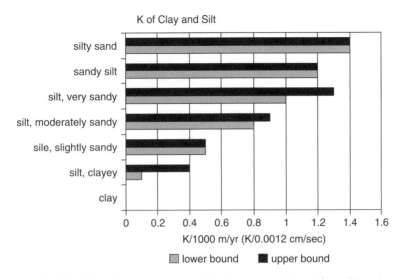

FIGURE 2.18. Hydraulic conductivity for silts and clays (from [8]).

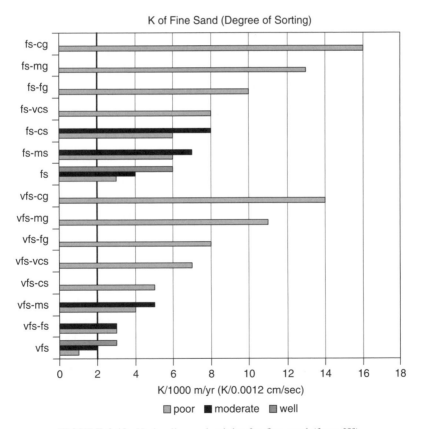

FIGURE 2.19. Hydraulic conductivity for fine sand (from [8]).

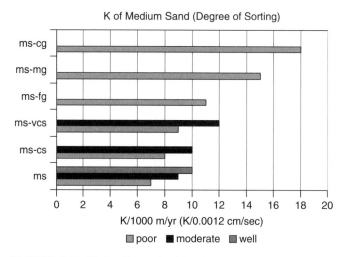

FIGURE 2.20. Hydraulic conductivity for medium sand (from [8]).

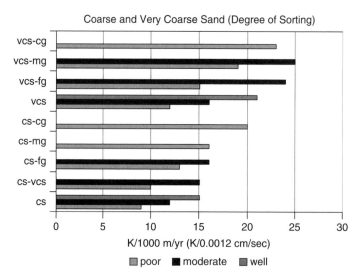

FIGURE 2.21. Hydraulic conductivity for coarse and very coarse sand (from [8]).

compared to friction losses as the water passes through the sand-filled portion of the column (think about the capillary-sized openings in the soil as compared to the relatively huge opening of the column above the soil). While the head at the manometer port may not exactly represent that at the surface of the sand, it is for practical purposes the same. Since we are interested in the head loss across the sand, and not the pressure per se, measurement of head in the column of water above the sand to determine the head at the top of the sand-filled portion of the column is quite acceptable.

The last question is more challenging. To answer it we must modify our experiment. Consider the setup illustrated in Figure 2.23. The sand is initially dry. We introduce water at the top at a small and controlled rate and allow the system to reach equilibrium. In

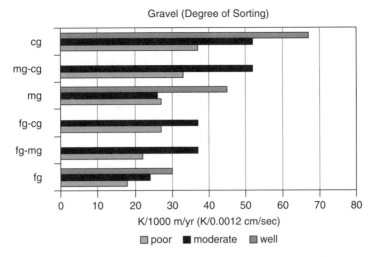

FIGURE 2.22. Hydraulic conductivity for gravel (from [8]).

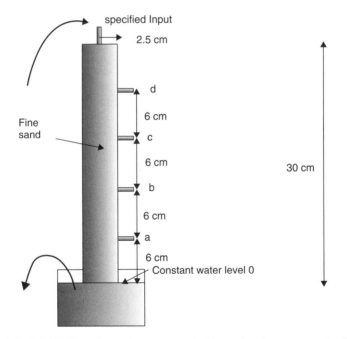

FIGURE 2.23. Experimental setup to study Darcy flow in unsaturated soil.

other words, we wait until the amount of water coming out the bottom of the column equals that being introduced at the top. The rate is set (i.e., limited) so that the sand column never becomes saturated.

At four locations along the length of the column we have installed pressure measuring devices. Because we are measuring water pressures that are below atmospheric due to the fact that the sand is not saturated, a special device is required.

At the end of the measuring tube that is imbedded in the sand, there is a small cup that is made of ceramic that allows water but not air to enter. In fact, the cup is made of the same basic kind of ceramic used in the experiment shown in Figure 2.10. It has pores that are of such a size that the surface tension associated with the water resident in the soil pores of the ceramic blocks the entry of air. In other words, the air entry pressure of the cup is sufficiently great that air does not enter under typical laboratory conditions. Of course, if sufficient suction is applied, air will eventually enter, but the cup is designed to preclude air invasion over the range of pressures of interest to us.

The opposite end of the measuring instrument, that is, the end external to the column, is attached to a pressure measuring device. The tension of the water in the sand generates a suction (or capillary) pressure, which is communicated to this device. This is a variant on an instrument generally described in the literature as a *tensiometer*.

Recall that we established earlier that the total head was the sum of the pressure head and the elevation head:

$$h(z) = h_p(z) + \zeta(z). \tag{2.28}$$

The pressure head h_p is determined from the measured pressure and Eq. (2.4). Thus, in our experiment, the total head at point a is

$$h(a) = h_p(a) + (a - 0), \tag{2.29}$$

and a similar calculation can be made for the other locations b, c, and d.

At this point, we have a measurement of the total head at the four locations in the sand column and the total rate of flow within the column can be measured. According to Darcy's law given earlier,

$$Q = -\frac{KA}{L}\Delta h, \tag{2.30}$$

where $\Delta h \equiv h_i - h_{i-1}$, $z_i > z_{i-1}$, $L = z_i - z_{i-1}$, and the coordinate axis is positive in an upward direction. Thus, if flow is down, that is, $Q < 0$, then $h_i > h_{i-1}$ results in the right-hand side of Eq. (2.30) being negative (downward flow), as we would expect.

At this point we have from our experiment values for all of the variables in Eq. (2.30) with the exception of K, the hydraulic conductivity. We record the various values that we have measured and repeat the experiment with an increased flow rate. In general, we have the following relationship, just as Henry Darcy did in 1855 [9]:

$$Q_i = -\frac{K_i A}{L}\Delta h_i \quad i = 1, \ldots, n, \tag{2.31}$$

where i is the experiment and n is the total number of experiments.

We now tabulate the information we obtain from two experiments, one with the column saturated and the other with the column unsaturated. To obtain saturated conditions, the experiment is modified such that water flows from the bottom of the column and discharges from the top. Using this experimental procedure, one assures that the column remains saturated (except for small quantities of entrapped air that can be removed using carbon dioxide gas prior to introducing the water). The results of these experiments are shown in Figure 2.24. The curve with the slope of 0.1167 is obtained under saturated

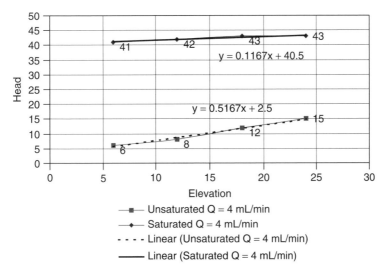

FIGURE 2.24. Head versus elevation measurements obtained using the unsaturated column flow apparatus shown in Figure 2.23.

TABLE 2.2. Information for Saturated–Unsaturated Darcy Experiment

Experiment	Q	A	dh/dx	K
Saturated flow	4.000 cm^3/min	19.63 cm^2	−0.1167	1.746 cm/min
Unsaturated flow	−4.000 cm^3/min	19.63 cm^2	0.51670	0.4011 cm/min

conditions. The unsaturated experiment results provide a slope of 0.5167. Using the information from our experiment and Eq. (2.30), we obtain the information presented in Table 2.2.[21]

It is evident from this table that the hydraulic conductivity, K, is much lower for the case of unsaturated flow. The reason lies in the observation that the more air that exists in a column of partially saturated soil, all other things being equal, the lower the hydraulic conductivity. In essence, the air isolated in the pore space is blocking the water from flowing through the porous medium.

As an example of the relationship between the degree of saturation and the hydraulic conductivity, we take a peek preview of what will be considered in more detail in Section 11.2.2. In Figure 2.25 we see the relationship between the saturation, the pressure, and hydraulic conductivity.

In the top panel of Figure 2.25 is plotted the hydraulic conductivity K versus the suction p_c.[22] Suction is a synonym for negative pressure in this example since the water pressure is lower than atmospheric, taken as reference. As the suction (or pressure less than atmospheric) increases, the hydraulic conductivity decreases. The reason that the

---

[21]A comprehensive discussion of the methodology for measuring saturated hydraulic conductivity is found in [10].

[22]One pascal (Pa) is the pressure resulting from the force of 1 newton (N) exerted over 1 square meter. One pascal equals 1.450377×10^{-4} pounds/in.^{2} and corresponds to the pressure of a column of water 0.1 mm in height.

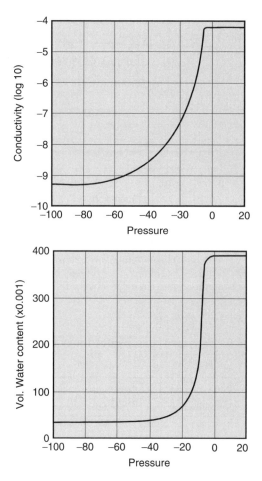

FIGURE 2.25. Relationship between water content, conductivity (m/s), and pressure (in cm of water) (adapted from [11]).

hydraulic conductivity decreases with increasing suction pressure is found in the lower panel of Figure 2.25. In this figure, which plots the volumetric water content θ_v versus the capillary pressure less than atmospheric, p_c (or the magnitude of the suction), one sees that the volumetric water content decreases as the suction increases. Recall that the water content is related to the saturation through the formula $\theta_v = S_w \times \varepsilon$, where S_w is the water saturation and ε is the porosity. Thus, given the porosity is constant with changes in pressure[23], the saturation decreases as the suction (negative gauge pressure, or pressure less than atmospheric) increases. We have seen this previously in Figures 2.10–2.13.

The suction and pressure are both aligned along the abscissa (horizontal axis), although the axis orientation is reversed because the suction increases as the pressure decreases. Taking the axis orientation into account, one observes that as the hydraulic conductivity increases, the water content, and therefore the saturation of water, also increases. Thus,

[23] In the case of sand, the assumption that the porosity does not change significantly over ranges in pressure normally encountered in the field is a reasonable one.

as observed earlier, the more air present in the soil, the lower the hydraulic conductivity, all other things being held constant.

2.5.1 Extended Forms of Darcy's Law

The Darcy experiment related the total discharge Q to the gradient in head, that is $\Delta h / l$. The proportionality factor was found to be KA, where K is the hydraulic conductivity and A is the cross-sectional area of the column through which the water is flowing.

We define the *specific discharge*, or *volumetric flux*, as the volume of fluid passing through a porous medium (fluid plus solid) with cross section of area A perpendicular to the flow per unit time, divided by the cross-sectional area of the sample. We denote this term by the symbol q, so that

$$q = \frac{Q}{A}. \tag{2.32}$$

We may also consider the Darcy experiment with a very fine spatial resolution of head measurements along the length of the column. In the limit, these measurements would be separated by a differential distance dl, with an associated change in hydraulic head dh. In this case, the Darcy equation tells us that the specific discharge is proportional to dh/dl, such that

$$q = \frac{Q}{A} \sim K \frac{dh}{dl}.$$

Because flow is always from regions of higher head to lower head, the equation for Darcy flow requires a negative sign, so that in the differential limit we write

$$q = -K \frac{dh}{dl}. \tag{2.33}$$

Equation (2.33) is a constitutive relationship; that is, it is determined through experiment. In this particular case, one can also show that this expression is equivalent to the momentum balance equation, given certain restrictive assumptions. The relationship between the generalized momentum balance equation for a fluid flowing in a porous medium has been derived by a number of authors. A classic paper on this topic was done by Hubbert in 1957 [4]. Therein he derived Darcy's law from the Navier–Stokes equation, which describes the motion of a viscous fluid. While less mathematically sophisticated than modern derivations (e.g., see Gray and O'Neill [12]), the fundamental concepts are present in Hubbert's paper. As a consequence of his derivation, he arrived at the following expression for the flow of fluid through a saturated soil:

$$\mathbf{q} = -(Nd^2)(\rho/\mu)g \cdot \nabla h \tag{2.34}$$

where N is a grain-shape factor, d is the diameter of randomly packed uniform spheres, ρ is the fluid density, g is the gravitational acceleration vector, and μ is the fluid viscosity.

The boldface notation implies vector quantities. In this case, the general flow vector $\mathbf{q} \equiv (q_x, q_y, q_z)$, where q_x, q_y, q_z are the components of the flow vector in the x, y, z coordinate directions.

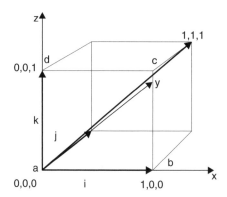

FIGURE 2.26. Definition sketch of the concept of the unit vector as used in vector calculus.

The term ∇h is the three-dimensional equivalent of the space derivative dh/dx. The operator ∇ is the gradient operator and is defined as

$$\nabla\,(\cdot) = \frac{\partial\,(\cdot)}{\partial x}\mathbf{i} + \frac{\partial\,(\cdot)}{\partial y}\mathbf{j} + \frac{\partial\,(\cdot)}{\partial z}\mathbf{k}, \qquad (2.35)$$

where $\mathbf{i}, \mathbf{j}, \mathbf{k}$ are the unit vectors in the three spatial coordinate directions. A geometrical interpretation of the unit vector is provided in Figure 2.26. A vector drawn from $(0, 0, 0)$ to $(1, 1, 1)$, that is, $\mathbf{l}$, is represented by the vector sum of vectors $\mathbf{i}, \mathbf{j}$, and $\mathbf{k}$. It will have a length of $\sqrt{1^2 + 1^2 + 1^2} = \sqrt{3}$. If instead of multiplying $\mathbf{i}, \mathbf{j}$, and $\mathbf{k}$ by unity, one multiplies by $\partial h/\partial x, \partial h/\partial y$ and $\partial h/\partial z$ respectively, the vectors in the x, y, and z directions would be stretched and the resulting vector sum would be a new vector replacing $\mathbf{l}$. The new vector would represent the three-dimensional gradient of h in the x, y, z plane as represented by ∇h in Eq. (2.35). We will shortly demonstrate these concepts with a concrete example.

Hubbert defined the product Nd^2 to be the permeability of the soil and gave it the symbol k, that is, $k \equiv Nd^2$. Note that the permeability is a function only of the geometry of the soil grains and their arrangement or packing. He represented the hydraulic conductivity K as

$$K \equiv (Nd^2)(\rho/\mu)g = \frac{k\rho g}{\mu}.$$

For porous media that have structure in their packing, the ease with which fluid can move through a medium may be different in different directions. Because hydraulic conductivity is a measure of the ability of the fluid to flow through the medium, this directional difference means that the hydraulic conductivity can have different values for different directions of flow. For the simple case when flow is equally easy in all directions, there is only one value of K, and the medium is called *isotropic*. When K changes with direction, the material is called *anisotropic*.

When hydraulic conductivity differs for different directions, we need to redefine the hydraulic conductivity so that it can have multiple values at any point in space. We do this by expanding the conductivity from a scalar to a matrix of values that we call a *tensor*. We denote the hydraulic conductivity tensor by the boldface symbol $\mathbf{K}$. Using this notation, Darcy's law is written in the form

$$\mathbf{q} = -\mathbf{K} \cdot \nabla h. \tag{2.36}$$

For a three-dimensional system, the tensor $\mathbf{K}$ will be a 3×3 matrix. To understand what is implied when one writes the hydraulic conductivity as a tensor, let us examine its elements. We begin with the simplest form of the tensor $\mathbf{K}$, that is, we assume a diagonal structure for it such that

$$\mathbf{K} = \begin{bmatrix} K_{xx} & 0 & 0 \\ 0 & K_{yy} & 0 \\ 0 & 0 & K_{zz} \end{bmatrix}. \tag{2.37}$$

In this case the three components of $\mathbf{K}$ denote hydraulic conductivity in each of the three coordinate directions. When introduced into Eq. (2.36) and expanded, this yields

$$\begin{Bmatrix} q_x \\ q_y \\ q_z \end{Bmatrix} = - \begin{bmatrix} K_{xx} & 0 & 0 \\ 0 & K_{yy} & 0 \\ 0 & 0 & K_{zz} \end{bmatrix} \begin{Bmatrix} \partial h/\partial x \\ \partial h/\partial y \\ \partial h/\partial z \end{Bmatrix}. \tag{2.38}$$

Multiplying the first row of the matrix in Eq. (2.38) by the column vector on its right-hand side, one obtains

$$q_x = -K_{xx}\frac{\partial h}{\partial x} + 0\frac{\partial h}{\partial y} + 0\frac{\partial h}{\partial z} \tag{2.39}$$

$$= -K_{xx}\frac{\partial h}{\partial x}. \tag{2.40}$$

Similar expressions can be obtained for the remaining two rows of the matrix.

Equation (2.39) states that the specific discharge in the x coordinate direction is proportional to the component of the head derivative in that direction, and that the proportionality constant is the diagonal element of the hydraulic conductivity matrix identified with that direction, namely, K_{xx}. However, since K_{xx}, K_{yy}, and K_{zz} need not be equal, the flux in different coordinate directions may be different even if the component of the head gradient in all directions were the same. The directional dependence of hydraulic conductivity is a common property in layered soils and is called anisotropy. It is often observed that the magnitude of the vertical component of hydraulic conductivity in a horizontally layered soil or rock can be several orders of magnitude less than the horizontal. Figure 2.27 illustrates the situation where, due to stratification or layering of sediments, the hydraulic conductivity in the x direction is larger than that in the z direction.

The form of the $\mathbf{K}$ matrix will change if the coordinate axes are not coincident with the directions represented by the maximum and minimum hydraulic conductivity values, where the maximum and minimum usually correspond to directions parallel and perpendicular to stratigraphic layering, respectively. Figure 2.28 illustrates just such a situation. The maximum values are along x' and z' and these directions are not coincident with the x and z coordinate directions. Under these circumstances, the $\mathbf{K}$ matrix takes the form

$$\mathbf{K} = \begin{bmatrix} K_{xx} & K_{xy} & K_{xz} \\ K_{yx} & K_{yy} & K_{yz} \\ K_{zx} & K_{zy} & K_{zz} \end{bmatrix}. \tag{2.41}$$

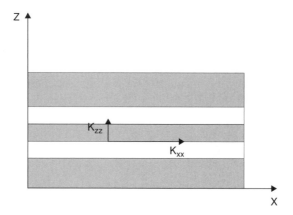

FIGURE 2.27. Illustration of the effect of lithology (layering of sediments in this case) on hydraulic conductivity when the coordinate axes are oriented such that they are in alignment with the layers.

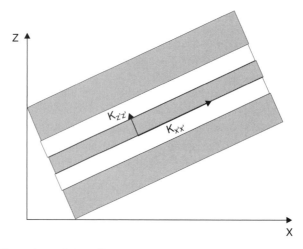

FIGURE 2.28. Illustration of the effect of lithology (layering of sediments in this case) on hydraulic conductivity when the coordinate axes are oriented such that they are not in alignment with the layers.

The off-diagonal elements, such as K_{xz}, can be interpreted as the coefficients describing the effect of a gradient in the z coordinate direction on flow across a face orthogonal to the z coordinate direction. If we now expand the first line in Eq. (2.36) using Eq. (2.41), we obtain

$$q_x = -\left[K_{xx}\frac{\partial h}{\partial x} + K_{xy}\frac{\partial h}{\partial y} + K_{xz}\frac{\partial h}{\partial z} \right],$$

which clearly shows that the flow in the x coordinate direction now depends on all three components of the head gradient.

In practice, one attempts to orient the coordinate axes such that they are collinear with (in line with) the major and minor components of the hydraulic conductivity tensor. The reason for this is to minimize the computational effort required to solve the equations that model the system. However, if the direction of the maximum and minimum values of hydraulic conductivity change their orientation in space, it is not easy to provide the required coordinate axes.

Before leaving this section, we note once again that while the dimensions of the specific discharge vector q are [L/T], this is not a measure of water velocity. The specific discharge is a measure of volume of water passing through a surface area A in time ΔT, divided by area A and ΔT, where the area A is the area of total porous medium, including void and solid. The actual velocity of water is determined by dividing by the area through which the water flows, as opposed to the total area. Because porosity is the ratio of void volume to total volume, the water area may be expressed as the total area multiplied by porosity. Thus we obtain for the velocity of a water particle $\mathbf{v}$ in a saturated porous medium the relationship $\mathbf{v} = \mathbf{q}/\varepsilon$.

2.5.2 Example of a Groundwater Flow Velocity Calculation in Two Dimensions

To illustrate the concept of the calculation of groundwater flow velocity in two dimensions, we now detail an example. Before doing so, we review a few basic geometrical concepts.

Consider the information presented in Figure 2.29. Flow enters the box uniformly along the right-hand side. It flows through the box to exit uniformly along the face on the left-hand side. From our understanding of Darcy's law, we would expect that a series of manometers located uniformly (at equal distances) along either the front or back sides would show uniformly decreasing water-level elevations from right to left. This is illustrated by the line connecting the head at point 1, call it h_1, and the head at point 3, call it h_3, which represents the water-surface elevation when moving from measuring point 1 to measuring point 3. The water level in each manometer is indicated by the top of the shaded portion of the manometer. The straight line we have used to connect these two points is an interpolation using the two measured water levels h_1 and h_3. However, it is a good approximation in this case since we would expect to find a uniform loss of energy (as reflected in the decrease in water-level elevations) due to frictional dissipation as the water moves uniformly through the tortuous pore pathways generated by the sand grains. In other words, as we noted in our discussion of Darcy's law, there is a change from potential energy as demonstrated by the difference in water levels to heat energy via friction as water moves through the soil.

We now return to Figure 2.29. Note that the water level in manometer 1 is the highest, since it is closest to the inlet end of the box. Manometer 2 has the next highest level since it is located a little further to the left of the inlet than manometer 1. Finally, manometer 3 exhibits the lowest water level, which is consistent with its location nearest the outlet.

In the experiment in Figure 2.29, we have located three manometers at points numbered 1, 2, and 3.

We have seen that a straight line connecting h_1 and h_3 is a reasonable representation of the water level between these two points. Let us now extend this concept and connect points 1 and 2 and points 2 and 3. This forms a triangular shape, which can now be used to define a plane (the darkest area shown in the figure). This plane is a reasonable representation of the water-level surface within the triangle. The smaller the triangle, the better is the representation because we are representing with a plane the water-level surface that usually has some curvature.

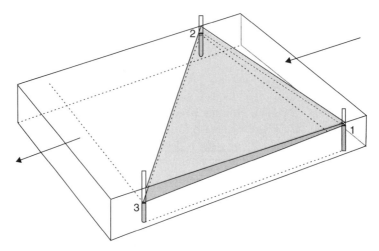

FIGURE 2.29. Experimental setup for the investigation of flow in two space dimensions. Ground-water moves from right to left through the rectangular box filled with sand. The water level in the sand is demonstrated by the elevations shown in the manometers.

Although, in general, we would not know the direction of groundwater flow, in this special case we know from the design of the experiment that it is from right to left. Because of this special flow pattern, we also know that, since the water level declines uniformly from right to left, lines denoting constant water levels must be parallel to the ends of the box. Let us now see how we would establish that this is indeed the case, if all we had to work with were the three water levels denoted as h_1, h_2, and h_3 in Figure 2.29.

We know the water level at manometer 2 has water-level elevation h_2. We also know that water-level elevation h_1 is higher than water-level elevation h_3. Since we have assumed that the water-level elevation between elevations 1 and 3 can be represented by a straight line between points 1 and 3, then *somewhere along this line there must be a point where the water-level elevation is equal to 2.* But where is it?

To determine this, we recall the equation of a straight line:

$$h(x) = a + bx, \tag{2.42}$$

where h is the water-level elevation, x is the distance along the line measured from point 3 and defined so that, at point 3, $x = 0$, and a and b are constants that we must determine from observations of h at two locations x. Let the distance between points 1 and 3 be denoted by L_{1-3}. We then determine the two coefficients as follows: at $x = 0$ we know that $h = h_3$. At $x = L_{1-3}$ we know that $h = h_1$. We can now substitute this information in Eq. (2.42) to obtain

$$h_3 = a + b \times 0 = a,$$

$$h_1 = a + b \times L_{1-3}.$$

We solve for a and b and find

$$a = h_3,$$

$$b = -(h_3 - h_1)/L_{1-3}.$$

After substitution of these results in Eq. (2.42), we have

$$h = h_3 + (h_1 - h_3) \times x/L_{1-3}.$$

From this equation, we determine the x-location where $h = h_2$ by setting the left-hand side equal to h_2 and solving for the position that we will call x_2:

$$x_2 = (h_2 - h_3)/(h_1 - h_3) \times L_{1-3}.$$

Recall that we are representing hydraulic head by a planar surface passing through the three values h_1, h_2, and h_3 at points $1, 2$, and 3, respectively. As noted earlier, within planar surfaces, lines of constant head values (so-called equipotentials) must fall on straight lines. Therefore along a straight line connecting point 2 and the location x_2 between points 1 and 3 will form an equipotential line, so that all head values along that line will equal h_2. We can derive similar equipotential lines for other values of h between the maximum (h_1) and minimum (h_3) values. Because of the properties of planar surfaces, all equipotential lines within this plane will be parallel to one another. Furthermore, for isotropic media, the direction of flow will be perpendicular to the equipotential lines, meaning that water flows in the direction of steepest descent of the hydraulic head surface.

It turns out that we can develop a simple, systematic methodology to determine the direction of maximum groundwater head gradient for any planar surface defined by three head measurements [13]. We do this by noting that the equation for a plane is given by

$$h(x, y) = a + bx + cy. \tag{2.43}$$

Since there are three undetermined coefficients in this equation, that is, a, b, and c, one must have three values of $h(x, y), x,$ and y to uniquely determine them.

Let us assume that the function $h(x_j, y_j), j = 1, 2, 3$, represents the hydraulic head measured at three points in an aquifer, all three points located at the same elevation. Further more, assume the three measuring point locations are not in a straight line and therefore form a triangle such as shown in Figure 2.30. Given this information, one can write

$$h(x_1, y_1) = a + bx_1 + cy_1, \tag{2.44}$$

$$h(x_2, y_2) = a + bx_2 + cy_2, \tag{2.45}$$

$$h(x_3, y_3) = a + bx_3 + cy_3. \tag{2.46}$$

This provides three equations in the three unknowns, a, b, and c. If we assume the triangle is numbered in counterclockwise order, such as seen in Figure 2.30, the solutions for b and c are given by

$$b = \frac{(h_1 - h_2)(y_2 - y_3) - (h_2 - h_3)(y_1 - y_2)}{(x_1 - x_2)(y_2 - y_3) - (x_2 - x_3)(y_1 - y_2)}, \tag{2.47}$$

$$c = \frac{(h_1 - h_2)(x_2 - x_3) - (h_2 - h_3)(x_1 - x_2)}{(y_1 - y_2)(x_2 - x_3) - (y_2 - y_3)(x_1 - x_2)}. \tag{2.48}$$

Having obtained these coefficients, we now return to the determination of the gradient.

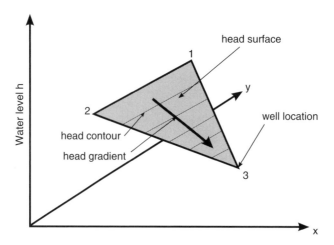

FIGURE 2.30. Nodal arrangement for the use of the algebraic approach to the calculation of gradients.

The groundwater gradient is given by

$$\nabla h = \frac{\partial h}{\partial x}\mathbf{i} + \frac{\partial h}{\partial y}\mathbf{j}, \tag{2.49}$$

where, as earlier, $\mathbf{i}$ and $\mathbf{j}$ are the unit vectors in the x and y directions, respectively. Differentiation of Eq. (2.43) yields

$$\frac{\partial h}{\partial x} = b \tag{2.50}$$

and

$$\frac{\partial h}{\partial y} = c. \tag{2.51}$$

Thus one has directly the x and y components of the gradient from which the resultant vector can readily be obtained. The velocity can then be determined by following the procedure outlined above (see Eq. (2.38)).

It is interesting to note that the algebraic formulation lends itself quite readily to determination of the gradient in three dimensions. A tetrahedron replaces the triangle in the three-dimensional formulation. Since the tetrahedron has four undetermined coefficients, four observations are required to uniquely define the four coefficients. Once obtained, the equation describing the tetrahedron is differentiated to obtain the three gradient components. From this information the gradient can be defined and the velocity calculated.

Finally, let us return to the calculation of specific discharge for the example of Figure 2.28. Consider the situation pictured in Figure 2.31. The numbers appearing at the four corners of the square are the coordinates. The [0, 1000] location is, for example, 1000 feet north of the [0, 0] location. The bracketed quantities represent the water levels observed

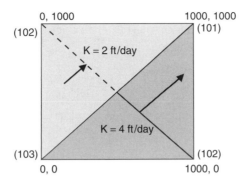

FIGURE 2.31. Example problem showing the calculation of the constant water-level line and the resulting velocity vector (not to scale).

in the wells located at the corners of the square. Let us now calculate the groundwater velocity.

We can use Eq. (2.47)–(2.51) to calculate velocities. In triangle number 1, use of Eq. (2.47) and (2.50) yields

$$\frac{dh}{dx} = b = -1/1000.$$

Similarly, Eq. (2.48) and (2.51) yield

$$\frac{dh}{dy} = c = -1/1000.$$

From these, we find that

$$q_x = -K_{xx} \times \frac{dh}{dx} = 0.004,$$

$$q_y = -K_{yy} \times \frac{dh}{dx} = 0.004.$$

Given a porosity of 0.25, we then have $v_1 = (0.016, 0.016)$, which is a vector that aligns along the 45-degree line and has magnitude 0.0226 ft/day. For the second triangle, we perform analogous calculations to find $v_2 = (0.008, 0.008)$.

2.5.3 Additional Concepts of Fluid Potential

It is now time to revisit the concept of fluid potential. Earlier, in our discussion of fluid potential and head, we showed that

$$h = \int_{z_0}^{z} dz + \int_{P_{atm}}^{p} \frac{d\pi}{g\rho(\pi)}, \tag{2.52}$$

where we have replaced p with π to emphasize the role of the dummy variable of integration in this relationship. One can introduce this expression into Eq. (2.36) to give

$$\mathbf{q} = -\mathbf{K} \cdot \nabla h = -\mathbf{K} \cdot \nabla \left(\int_{z_0}^{z} dz + \int_{P_{atm}}^{p} \frac{d\pi}{g\rho(\pi)} \right). \tag{2.53}$$

To further evaluate this expression, we need to introduce a mathematical relationship that describes how to differentiate an integral. The vehicle is called Leibnitz's rule and is written

$$
\nabla \int_{a(\mathbf{x})}^{b(\mathbf{x})} f(\mathbf{x}, \boldsymbol{\xi})d\xi = \int_{a(\mathbf{x})}^{b(\mathbf{x})} \nabla f(\mathbf{x}, \boldsymbol{\xi})d\xi + f(\mathbf{x}, b(\mathbf{x})) \cdot \nabla b - f(\mathbf{x}, a(\mathbf{x})) \cdot \nabla a. \quad (2.54)
$$

The utilization of Leibntiz's rule in conjunction with Eq. (2.53) gives

$$
-\mathbf{K} \cdot \nabla \left(\int_{z_0(\mathbf{x})}^{z(\mathbf{x})} dz + \int_{p_{atm}(\mathbf{x})}^{p(\mathbf{x})} \frac{d\pi}{g\rho(\pi)} \right)
$$

$$
= -\mathbf{K} \cdot \left[\int_{z_0(\mathbf{x})}^{z(\mathbf{x})} \nabla(1)dz + (1)\nabla z - (1)\nabla(z_0) \right]
$$

$$
-\mathbf{K} \cdot \left[\int_{p_{atm}(\mathbf{x})}^{p(\mathbf{x})} \nabla \frac{1}{g\rho(\pi)}d\pi + \frac{1}{g\rho(p)}\nabla p - \frac{1}{g\rho(p_{atm})}\nabla p_{atm} \right]
$$

which yields, assuming $z_0 = 0$ and $p_{atm} = 0$ and using Eq. (2.53),

$$
\mathbf{q} = -\mathbf{K} \cdot \left(\nabla z + \frac{1}{g\rho(p)}\nabla p \right) \quad (2.55)
$$

or

$$
\mathbf{q} = -\frac{\mathbf{K}}{g\rho(p)} \cdot \left[g\rho(p)\nabla z + \nabla p \right], \quad (2.56)
$$

which is the pressure-based form of Darcy's law. In obtaining Eq. (2.55), two points are worthy of mention: (1) z in this equation is the elevation and not necessarily the z coordinate value and (2) the integrand in the pressure integral involves the dummy variable of integration and is not a function of $\mathbf{x}$.

2.6   FLUID FLOW AND MASS AND ENERGY FLUXES

The specific discharge is not the fluid velocity. As noted earlier, it is the average flow across a face that contains both pores and solids. To obtain the average pore velocity in a saturated porous medium, one must take into account the fact that the fluid moves only through the pore space. Thus, in an average sense, the pore space associated with the cross section would be εA. The *average pore velocity* $\mathbf{v}$, that is, the average velocity in the pores in the cross section A, is

$$
\mathbf{v} = -\frac{\mathbf{K}}{g\rho(p)\varepsilon} \cdot \left[g\rho(p)\nabla z + \nabla p \right]. \quad (2.57)
$$

Because the porosity is always less than one, the average pore velocity is always larger than the specific discharge. In fact, the smaller the porosity, the larger the pore velocity, all other things being held constant.

   To this point we have studied only the movement of the bulk water phase. We have not considered the possibility that dissolved constituents might exist and that their movement might differ from that of the bulk water phase. In the following experiment we will address this issue.

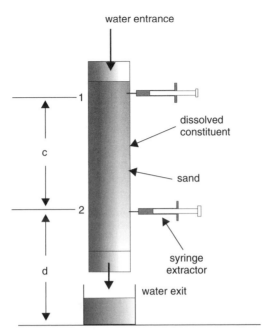

FIGURE 2.32. As the dissolved constituent moves down the column, it is carried by both convection and dispersion. The convection moves the center of mass and the dispersion smears the concentration front.

2.6.1 Convection, Diffusion, and Dispersion

Consider the column of sand shown in Figure 2.23. Let us assume that we are adding water to the column at a rate such that the column remains saturated.[24] In addition, we will place multiple ports along the length of the column that will allow us to take water samples (see Figure 2.32 for the case of two sampling ports). We will also collect water samples periodically at the base of the column.

At an arbitrary time t_0, we add some dye along with some ordinary table salt to the water being introduced into the column. The salt will permit us to measure its concentration through the change that it causes in the electrical conductivity of the solution. The dye will allow us to see qualitatively the progress of the dye concentration front. We can extract the liquid for testing either via the syringe or at the base of the column. We are assuming the concentration of the salt is indicative of the concentration of the salt–dye mixture[25].

To know when to sample and how often, it would be helpful to know the approximate speed of the concentration front. A reasonable approximation to the speed of the front is the speed of the water. From our earlier experiments, we have the information necessary to make this calculation.

From those investigations that used this sand-filled column we know that the discharge from the end of the column will be approximately 0.1 cm^3/s. We measure the diameter of our column and find it to be 5 cm. Dividing the discharge by the cross-sectional

---

[24]This is not a requirement of our approach, but it makes the experiment easier to analyze.

[25]In actual fact, the dye may move slightly more slowly than the salt because of a phenomenon know as *retardation*.

area of the column, we obtain the specific discharge, that is, $q = Q/A = 0.1/(\pi r^2) = 0.1/(3.14 \times (2.5)^2) = 0.005$ cm/s.

However, as we just discussed, the velocity of the water is not the specific discharge, but rather the specific discharge divided by the porosity. Preliminary analysis showed that for the sand we used for our experiments the porosity was 20% (or 0.2). Using this value, we obtain a pore velocity v of $0.005/0.20 = 0.025$ cm/s. One may then expect that a molecule introduced at the top of the column will move 3 cm in $3.0/0.025 = 120$ s. This gives us helpful insight into when we must take measurements if we are to capture the concentration profile as it moves along the length of the column.

Let us assume that the salt concentration of the solution entering the column is c_0 and that this corresponds to an electrical conductivity value of EC_0. Then by measuring the electrical conductivity at the ports and at the exit, one can determine the relative concentration, that is, $EC/EC_0 = c/c_0$, where EC and c are the measured electrical conductivity values, and the calculated concentration values, respectively. The salt concentration can now be computed as

$$c = \frac{c_0}{EC_0} EC \quad \text{mg/L.} \tag{2.58}$$

The results of the experiment are found in Figure 2.33. Each curve represents the concentration of the salt compared to the reference concentration c_0 as measured along the length of the column at a specific time.

The first thing one notices is that the concentration is highest at the top of the column and decreases toward the bottom. At the top the concentration has a value of c_0, as one would expect. As one moves down the column the concentration decreases.

An important concentration value to note is that of c/c_0 of 0.5. Based on the time the measurement was taken and the relationship $d = vt$, where d is the distance traveled according to the fluid-flow velocity, we determine that the observed location of the 0.5 value of concentration at time t coincides approximately with the position where we would have expected the average pore-water velocity to have moved a water molecule.

Another interesting thing to note is the change in shape of the concentration profile through time. At early time, for example, at $t = 6.0$ minutes, the slope of the concentration profile is steep, that is, the concentration changes rapidly as one moves along the column. At later time, for example, $t = 17.0$ minutes, the concentration change with distance is not quite as steep. It appears that as time increases, the slope of the concentration front decreases, such that spatial changes in c become more gradual.

To understand why one observes a decrease in the slope of the concentration front as time elapses, it is necessary to realize that the movement of contaminants in groundwater can be attributed to three distinctly different mechanisms. One is *convection* (or *advection*), another is *diffusion,* and the third is *dispersion.* These concepts can be best understood via an analogy.

Consider that you are standing on a bridge looking down at a rapidly flowing river. You observe that there is a canoe approaching. On board is a park ranger who has been asked to study the small-scale water velocity behavior of the stream. As he approaches the bridge he rests his paddle in the canoe while he releases a quantity of dye into the stream. You happen to have the appropriate instrumentation with you so you decide to measure the change in concentration of the dye in the river as it moves under the bridge. You observe that two physical phenomena are responsible for the concentrations you are observing. The first is the average velocity of the stream as represented by the floating

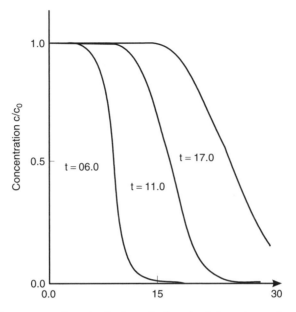

FIGURE 2.33. The concentration of salt along the length of the column changes as a result of both convection and dispersion. Note that elapsed time is in minutes.

canoe. It is clear that the mass of dye is moving more or less at the same velocity as the average velocity of the water in the stream. The second observation is that the dye is spreading outward around its more concentrated center of mass.

The movement by virtue of the average stream velocity is known as *convection*. The spreading of the dye is due primarily to the small-scale variability in the stream velocity relative to the average stream velocity. Spreading via the small-scale velocity variations we denote as *dispersion*. Molecular diffusion also causes spreading but generally plays a small role relative to dispersion. Molecular or *Fickian diffusion* is the movement of the solute due only to the existence of a concentration gradient. It is the mechanism that causes a drop of dye to spread in a beaker when the water in the beaker is at rest. The combination of these three phenomena, namely, convection, dispersion, and diffusion, gives rise to the overall behavior of the dye as viewed from the bridge. If it were not for the turbulence and molecular diffusion, the concentration in the stream would be either 0 or 1. Diffusion and dispersion cause the concentration to spread, so that from our point of view on the bridge, the concentration rises gradually and then decreases as the dye moves beneath us. It is not a step function (abrupt change) as would be the case in the absence of the dispersive phenomenon[26].

Returning to our column experiment involving the introduction of salt as the dissolved substance, we observe that the same three phenomena are at work in flow through porous media. The average groundwater flow velocity carries the dye along. This average velocity can be determined approximately by examining the 0.5 concentration location, since this location is the best indicator of the distance the dissolved constituent would have traveled in the absence of dispersion. The reason the 0.5 value is a good indicator is that the

---

[26]In actual fact, there are two step functions involved: one is from zero to one upon the arrival of the dye and the other is from one to zero as the dye passes by.

phenomena of diffusion and dispersion tend to be symmetric in the sense that the mass apparently lost upstream of the 0.5 value is found downstream of this value. Transport by virtue of fluid convection is called convective transport.

The spreading found in the movement of the salt through the column, and illustrated by the decreasing slope in the concentration profile as time evolves, is due primarily to small-scale velocity components in the pores that are different from that of the average velocity. We refer to this as mechanical dispersion. As in the case of our river flow analogy, the other phenomenon at work is Fickian diffusion.

If the various physical–chemical vehicles available for the movement of a dissolved compound are combined into an equation, we can write,

$$q_c = q_v + q_D + q_F, \tag{2.59}$$

where q_c is the *total solute transport*, q_v is the transport due to convection, q_D is transport via dispersion, and q_F is transport attributable to Fickian diffusion. In a later chapter we will examine the experimentally determined equations that describe the various terms on the right-hand side of Eq. (2.59).

In Figure 2.33 we plotted the change in concentration along the length of the column as a function of time; that is, we provided three space-dependent concentration profiles, one for each of the three times when measurements were made. Another way to view these data is to examine the change in concentration at specific locations along the column as a function of time. The results for a specific location along the column are given in Figure 2.34. Note that the curve rises as the concentration front approaches the observation point and appears similar in shape to, although a mirror image of, the profile presented in Figure 2.33. This type of curve is usually called a breakthrough curve.

How could you use this information to determine the average velocity of the water in the column? The answer lies in the earlier observation that the 0.5 concentration location represents the center of mass of the concentration front and the location of salt particles that are moving at the average water velocity in the porous medium. Thus, if one determines the location of the 0.5 concentration relative to the source location and also the amount of elapsed time since the salt was introduced, one can determine the average velocity by dividing the distance traveled by the time of travel.

Before moving on to the next topic, a brief discussion of the units of measure used in water quality analysis is appropriate. Concentration is generally expressed in terms of mass per unit volume of solution. For example, grams per cubic centimeter of fluid. If, for example, one had a mass of 10 g of salt in a solution of 1 liter, the concentration could be expressed as a concentration of 10 g/1000 cm^3 or 0.01 g/cm^3. Alternatively, one could express the concentration as the number of grams of salt per mass of solution. In the above example this would yield 10 g/1010 g or 0.0099 (assuming a density of 1 g/cm^3 for water). Clearly, for dilute solutions, these two measures are very nearly equal.

In the field, one normally rewrites the ratio 10 g/1010 g as 10,000 mg/1,010,000 mg which is read as ten thousand parts per million (ppm), although this is not formally correct. It is also common to read this concentration as 10,000,000 μg/1,010,000,000 μg[27] ten million parts per billion (ppb), because organic contaminants are often considered important in concentrations of only a few parts per billion.

---

[27]One microgram, μg, is one millionth of a gram.

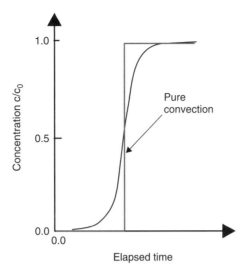

FIGURE 2.34. The concentration at a particular location responds to a continuous source of strength c/c_0.

2.6.2 The Phenomena of Adsorption and Retardation

Retardation is a phenomenon whereby dissolved ions or compounds adhere to soil particles. To understand the importance of this phenomenon on groundwater flow and transport, consider the experiment presented schematically in Figure 2.35. The burette shown in this figure has a small amount of glass wool placed inside at its tip. The burette has then been filled with activated carbon to near the top, where a second glass wool plug has been inserted. The burette empties into a small beaker wherein a total dissolved solids sensor is located. The beaker is chosen small enough such that the concentration measured in it is a reasonable representation of the concentration leaving the burette. When the beaker is full, it overflows into a larger beaker that contains it.

The experiment requires that we first saturate the carbon-filled burette with distilled water and adjust the fluid input at the top and output at the bottom to permit flow through the column while maintaining it at a fully saturated state. Now we observe, as a reference, the total dissolved solids (TDS) measurements on the TDS meter that we have placed in the beaker at the outlet of the burette.

Next, we prepare, as we did earlier, an aqueous solution of sodium chloride (common table salt) to which we add enough methylene blue to form a dark blue solution. Having prepared the solution, we change the fluid source entering the burette from the distilled water to the salt solution containing the methylene blue coloring. One can observe the blue solution entering the column. Monitoring of the TDS meter will indicate the point in time when the salt solution first exits the burette. Using a 50 ml burette, the time required to move through the column will be approximately 4 minutes.

Even though the methylene blue plus saline solution entered the burette together and the mixture is obviously blue in color, the solution, heralded by the readings on the TDS meter, appears to exit the burette clear. There is no trace of the dye when the salt begins to exit. The dye has been adsorbed on the activated charcoal. Sodium chloride, being a conservative tracer, is not adsorbed and therefore is a good measure of the pore velocity of the water. Assuming the solution is continuously added, in a day or two the blue

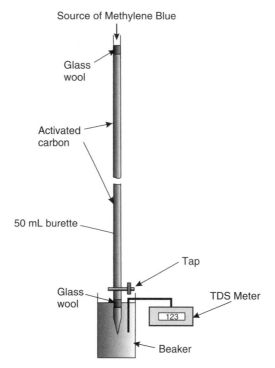

FIGURE 2.35. Diagrammatic representation of the apparatus setup to illustrate the phenomenon of adsorption. A methylene-blue-dyed saline solution is passed through a carbon-filled burette. The arrival time of the sodium chloride is determined by the total dissolved solids meter and the arrival of the methylene–chloride is determined by a color change in the effluent water.

coloring will appear at the exit of the burette. The apparent rate of movement of the water (and salt) is much different from that of the blue coloring. The apparent progress of the coloring has been retarded.

The *retardation coefficient R* is a measure of the degree to which this phenomenon takes place. The effect of retardation is to reduce the apparent velocity of a dissolved compound. For example, referring to Eq. (2.59), it would appear that there is a coefficient R such that

$$q_c = \frac{q_v + q_D + q_F}{R}, \tag{2.60}$$

where R is always greater than one[28]. Indeed, we will see later that such a factor exists and that it inhibits the apparent velocity of the solute.

Retardation depends on the type of soil and the dissolved species. The most adsorptive soils, all other things being equal, are those containing organic carbon. In addition, clay exhibits adsorption because clay particles are made up of clay minerals and have, by their mineralogical makeup, many sites available to adsorb ions. Sands, on the other hand, are generally not very adsorptive unless organic carbon is present. Organic carbon is very effective in retarding many compounds.

---

[28]The possibility that R can be smaller than one has been associated with the phenomenon of ion exclusion.

The simplest form of the retardation coefficient for dissolved ions, such as calcium, is given by

$$R = 1 + \frac{1 - \varepsilon}{\varepsilon} \rho_s K_d, \tag{2.61}$$

where ρ_s is the *density of the soil grains*[29], and K_d is the *distribution coefficient*.

The value of K_d is obtained from an experiment conducted using a specific soil and a specific solute. The experiment, in its simplest form, involves placing a soil sample of known bulk density (mass of grains per total volume of medium) and volume in a solution with a known concentration of the ion of interest. The amount of the ion bound to the substrate, per unit mass of substrate, is denoted by F. It is found by experiment to be a function of the concentration of the substance in the solution.

One now plots the value of F versus the concentration in the solution surrounding the soil grains, c_s. The parameter K_d is determined as the derivative dF/dc_s. If the assumed relationship is of the form $F = k_1 + k_2 c_s$, it is called a *linear isotherm*[30]. The derivative in this case is the slope of this straight line. Thus we can obtain directly $K_d = k_2 = \Delta F / \Delta c_s$. From this relationship, K_d is easily determined.

More complicated relationships have been proposed. Langmuir [14], for example, suggested the nonlinear equilibrium isotherm given by

$$F = \frac{k_3 c_s}{1 + k_4 c_s}. \tag{2.62}$$

The logic behind this formulation can be justified by either equilibrium or kinetic considerations. If one divides both sides of Eq. (2.62) by c_s, one obtains

$$\frac{F}{c_s} = \frac{k_3}{1 + k_4 c_s}$$

or

$$\frac{c_s}{F} = \frac{1 + k_4 c_s}{k_3} = \frac{1}{k_3} + \frac{k_4}{k_3} c_s.$$

Therefore, a plot of c_s/F versus c_s will provide, via the intercept and the slope, the coefficients k_3 and k_4.

Other forms of isotherms are known and used. Each is a special form of the relationship

$$E(c) = \frac{\rho_b \alpha c^\beta}{(1 + \gamma c)^2}, \tag{2.63}$$

where, for the linear case $\beta = \gamma = 0$, for the so-called *Freundlich isotherm* $\gamma = 0$ and for the Langmuir isotherm $\beta = 0$. Once again, it should be emphasized that α, β, and γ are experimentally determined coefficients.

For the case of organic compounds in solution, such as trichloroethylene, one again assumes a linear relationship, in this case given by

$$F = k_1 + K_p c'_s, \tag{2.64}$$

---

[29]Density of the soil grains is the ratio of the mass of the solid particles in a soil sample to the volume of soil grains.

[30]The term isotherm implies measurements are done at a constant temperature.

where c_s' is expressed as mass of solute per unit mass of water. Given this definition, the retardation coefficient becomes

$$R = 1 + \frac{1 - \varepsilon}{\varepsilon} \frac{\rho_s}{\varrho} K_p. \tag{2.65}$$

While the value of K_p may be determined much as was K_d, there is an alternative strategy for hydrophobic compounds (compounds that do not like water). By employing what is known as the hydrophobic theory, one can use the relationship

$$K_p = K_{oc} f_{oc}, \tag{2.66}$$

where K_{oc} is a dimensionless distribution coefficient defined for a soil made up of only organic material. The organic carbon ratio f_{oc} is the dimensionless ratio of solid organic carbon to the dry weight of the soil.

There exist relationships between K_{oc} and K_{ow}, the latter being the dimensionless partition coefficient between water and octanol. One proposed relationship is [15]

$$K_{oc} = 0.411 K_{ow} \tag{2.67}$$

and another is

$$\log K_{oc} = a \cdot \log K_{ow} + b, \tag{2.68}$$

where values of a and b can be found in the literature [16] along with values of K_{ow} for selected compounds. Thus by measuring only the organic carbon ratio, it is possible to estimate K_p for a given organic compound. This leads naturally to the estimation of the retardation coefficient using Eq. (2.65).

2.7 SUMMARY

The focus of this chapter has been a description of the fundamental physics that dictate the behavior of groundwater flow and the transport of dissolved constituents. The chapter introduces the concepts of pressure, hydraulic head, and fluid potential, three of the most fundamental concepts in the physics of flow through porous materials. Since the shallow subsurface pore space is occupied by both water and air, we discuss the concept of saturation, which is a vehicle for quantitatively describing the air-water mixture. With these concepts in hand, we proceeded to describe Darcy's classical flow experiment and the ensuing law derived therefrom. The methodology required to determine the velocity of groundwater is also described and illustrated. The chapter closes with a discussion of the physical processes that dictate the behavior of dissolved species in a flowing groundwater system.

2.8 PROBLEMS

2.1. A sand and gravel outwash formation found on Cape Cod, Massachusetts is estimated to have a hydraulic conductivity of 120 meters per day. What is the intrinsic permeability of this formation? Express your answer in units of m^2 and darcies.

2.2. Consider a column setup analogous to the Darcy experiment, such that a column of length L is filled with water-saturated sand. The hydraulic head at the top of the column is h_{top}, the hydraulic head at the bottom of the column is h_{bot}, the cross-sectional area is A, the hydraulic conductivity of the sand is K, and a steady-state flow of water through the column is established. (a) For a homogeneous sand, plot hydraulic head as a function of location along the column (in this case of a vertically oriented column, plot h versus z). You should also write the equation for $h(z)$. (b) Next, consider the same physical setup as that used in part (a), except that the information you are given is L, A, h_{top}, K, and the volumetric flow rate Q (volume per time) through the column. Again plot h as a function of z, and write the equation for $h(z)$. You should also write the specific equation for h_{top} as a function of L, A, h_{top}, K, and Q. (c) Finally, consider a column whose top one-third and bottom one-third are filled with sand of conductivity K, and whose middle one-third is filled with silt having conductivity $K/10$. Repeat parts (a) and (b) for this case.

2.3. Consider a layered formation, in which layers are homogeneous with constant thickness. Assume the layering is aligned with the horizontal direction, denoted by the spatial coordinate x, while the vertical coordinate is denoted by z (see Figure 2.27). Show that the effective hydraulic conductivity of this layered system in the direction parallel to the layering is given by the weighted arithmetic average, that is,

$$K_{xx}^{eff} = \frac{\sum b_i K_i}{\sum (b_i}.$$

where the thickness of layer i is denoted by b_i, the hydraulic conductivity of layer i is K_i, and the summation is taken over all layers.

2.4. For the same layered system as in Problem 2.3, show that the effective hydraulic conductivity in the direction perpendicular to the layering is given by the weighted harmonic average,

$$K_{zz}^{eff} = \frac{\sum b_i}{\sum (b_i/K_i)}.$$

where b_i, K_i, and the summation all have the same meaning as in Problem 2.3.

2.5. The hydraulic conductivity for an aquifer, described in two spatial dimensions, is given by the following tensor

$$\mathbf{K} = \begin{bmatrix} 25 & 6 \\ 6 & 20 \end{bmatrix},$$

where the units are in feet per day. (a) Is the aquifer isotropic or anisotropic? Explain. (b) Is the aquifer homogeneous or heterogeneous? Explain. (c) If the hydraulic head gradient is given by $\partial h/\partial x = 4$ and $\partial h/\partial y = 6$, determine the volumetric flux vector $\mathbf{q} = (q_x, q_y)$.

2.6. Consider a geological formation that has a direction preference to flow, such that the direction of maximum permeability is at an angle θ relative to the x-axis, in the xz plane (e.g., see Figure 2.28). Assume that material properties do not change in the

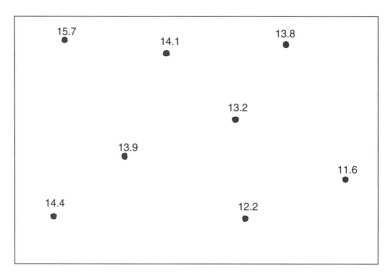

FIGURE 2.36. Location of wells in Problem 2.7

y-direction, so that the system can be treated as two-dimensional. For a hydraulic gradient $(-\nabla h)$ at an angle α relative to the *x*-axis in the xz plane, determine the direction of fluid flow. You should express the flow direction by the angle β, measured relative to the *x*-axis. The direction of flow should be expressed as a function of α, θ, and the directional values of the hydraulic conductivity. Identify the conditions under which the flow direction aligns with the (negative of the) hydraulic gradient; that is, flow vector **q** is parallel to $-\nabla h$.

2.7. Consider a horizontal aquifer of uniform thickness, which exhibits essentially horizontal flow. Eight wells are screened within this aquifer, and measurements of water levels have been taken in all eight of these wells. Figure 2.36 shows the location of the wells and their associated water levels, with the water levels reported in meters above mean sea level. (a) From the figure, draw contours of the hydraulic head, using a contour interval of 1 meter. (b) Based on these contour lines, indicate the direction of the (negative) hydraulic head gradient, $-\nabla h$, for this system. (c) Based on the head gradients, estimate the directions of flow for this system, and describe qualitatively the flow system. Explain any assumptions you have made (*Hint:* Recall Problem 2.6 that relates the direction of **q** and the direction of $-\nabla h$.)

2.8. For the system described in Problem 2.7, use the triangulation procedure of Section 2.5.2 to estimate the directions of flow. How do these estimates compare to those from Problem 2.7?

2.9. Explain how you would use the flow information from either Problem 2.7 or Problem 2.8 to estimate groundwater "travel times" (i.e., the time it takes for water to flow from an initial location to another location along a flow path). What other information is needed to make this calculation?

2.10. If a contaminant is moving along a flow path in the flow system analyzed in Problems 2.7–2.9, with the contaminant introduced at a particular time, call it $\underline{t}^*$, at a particular location (say, in the upper left of the domain), explain how the subsequent transport of the contaminant would evolve in the cases of (a) convection only, (b) convection and dispersion, (c) convection and retardation, and (d) convection, dispersion, and retardation.

BIBLIOGRAPHY

[1] M. K. Hubbert, The theory of groundwater motion, *J. Geol.* **48**:785, 1940.

[2] J. G. Guarnaccia and G. F. Pinder, *NAPL: Simulator Documentation*, National Risk Management Research Laboratory, U.S. Environmental Protection Agency, EPA/600/SR-97/102, 1997.

[3] S. Finsterle, T. O. Sonnenborg, and B. Faybishenko, Inverse modeling of a multistep outflow experiment for determining hysteretic hydraulic properties, *Proceedings, TOUGH Workshop '98*, May 4–6, Lawrence Berkeley National Laboratory, Berkeley, CA, 1998, p. 250.

[4] M. K. Hubbert, Darcy's law and the field equations of the flow of underground fluids, *Bull. Assoc. Hydrol Sci.* **5**:24, 1957.

[5] P. Darcy, *Henry Darcy: Inspecteur Général des Ponts et Chaussées, 1803–1858.* Imprimerie Darantiere, Dijon, 1957.

[6] J. R. Philip, Desperately seeking Darcy in Dijon, *Soil Sci. Soc. Am. J.* **59**:319, 1995.

[7] G. Brown, "Henry Darcy and His Law," http://biosystems.okstate.edu/ darcy/index.htm, 1999.

[8] C. Yu, C. Loureiro, J.-J. Cheng, L. G. Jones, Y. Y. Wang, Y. P. Chia, and E. Faillace, *Data Collection Handbook to Support Modeling Impacts of Radioactive Material in Soil*, Environmental Assessment and Information Sciences Division, Argonne National Laboratory, Argonne, IL, 1993.

[9] H. Darcy, *Les Fontaines Publiques de la Ville de Dijon*, Victor Dalmont, Paris, 1856.

[10] EPA, SW846, "Method 9100, Saturated Hydraulic Conductivity, Saturated Leachate Conductivity, and Intrinsic Permeability," http://www.epa.gov/epaoswer/hazwaste/test/9100.pdf, 1986.

[11] "SEEP-W Version 5," http://home.geo-slope.com/webhelp/sep/sephlp.htm, 2002.

[12] W. G. Gray and K. O. O'Neill, On the general equations for flow in porous media and their reduction to Darcy's law, *Water Resour. Res.* **12**(2):148, 1976.

[13] G. F. Pinder, M. Celia, and W. G. Gray, Velocity calculation from randomly located hydraulic heads, *Groundwater* **19**(3): 262, 1981.

[14] I. Langmuir, Constitution and fundamental properties of solids and liquids, *J. Am. Chem. Soc.* **38**:2221, 1916.

[15] S. W. Karickhoff, Semi-empirical estimation of sorption of hydrophobic pollutants on natural sediments and soils, *Chemosphere* **10**(8): 833, 1981.

[16] W. J. Lyman, W. F. Reehl, and D. H. Rosenblatt, *Handbook of Chemical Property Estimation Methods*, McGraw-Hill, New York, 1982.

CHAPTER 3

THE GEOLOGIC SETTING

In the previous chapter we were introduced to some of the fundamental properties and processes associated with groundwater flow and transport. In this chapter we investigate the geologic framework within which groundwater professionals operate. The focus is on the nature of geologic deposits, their mechanism and history of formation, and their characteristics. This information, when combined with that provided earlier, sets the stage for a discussion of groundwater occurrence and its behavior.

Hydrological properties of materials are normally dependent on the geologic environment in which they were created. For example, *clay*[1] *particles* have a very small grain size and an associated very small pore space (although the overall porosity of clay can be quite high!). Because of the resulting large ratio of pore surface area to volume, there are substantial friction losses as groundwater moves through clay deposits. As a result, clay is a low hydraulic conductivity material. However, because of its high porosity, it can store significant quantities of water.

The number of geologic materials, environments, and landforms is enormous. We will not attempt to visit all of them, but rather we will focus on a few of the more relevant. The approach we will take is to catalog geologic materials as *unconsolidated deposits, consolidated deposits, metamorphic rocks,* and *igneous rocks.* The genesis of and relevance of each to groundwater flow and transport will be discussed.

3.1 UNCONSOLIDATED DEPOSITS

3.1.1 Clastic Sedimentary Environment

A *sedimentary environment* can be thought of as one in which all of the deposited material has been either precipitated out of solution or transported from elsewhere. *Clastic*

---

[1]Note that the term clay is applicable to both a grain size and a mineral.

Subsurface Hydrology   By George F. Pinder and Michael A. Celia
Copyright © 2006 John Wiley & Sons, Inc.

materials are those that have been transported from elsewhere. Cobbles, boulders, sand, silt, and clay constitute materials that are normally found in a clastic sedimentary environment. Such materials are normally transported by water. However, this is not universally the case. There are materials that have been and are being moved by wind to form *loess* and *dune* deposits, by gravitational forces to form *talus* deposits, and by ice to form *glacial* deposits (which we will discuss at length in Section 3.1.3).

In general, when one encounters granular materials such as indicated above, they can be associated with specific geologic environments. The energy associated with an environment dictates the grain size of the sediments found there. A high-energy environment will transport coarse sediments, such as coarse sand and gravel, while a quiescent, low-energy environment will favor the deposition of finer grained materials. A qualitative representation of the range of grain sizes that one might expect to encounter in various clastic environments is presented in Figure 3.1. To get a sense of the grain size range represented by the terms clay, silt, sand, and gravel, the reader should return to page 16 and view Table 1.3 or Figure 3.3 below.

While deposits such as those identified in Figure 3.1 are often found at the earth's surface in their original depositional environment, this is not always the case. Energy environments can and do change. Thus a low-energy environment may become a high-energy environment if there is local geologic uplift of the land surface such as occurs in mountain building. When this happens, the finer sediments can and often are eroded away to be deposited elsewhere in areas that have been transformed into lower energy environments. Alternatively, coarser materials may be transported to and deposited upon the finer grained sediments. Thus materials deposited in one energy environment can be buried by younger sediments deposited in a different energy environment. As a consequence, when one examines a sequence of deposits through drilling or excavation, *deposits indicative of historically different energy environments may be, and generally are, encountered.*

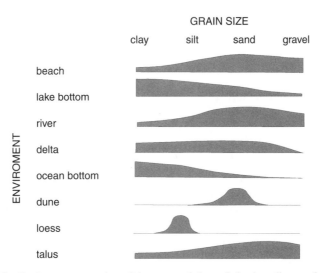

FIGURE 3.1. Qualitative representation of the range of sizes of clastic sediments found in selected environments.

The term "younger" in the sense used in the preceding paragraph needs further explanation. If a series of sediments has not been disturbed, the topmost layer is always the youngest and the lowermost layer is always the oldest. Such a sequence of sediments is said to conform to the tenants of the *law of superposition*.

An understanding of the *distribution of sediments either areally or vertically* is very important in contaminant transport. Because coarse sediments tend to be more permeable, groundwater tends to move more rapidly through these geologic horizons. Therefore, if a coarse horizon is found to extend over significant distances, one might reasonably expect, all other things being equal, that contaminants would move more rapidly and over greater distances in these deposits. On the other hand, the existence of a low-permeability, finer grained deposit may inhibit movement of contaminants. The role of inhibition is particularly important when the finer grained deposit is areally extensive and overlain by more permeable deposits. In this instance the role of the low-permeability layer (often referred to as either a *confining bed, aquitard,* or even *aquiclude*) in limiting vertical migration of contaminants can be important.

Groundwater professionals have tried for decades to relate hydraulic conductivity to grain size (e.g., see the discussion found on page 86). While this task would appear to be rather straightforward, it has been found that this correlation is not easily established. Meinzer, one of the giants in the field of hydrology, stated the situation as follows [1]:

> In the indirect laboratory methods the permeability[2] is computed from the mechanical composition and porosity.... Indirect methods of computing permeability are useful for some purposes but have not always given consistent results and are not in general to be recommended. Direct tests of permeability require no more work and are generally more satisfactory.

Nevertheless, numerous papers have been written regarding these relationships. In his 1927 paper, Kozeny [2] presents a model based on the analogy between a porous soil and an ensemble of channels of the same length, but with different cross sections. He then solves the Navier–Stokes equations for all the channels as defined along a cross-sectional plane. The resulting expression is

$$S^2 = \frac{c\varepsilon^3}{k},$$

where S is the specific surface of the channels, c is a shape factor that has different values depending on the shape of the capillary, ε is the porosity, and k is the intrinsic permeability. Kozeny's work was extended by others who added additional complexity to his original conceptual model (e.g., see Leverett [3]).

A typical heuristic relationship was published by the Illinois State Water Survey and is reproduced in Figure 3.2. This figure shows the relationship between the effective grain size[3] and the hydraulic conductivity. To relate this to the grain size distribution curve, we reproduce the example presented in Chapter 1 (see Figure 3.3).

---

[2]Permeability as used here was defined by Meinzer as "the rate of flow of water at 60 °F, in gallons a day, through a cross section of 1 square foot, under a hydraulic gradient of 100 percent [1]. Therefore the *permeability* of Meinzer is the same property we define as *hydraulic conductivity.*

[3]Recall from Chapter 1 that the effective grain size is defined as the smallest grain size that is larger than 10% of the sample by weight.

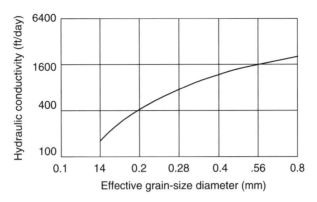

FIGURE 3.2. Relationship between effective grain size and hydraulic conductivity (adapted from [5]).

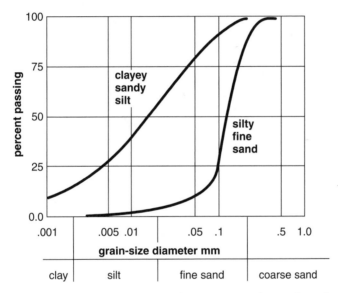

FIGURE 3.3. Grain size distribution curve showing percent passing a given sieve size versus sieve size diameter. Also illustrated is grain size nomenclature for the given range of grain sizes (from [6]).

The more common way to present hydraulic conductivity values is via a tabulation such as is presented in Table 3.1.

While we cannot compare Figures 3.3 and 3.2 with Table 3.1 directly, one can see through extending the curve appearing in Figure 3.2 that the values for these representative samples are consistent with the relationship developed by the Illinois State Water Survey.

The grain size distribution and concomitantly the hydraulic conductivity of a clastic deposit depend very much on the origin of the deposit. For example, in Figure 3.4 one observes a gravel deposit located near its point of origin. It has a wide range of grain sizes, making it a *well graded* or *poorly sorted* deposit. Because the material pictured

TABLE 3.1. Representative Values of Hydraulic Conductivity

Geologic Classification	K (cm/s)	K (ft/day)
Clay	10^{-9}	10^{-5}
Very fine sand	8.5×10^{-3}	2.4×10
Medium sand	2.2×10^{-1}	6.2×10^2
Coarse sand	2.7	7.8×10^3
Gravel	3.7×10	1.05×10^5

Source: Davies and DeWiest [4].

FIGURE 3.4. Gravel deposit near Avon, Ontario, Canada generated during the last ice-age near a glacier where there was high energy.

here was deposited near its source, it did not have adequate opportunity to separate out into different grain sizes. Based on our earlier discussion of grain size distribution, one recognizes that this deposit will have a lower porosity than a more uniform deposit and therefore will have a lower hydraulic conductivity than would be expected for a sediment made up solely of the coarser fraction of this deposit.

In Figure 3.5 we observe gravels in a very different environment. In this case the source of material was an enormous river of water generated when an ice dam collapsed during the last ice age. The rapidly moving water eroded material and carried it great distances. The distance of travel allowed the sorting of material such that this deposit is remarkably well sorted (or poorly graded). Due to the large pores generated by this uniform gravel, the resulting deposit has exceedingly high hydraulic conductivity.

In Figure 3.6 is shown a river alluvium sand deposit. The sand is relatively uniform due to its mode of deposition. Notice the layering that varies from nearly horizontal to dipping.

FIGURE 3.5. Very well sorted (poorly graded) coarse gravel deposited near Hanford, Washington during an enormous flood associated with the last ice age.

FIGURE 3.6. River alluvium deposit of fine to medium sand. This sediment illustrates the phenomenon of cross bedding.

The dipping beds are called *cross beds* and demonstrate, in this case, a depositional source to the right of the picture. Sand, carried by the moving water, traversed from the right of the picture to the left and then was deposited as a new dipping layer as the deposit extended from right to left.

Because this sand deposit is relatively well sorted (poorly graded), it would enjoy a high hydraulic conductivity were it not for the fact that the grain size is fine to medium. The small grain size results in a small pore size and this, in turn, makes this deposit less permeable than the uniform gravel deposit shown in Figure 3.5.

Finally, to give a sense of the scale of some of the geologic features associated with clastic deposits, we provide in Figure 3.7 a photograph of sand dunes in eastern Washington State. These barchan dunes are crescent shaped with horns pointing downwind. The windward slope is gentle relative to the leeward slope (located inside the horns). Such dunes are typically about 100 ft high and 1000 ft from one tip of the horn to the other. While the hydraulic conductivity of such deposits would be considerable, they are of relatively little hydrogeologic interest because they are normally formed above the water table and are therefore in the vadose zone. Consequently, they are of very little interest from a water supply perspective. However, they can be important aquifers if they are buried and below the water table.

You will note that in Table 3.1 we used two sets of units to describe hydraulic conductivity. In fact, there are many other sets of units that are used routinely by groundwater professionals to describe this parameter. In Table 3.2 we present the conversion from one set of units to another.

To use this table, the unit provided is multiplied by the appropriate conversion value for the unit desired. As an example, consider the problem of changing a hydraulic conductivity value of 4.0 ft/day into the equivalent value in cm/s. The calculation is as follows:

$$1 \text{ ft/day} = 3.528 \times 10^{-4} \text{ cm/s}$$

$$4 \times 1 \text{ ft/day} = 4 \times 3.528 \times 10^{-4} \text{ cm/s} = 1.411 \times 10^{-3} \text{ cm/s}.$$

It is also interesting to note, as we will discuss later, that fine-grained deposits tend to hold a disproportionately large amount of water and, in some circumstances

FIGURE 3.7. Barchan sand dunes in eastern Washington State.

TABLE 3.2. Conversion Table for Hydraulic Conductivity

	Hydraulic Conductivity Conversions				
	g/day/ft^2	m/day	cm/s	ft/day	ft/s
g/day/ft^2	1.000	4.070×10^{-2}	1.852×10^{-8}	1.337×10^{-1}	1.547×10^{-6}
m/day	24.54	1.000	1.157×10^{-3}	3.280	3.780×10^{-5}
cm/s	1.636×10^{-3}	6.667×10^{-5}	1.000	2.187×10^{-4}	2.532×10^{-9}
ft/day	7.480	3.050×10^{-1}	3.528×10^{-4}	1.000	1.157×10^{-5}
ft/s	6.463×10^5	2.633×10^4	30.48	8.640×10^4	1.000

contaminants, in storage. These sediments may release their fluids, including the contaminants, over long periods of time. The consequent slow release of groundwater from these sediments into more permeable formations often tends to make the rapid remediation of contaminated sites difficult, time consuming, and, in some instances, nearly impossible.

We also learned in Chapter 1 that contaminants can be slowed in their movement via the process of retardation. Recall that a key factor in the retardation of ions was the amount of clay present. We also noted earlier that the retardation of organic solvents is influenced by the amount of organic material present. Thus sediments with more organic material, such as peat, will tend to retard organic contaminants and therefore slow down their rate of migration.

3.1.2 Precipitate Sedimentary Environment

Under favorable conditions, chemical precipitates can form and become deposited on the bottom of a quiescent water body. The Great Salt Lake is an example where a supersaturated solution of halite (i.e., common table salt) is found and is being precipitated to form a salt deposit. Calcium carbonate and calcium magnesium carbonate also precipitate under suitable environmental conditions. In addition, it is possible to form a sedimentary deposit via the accumulation on the ocean floor of the *calcium carbonate remains of aquatic plants and animals*.

In general, unconsolidated precipitates would not be catalogued as permeable porous media and therefore would play a minor role in groundwater flow and transport. However, when they form rocks, as will be discussed in Section 3.2, they often are important.

3.1.3 Glacial Environments

Over the northern third of the United States, most of Canada, and in areas of higher elevation elsewhere on the continent, the topography of surficial landforms has been impacted by processes identified with *Pleistocene glaciation*. During Pleistocene time, massive continental ice sheets moved southward from polar regions to form enormous ice thicknesses. For example, in the Vermont area glacial ice was more than a mile in thickness. The extent of Pleistocene glaciation is shown in Figure 3.8.

The term Pleistocene is associated with a particular period in Earth history. It is one of two *epochs* or *series* that make up the *period* or *system* termed *Quaternary*. The Quaternary period extends from 1.8 million years ago to the present. The other epoch

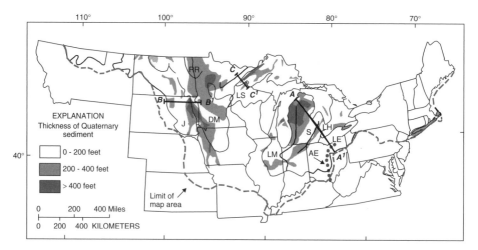

FIGURE 3.8. The dashed line denoted as the "limit of map area" represents the furthest extent of the glacial ice. The various letter pairs, such as DM, refer to specific ice lobes, in this case the Des Moines lobe. The location of these lobes is shown by the bold solid line. The straight lines with the same letter on each end (one end has a superscript 1) represent locations where cross sections have been generated. Note the thickness of the deposits shown by the shaded areas (from [7]).

TABLE 3.3. Epochs and Periods that Constitute the Segment of the Geologic Time Scale Known as the Cenozoic Era[a]

Era	Period or System	Epoch or Series
	Quaternary 0–1.8	Holocene 0–0.008
		Pleistocene 0.008–1.8
		Pliocene 1.8–5.3
Cenozoic 0–65		Miocene 5.3–23.8
	Tertiary 1.8–65	Oligocene 23.8–33.7
		Eocene 33.7–55.5
		Paleocene 55.5–65

[a] All time is in terms of millions of years before the present.
Source: Adapted from [8].

that is found in the Quaternary period is the Holocene, which extends from 8000 years ago to the present. The Pleistocene epoch therefore extends from 1.8 million to 8000 years before the present. To make life more complicated, it turns out that the Quaternary period is one of two that, in combination, constitute the *Cenozoic era*, which extends from 65 million years ago to the present. This portion of the *geologic time scale* is found in Table 3.3.

As a point of reference in imagining the extent of the continental glaciation, note that the Missouri and Ohio rivers roughly represent the southern margin of the ice sheet during its most southward advance. In addition to the continental ice sheet, valley glaciers expanded and advanced in more mountainous areas beyond the ice sheet margin.

In some areas the slowly moving glacial ice carved out and transported surficial deposits such as described in Section 3.1.1 and redeposited them far from their source.

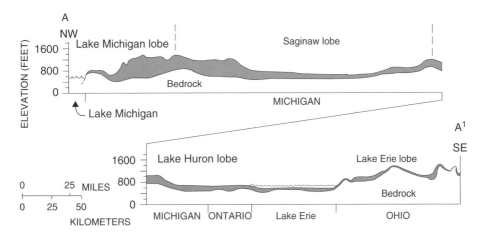

FIGURE 3.9. Cross section along the line A–A^1 in Figure 3.8. The shaded area represents the Quaternary deposits. Note that these deposits are found at all elevations, unlike sediments deposited by water. The thickness of deposits shown in this figure can be compared with those found in Figure 3.8 (from [7]).

FIGURE 3.10. Outcrop of schist, a very hard metamorphic rock, that has been carved by continental glaciation to form smooth surfaces of unusual topology.

For example, much of the fertile soil of the midwestern United States once resided in Canada and was transported via ice, wind, and water associated with continental glaciation. The thickness of the deposits that are associated with glaciation can be large, often in excess of 400 ft, as is illustrated in Figures 3.8 and 3.9. We discuss the nature of these materials in more detail in the following sections.

Glacial ice has enormous erosive potential. Rocks of great hardness can be and have been reshaped by ice during the latest ice age. Figure 3.10 shows an outcrop of schist, a very hard and dense metamorphic rock, that has been molded by a continental glacier during the last ice age. Note that the resulting rock surface is smooth and can exhibit unusual topography. A close-up view of the rock illustrated in this picture is found in Figure 3.23.

TABLE 3.4. Landforms and Materials Arising Out of Glacial Activity.

Glacial Deposits	
Stratified	Unstratified
Outwash Deposits	Basal Till
Varved Clay	Ablation Till
Deltaic Deposits	Lateral Moraine
Esker Deposits	Medial Moraine
Cravasse Deposits	Terminal Moraine
Kame Deposits	Ground Moraine
Kettle Deposits	Drumlin

Many of the important groundwater quality and quantity problems are located in the northeastern United States and the Upper Midwest. Inasmuch as the relevant aquifers in these areas are often made up of materials deposited by the glaciers, it is important to understand the basic mechanisms at work in glaciation and the nature of the resulting geologic deposits.

The deposits formed via glacial action can be classified as stratified and unstratified. Table 3.4 shows the cataloging protocol we will use in the following sections.

Unstratified Deposits   The material scraped from the earth's surface by the glacier and later directly deposited in various landforms (without transport by water) is called *till*. Till is very unusual inasmuch as it typically exhibits no grain size sorting. In other words, mechanical analyses would show a wide distribution of grain sizes. The resulting deposits are therefore *poorly sorted* and *well graded*.

The poor sorting is due to the fact that there is no mechanism for sorting the materials as there is in water-transported or wind-transported materials. As a consequence, in the same sample of till, one might find every grain size from clay to boulder. In Figure 3.11 we see where two valley glaciers converge. Each of the two valley glaciers has *lateral moraines* upstream of their confluence. After the glaciers merge, two of the lateral moraines combine to form a *medial moraine*. We will discuss moraines in more detail shortly.

In Figure 3.12 we see till being deposited at the toe of the Tasman glacier in New Zealand. The glacier at this point consists of a mixture of glacial ice and crushed rock fragments. All grain sizes are represented and the individual rock fragments do not, in general, exhibit the rounding effects identified with water-transported materials. The pool shown in this figure consists of glacial meltwater attributable to the melting ice. Rock dust is held in suspension in this water, making it, for all practical purposes, undrinkable. On the top of the ice is found the ablation till, which we will discuss momentarily.

The composition of the till, both mineralogically and in terms of grain size, depends on the source of the material that was available to the glacier as it flowed under gravity from its source area to its furthest extent. For example, glacial ice moving through the Great Lakes basins tended to erode fine-grained materials. As one would expect, this generated a till that exhibits an abundance of clay-sized particles.

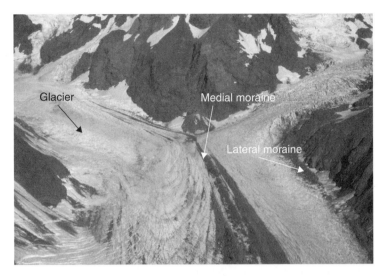

FIGURE 3.11. Confluence of two glaciers on Mount Cook in New Zealand. Note the existence of lateral moraines where the glaciers meet the rock walls and the formation of a medial moraine where two lateral moraines merge along the midline of the new larger glacier.

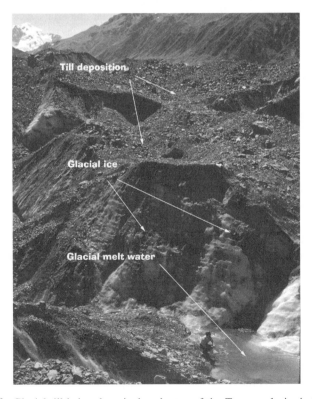

FIGURE 3.12. Glacial till being deposited at the toe of the Tasman glacier in New Zealand.

Till is categorized as *ablation till* and *basal till*. Ablation till is that deposited via transport by the ice to the surface of the glacier. Through melting, this material is released and forms a relatively loose and friable deposit as is evident in Figure 3.12. In other words, it appears similar to water-transported deposits of sand and gravel, but for its poor sorting and a lack of rounding of individual particles.

Basal till, on the other hand, is deposited beneath the glacier. One could envision this process as similar to using a knife to spread butter on a slice of bread (the glacier would be the knife, the till the butter and the rock underlying the glacier the bread). Basal till can be extremely hard with characteristics similar to rock. Both ablation and basal till deposits are found over much of the area of the United States that has experienced glaciation (see Figure 3.8). While ablation till can be important from a water supply and contaminant-transport perspective, basal till is generally of relatively low permeability and of lesser interest.

A boulder that has been deposited on subsurface material different from the material of origin of the boulder is called an *erratic*. Erratics can be enormous, often measuring more than 10 feet in diameter. A trail of erratics leading from the source area of the boulders to their final resting place is called a *boulder train*. It can be used to determine the direction of movement of the ice in the neighborhood of the boulder train. While of interest from a geologic point of view, erratics and boulder trains are of little importance hydrogeologically.

Several characteristic landforms are generated by glacial deposition and erosion. *Moraines* are accumulations of poorly sorted granular materials carried by a glacier and subsequently deposited on the earth's surface and are the most commonly encountered. The *lateral moraine,* mentioned earlier and shown in Figure 3.11, is formed where the glacial ice comes in contact with the valley wall containing it. This kind of moraine is found as a long, narrow ridge located at the ice–rock contact.

As described earlier and again illustrated in Figure 3.11, medial moraines are found on the coalescing edges of two glaciers at the point where they come together. In the center of the combined river of ice, the two ridges of material that formed two of the four lateral moraines associated with the two coalescing glaciers come together to form a moraine that is now somewhere in the interior of the ice flow rather than on the edges. Of course, there are still two lateral moraines located along the contact between the new combined glacier and the valley walls.

Terminal moraines occur along the ice margin at the furthermost advance of the ice. They are located where the downstream edge of the melting ice remained stationary for a prolonged period because its rate of melting is in equilibrium with the rate of flow of ice from the source areas. In other words, the ice is flowing to the ice margin at the same rate as the ice is melting at the margin. At this point, rock debris is released from the ice as the ice melts, something along the lines of a conveyer belt, leaving a deposit of till. Such moraines can be hundreds of feet in height. In the case of a valley glacier, end moraines have a crescent shape that points downstream.

Continental glaciers, on the other hand, tend to leave terminal moraines that are more irregular and can extend miles across the landscape. As the glacier recedes, it can also form recessional moraines at points where the ice front pauses long enough in a state of equilibrium to form a moraine.

Finally, it is important to mention *ground moraine*, which consists of widespread deposits of till that are laid down as the ice retreats. Consequently, they are found over vast areas.

Till can have a *wide range of hydraulic conductivity* characterized by significant changes in material properties over short distances. As such, they are very nonhomogeneous[4] (or heterogeneous) with respect to hydraulic conductivity in contrast to deposits such as dune sand that exhibit uniform homogeneous properties.

Because of the nonhomogeneous nature of till deposits, the analysis of the movement of water and contaminants in till is a very challenging problem. The nature of the challenge is not so much in the determination of the direction and speed of groundwater once the geologic materials are adequately mapped, but rather in the mapping process itself. Inasmuch as there is, in general, a lack of continuity of deposits of a given grain size in till, it is difficult to interpolate material properties between subsurface boring locations. Thus it is necessary to have a large number of boring locations to adequately define hydrodynamic properties, such as hydraulic conductivity, in till. Since boring into the subsurface tends to be expensive, and consequently a limited number of borings can be realized in a given area, till is not, as a general rule, very well characterized hydrogeologically.

Stratified Deposits   Deposits of materials derived from glacial ice, but water transported, are collectively called *glacial drift*. *Outwash* is a form of glacial drift deposit that is generated when the meltwater from the ice encounters and entrains materials being transported by the ice and deposits them down-valley from the ice margin. Streams and rivers emanating from the ice margin tend to be very fast moving and constitute a high-energy environment. As a result, they tend to carry coarse sediments that, upon deposition, become glacial-fluvial deposits, that is, stream-carried deposits. Outwash is characterized by coarse material interbedded and intermixed with finer grained materials such as silts and clays. A relatively permeable aquifer, outwash deposits are often used as a source of groundwater for domestic, municipal, and industrial supplies.

In Figure 3.13 meltwater is being generated at the toe of the Franz Joseph glacier in New Zealand. Notice in this photograph that the ice is in retreat as is deduced from the fact that the now visible bedrock has been molded by the ice when it was located further down the valley. Debris from the ice is carried by the meltwater downstream, where it is deposited as outwash. In Figure 3.14 one observes the outwash plain generated at the toe of the Tasman glacier. This photo was taken a short distance down the valley from that presented in Figure 3.12.

Where glacial lakes are found, a clay deposit unique to this environment is often formed. Characteristically, these clay deposits are made up of alternating layers of finer grained and coarser grained materials. In cross section, these materials appear striped. They are called *varved clays*. It is believed that these alternating layers represent summer and winter sedimentation. As might be expected, glacial lake clays are often found in areas near the terminus of the glacier. Some of the glacial lakes formed during the last Pleistocene glaciation were enormous, some even larger than the current Great Lakes.

---

[4]The terms *inhomogeneous, heterogeneous,* and *nonhomogeneous* are often used interchangeably.

FIGURE 3.13. Meltwater generated at the toe of the Franz Josef glacier in New Zealand.

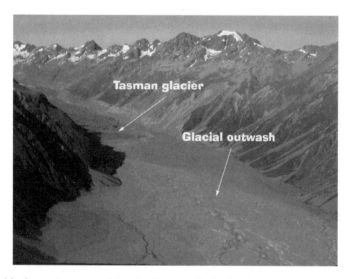

FIGURE 3.14. Outwash generated by the Tasman glacier in New Zealand. Note the toe of the glacier to the left in the photograph.

Because lake clays have relatively low hydraulic conductivity, they are important primarily in their role as barriers to the vertical migration of contaminants in layered multiaquifer systems.

At the point where a high-energy water body, such as a river, enters a quiet, low-energy environment, such as a lake, deposition of coarse-grained material takes place. These deposits have characteristically dipping bedding and are called *deltaic deposits*. Outwash and deltaic deposits associated with glacial discharge are similar in terms of

FIGURE 3.15. Glacial deltaic deposits characterized by their poor sorting and cross bedding. Photo taken near Avon, Ontario, Canada.

grain size. However, *deltaic cross bedding*[5] can be used to distinguish one from the other (Figure 3.15).

Deltaic deposits can constitute important aquifers if they are large and are made up of relatively coarse materials. Glacial-lake deltas often satisfy these requirements.

Several other landforms are associated with glacial activity. One of the most interesting is the *esker*. Formed by streams running through tunnels in stagnant ice, these ridges of stratified gravel (often called stratified drift) are revealed when the surrounding ice melts away. They can be from 10 to more than 50 feet in height and can extend for miles. An example is shown in Figure 3.16. Since they were deposited in ice channels by water under pressure, the transporting water could run uphill as well as down, much as does a hose draped over a chair in the garden. Consequently, the esker deposits can follow the land topography even though it is uphill in the direction of original glacial-water flow.

Although striking land features, they are relatively unimportant from a hydrogeological perspective. Their lack of importance stems from the fact that, although they are highly permeable deposits, they normally project above the neighboring landscape and therefore are unsaturated. Lacking saturation, they cannot be effectively exploited as a water supply. However, research has been conducted on using eskers as hot-water energy repositories. Groundwater that is heated during the summer months is injected into the esker to be pumped later and "mined" for its thermal energy content.

Crevasse fillings are similar to eskers in many respects but tend to be straighter. Their geometry reveals the fact that they were formed by material collecting in crevasses in the ice. They were subsequently left as ridges as the ice melted, much as were eskers.

A close relative to the crevasse deposit is the *kame*. A stratified mound of irregular shape, it was probably formed by debris collecting in openings in stagnant ice. A *kame*

---

[5]Cross bedding occurs when sediments are deposited on a slope at the leading edge of the deposit. In a cross section taken along the direction of flow, this bedding has a negative slope in the downstream direction.

FIGURE 3.16. An esker ridge, deposited into a former glacial lake with a kettle located in the southwest quadrant of the photograph. Location is in northwestern Manitoba (from Canadian Landscapes Photo Collection [9]).

terrace is found between the wasting ice and the valley wall. As the name implies, this material is normally left as a terrace along the side of a valley wall once the ice has melted.

Drumlins are landforms found largely in New England and southeastern Canada. They are ridges of an elliptic shape with the long axis of the ellipse oriented in the direction of ice flow. Because they tend to appear in clusters, localities wherein they are found are often referred to as *drumlin fields*. They are characteristically streamlined, but steeper on the upstream end, that is, the direction from which the ice flowed. Typically, drumlins are 25 to 200 feet in height and are composed of till. Due to the fact that drumlins are normally made up of till and elevated relative to surrounding topography, they are typically relatively unimportant from a hydrogeological perspective.

When an ice block becomes isolated during the retreat of a glacier, it can sometimes be surrounded by till, or even buried. As the ice melts, it leaves a hole, or depression, in the till-dominated landscape. This depression is called a *kettle*. It sometimes becomes a lake or swamp and fills with organic matter. Kettles can be miles in diameter and are often recognized by the rich organic content of the soil occupying them. *Peat bogs* found in the northeastern United States are sometimes the remnants of kettles. An example of a kettle is found in Figure 3.16.

3.2 CONSOLIDATED ROCKS

When sedimentary deposits are buried for long periods of time, the materials consolidate and become cemented together to form consolidated rocks. At this point the deposits lose

SEDIMENTARY ROCKS			
ORIGIN		TEXTURE	PARTICLE SIZE COMPOSITION	ROCK NAME

ORIGIN		TEXTURE	PARTICLE SIZE COMPOSITION	ROCK NAME
CHEMICAL	DETRITAL	CLASTIC	GRANULAR	CONGLOMERATE
			SAND	SANDSTONE
			SILT AND CLAY	MUDSTONE OR SHALE
	INORGANIC	CLASTIC AND NONCLASTIC	CALCITE	LIMESTONE
			DOLOMITE	DOLOSTONE
			HALITE	SALT
			GYPSUM	GYPSUM
	BIOCHEMICAL		CALCITE	LIMESTONE
			PLANT REMAINS	COAL

FIGURE 3.17. Sedimentary rock chart (adapted from [10]).

much of their pore space and the resulting porosity can be attributed, at least in large part, to fractures and other pathways associated with dissolution. The consolidated rocks derived from the unconsolidated deposits described previously are summarized below (see Figure 3.17).

In Figure 3.17 the first column represents the geologic process that formed the rock. The term *detrital* appearing in this column denotes rocks created from the *erosion* or *weathering* of other rocks. The texture or appearance of the rock created as a result of detrital processes is denoted as *clastic*, which means made up of rock and mineral fragments. Particle size information for each rock type is given in the third column and the rock name is found in the fourth. While this tabulation is somewhat simplified, it forms a useful template for cataloging sedimentary rocks.

In general, consolidated rocks are less productive as aquifers than unconsolidated rocks, primarily because they have lower permeability. On the other hand, while the specific discharge for consolidated rocks may be lower than that of unconsolidated rocks, consolidated rocks often have low porosity, which, for the same specific discharge, results in relatively high pore velocity. The reason for this rapid movement can be seen via Eq. (3.1),

$$\mathbf{v} = \frac{\mathbf{q}}{\varepsilon}, \tag{3.1}$$

written for saturated media, where $\mathbf{v}$ is the average pore velocity, $\mathbf{q}$ is the specific discharge, and ε is the porosity. As ε becomes small, as it is in fractured rocks, all other things being equal, the average pore velocity increases. Thus contaminants can move long distances in short times in consolidated rocks with significant *secondary permeability*, that is, permeability due to such features as fractures, faults, or solution cavities.

Secondary permeability is illustrated in Figure 3.18, where it is evident that water percolating along fractures in the consolidated sedimentary rock has dissolved the rock

FIGURE 3.18. Secondary permeability observed in limestone in western Kentucky.

and produced enormous secondary permeability. In Figure 3.19 we see how secondary permeability facilitates the transmission of large quantities of water significant distances in relatively short times.

Because flow due to secondary permeability tends to be directional in nature, there is often significant anisotropy, that is, preferential flow direction (see page 87 for a discussion of anisotropy).

The importance of secondary permeability also gives rise to the concept of a *double porosity*[6] conceptual model of the aquifer system. In such a conceptual model there are two sets of permeabilities and two sets of porosities, one associated with the secondary permeability features, such as the fractures, and the other associated with the *primary permeability* features, namely, the host rock (more on secondary permeability can be found on page 19). The host rock in this instance can be thought of as consisting of blocks, which are, in turn, surrounded by fractures. The primary and secondary flow systems are each described by a separate flow equation with its own set of parameters, boundary conditions, and hydrological stresses. The two systems are coupled one to the other by a leakage term that describes the movement of water (and contaminants) from the blocks to the fractures and vice versa. Needless to say, the determination of this leakage term is not an easy task and constitutes a significant challenge to groundwater professionals using the double porosity model.

---

[6] *Double porosity models* are also known as *dual porosity models*.

FIGURE 3.19. Spring emanating from secondary permeability in a limestone formation in Kentucky.

3.3 METAMORPHIC ROCKS

When consolidated sediments, and indeed also igneous rocks (which will be discussed in Section 3.4), are subjected to intense pressure and heat, generally due to phenomena associated with tectonic-plate dynamics (described in detail in Section 3.5), the minerals that constitute the rock recrystallize to form a much more dense rock. The geologic phenomenon taking place is called metamorphism and the new rocks are called metamorphic rocks. The resulting metamorphic rocks are made up of new minerals in equilibrium with the new conditions under which they now formed. While the molecular composition of the resulting rock is normally the same as the original consolidated rock, the minerals that form it are generally quite different. Some typical metamorphic rocks and the host rock from which they formed are listed in Figure 3.20. The first column is the host rock from which the metamorphic rocks listed in the ensuing columns formed. The last three of these rocks, namely, basalt, granite, and rhyolite, are igneous rocks that will be considered shortly.

Across the top of the chart are listed the *metamorphic zones*. These names correspond to the different pressure–temperature environments to which the host rocks have been subjected. The lowest grade of metamorphism is the *chlorite zone*, which corresponds to relatively low temperatures and pressures. The zone corresponding to the highest temperatures and pressures is the *sillimanite zone*. The names of the zones are associated

ORIGINAL ROCK	METAMORPHIC ZONE				
	CHLORITE	BIOTITE	ALMANDITE	STAUROLITE	SILLIMANITE
	METAMORPHIC ROCK				
SHALE	SLATE	BIOTITE PHYLLITE	BIOTITE-GARNET PHYLLITE	BIOTITE-GARNET STAUROLITE SCHIST	SILLIMANITE SCHIST OR GNEISS
CLAYEY SANDSTONE	CLAYEY SANDSTONE	QUARTZ-MICA SCHIST	QUARTZ-MICA-GARNET-SCHIST		
QUARTZ SANDSTONE	QUARTZITE				
LIMESTONE DOLOMITE	LIMESTONE DOLOMITE	MARBLE			
BASALT	CHLORITE-EPIDOTE-ALBITE-SCHIST		ALBITE EPIDOTE AMPHIBOLITE	AMPHIBOLITE	
GRANITE	GRANITE	GRANITE GNEISS			
RHYOLITE	RHYOLITE	FINE-GRAINED BIOTITE GNEISS			

FIGURE 3.20. Chart of metamorphic rocks (adapted from [10]).

with the indicator minerals that are created in the various geochemical environments. For example, sillimanite, a mineral, is formed under the temperatures and pressures associated with the sillimanite zone.

Different minerals form from different host rocks. The composition of the host rock dictates to some degree the composition of the metamorphic rock. Limestone, for example, will form marble under conditions similar to those that will form quartzite from sandstone. It is interesting to note that while a fracture plane through a sandstone will tend to follow grain boundaries, similar fractures in quartzite will cut through grains. This is due to the fact that, during metamorphism, the grain boundaries, for all intents and purposes, fuse and are no longer physically distinct[7]. Commonly encountered metamorphic rocks are *slates*, *quartzite, marble*, and various forms of *gneiss*. *Fossils*, preserved forms of ancient life often found in consolidated rocks, are normally absent from metamorphic rocks, all evidence of their existence having been destroyed in the process of metamorphism.

Because metamorphic rocks are exceedingly dense, they have virtually no primary porosity or permeability. Thus secondary permeability due to fractures and faults constitutes the principal mechanism for groundwater flow. Porosity and permeability are therefore very low in these rock units. However, as in the case of consolidated rocks, groundwater velocities can be quite high due to the low porosity and the relationship presented in Eq. (3.1). In visualizing the secondary permeability in consolidated rocks, one should keep in mind that fractures occur at all scales. Thus very small-scale fractures

---

[7]It is, however, possible to see the grain boundaries with the aid of specialized microscopes trained on specially prepared rock specimens.

FIGURE 3.21. Folded and fractured metamorphic rock of Precambrian age. While the rock blocks are of low permeability and porosity, the factures have a high permeability and low porosity that facilitates the movement of contaminants over long distances.

can act as conduits of flow just as can larger-scale dislocations, although the smaller-scale fractures may, on the whole, be less conductive.

A folded metamorphic rock is shown in Figure 3.21. Note that although the matrix of this rock is of very low porosity and permeability, there are numerous fracture planes associated with the rock fold. The fracture planes provide conduits for rapid movement of water, although the storage capacity of the overall rock mass is small. The consequence of this, as discussed earlier, is that in rocks of this kind contaminants can move large distances rapidly.

A second example of a fractured metamorphic rock is found in Figure 3.22. The rock illustrated here is slate and is characterized by vertical, closely spaced fractures. The flow and transport characteristics of this rock are similar to the rock shown in Figure 3.21. A third and final example is given in Figure 3.23. This magnetite-bearing chlorite-muscovite-albite schist is from the Underhill Formation of Lower Cambrian age collected in Vermont (see Section 3.5 for a discussion of geologic time and the nomenclature associated with it). This sample is very dense with relatively few significant fractures.

3.4 IGNEOUS ROCKS

Igneous rocks are formed directly from a *liquid* or *melt*. Such liquids may develop due to the complete melting of existing rocks due to mountain building. Alternatively, the melt may be derived directly from the earth's *mantle*, having been convected upward through the earth's crust to near the surface.

Igneous rocks tend to be cataloged according to grain size (or more specifically crystal size) and mineral composition. This concept is illustrated in Figure 3.24. Along the vertical axis is plotted the grain size. The horizontal axis is used to represent the percentage of dark (generally iron-bearing) minerals.

FIGURE 3.22. The metamorphic rock illustrated here is slate. It is characterized by closely spaced vertical fractures that have flow and transport properties similar to those associated with the preceding example in Figure 3.21.

FIGURE 3.23. Magnetite-bearing chlorite-muscovite-albite schist of the Underhill Formation of Lower Cambrian age located near Burlington, Vermont.

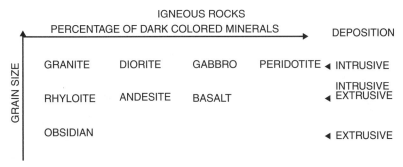

FIGURE 3.24. Igneous rocks defined by mineralogy and grain size.

The mode of rock formation is indicated on the right-hand side of the figure. There are two basic types of igneous rocks, extrusive and intrusive. *Extrusive rocks* are formed when liquid magma from the mantle reaches the earth's surface and forms lava. Because it cools quickly, it tends to form very small mineral crystals. One should keep in mind that the lava also extrudes at the bottom of the ocean, especially where, due to continental drift associated with plate tectonics, rifts occur in the crust of the earth[8] (rifts due to continental drift are discussed in Section 3.5).

Lava has three basic forms. *Ash* is the material that, when extruded during volcanic activity, is carried via the wind to form ash cones. *Basalt* is a liquid form of the extruded material that flows overland to form large volcanic beds. Basaltic flows can be either very dense and ropey, in which case they are called *pahoehoe*, or composed largely of coarse, porous-block beds and known as *ah ah*. The dense beds can also be quite porous, but the porosity may be irrelevant to groundwater flow because it is often composed of *isolated air bubbles* (see Figure 1.23 for an example of this phenomenon).

Enormous areas of the northwestern United States are covered by basaltic flows. Often the basaltic flows consist of layers of dense basalt interbedded with relatively permeable basalt. The permeable basalt layers are formed where one basaltic flow encounters and overrides an earlier one. In this circumstance, the surface of the older flow tends to be irregular due to the cooling of the lava, a phenomenon similar in concept to ice blocks forming on the surface of a lake or stream. The permeable interbeds generally form good aquifers that yield considerable volumes of water to wells. The dense basalt is capable of groundwater flow primarily through secondary permeability.

Figure 3.25 illustrates extrusive igneous rocks in the form of lava. As is evident from this photograph, the upper layer of basaltic rock is broken and irregular. It is easy to visualize that when this material is buried, it may form a very permeable interbed.

In Figure 3.26 one sees the mold of a tree. Now, where only a hole is found, a tree once stood. The ash from the Kilauea Iki vent on the island of Hawaii buried the tree, which was destroyed by the heat of the ash. This is only one of many unusual openings that are found in volcanic terrain. Since the void space created by the demise of the tree is not continuous beyond a few feet, it is not an important hydrogeological feature. However, many volcanically derived openings are continuous and important from a groundwater flow perspective. On such example is a *lava tube*, now empty, that once conducted lava over extensive distances (see Figure 3.27).

---

[8]The crust of the earth can be thought of as the hard rock shell that surrounds the molten interior of the planet. It is thickest (20–50 km) under continents and thinnest (7–8 km) under the oceans.

FIGURE 3.25. Lava flows on the island of Hawaii. Note the volcanic vents in the distance and the irregular surface of the cooled lava field.

FIGURE 3.26. Mold of tree. The original tree was destroyed by the heat of the ash that was ejected from the Kilauea Iki vent in Hawaii.

FIGURE 3.27. Lava tube on the island of Hawaii. View is from inside the lava tube looking out. Note the handrails for scale.

Intrusive rock is the second kind of igneous rock. Rocks of this family are very dense and often have large crystals, consistent with what one might expect given an environment of slow crystallization far beneath the surface of the earth. Those rocks that form at great depths may eventually be exposed at the earth's surface by erosion. Such rocks are often *granites*. However, other coarse-crystalline rocks called by different names because of their grain size or mineral composition are also widely found. In general, granitic rocks have virtually no primary porosity or permeability. Secondary permeability and porosity are due to fracturing and faulting. Because intrusive rocks normally appear on the earth's surface through the erosion of overlying rock units, the rocks that form the core of mountain chains can, through erosion of the overlying rock units, eventually make their way to the earth's surface. Such intrusive rocks are often found as *veins* in metamorphic rocks, where molten rock has intruded along preexisting fractures, cooled, and crystallized.

Figure 3.28 shows a granite outcrop at a quarry in Barre, Vermont. Vertical and horizontal jointing is especially well developed at this site and evident because of the tendency of the rock to cleave along the joint planes. Typical of granitic bodies, fracture planes make up virtually all of the permeability in this rock.

In Figure 3.29 a piece of the host sedimentary rock into which the granite magma was intruding was dislodged from the sedimentary rock and incorporated into the magma. As the granite melt solidified, the piece of host rock was entrapped. The entrapped piece of rock is called a *xenolith*.

FIGURE 3.28. Granite of Devonian age formed during mountain building from the partial remelting of Silurian–Devonian sedimentary rocks. Photograph taken at the Rock of Ages quarry in Barre, Vermont.

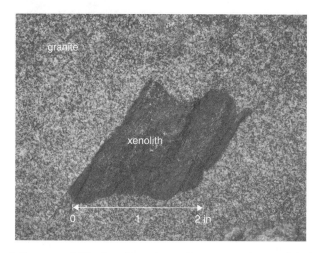

FIGURE 3.29. A fragment of Silurian–Devonian sedimentary rock was dislodged and incorporated into the granite melt during its intrusion. The granite solidified around the piece of host rock to form a *xenolith*.

Intrusive rocks can also extrude along fractures and faults that intersect with the land surface. Molten rock will travel along such fractures, whether horizontal or vertical. As the molten rocks approach the earth's surface, they form dense rock masses that are generally referred to as *basaltic* intrusive rocks.

When the fractures occupied by these rocks are vertical and crystallization occurred in the subsurface, the consequent intrusive rocks are called *dikes*. When the filled fractures are horizontal, the resulting rock bodies are called *sills*. In either case, the contact between the basalt and the host rock can be very permeable whenever a zone of cooling between the molten and host rock exists and thermally induced fracturing has occurred. As was the case for extrusive rocks, such zones can represent important zones of secondary permeability. Dissolved and nonaqueous contaminants can travel significant distances rapidly through these very permeable zones.

Various kinds of intrusive igneous bodies are illustrated in Figure 3.30.

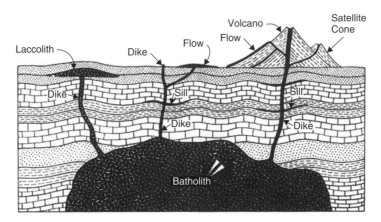

FIGURE 3.30. Diagrammatic representation of intrusive and extrusive igneous bodies. The batholith is the source of the molten rock that forms the various intrusive and extrusive bodies represented in this diagram (from [10]).

3.5 GEOLOGIC TIME

Over the approximately 4.5 billion years that the Earth has been in existence, the Earth's crust has undergone enormous change. In some periods large parts of North America were volcanically quiescent, giving rise to the creation of inland seas that covered much of the continent. In such an environment, chemical precipitates accumulated on the sea floor and limestone, dolostone, and salt beds were created. During and subsequent to periods of mountain building, erosion of the uplifted rocks gave rise to clastic sedimentary rocks such as sandstones and conglomerates. This example illustrates how a knowledge of depositional environments active at a given point in geologic time permits a more informed interpretation of limited geologic information. In addition, the fossil record can provide insight into the relative age of different geologic horizons, further assisting the groundwater professional in his/her efforts to establish the nature and extent of aquifers and aquitards.

In Table 3.3 we introduced the concept of geologic time in the context of Pleistocene glaciation. However, the *Cenozoic era* described in that table represents only a small fraction of the time since the Earth was formed. The three eras that make up geologic time, from youngest to oldest, are *Cenozoic* (recent life), *Mesozoic* (middle life), and *Paleozoic* (ancient life). All time prior to the Paleozoic era is bundled into the *Precambrian eon*.[9] . The Precambrian represents approximately 90% of geologic time (see Figure 3.31) and is subdivided into the *Proterozoic* (early life), *Archaean* (ancient or primitive), and *Hadean* (hades-like) *eras*.

Because a knowledge of the geologic time scale plays a key role in the concept of rock correlations—that is, the extrapolation of information regarding a rock formation from a limited number of rock samples—we will now briefly discuss the geologic time scale. However, we will not dwell on this topic because, although it sets the stage for understanding geologic processes, it is in the end the hydrodynamic characteristics of rocks rather than their age that are of primary importance to us.

[9]The nomenclature adopted here for the geological time scale is that of the U.S. Geological Survey; however, other interpretations are also found in the literature [8].

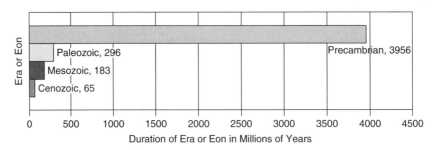

FIGURE 3.31. Graphical representation of the length of time identified with each of the three geological eras and the Precambrian eon.

3.5.1 The Hadean Era

We begin Earth's history with the oldest era in the Precambrian eon, the *Hadean era*. The beginning of this era coincides with the formation of the Earth from dust and gas orbiting the Sun about 4.6 billion years ago and ends about 1 billion years ago. The Earth's surface during this period was made up of liquid rock created by lava flows. The atmosphere was composed of carbon dioxide, water vapor, and traces of nitrogen and sulfur compounds [11]. Incoming objects from space frequently reached the molten Earth's surface. It is believed by some that a large asteroid struck the Earth during this era and resultant ejected molten rock formed the Moon.

No rocks have been found that are representative of this era. Thus, if there was any life on Earth at that time, no evidence of it has been preserved.

3.5.2 The Archaean Era

We now move on to the *Archaeon era* at which time the Earth is about 3.6 billion years old. It is covered by a global ocean whose bottom is formed by solidified lava. Chains of islands appear in the ocean and are moved about by convection cells in the molten rock beneath the ocean floor. When the islands Collide they coalesce to form a larger land mass. Eventually this accretion will be responsible for the formation of continents. The movement of land masses by virtue of thermal convection in the Earth's mantle is the beginning of a phenomenon that continues to this day and is called *plate tectonics*.

The atmosphere has evolved and now consists primarily of nitrogen. There are clouds and rain similar to what we find today. The carbon dioxide that was characteristic of the atmosphere during the Hadean era has combined with other available chemical species to form limestone that is found at the bottom of the oceans [11].

Life on earth is believed to have begun in the oceans at the beginning of this era. The oldest known fossils are those of bacteria dated at about 3.5 billion years before the present [11].

3.5.3 Proterozoic Era

The *Proterozoic era* starts about 2.6 billion years before the present. The drifting islands mentioned earlier have coalesced in part to form continents. At this time there are two supercontinents, one in the northern hemisphere and the other in the southern. The Earth is very cold. Although the precise location of the supercontinents is not known, they were probably located near the north and south poles. It is hypothesized that their polar locations are responsible for the fact that they appear to have been largely covered with glaciers, possibly similar to what was observed in the Pleistocene.

Life on Earth does not evolve dramatically and is still found only in the ocean and is largely bacteria. However, about midway through this era *single-celled creatures* began to form and near the end of the era soft, *multicelled creatures* are seen.

Although the atmosphere is still largely nitrogen, the biological activity attributable to the primitive sea life has created small amounts of oxygen [11].

3.5.4 Paleozoic Era

The *Paleozoic era* heralds the end of the Precambrian eon. The Paleozoic era is made up of six major periods (see Table 3.5). The oldest of these is the *Cambrian period*, which existed between 544 and 505 mybp[10]. It is named after Cambria, the name the Romans gave to Wales in the United Kingdom, where the rocks of this period were first studied [8]. Table 3.5 catalogs the various periods and epochs that constitute the Paleozoic era.

Cambrian Period At the dawn of the *Cambrian period* the southernmost supercontinent, called *Gondwanaland*, remained in the southern hemisphere. Over geologic time, Gondwanaland would evolve to form what are now South America, Africa, Madagascar, India, Australia, and Antarctica. The northernmost supercontinent split during

TABLE 3.5. Subdivision of the Paleozoic Era

Era	Period or System	Epoch or Series	
Paleozoic 544–248	Permian 286–248		Late or Upper
			Early or Lower
	Carboniferous 360–286	Pennsylvanian	Late or Upper
			Middle
			Early or Lower
		Mississippian	Late or Upper
			Early of Lower
	Devonian 410–360		Late or Upper
			Middle
			Early or Lower
	Silurian 440–410		Late or Upper
			Middle
			Early or lower
	Ordovician 505–440		Late or Upper
			Middle
			Early or Lower
	Cambrian 544–505		Late or Upper
			Middle
			Early or Lower

[10]The acronym mybp stands for million years before present.

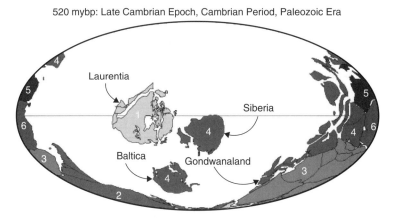

FIGURE 3.32. Location of continents (named) and future continents (numbered) during the Late Cambrian epoch. The numbers indicate 1 = North America, 2 = South America, 3 = Africa, 4 = Eurasia, 5 = Australia and 6 = Antarctica (from [12]).

this period to form three continents, Laurentia, Baltica, and Siberia (see Figure 3.32). Laurentia would evolve into what is now North America, and Baltica would become northern Europe. While the southernmost polar regions remained cold, those continents near the equator, especially Laurentia, were tropical and flooded by a shallow tropical sea. Many hardy species lived in these seas (e.g., *brachiopods, trilobites*, and *graptolites*) and thick deposits of limestone were formed from those members of the population that were shelled marine animals. Fossil remains of brachiopods, trilobites, and graptolites are presented in Plate 3.1 [13].

Sandstone formations of Cambrian age are found in North America, evidence of the erosion of rocks exposed above sea level during this period.

Seaweed was found in the oceans and lichens were found on land. The whole animal kingdom except vertebrates developed during this period.

Ordovician Period The *Ordovician period* spans the interval between 505 and 440 mybp. It is named after a Celtic tribe called the Ordovices. During this period, Gondwanaland remained in tact, drifted over the south pole, and was in the grips of a major glaciation. North America straddled the equator and western and central Europe were separated from the rest of Eurasia (Figure 3.33). Baltica and North America converged, giving rise to the creation of a mountain chain that stretched from Norway to Scotland to Ireland to Greenland and finally to northeastern North America (pre-Appalachian range). Erosion of the mountain chains gave rise to extensive clastic deposits, such as sandstone and conglomerates, in North America.

At this time the climate was warm, at least in the area of the tropics, and North America was covered by shallow tropical seas. Life forms diversified dramatically. In fact, the diversity of marine life reached a peak during the Ordovician period. The *corals* (see Plate 3.1) first appeared in this period as did the *bivalve mollusks* and planktonic graptolites. Among the most successful of the life forms that developed in the Cambrian that continued into the Ordovician are the brachiopods.

Primitive vascular plants began to appear on land, which had heretofore been nearly barren but for lichens.

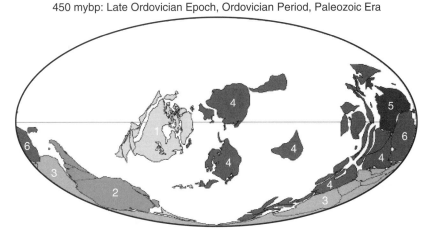

PLATE 3.1. Panel 1 shows Silurian *Halysites* chain coral from Louisville, Kentucky; panel 2 shows a graptolite, Ordovician, Danville, Kentucky; panel 3 shows a Silurian *Gravicalymene celebra* trilobite, Bardstown, Kentucky; and panel 4 shows an Ordovician *Hebertella* brachiopod, Owingsville, Kentucky (with permission of R. Schrantz [13]).

FIGURE 3.33. Location of land masses during the Late Ordovician epoch. The numbers indicate 1 = North America, 2 = South America, 3 = Africa, 4 = Eurasia, 5 = Australia, and 6 = Antarctica (from [12]).

At the end of the Ordovician period there was a mass extinction of tropical marine faunas. While there are many hypotheses regarding why this took place, the most probable is that the Earth cooled, with concomitant cooling of the oceans. Those organisms unable to adapt vanished.

Silurian Period   The *Silurian period* is named after a Celtic tribe called the Silures. It spans the period from 440 to 410 mybp. The continents during the Silurian were arranged much as they were during the Ordovician period (see Figure 3.34) with large areas of North America flooded by shallow seas. Large coral reefs and algae were abundant and major evaporite basins existed (e.g., the Michigan basin). Due to the erosion of the pre-Appalachian mountains, the early Silurian deposits in the eastern United States are sandstone and conglomerate. In the West, marine limestone was deposited.

The fossil record indicates that trilobites were still numerous in the sea and primitive fishes appear in increased numbers. Notably, scorpion fossils are found during this period. These fossils record what may have been the first animals to live on land and breath oxygen from the air.

Devonian Period   The *Devonian period* extends from 410 to 360 mybp. It is named after Devonshire, England, where rocks of this period were first recorded. During the Devonian the continents were moving together as seen in Figure 3.35. The separating oceans were shrinking, allowing freshwater fauna to migrate from the southern continent to those in the north.

Much of North America was submerged under shallow marine seas during this period. As a result, there are extensive deposits of evaporites, limestone, and dolostone, some of which is very fossiliferous. However, along the east coast of North America, mountain building was especially active. Erosion of these mountains produced large clastic formations, particularly well known is the Catskill deltaic deposits that resulted in clastic sedimentation far to the west of the mountains. At this time there were also mountains being formed along a line stretching from British Columbia, Canada, to Nevada.

420 mybp: Middle Silurian Epoch, Silurian Period, Paleozoic Era

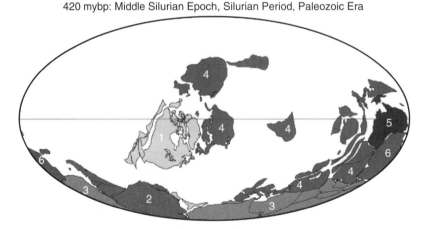

FIGURE 3.34. Location of continents during the Middle Silurian period. The numbers indicate 1 = North America, 2 = South America, 3 = Africa, 4 = Eurasia, 5 = Australia, and 6 = Antarctica (from [12]).

390 mybp: Middle Devonian Epoch, Devonian Period, Paleozoic Era

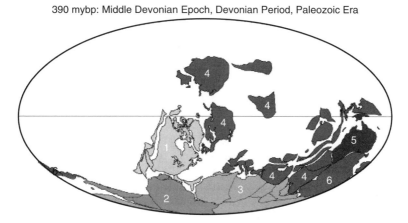

FIGURE 3.35. Location of continents during the Middle Devonian epoch. The numbers indicate 1 = North America, 2 = South America, 3 = Africa, 4 = Eurasia, 5 = Australia, and 6 = Antarctica (from [12]).

Graptolites died out during the Early Devonian and trilobites were much less common than during the Cambrian period. The two-shelled brachiopods were at their greatest diversity and corals were well represented. *Placoderm fish*, the most primitive of the vertebrates with jaws, were found as were sharks with their cartilaginous skeletons. The Devonian also saw the introduction of the bony fishes. The oldest known amphibians and insects were also found in the Devonian. Land plants likewise began to flourish during this period, forming the first forests.

At the end of the Devonian, there was a mass extinction in the oceans. Terrestrial flora were essentially unaffected. The major reef builders, who earlier in the period were responsible for building the largest reefs in geologic history, were the most affected. So severe was their decline that major reef building did not occur again in this era. Among the marine invertebrates, 70% did not survive into the next period.

The reasons for this mass extinction are not certain, but it is believed that the cause was global cooling, similar to what possibly occurred in the Ordovician period.

Carboniferous Period The *Carboniferous period* is composed of the *Mississippian* and *Pennsylvanian* periods and extends from 360 to 286 mybp. The Mississippian and Pennsylvanian periods are named after locations where the type of rock widespread. The type rocks, in this case, are those used to first describe and identify the rocks indicative of the period.

Near the end of the Carboniferous period three land masses came together to form a supercontinent called *Pangea*, from the Greek meaning "all the earth" (see Figure 3.36). During this period, due to changes in sea level resulting from either tectonic deformation or the formation of glaciers in Gondwanaland, there were numerous transgressions and regressions of the shallow seas that inundated large portions of North America. Mountain building took place in the Appalachian region of the United States, extending into the Maritime provinces of Canada to the north and to Oklahoma in the south. During periods of shallow sea transgression, shallow marine carbonates were deposited. In addition, in response to the erosion of the uplifted mountain chains, alluvial and deltaic deposits accumulated in the eastern half of North America.

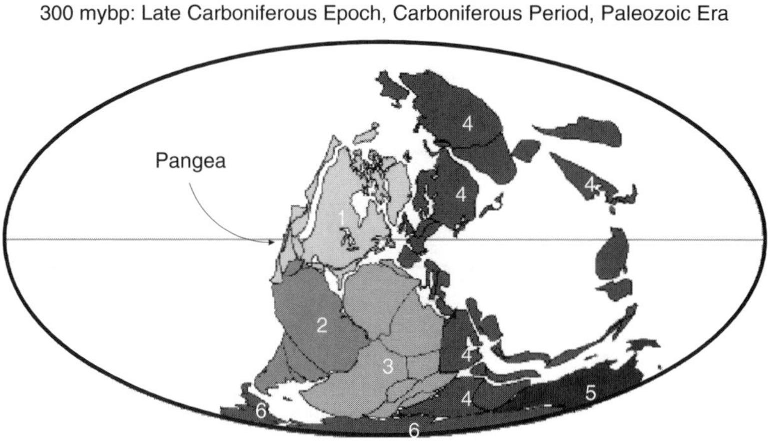

300 mybp: Late Carboniferous Epoch, Carboniferous Period, Paleozoic Era

FIGURE 3.36. Location of supercontinent Pangea and future continents at the end of the Carboniferous period. The numbers indicate 1 = North America, 2 = South America, 3 = Africa, 4 = Eurasia, 5 = Australia, and 6 = Antarctica (from [12]).

The Carboniferous shallow seas supported abundant life. Brachiopods, *cephalopods*, *pelecypods*, *gastropods* (snails), crinoids, corals, and *bryozoa* were represented (see Plate 3.2 [13] for photos of some of these fossils). Although trilobites were present, they were nearing the end of their long history. Sharks were found in both fresh and salt-water environments. Amphibians were represented, as were the first small reptiles. Large insects, such as dragonflies and cockroaches, were also present [14].

Perhaps best known for its abundant plant life, the Carboniferous period was characterized by widespread, swampy coastal forests with trees that reached heights of 30 m. The first seed-bearing plants appeared during the Middle Carboniferous. As a consequence of this environment, the world's leading deposits of coal were formed. Contrary to one's intuition, coal deposits with their considerable secondary permeability due to fractures constitute significant groundwater reservoirs.

Permian Period The Permian period extends from 286 to 248 mybp. The name is derived from the province of Perm in Russia where rocks of this age were first studied [8]. The supercontinent *Pangea* stretched from the north pole to the south pole as seen in Figure 3.37. The southern regions of Pangea, which later formed southern South America, southern Africa, Antarctica, India, southern India, and Australia, were glaciated during this period. By the mid-Permian, a mountain range extending into North America blocked moisture from the equatorial regions and desert-like conditions developed. As would be expected, due to the existence of the supercontinent, deposits throughout the world have some common characteristics. At this point a single tropical ocean occupied 75% of the Earth's surface. In response to the desert conditions, dune sands, evaporites, redbeds, and calcic soil zones developed.

Brachiopods became highly specialized during the Permian, thereby forecasting their demise. By the end of the Permian, only 20% of the families that existed at the beginning of this period survived. A similar fate befell the bryozoans and the ammonoids. Insects made significant advances over those of the Carboniferous period. Plants adapted to accommodate to the relatively dry conditions encountered in the Permian. Amphibians

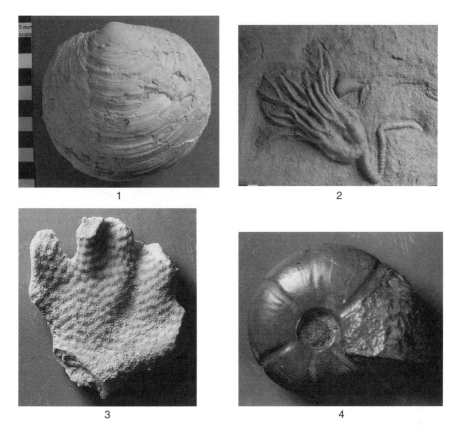

PLATE 3.2. Panel 1 is a silicified, Devonian pelecypod, Clark Country, Indiana; panel 2 is an Ordovician *Archaeocrinus* crinoid, Garrard County, Kentucky; panel 3 is an Ordovician *Constellaria* bryozoan, Frankfort, Kentucky; and panel 4 is a Mississippian *Muensteroceras* ammonoid, Morehead, Kentucky (with permission of R. Schrantz [13]).

were common, but the reptiles made the most significant advances. Of greatest importance were those mammal-like reptiles that would evolve in the Triassic into mammals [15].

3.5.5 Mesozoic Era

The Mesozoic era, extending from 248 to 65 mybp, has three periods (see Table 3.6). From oldest to youngest these are the Triassic, the Jurassic, and the Cretaceous.

Triassic Period The *Triassic period* extends from 248 to 213 mybp. The name is derived from the threefold division of the rocks of this age found in Germany [8]. The continental collisions that started in the Devonian and led to the supercontinent Pangea continued into the late Triassic. The state of the Earth in the early Triassic is shown in Figure 3.38. The Triassic was a period of relative calm from a tectonic perspective. The various continents formed rigid, relatively stable, continental blocks. By the close of the Permian, the geography of the lands and of the seas were similar to what we see today [16]. In the Early Triassic, the seas retreated from eastern North America to a

240 mybp: Lower Triassic Epoch, Triassic Period, Mesozoic Era

FIGURE 3.37. Location of the supercontinent Pangea and the future continents at the middle of the Permian period. The numbers indicate 1 = North America, 2 = South America, 3 = Africa, 4 = Eurasia, 5 = Australia, and 6 = Antarctica (from [12]).

TABLE 3.6. Subdivision of the Mesozoic Era

Era	Period or System	Epoch or Series
	Cretaceous 145–65	Late or Upper Early or Lower
Mesozoic 248–65	Jurassic 213–145	Late or Upper Middle Early or Lower
	Triassic 249–213	Late of Upper Middle Early or Lower

240 mybp: Lower Triassic Epoch, Triassic Period, Mesozoic Era

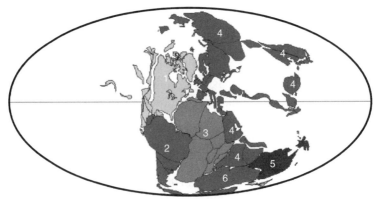

FIGURE 3.38. Location of continents during the early Triassic period. The numbers indicate 1 = North America, 2 = South America, 3 = Africa, 4 = Eurasia, 5 = Australia, and 6 = Antarctica (from [12]).

north–south line through central Nevada. To the west of this line one encounters marine limestone, shale, a volcanic sequence, and again limestone. To the east of this line clastic deposits are predominant.

The reduction in the shallow marine environment during this period gave rise to the extinction of a large portion of the marine fauna. Many families of corals, echinoderms (spiny-skinned invertebrates such as starfish), arthropods (insects), mollusks (includes snails, octopuses and squid), bryozoans, and brachiopods became extinct [16]. Ammonoids are particularly abundant during the Triassic, especially interesting since only one group survived the Paleozoic.

Land animals were unaffected by the extinction in the seas. Of special note was growth of the vertebrate groups that would dominate in later periods. Land plants suffered during this period, probably because of adverse climatic conditions. In the Late Triassic, however, a relatively uniform flora of land plants was found throughout the world [16].

Jurassic Period   Named after the Jura mountains located between France and Switzerland where rocks of this age were first studied, the Jurassic extended from 213 to 145 mybp. Just as Pangea had been built piecewise through the agglomeration of smaller continents, so also did Pangea subdivide (Figure 3.39). As described by Scotese, 1997 [17]:

> The supercontinent of Pangea…was subdivided into smaller continental blocks in three main episodes. The first episode of rifting began in the Middle Jurassic, about 180 million years ago. After an episode of igneous activity along the east coast of North America and the northwest coast of Africa, the Central Atlantic Ocean opened as North America moved to the northwest…. This movement also gave rise to the Gulf of Mexico as North America moved away form South America. At the same time, on the other side of Africa, extensive volcanic eruptions along the adjacent margins of east Africa, Antarctica, and Madagascar heralded the formation of the western Indian Ocean.

> During the Mesozoic North America and Eurasia were one landmass, sometimes called Laurasia. As the Central Atlantic Ocean opened, Laurasia rotated clockwise, sending North

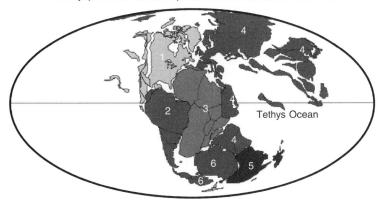

180 mybp: Middle Jurassic Epoch, Jurassic Period, Mesozoic Era

FIGURE 3.39. Locations of continents in the Middle Jurassic period and the Tethys Ocean. The numbers indicate 1 = North America, 2 = South America, 3 = Africa, 4 = Eurasia, 5 = Australia, and 6 = Antarctica (from [12]).

America northward, and Eurasia southward. Coals, which were abundant in eastern Asia during the early Jurassic, were replaced by deserts and salt deposits during the Late Jurassic as Asia moved from the wet temperate belt to the dry subtropics. This clockwise, see-saw motion of Laurasia also led to the closure of the wide V-shaped ocean, Tethys, that separated Laurasia from the fragmenting southern supercontinent, Gondwana.

In North America the Canadian Shield, a very stable continental land mass, was bordered to the north and west by seas that received sediments eroded from the shield areas. While in general the Jurassic was relatively passive from a tectonic perspective, mountain building did occur along the western margin of North America. Resulting high mountains were eroded to provide clastic material that was deposited to the east of the mountains.

Reptiles were in abundance during the Jurassic. On land, the dinosaurs such as *Allosaurus*, *Brontosaurus*, and *Stegosaurus* were dominant. In the sea, *Plesiosaurus* and *Ichthyosaurus* and crocodiles such as *Stenosaurus* were found locally. In the air, pterosaurs represented by the likes of Pterodactylus were found. Fossil remains of the first bird, *Archaeopteryx*, were found in the Late Jurassic. All groups of modern fishes were represented in the Jurassic along with amphibians such as frogs and toads. Evidence of small mammals has also been found.

The most important marine invertebrates were the ammonites. Both gastropods and pelecypods are also abundant in the fossil record. Echinoids are abundant, but the crinoids and brachipods are represented by only a few groups.

Land invertebrates are represented by such insects as flies, butterflies, and moths [18].

Cretaceous Period   The name of this period is derived from *creta*, the Latin word for chalk, which is indicative of rocks of this period located along the cliffs of the English Channel. The breakup of Pangea that started in the Jurassic continued into the Cretaceous. As described by Scotese [17], the events took place as follows:

> The second phase in the breakup of Pangea began in the early Cretaceous, about 140 million years ago. Gondwana continued to fragment as South America separated from Africa opening the South Atlantic, and India together with Madagascar rifted away from Antarctica and the western margin of Australia opening the Eastern Indian Ocean. The South Atlantic did not open all at once, but rather progressively "unzipped" from south to north. That is why the South Atlantic is wider to the south.

> Other important plate tectonic events occurred during the Cretaceous period. These include: the initiation of rifting between North America and Europe, the counter-clockwise rotation of Iberia from France, the separation of India from Madagascar, the derivation of Cuba and Hispaniola from the Pacific, the uplift of the Rocky Mountains, and the arrival of exotic terranes (Wrangellia, Stikinia) along the western margin of North America.

As can be seen from Figure 3.40, the location of the continents in the Late Cretaceous begins to resemble what we see today.

The climate during the Cretaceous mimicked that of the Triassic and Jurassic, being much warmer than today. Because sea levels were 100 to 200 m higher than they are today, shallow seas covered large areas of the continents and chalk deposition was widespread. Extensive deposits of marine shale, sandstone, and conglomerate along the Pacific coast of North America reflect the encroachment of the Pacific Ocean onto the continent. The Pacific and Atlantic oceans were connected by a narrow passageway west

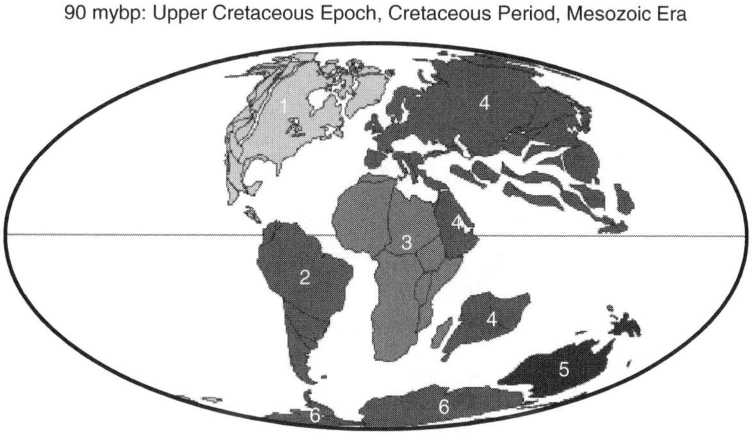

90 mybp: Upper Cretaceous Epoch, Cretaceous Period, Mesozoic Era

FIGURE 3.40. Location of the continents in the Late Cretaceous. The numbers indicate 1 = North America, 2 = South America, 3 = Africa, 4 = Eurasia, 5 = Australia, and 6 = Antarctica (from [12]).

of Mexico City. During the Late Cretaceous the sea intruded on the land along the east coast of North America and depositing marl[11], clay, and sand.

The close of the Cretaceous was marked by the withdrawal of the seas from the continents. It also signaled the extinction of the dinosaurs, flying reptiles, and huge marine reptiles. The ammonites and other groups of marine mollusks also became extinct during this period [19]. The reason for this extinction is still in question, but the prevailing hypothesis is that a large meteor struck the Earth in what is now the Gulf of Mexico and caused a catastrophic change in climate.

This brings to a close the Mesozoic era. The remaining era, the Cenozoic, was addressed briefly earlier in Section 3.1.3 and is discussed further in the next section.

3.5.6 Cenozoic Era

The *Cenozoic era* extends from 65 to 0 mybp and is made up of the Tertiary and Quaternary periods.

Tertiary Period   The *Tertiary period* began with a warm climate, but at the end of the period temperatures began to drop as a warning of the extensive glaciation that was soon to come. During this time mammals diversified rapidly. Marsupials, insectivores, bears, hyenas, dogs, cats, seals, walruses, whales, dolphins, early mastodons, hoofed mammals, horses, rhinoceroses, hippopotamuses, oreodonts, rodents, rabbits, monkeys, lemurs, apes, and yes humans (*Australopithecus*) entered the scene.

The continents were continuing to drift and collide during this era (see Figure 3.41). Scotese describes this period of activity as follows [17]:

> The third, and final phase in the breakup of Pangea took place during the early Cenozoic. North America and Greenland split away from Europe, and Antarctica released Australia which, like India 50 million years earlier, moved rapidly northward on a collision course

---

[11]Marl is a soft, friable material that normally contains abundant shell fragments.

30 mybp: Oligocene Epoch, Quaternary Period, Cenozoic Era

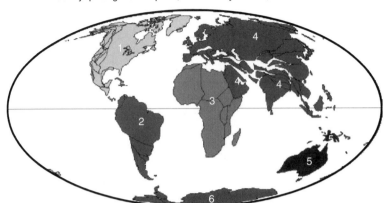

FIGURE 3.41. Location of the continents during the middle of the Tertiary period. The numbers indicate 1 = North America, 2 = South America, 3 = Africa, 4 = Eurasia, 5 = Australia, and 6 = Antarctica (from [12]).

with southeast Asia. The most recent rifting events, all taking place within the last 20 million years include: the rifting of Arabia away from Africa opening the Red Sea, the creation of the east African Rift System, the opening of the Sea of Japan as Japan moved eastward into the Pacific, and the northward motion of California and northern Mexico, opening the Gulf of California.

Though several new oceans have opened during the Cenozoic, the last 66 million years of Earth history are better characterized as a time of intense continental collision. The most significant of these collisions has been the collision between India and Eurasia, which began about 50 million years ago. During the Late Cretaceous, India approached Eurasia at rates of 15–20 cm/yr—a plate tectonic speed record. After colliding with marginal island arcs in the Late Cretaceous, the northern part of India, Greater India, began to be subducted beneath Eurasia raising the Tibetan Plateau. Interesting, Asia, rather than India, has sustained most of the deformation associated with this collision. This is because India is a solid piece of continental lithosphere riding on a plate that is primarily made up of stronger oceanic lithosphere. Asia, on the other hand, is a loosely knit collage of continental fragments. The collision zones, or sutures, between these fragments are still warm, and hence, can be easily reactivated. As India collided with Asia, these fragments were squeezed northwards and eastwards out of the way, along strike-slip faults that followed older sutures. Earthquakes along these faults continue to the present-day.

The collision of India with Asia is just one of a series of continental collisions that has all but closed the great Tethys Ocean. From east to west these continent–continent collisions are: Spain with France forming the Pyrenees Mountains, Italy with France and Switzerland forming the Alps, Greece and Turkey with the Balkan States forming the Hellenide and Dinaride mountains, Arabia with Iran forming the Zagros Mountains, India with Asia, and finally the youngest collision, Australia with Indonesia.

Sedimentary rocks of this period include clastic sediments such as sandstones, marls, mudstones, and conglomerates, as well as limestones. The clastic nature of the rocks is consistent with the reduction in the size of the seas, which had reached their greatest extent during the Cretaceous.

Quaternary Period   The *Quaternary period* includes the *Pleistocene epoch* that stretched from 0.008 to 1.8 mybp. Deep ocean sediments reveal that during this period of time there were more than 16 glaciations. Each glaciation consisted of a warm period (interglacial) and a cold period, which, in combination, lasted approximately 100,000 years. During the cold periods ice sheets covered most of Canada and the northern margin of the United States. The last two glaciations are called the *Illinoian* and the *Wisconsin*. Since each glaciation tends to destroy the evidence of the preceding, most of the existent information pertains to the Illinoian and the Wisconsin. The most recent, the Wisconsin glaciation, started about 75,000 years ago and came to a close about 10,000 years ago [20].

The warm period between the Illinoian and Wisconsin is called the *Sangamon interglacial*. During this time the sea level rose to about 16 feet above where it is today. The higher sea level resulted in the erosion of a new shoreline that still is visible in some areas.

Although there is ongoing controversy about when *Homo sapiens* first arrived on the scene, it seems that they lived in Africa 200,000 years ago and first appeared in Europe and in South Africa about 50,000 years ago. The observation that large mammals begin to decline with the arrival of *Homo sapiens* suggests that the *Homo sapiens* may have been responsible for the demise of the large mammals. The fossil remains of large mammals that exhibit evidence of butchery and the discovery of spearheads in the same location as the bones support this hypothesis.

3.6   FIELD INVESTIGATION

3.6.1   Near-Surface Investigation

Let us now leave the discussion of the history of the Earth and return to the more practical matter of quantifying the distribution of geologic materials. An important responsibility of the groundwater professional is the design and execution of the field investigation. Such an investigation has many elements. As a first step, the groundwater professional often employs available topographical or geologic maps to identify available soil and rock exposures. Road cuts, quarries, gravel pits, man-made excavations, and recently eroded lake, stream, or river banks often provide access to the near-surface subsurface environment. If the site is large, a reconnaissance investigation by automobile or even by aircraft may provide a quick overview and initial screening. The result of such an initial investigation may be the prioritization of sites that warrant a more detailed on-site examination.

When a geologic section is revealed, such as the gravel face illustrated in Figure 3.4, the field investigator proceeds to clear the face of the exposure of debris and vegetation and to examine the undisturbed geologic materials. A record is made of the materials that are encountered in a field notebook augmented by photographs and soil samples where appropriate. Any extraordinary occurrences, such as seeps, unusual coloration and staining, and unusual odor, are documented for future reference.

The tools readily available to the groundwater professional in the field for a near-surface investigation include the *portable shovel*, the *hand auger*, the *geologist's pick*, and the *soil pick*. The soil pick is pictured in Figure 3.4 and the geologist's pick can be seen in Figure 3.21. The hand auger, in essence, is a wood drill auger of approximately 2.5 cm diameter attached to a metal rod, which, in turn, is connected to a second metal rod that forms a T joint and serves as a handle. Used routinely in soil investigations, the hand auger can be employed to investigate soil at depths that generally range from 1 to 3 meters, depending on the nature of the soil deposit. Attachments to the hand auger shaft that replace the auger are available that will allow the user to press a coring device into the base of the hole to obtain a relatively intact sample.

An alternative strategy for near-surface investigations using an auger is to employ a post-hole drilling device. Although hand-held, this gasoline powered auger can be used to drill and sample tens of feet beneath the surface under optimal soil conditions.

3.6.2 Deep Subsurface Investigation

While the investigation of the near-surface is relatively straightforward and inexpensive, knowledge of the deep subsurface tends to require more sophisticated tools and concomitantly additional expense. The most economical tool for intermediate-depth subsurface investigations is the *truck-mounted auger*. A larger version of the hand auger, the truck-mounted auger can investigate unconsolidated deposits quickly and inexpensively to significant depths. Powered by an internal combustion engine, the truck-mounted auger penetrates the subsurface and in the process brings to the surface *soil cuttings* not unlike the hand auger. As the auger drills deeper into the ground, additional auger flights are

FIGURE 3.42. Truck-mounted auger.

added at the surface to the string of tools resident in the boring. The same protocol is used if a post-hole digger is used.

The groundwater professional notes the materials being brought to the surface and records the properties of the observed soil as well as the apparent depth from which it is being removed. A truck-mounted auger is shown in Figure 3.42.

It is often helpful to be able to obtain an *undisturbed sample of soil* from the bottom of an auger hole without the removal of the auger flights. The tool for accomplishing this task is the *hollow-stem auger*. With this equipment it is possible to insert a sampler into the

FIGURE 3.43. A split spoon sampler containing a clay-soil core.

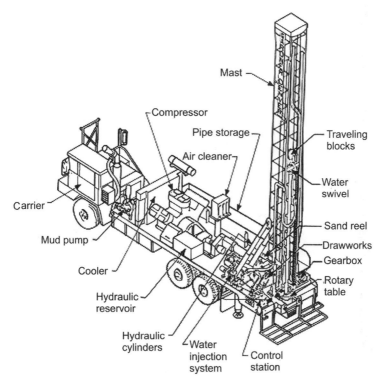

FIGURE 3.44. Components of a drilling rig (from [21]).

auger stem and, after hammering the core barrel into the ground, retrieve it along with the enclosed sample. Upon arriving at the surface, the core barrel is split apart and the sample is examined and further sampled as appropriate. A core barrel that has been split open (*a split spoon sampler*) revealing a clay sample is shown in the photograph in Figure 3.43.

While it is desirable to use an auger when possible, it is often necessary to employ a more complex technology known as a *drill rig*. Many drill rig designs are currently in operation as part of groundwater investigations.

The classic drill rig employs a drill bit that is connected to the ground surface via a string of hollow *drilling rods* (see Figure 3.44). The drill rods are rotated by a turntable that is powered by an internal combustion engine. Rotation of the turntable, in turn, causes a rotation of the *drill bit* located at the bottom of the boring. To remove the *drill cuttings* from the bottom of the hole, a viscous fluid mixture of water and drilling fluid additive, often called *drilling mud*, is forced through the hollow drill stem, thence through holes in the bit, and thence to the surface via the orifice located between the walls of the boring and the external surface of the drill rods. The drilling fluid coats the walls of the boring and assists in preventing the spontaneous collapse of the boring walls. Occasionally the walls of the boring do collapse and great effort must be expended in attempting to retrieve the drilling tools resident in the hole at the time of the collapse.

Since the entire string of drilling tools must be removed in order to obtain an undisturbed sample from the bottom of the hole, such sampling is less frequent than when a hollow-stem auger is used.

FIGURE 3.45. Mud rotary drill rig.

FIGURE 3.46. Tri-cone roller bit. The bit in this photograph will provide a 17.5 inch diameter boring (photo from [21].)

The groundwater professional relies heavily on the cuttings being brought to the surface by the drilling fluid to identify the characteristics of the soil being encountered at depth. As might be expected, the ability to interpret this inexact form of information requires considerable training and experience.

A close-up photograph of a drilling rig is presented in Figure 3.45. Visible is the drill rod, the turntable that turns the drill rod, the drilling fluid, and the screen that separates out the cuttings. In Figure 3.46 a photograph of one kind of bit, called a tri-cone roller bit, is presented. The number of teeth on the bit is increased as the soil or rock being drilled becomes harder and more difficult to drill.

A new technology has recently been developed for the investigation of the subsurface at intermediate depths. Commonly known as a *cone penetrometer*, this device consists of a metal shaft connected to a cone-shaped probe that is pressed into the subsurface by the weight of the vehicle that is used to drive the cone penetrometer into the ground. The probe may contain very sophisticated electronics such that water pressure and, in some instances, water quality can be determined in situ. While an attractive technology when applicable, it is restricted to relatively shallow depths, generally tens of feet, and its applicability to a particular field situation obviously depends on the nature of the soil being penetrated.

3.7 THE HYDROGEOLOGICAL RECORD

It should be evident at this point that a general knowledge of the geology of an area of interest is important. To understand the geology often requires that one organize available information into forms usable to the groundwater professional. Two organizational forms often encountered are the *geologic cross section* and the *contour map.*

3.7.1 The Cross Section

The *cross section* and its close cousin the *fence diagram* are generated directly from subsurface information. The most common form of subsurface information is the *well log* or *boring log*. These logs record the information obtained via the deep subsurface investigation techniques described in the preceding section. An example of a hypothetical boring log is found in Figure 3.47.

The well or boring log normally has a standardized format. The heading of the log tells who is recording the log, which site is being investigated, which boring is being recorded, and the project number.

The leftmost column records the depth of the hole. Thus as one moves down the log, the information recorded is coming from ever greater depth.

The second column indicates the locations where samples of the soil were taken. For example, a sample identified as MT16 was taken at a depth of 5 feet. Column three tells us more specifically that the sample was taken from a depth of 4.7 to 5.3 feet and therefore is approximately 6 inches long. In the next column to the right, at a location opposite to the sample elevation, there are two numbers in the column marked TPH-P, TPH-E. The top number of each pair represents the *total purgeable hydrocarbon concentrations* and the lower one describes the *total extractable hydrocarbons*. The *purgeable* compounds are *volatile* as compared to the *extractable*. As a result, different methods are used for their analyses. The concentrations are presented in milligrams per liter. Sometimes this information is expressed as parts per million by weight[12].

The fifth column is labeled TLC results. TLC is the acronym for *thin-layer chromatography*. It is a method of chemical analysis that can be used in the field to get a rough estimate of the contaminant concentration in a sample. It is very helpful in determining the relative level of concentrations between samples. Thus it can be used to guide the groundwater professional in determining which samples should be sent to the laboratory.

Now let's examine the column on the right. It is the actual elevation of the point of observation. It is similar in concept to the information that is in the first column on the left, except that the ground elevation of the boring location is now taken explicitly into account. The reason that it is important to have the elevation, as distinct from just the depth, is that groundwater professionals are often interested in the correlation of information from one boring to the next. The description of the material encountered at depth is used to prepare cross sections and even contour maps of the top and bottom surfaces of different soil horizons. To make these correlations, it is the actual elevations above a common datum or reference elevation, for example, elevation above mean sea level, rather than their depths below land surface that are important.

For example, if you were interested in building a tunnel that was horizontal through a mountain, it would be the elevation of the tunnel roadway that would be important in the design, not the depth below the land's surface. Of course, if you knew the elevation of the land's surface, you could easily determine the elevation of the roadway given knowledge of its depth. The results of that elevation calculation are recorded in this column.

In the sixth column is a graphical presentation of the soil that was encountered as the drilling equipment made its way into the subsurface. It is the recorded opinion of the groundwater professional as to the nature of the soil (or rock). All of the information

---

[12]Note that while these two measures are often used interchangeably, they are not formally equivalent since the weight of a liter of solution containing dissolved ions would not be exactly 1 kilogram.

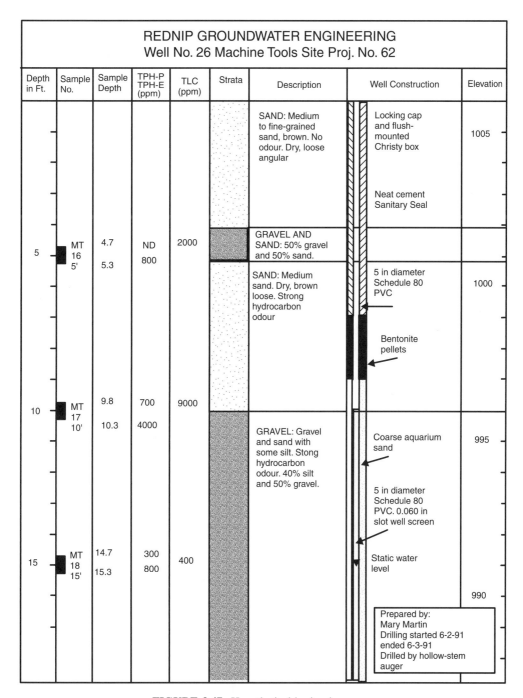

FIGURE 3.47. Hypothetical boring log.

available to the person making the log is synthesized in arriving at the interpretation presented in this column. This includes *split spoon samples*, *cuttings* brought to the surface by the drilling equipment, and information from the drillers as to how the drilling equipment is responding as they penetrate the different soil horizons. This kind of geologic information is sometimes called a *stratigraphic log* because it describes the nature of the *geologic horizons* that are being encountered at depth. The associated verbal description of the material is given in column seven.

As one can readily see, there are two distinctly different printed patterns in column six. One is fine grained and represents sand. One can see that this is the case by examining the information in column seven. The other is made up of open ellipses and circles. This represents gravel. Once again this can be confirmed by looking at column seven.

Now let us turn our attention to the top of the column dedicated to well construction information. As you may have guessed, this column provides a detailed description of how the well was constructed or, as it is often described, completed. As in the case of the geologic data, well construction information is input both visually and in words.

Before proceeding, we will take a slight diversion to describe briefly how a well is constructed. The hole created by the auger or other drilling device is normally larger than the desired diameter of the well. Thus there is a space between the wall of the boring and the outside of the pipe that will be placed in the hole and through which the water will be removed or water-level measurements taken. This space is sometimes referred to as the *annulus* and it is eventually filled with various materials selected to achieve specific purposes.

Now, returning to the log (Figure 3.47), the first thing indicated is that the top of the well is at land surface and is accessed via a locking cap. We also note that the pipe that is placed in the well is plastic pipe made of PVC with a diameter of 5 inches. Notice, however, that below a depth of about 10 feet, the pipe is replaced by 0.06 in slot *well screen*. The role of the well screen is to allow water to enter the well via the screen and then to be removed from the well. The screened area is represented by the horizontal lines in this diagram.

Next, note that the upper 7 feet of annulus that would normally exist between the pipe that forms the wall of the well and the original wall of the boring is, in this instance, filled with cement. This is put in place to assure that water from the ground surface around the borehole does not find its way into the well.

The black area below the cement represents a section of the annulus that is filled with bentonite clay. *Bentonite clay* forms a very effective water seal. It therefore prevents movement of water vertically across the portion of the annulus that it occupies.

Now notice that below the bentonite seal the annulus is filled with coarse *aquarium sand*. The reason for this is to assure that water moving from the soil into the annulus and then into the well via the well screen is unimpeded by fine-grained materials. It serves the purpose of filtering out any fine sand, silt, or clay particles that might otherwise enter the well and degrade the quality of the water being extracted.

It is important to note that water will enter the well along the entire length of the zone filled with the coarse sand. In other words, water will enter the well from the soil adjacent to the small section of sand-filled annulus above the well screen as well as from the soil directly opposite the well screen. It should now be apparent that the bentonite clay seal is needed to prevent water from entering the well from higher elevations that may not be of scientific interest.

There are two reasons it is very important that we know as precisely as possible the section of soil that is contributing water to the well. The first is that we want to know the soil zone that is contributing contaminated water, if indeed it is encountered. The second reason is the need to measure the water pressure or more commonly the hydraulic head (see Section 2.3 for an in-depth discussion of hydraulic head). It is important to know the elevation at which the pressure (or head) measurement is being taken because the hydraulic head value represented by the water level in the well will be the average of the hydraulic head values that are found along the length of the soil column that is adjacent to the sand-filled annulus. This is especially important when a knowledge of flow in the vertical direction is needed.

Let us now see how the information contained in the boring log is employed in generating a cross section. The procedure is the following. A line is drawn on a base map that is generally close to wells for which there are logs (for e.g., line A–B in Figure 3.48).

Information from these wells is now projected horizontally to the line A–B. When this depth information is plotted, it will appear as a series of logs such as shown for the four wells illustrated in Figure 3.48. The groundwater professional now interpolates between the known boring-log information using all of the knowledge at his/her disposal and presents the results as a section through the earth. One representation of a cross section that corresponds to that selected by the line A–B in Figure 3.48 is given in Figure 3.49. The two geologic horizons encountered in the well logs for the indicated wells are illustrated by the two patterns appearing on this cross section.

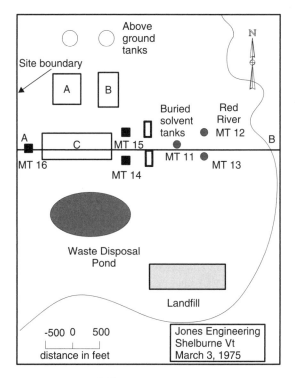

FIGURE 3.48. Location of cross section.

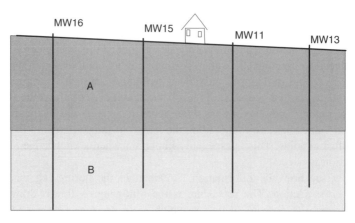

FIGURE 3.49. Typical cross section as identified in the preceeding map of Figure 3.48.

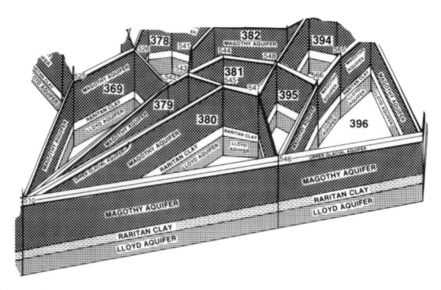

FIGURE 3.50. Fence diagram of the geology of Long Island, New York. The larger numbers refer to the elements defined by the vertical panels and the smaller numbers identify nodes in a finite-element model.

The *fence diagram* is a three-dimensional variant on the cross section and is made up of several concatenated cross sections connected together in a pattern resembling farmers' fences as seen from an airplane. Thus the name fence diagram. A fence diagram generated to describe the geology on Long Island, New York is presented as Figure 3.50.

A variant on the fence diagram that has come into popular use with the advent of computer graphics is the block diagram. A typical block diagram is seen in Figure 3.51. In this figure we see not only geologic formations such as the Blackwater Draw and Ogallala, but also the saturation profile. A perched water lens is found above the Fine-Grained Zone due to the resistance of this zone to vertical migration. Below the Fine-Grained Zone is a continuation of the Ogallala with the lower portion saturated. The saturated portion

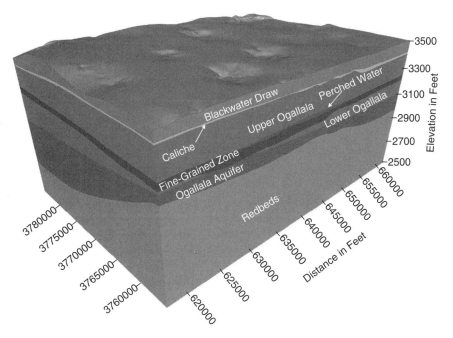

FIGURE 3.51. Block diagram of the Pantex site in Texas, USA. The site is characterized by several formations that give rise to a perched water zone above the regional Ogallala aquifer (by permission of SAIC Corporation).

is denoted as the Ogallala Aquifer. The horizontal scale is in terms of the Texas State coordinate system. The surface depressions are playa lakes.

3.7.2   The Contour Map

While the above analysis gives insight into the stratigraphy of a geologic sequence and, to a certain degree, the extent of various geologic horizons, this is not the specific information required by a hydrogeologist, especially with respect to the needs of a groundwater simulation program. Rather, the computer needs to know the elevation of the interfaces between the various geologic horizons. Thus the information generated at discrete boring locations must then be interpolated to established a three-dimensional surface.

Several forms of interpolation exist that are applicable to this problem. The simpler algorithms use a linear or quadratic spatial interpolation technique. A popular and more sophisticated approach that incorporates, to some degree, the uncertainty inherent in interpolating over large spatial areas is called *kriging* (see Section 7.7.4 for a discussion of kriging).

The resulting surface may be presented as a block diagram or may be contoured prior to being input into a groundwater flow model. A hypothetical contour map is presented in Figure 3.52. In this instance, the quantity being contoured is the water table. Note that each line represents the locus of points with equal elevations. The elevations are indicated by the numbers associated with each contour line. Thus, if one were to walk along any one of the contour lines, the water-table elevation should not change.

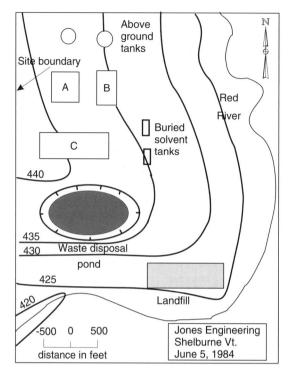

FIGURE 3.52. A contour map of the water table at the hypothetical Machine Tools site. The number associated with a line indicates the line elevation.

3.8   THE MEASUREMENT OF STATE VARIABLES

In the course of a hydrogeological investigation, there are a number of state variables that will be measured. Among the most common are water-level elevations and solute concentrations.

3.8.1   Water-Level Measurements

The objective of water-level measurements is the definition of the hydraulic head surface. In many instances this is well represented by the water table. The classical approach to measuring water levels involves the use of a *chalked tape*. In applying this method one first applies carpenters' chalk to the first few feet of a metal tape. A weight is attached to the tape and it is lowered into the well. By listening carefully, one can hear when the weight encounters the water level in the well. The tape is then lowered another few inches to be sure that the water wets the lower portion of the chalked tape. The tape is then held against a measuring or reference point at the top of the well casing and the distance from the end of the tape to the measuring point is recorded. Let us identify this measurement with the letter A. The tape is then removed from the well.

To obtain the depth to water, one first records the highest wetted point on the tape; let us call this value B. By subtraction of this value B from value A, one obtains the depth to the water in the well relative to the measuring point location where the value A was obtained. The elevation of the water level in the well is now calculated by subtraction of the depth to water from the elevation of the reference point at the top of the casing where the measurement A was taken. Water-level accuracies within a tenth of an inch can be achieved using this classical approach.

A variant on this method involves the use of an electric tape. The electric tape consists of a role of paired wires connected at one end to a pair of electrodes and at the other end to a power source and a signal light or audible alarm. The distance from the end of the electrode is recorded on the tape with markings analogous to a standard steel tape. The electrodes at the end of the tape are separated from one another by a short distance. When in air, these electrodes do not record the very low air conductivity between them. When they are immersed in water, the connection between the two electrodes is complete, current moves between and through the electrodes and therefore between the paired wires. The completion of this circuit either illuminates the light or sounds the alarm.

To measure water levels with an electric tape, one lowers the paired set of electrodes into the well until either the illuminated light or the audible alarm signals that the probe has encountered water. The depth to water is determined by recording the distance from the tip of the electrodes to the point on the electric tape that is adjacent to the well casing elevation reference point. One then proceeds as in the case of the chalked tape method.

Recent advances in electronics have resulted in an ability to record water levels continuously or automatically at specified time intervals. The general concept is to use a *pressure transducer*. This device changes mechanical energy, in this case deformation of a mechanical device due to pressure, to electrical energy. The transducer is located below the water table at a specified depth in the well relative to the measuring point. The pressure at that depth is determined by the transducer and the information is communicated to a recording device on the surface, which is generally, but not always, located near the well. From the pressure measurement it is possible to compute the height of the column of water above the transducer by using the relationship $p = \rho g h$, where p is the pressure relative to atmospheric, ρ is the fluid density, g is gravitational acceleration, and h is the height of the water column above the transducer. From this information the depth to the water-level surface in the well, and therefore the water-level surface elevation, is easily calculated.

3.8.2 Solute Concentration Measurements

An essential element of modern contaminant groundwater hydrology is the accurate determination of groundwater quality. In obtaining groundwater samples, one must be sensitive as to how well the sample represents the resident groundwater. Particularly in the case of measuring volatile organic contaminants, such as *trichloroethylene*, it is very important to obtain a sample of resident aquifer water rather than standing water in the monitoring well. To achieve this, one must remove the standing water in the well and induce water resident in the adjacent soil formation to enter the well. This is normally achieved by removing three to ten well-bore volumes of water from the well before a sample is collected. In estimating the volume of water required to evacuate a well casing, one should take into account that water is entering the well from the formation at the same time as it is being removed from the well. Thus the higher the hydraulic conductivity of the soil adjacent to the well screen, the more water that must be removed before one is certain that only formation water is being tested.

The water sample may be collected using either a *bailer* or a *pump*. The bailer is generally a hollow tube with a valve on the bottom that is lowered into the well until it resides at the desired elevation below the water surface. It is then removed. A valve at the bottom of the bailer is designed such that the water resident in the bailer at the specified depth is not allowed to exit through the bottom of the bailer as it is removed. Thus the water retrieved at the surface from the bailer is indicative of the water at the desired elevation.

The advantage of using a bailer is its portability and simplicity in field application. One disadvantage is that it is not practical for removing large volumes of water, such as might be required in evacuating a well casing. A second disadvantage is that one may lose volatile organic compounds (VOCs) through evaporation when transferring the water sample from the bailer to the sample bottle.

An alternative strategy is to use some form of pump. Among the various pump designs, the *suction-lift* and *submersible pumps* are the most commonly encountered. The suction-lift pump generates a negative pressure, relative to atmospheric, in a pipe that resides, in part, beneath the water surface in the well. Under these circumstances, atmospheric pressure forces water through the pipe to the surface. These readily available and inexpensive pumps have the advantage that they are relatively portable. The main disadvantages of the suction pump are that it can be used only for water levels within about 20 feet of the ground surface and, since a negative pressure is involved, some dissolved gases will be lost.

The submersible pump, in contrast, pushes water to the surface using a down-hole rotor and stator configuration. It has the advantages of being able to pump from greater depths and to retain a positive pressure on the water sample. The primary disadvantage of this approach is that most submersible pumps do not come in small diameters and therefore cannot be used in small-diameter wells. Most submersible pumps are too large, for example, to be used in the commonly encountered 2 inch diameter monitoring well.

Several visual formats exist for presenting groundwater concentrations. The most common is the contour diagram representing concentration. Although concentrations do not form a smooth surface, groundwater professionals often prefer to represent concentrations using a contour map. The resulting contours can often provide a helpful, albeit fuzzy, picture of the contaminant distribution. A contour map of the concentration of tetrachloroethylene (PCE) in the San Fernando Valley of California is provided in Figure 3.53.

3.9 SUMMARY

The chapter begins with a discussion of the role of geology and geologic processes in the evolution of the subsurface environment, especially the type and distribution of pore space. Nomenclature commonly encountered in descriptions of the subsurface and its properties are presented within the context of physical and historical geology. In the discussion of the history of the Earth, its tectonic evolution and the evolution of life on Earth are considered. The methodology used to characterize the subsurface is briefly described and the techniques employed in presenting the resulting data in visual form are discussed. Finally, the methodology used to measure the hydrodynamic and geochemical properties of groundwater systems is introduced.

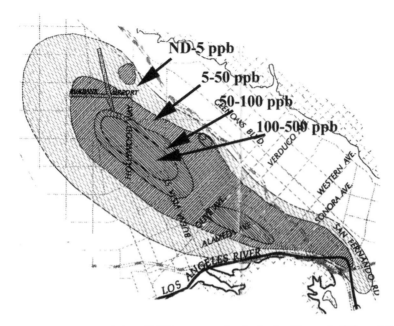

FIGURE 3.53. Tetrachloroethylene (PCE) concentration in the San Fernando Valley of California, Spring 1993.

3.10 PROBLEMS

3.1. Deltaic deposits found in a glacial environment generally are well graded (poorly sorted). Keeping in mind the concept that water in a high-energy environment (fast moving) can carry coarser grained materials, explain why deltaic deposits are often found at the location where rivers enter quiet bodies of water such as lakes.

3.2. In Figure 3.6 a cross-bedded sand deposit is shown. From which direction did the water flow during the deposition of the foresets (materials deposited on an incline in this figure) and why? What is the sedimentary relationship between the foresets and the horizontal underlying deposits; that is, why are the two sets of beds found deposited at different angles (sloping beds overlying horizontal deposits)?

3.3. Glacial till contains a wide range of grain sizes (well graded or poorly sorted). Explain why this is the case for till in contrast to sand dunes, which have a very uniform grain size.

3.4. Explain the origin of medial moraines.

3.5. Diamond deposits are associated with peridotite deposits. Why would diamonds be more likely to form in peridotite than in basalt even though they are both igneous rocks? (*Hint*: Diamonds are formed at very high temperatures under very great pressures).

3.6. A pre-Appalachian range of mountains formed in the Ordovician period along a line that extends from Norway to northeastern North America. Considering the continental movements presented in Figures 3.33 and 3.34, present an explanation for the location of these mountains.

3.7. It is observed that during periods of animal and plant extinctions (such as occurred at the end of the Ordovician period) the more evolutionarily advanced life forms tended not to survive in comparison with the more primitive forms. Why do you think this would be the case?

3.8. Both a truck-mounted auger and a rotary drilling rig are used to drill borings in the subsurface. Describe when one would be superior to the other and vice versa.

3.9. The fence diagram presented in Figure 3.50 is based on information obtained from boring logs found at the locations indicated by the vertical lines in this figure. However, the lines connecting the boring locations and thus forming the "fence lines" are not unique; that is, a different set of lines could have been used to connect these borings. What design criteria do you think were used in arriving at the selected connection strategy?

BIBLIOGRAPHY

[1] O. E. Meinzer (ed.), *Hydrology*, Dover Publications, Mirieola, NY, 1942.

[2] J. Kozeny, Ober kapillare Leitung das Wassers im Boden, *S. Ber. Wiener Akad. Abt. IIa* **137**: 271, 1927.

[3] M. C. Leverett, Capillary behavior in porous solids, *Pent. Trans. AIME* **142**:161, 1941.

[4] S. N. Davis and R. J. M. DeWeist, *Hydrogeology*, John Wiley & Sons, Hoboken, NJ, 1966.

[5] K. E. Anderson, *Ground Water Handbook*, National Groundwater Association, Dublin, OH, 1991.

[6] G. de Marsily, *Quantitative Hydrogeology*, Academic Press, San Diego, CA, 1986.

[7] D. R. Soller, *Text and References to Accompany Map Showing the Thickness and Character of Wuaternary Sediments in the Glaciated United States East of the Rocky Mountains*, U. S. Geological Survey Bulletin 1921, 1992.

[8] U. S. Geological Survey website, http://vulcan.wr.usgs.gov/Glossary/geo_time_scale.html, 2002.

[9] Geological Survey of Canada, "Esker Ridge, with Side Deltas Deposited into a Former Glacial Lake, Northwestern Manitoba," Canadian Landscapes Photo Collection, http://gsc.nrcan.gc.ca/landscapes/details_e.php?photoID=460.

[10] L. D. Leet and S. Judson, *Physical Geology*, Prentice Hall, Englewood Cliffs, NJ, 1959.

[11] C. Kreger, "Exploring the Environment," http://www.cotf.edu/ete/modules/msese/earthsysflr/cambrian.html, 2000.

[12] USGS, "USGS in the Parks, Color-coded Continents," http://wrgis.wr.usgs.gov/docs/parks/pltec/scplseqai.html.

[13] R. N. Schrantz, "Photographs of Fossils Found on KPS Fieldtrips," http://www.uky.edu/OtherOrgs/KPS/pages/fossilphoto.html.

[14] W. A. Pryor, Carboniferous, in *McGraw-Hill Encyclopedia of Science and Technology*, Vol. 2, McGraw-Hill, New York, 1982.

[15] C. A. Ross and R. P. Ross, Permian, in *McGraw-Hill Encyclopedia of Science and Technology*, Vol. 10, McGraw-Hill, New York, 1982, p. 32.

[16] B. Kummel, Triassic, in *McGraw-Hill Encyclopedia of Science and Technology*, Vol. 14, McGraw-Hill, New York, 1982, p. 90.

[17] C. R. Scotese, "Paleomar Project," http://www.scotese.com/moreinfo9.htm, 2002.

[18] E. G. Nelson, Jurassic, in *McGraw-Hill Encyclopedia of Science and Technology*, Vol. 7, McGraw-Hill, New York, 1982, p. 458.

[19] W. A. Cobban, Cretaceous, in *McGraw-Hill Encyclopedia of Science and Technology*, Vol. 3, McGraw-Hill, New York, 1982, p. 719.

[20] Province of Nova Scotia website, "Fossils of Nova Scotia," http://museum.gov.ns.ca /fossils/geol/quat.htm.

[21] F. G. Driscoll, *Groundwater and Wells*, 2nd ed., Johnson Division Publication, St. Paul, MN, 1986.

CHAPTER 4

WATER MOVEMENT IN GEOLOGICAL FORMATIONS

In Chapter 1 we introduced the various ways in which water can occur in the subsurface. In a nutshell, water occurs primarily as a liquid, which, depending on the nature of its occurrence, may be mobile or immobile. Pendular water and adsorbed water are typically immobile. Capillary water and water occupying the saturated zone are typically mobile or have the potential to be mobile.

The zone above the water table—shown in Figure 1.25 as the elevation at which the pressure in the water phase is atmospheric—we denote as the *vadose zone*. In this zone we have both saturated flow in the capillary fringe and unsaturated flow above the capillary fringe. Below the water table we have saturated flow, assuming no occurrence of trapped air or nonaqueous phase liquids below the water table.

In this chapter we discuss the equations that describe saturated flow in the subsurface. Unsaturated flow is addressed in Section 11.3. The reason for this examination is the need to develop tools to simulate subsurface water flow. Models are needed because it is impractical, in general, and impossible in some instances, to study subsurface water flow behavior using only physical models.

We focus primarily on the form of the equations and their interpretation rather than the details of their development. Nevertheless, to set the stage, we provide a peak into how one would develop these and related equations using principles based on the physics of flow in porous media. Those readers who are not interested in the theoretical under-pinnings of these equations may wish to proceed to Section 4.3.1.

4.1 CONSERVATION OF FLUID MASS

Problems such as we will encounter in this book can be solved provided we have the following:

1. One or more conservation equations (often called balance equations)
2. A suitable number of constitutive relationships (experimental relationships among variables of interest)
3. Parameter estimates for these constitutive relationships (material properties)
4. Boundary conditions
5. Initial conditions
6. Source functions

Each of these items will be considered in due course. To begin, let us examine the *conservation of mass equation*.

There are many ways to develop the conservation of mass equation. The approach we will use employs the nomenclature in Figure 4.1. Consider the fluid element with volume V and surface S that is associated with a fluid body, in our case water. The fluid-element shape is arbitrary, but the volume is small relative to the overall fluid body.

The overall mass of fluid in this volume, let us call it M, is given by summing the mass in each *incremental volume* dV. If fluid density ρ is defined as the mass of fluid per volume of fluid [M/L^3], then the mass of fluid in incremental volume dV is $\rho\,dV$, and the mass within the entire volume V is

$$M = \int_V \rho\,dV. \tag{4.1}$$

Consider the possibility that mass is moving out of the control volume through the surface we have denoted as S. If the fluid velocity perpendicular to the incremental surface area dS is given by v_n, the *outward directed fluid velocity*, the mass moving through the incremental surface dS is $\rho v_n\,dS$. The total (net) rate of mass leaving the control volume via flow across its boundary is then given by

$$F_n = \int_S \rho v_n\,dS. \tag{4.2}$$

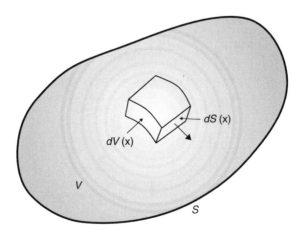

FIGURE 4.1. Control volume V, incremental volume dV, control volume surface S, and incremental control volume surface dS.

Note that v_n can be related to the *outward directed unit vector*, which we denote **n**, through the scalar product relationship for vectors, that is, $v_n = \mathbf{v} \cdot \mathbf{n}$, where **v** is the fluid velocity.

In the absence of internal sources or sinks of mass, the rate of change of mass in our control volume is balanced by the net amount moving through the surface S. Thus we can write that

$$\frac{\partial}{\partial t} M = -F_n. \tag{4.3}$$

The negative sign is due to the fact that if the left-hand side of Eq. (4.3) is positive, then the amount of mass must be increasing; that is, there must be a net inward flux of fluid mass. Since F_n, by definition, is the net outward flux, that is, the amount leaving, then the sign in Eq. (4.3) must be negative.

Let us substitute Eq. (4.1) and (4.2) into (4.3) and subsequently apply *Gauss's theorem*, which states that

$$\int_V (\mathbf{\nabla} \cdot \mathbf{F}) \, dv = \int_S \mathbf{F} \cdot \mathbf{n} \, dS,$$

where **F** is a vector function, and **n** is again an outward directed normal vector to the surface S enclosing the volume V. Note that $F_n = \mathbf{F} \cdot \mathbf{n}$. The relationship $\mathbf{\nabla} \cdot \mathbf{F}$ is called the divergence of **F**. In essence, Gauss's theorem converts the volume integral of the divergence of a vector field into a surface integral involving the normal component of the vector **F** over the surface enclosing V, that is, S. We observe that if vector **F** corresponds to the vector $\rho\mathbf{v}$, then Gauss's theorem tells us that

$$\int_S [\rho\mathbf{v} \cdot \mathbf{n}] \, dS = \int_V [\mathbf{\nabla} \cdot (\rho\mathbf{v})] \, dV.$$

Therefore the right-hand side of Eq. (4.3) may be replaced by

$$\int_V [\mathbf{\nabla} \cdot (\rho\mathbf{v})] \, dV.$$

Therefore we have the mass balance equation

$$\frac{dM}{dt} = \frac{d}{dt} \left(\int_V \rho \, dV \right) = \int_V [\mathbf{\nabla} \cdot (\rho\mathbf{v})] \, dV.$$

Because the volume V is not changing with time (by construction), V is not a function of time and therefore the time derivative can be exchanged with the spatial integration over V, yielding

$$\int_V \left[\frac{\partial \rho}{\partial t} + \mathbf{\nabla} \cdot (\rho\mathbf{v}) \right] dV = 0. \tag{4.4}$$

Because the volume V is arbitrary, restricted only to be a volume that is fixed in space, for Eq. (4.4) to hold in general (i.e., for any V) the integrand must vanish. The result is the *mass-conservation equation* for our fluid:

$$\frac{\partial \rho}{\partial t} + \mathbf{\nabla} \cdot (\rho\mathbf{v}) = 0. \tag{4.5}$$

There is an alternative form of this equation that you may find in the literature. To derive it, it is convenient to expand the second term in Eq. (4.5) to yield

$$\frac{\partial \rho}{\partial t} + \rho \nabla \cdot \mathbf{v} + \mathbf{v} \cdot \nabla \rho = 0, \tag{4.6}$$

which can be rewritten

$$\frac{D\rho}{Dt} + \rho \nabla \cdot \mathbf{v} = 0,$$

where the operator $D(\cdot)/Dt$ is called the *substantial derivative* defined as

$$\frac{D(\cdot)}{Dt} \equiv \frac{\partial(\cdot)}{\partial t} + \mathbf{v} \cdot \nabla(\cdot),$$

and the quantity denoted by $(\cdot)$ represents any scalar quantity (like density).

Recall from basic calculus that the definition of a partial derivative is the derivative taken while holding all other independent variables fixed. Therefore, by definition, $\partial \rho / \partial t$ is the change of density with respect to time, holding (x, y, z) fixed, which is to say, holding the spatial location fixed. Conversely, the substantial derivative $D\rho/Dt$ is the time rate of change of density defined in a moving coordinate system, where the velocity of the moving coordinate system is equal to the velocity of the fluid. This is perhaps best explained through a particularly effective analogy provided by Bird et al. [1]:

> Suppose we get into a canoe, and not feeling energetic, we simply float along counting fish. Now the velocity of the observer is just the same as the velocity of the stream, $\mathbf{v}$. When we report the change in fish concentration with respect to time, the numbers depend on the local stream velocity. This derivative is a special kind of total time derivative and is called the "substantial derivative" or sometimes (more logically) the "derivative following the motion."

By extension one can use this analogy to describe the meaning of the partial derivative, $\partial(\cdot)/\partial t$. Assume that you now select your observation point to be on a bridge crossing the stream containing the fish. You now count the fish from this fixed point. The change in fish concentration as now reported from this fixed location is the *partial time derivative*.

The mass balance equation presented as Eq. (4.5) has counterparts that describe the conservation of energy, momentum, and entropy. Since the methodology to develop these equations is similar to that presented for mass, we will not develop it. Rather, we will now proceed to the more interesting (from our point of view) case of the conservation equations for flow through soil, or more generally through a porous medium.

4.2 CONSERVATION OF FLUID MASS IN A POROUS MEDIUM

We have established in Eq. (4.5) an expression that describes the conservation of mass at any point in a fluid. To extend this concept to a porous medium like soil, we need some new tools, both mathematical and conceptual. Consider first the information provided in Figure 4.2. This arbitrary sample of poorly sorted gravel (i.e., a sample of the size of this picture) is composed of material with a range of grain sizes and its grain size composition varies spatially.

FIGURE 4.2. Representative elementary values (REVs) used to evaluate porosity.

We begin by drawing a circle with its center located at the dot numbered 1 in this figure. Although we use a circle of radius r, any shape would do. To make our analysis consistent with a three-dimensional formulation, assume the circle is the visible projection of a cylinder embedded in the gravel so we have a sample of the porous medium analogous in its dimensions $[L^3]$ to the development in Section 4.1. For now we assume the volume used to define porosity is sufficient to be representative of the material in the vicinity of the point numbered 1, so that the volume corresponds to a representative volume, or a *representative elementary volume* (REV).

Imagine that you want to measure the porosity in the sample volume (i.e., the REV). You collect the sample and make a determination that it has a porosity ε_1 and record this value. Since this is a hypothetical example, let us assume that after you take the sample, you replace it exactly as it was, thereby returning the deposit to its original state. Now move your cylinder to the location identified with the center point number 2 and repeat the sampling and analysis procedure, thereby obtaining a porosity determination ε_2. Replace the sample as before and continue this protocol at circles centered at 3, 4, and 5.

In Figure 4.3 are plotted the values of porosity obtained from the five samples as located in Figure 4.2. While it is evident that the material shown in Figure 4.2 is heterogeneous, we find that when the averaging volume is shifted only slightly, the value of the porosity does not change significantly. In the limit of very small shifts of the centroid of the averaging volume, the resulting curve of porosity as a function of spatial location is smooth, like that shown in Figure 4.3. This is the nature of volume averaging and is one of the reasons for using a REV concept to represent the porosity parameter. The smoothness of this curve is important because it allows the function representing the porosity to be differentiated in a formal mathematical sense. Differentiation of parameters such as porosity, or of state variables such as pressure, is helpful to make equations like Eq. (4.5) meaningful. A discontinuous functional representation of parameter and state variables may be mathematically tractable, but not convenient.

In summary, it appears that by taking moving averages of key variables such as the parameter and state variables, one can work with continuous functions that are convenient from a mathematical perspective.

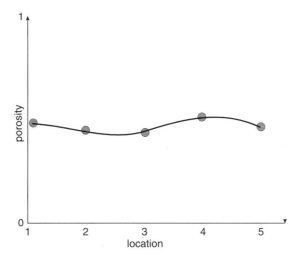

FIGURE 4.3. Porosity values obtained from samples selected at locations identified in Figure 4.2.

But, you ask, what happens as the REV is reduced in size? Eventually the control volume should contain either a pore or a solid, making the definition of porosity meaningless. The answer is apparent in the use of the word "representative" in the REV acronym. The sample must contain a statistically meaningful sample of pores and solids to meaningfully represent the parameter porosity.

The argument defending this concept is made using Figure 4.4. When small sample sizes are selected, the porosity determination is quite erratic. In fact, as one approaches zero volume for the REV, the porosity becomes either 1 or 0, depending on whether the REV is occupied by void space or a grain. As one moves to larger sample sizes, the estimate of porosity stabilizes, in this case around a value of 0.5. To the right of the REV size identified with the point A in Figure 4.4, one has a stable estimate of porosity. However, if one were to use a very large REV size, another problem arises. The REV may be so large that larger scales of heterogeneity are encountered that mask the smaller-scale spatial variability that should be present in the estimate of the porosity. In Figure 4.2 it

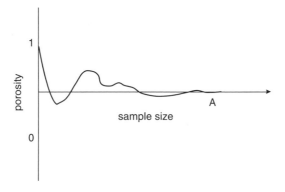

FIGURE 4.4. The dependence of the value of computed porosity on the size of the sample used to make the determination.

is clear that there is natural variability and that averaging over a large REV may mask that variability. Thus there is a range of appropriate averaging volume sizes depending on the parameter or state variable being considered and on the geologic material. For an equation such as mass conservation to hold formally, the REV must be of such a size that all the parameter and state variables in the equation are suitably represented.

Assuming a suitable REV can be defined, let us now see how one uses this concept in developing a conservation of fluid mass equation for a porous medium. It is helpful to consider first the definition sketch provided in Figure 4.5. The circle is the REV and the point B is the center of the REV.

To this point, we have not considered the problem of determining in which phase a given point, say, A in Figure 4.5, is located. It could be in the fluid phase or the solid phase. We know where it is with respect to the origin of the global coordinate system **x**, but we do not know if that particular point is in the solid or fluid phase. To assist in this endeavor, we introduce a very important concept, the *phase-distribution function*, $\gamma_\alpha(\mathbf{r}, t)$, $\alpha = s, l$.[1] In this notation, the subscript α denotes the phase identified with the phase-distribution function. In our case there are two phases, the solid s and the liquid l so there are two phase-distribution functions.

The phase-distribution function can have only one of two values. It is either one or zero. For example, if the point of interest, say, $(\mathbf{r}_0, t_0)$, is in the solid phase, then $\gamma_s(\mathbf{r}_0, t_0) = 1$. Since the solid and liquid cannot occupy the same point in space, by definition, $\gamma_l(\mathbf{r}_0, t_0) = 0$. If, on the other hand, the point of interest lies in the liquid phase, then $\gamma_l(\mathbf{r}_0, t_0) = 1$ and $\gamma_s(\mathbf{r}_0, t_0) = 0$. We can write this formally as

$$\gamma_l(\mathbf{r}, t) = \gamma_l(\mathbf{x} + \boldsymbol{\eta}, t) = \begin{cases} 1 & \mathbf{r} \in dV_l \\ 0 & \mathbf{r} \in dV_s \end{cases}, \tag{4.7}$$

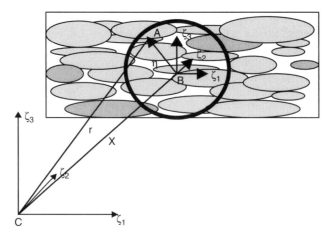

FIGURE 4.5. Relationship between local ξ and global ζ coordinates. The circle is the REV and the oval patterns represent tea leaves in a tea bag.

[1]We allow $\gamma_i(\mathbf{r}, t)$ to be a function of time or they may change their state because the phases may move over time (e.g., liquid water may freeze and thereby change from the liquid to the solid phase).

where the global coordinate $\mathbf{r}$ is defined as the coordinate that represents the sum of the coordinate identified with the center of the REV. that is, $\mathbf{x}$, and the *local coordinate* ξ that describes the position of a point relative to the REV center $\mathbf{x}$ that is associated with the REV.

We are now at the point where we can use our newly defined coordinate system and the phase-distribution functions to obtain the mass-conservation equation for the porous medium. We use averages defined over the REV as point values, identified with the center point of the REV. These averaged values are then used in the same way we used point values earlier to derive Eq. (4.5). Using explicit averages over the REV in conjunction with the control-volume concept then yields the following equation, written for the liquid phase:

$$\int_V \frac{1}{dV} \int_{dV} \left(\frac{\partial \rho}{\partial t} + \nabla \cdot (\rho \mathbf{v}) \right) \gamma_l \left(\mathbf{x} + \boldsymbol{\eta}, t \right) dv_\xi \, dV = 0. \tag{4.8}$$

Equation (4.8) states that we are identifying that portion of the REV that contains the liquid (i.e., we are multiplying by the phase-distribution function for the liquid phase), then integrating over the REV and dividing by the REV volume to derive the averaged quantities. Finally, we are integrating over the entire porous-medium control volume V.

Equation (4.8) may be formally correct, but it is not very useful. In general, we do not know the functional form of the phase-distribution function and so the proposed integration over dV, even if computationally feasible, is a practical impossibility. So how do we transform Eq. (4.8) into something useful?

While the path in the case of multiple fluid phases with dissolved species and phase interactions is very complex, for our simple case of one fluid and one solid we can proceed in a relatively straightforward fashion. Let us define the *volume average* $\langle \cdot \rangle_\alpha \left(\mathbf{x}, t \right)$,

$$\langle \cdot \rangle_\alpha \left(\mathbf{x}, t \right) \equiv \frac{1}{dV} \int_{dV} (\cdot) \left(\mathbf{x} + \boldsymbol{\eta}, t \right) \gamma_\alpha \left(\mathbf{x} + \boldsymbol{\eta}, t \right) dv_\xi, \tag{4.9}$$

where in this case the quantity defined by $(\cdot)$ must be an extensive quantity; that is, it must be summable (it depends on the amount of material present). The requirement that the quantity be extensive makes sense since we are integrating over it and integration is, in essence, a form of summation. The volume average can be thought of as the total quantity of $(\cdot)$ in the REV per unit volume of the REV. Moreover, this is a moving average because the average quantity is a function of the global coordinate ζ, not the local coordinate ξ. In essence, the effects of the size, shape, and distribution of the pores have been absorbed into the average through the integration process. In this sense, Eq. (4.9) is a mathematical representation of the REV averages used in Figure 4.2. While Eq. (4.9) may be used to define averages of variables, we may also use it to average entire equations. In this way we can, in effect, move from the world of the fluid particle to the world of the porous medium. The porous medium scale is the one at which practical calculations and measurements can be made, because averages are defined over many pores and grains and can correspond to the lengths over which instruments and devices measure porous media properties. Through combination of Eq. (4.9) and (4.8), we obtain for the fluid phase ($\alpha \equiv l$)

$$\int_V \left\langle \frac{\partial \rho}{\partial t} + \nabla \cdot (\rho \mathbf{v}) \right\rangle_l dV = 0,$$

which, for the relatively simple case we are considering wherein there is no phase change, can be written

$$\int_V \frac{\partial \langle \rho \rangle_l}{\partial t} + \nabla \cdot \left(\langle \rho \rangle_l \overline{(\mathbf{v})}^l \right) dV = 0, \tag{4.10}$$

where we have used the definition provided below:

$$\overline{(\cdot)}^\alpha (\mathbf{x}, t) \equiv \frac{1}{\langle \rho \rangle_\alpha (\mathbf{x}, t) dV} \int_{dV} \rho (\mathbf{x} + \boldsymbol{\eta}, t) \ (\cdot) (\mathbf{x} + \boldsymbol{\eta}, t) \ \gamma_\alpha (\mathbf{x} + \boldsymbol{\eta}, t) \, dv_\xi$$

and, in our case, $\alpha \equiv l$.

We now can argue, as we did in the case of the point fluid mass-conservation equations, that because the volume V was arbitrary the integrand must be zero, which yields

$$\frac{\partial \langle \rho \rangle_l}{\partial t} + \nabla \cdot \left(\langle \rho \rangle_l \overline{(\mathbf{v})}^l \right) = 0. \tag{4.11}$$

However, Eq. (4.11) is a little problematic. The volume average density of the fluid, $\langle \rho \rangle_l$, has a strange definition. According to Eq. (4.9), this quantity is the mass of fluid divided by the REV, and the REV includes both the solid and the liquid. The quantity normally identified with the fluid density would be the mass of liquid divided by the volume of liquid. To rearrange Eq. (4.11) into a more useful form, we introduce a third type of average, the *intrinsic volume average, $\langle \cdot \rangle_\alpha^\alpha (\mathbf{x}, t)$,* given by

$$\langle \cdot \rangle_\alpha^\alpha (\mathbf{x}, t) \equiv \frac{1}{dV_\alpha (\mathbf{x}, t)} \int_{dV} (\cdot) (\mathbf{x} + \boldsymbol{\eta}, t) \gamma_\alpha (\mathbf{x} + \boldsymbol{\eta}, t) \, dv_\xi, \tag{4.12}$$

where dV_α is the volume of the α phase in the REV. The average defined by Eq. (4.12) is just what we needed. If one replaces the $\cdot$ with ρ, we see that the mass is being averaged over the volume of fluid, precisely as is normally encountered in practice.

It is easy to show that

$$\langle \cdot \rangle_\alpha (\mathbf{x}, t) = n_\alpha \langle \cdot \rangle_\alpha^\alpha (\mathbf{x}, t), \tag{4.13}$$

where n_α is defined as dV_α / dV and is called the *volume fraction* of phase a. Through combination of Eq. (4.13), (4.12), and (4.11), we obtain

$$\frac{\partial \varepsilon \langle \rho \rangle_l^l}{\partial t} + \nabla \cdot \left(\varepsilon \langle \rho \rangle_l^l \overline{(\mathbf{v})}^l \right) = 0, \tag{4.14}$$

where we have substituted the porosity for n_l since, in this instance, where the soil is saturated, they are the same. Equation (4.14) is the generally accepted form of the *conservation of fluid mass equation for a single fluid occupying a porous medium.* Note that the combination $\varepsilon \overline{(\mathbf{v})}^l$ is what we have called the specific discharge $\mathbf{q}$ as discussed in Section 2.6.

Equation (4.14) appears similar to the fluid mass balance presented as Eq. (4.5). The similarity is to be expected since both represent the concept of the conservation of mass, albeit Eq. (4.14) is written for a porous medium. However, Eq. (4.14) includes the porosity, and the averages are over the REV, which is much larger than the averaging volume used to define pointwise fluid properties like those used in Eq. (4.4). In general, as we average over larger volumes, we tend to introduce additional parameters and terms into our equations.

In the remainder of this book we will simplify our notation for equations describing the physics of flow in porous media. Hereafter, variables used in porous-medium equations and denoted by their commonly used symbols will be assumed to represent the appropriately averaged quantities. Thus, for example, we will typically write Eq. (4.14) as

$$\frac{\partial (\varepsilon \rho)}{\partial t} + \nabla \cdot (\varepsilon \rho \mathbf{v}) = 0. \tag{4.15}$$

4.3 GROUNDWATER FLOW EQUATIONS

4.3.1 The Governing Equation

Examination of Eq. (4.15) reveals that we have one equation in four unknowns—ρ and the three components of $\mathbf{v}$ (assuming porosity is known). To make the system of equations solvable, we will need to add information. One piece of information is the constitutive (or experimental) relationship attributable to Darcy. Recall from Eq. (2.36) that the specific discharge $\mathbf{q}$ can be obtained from Darcy's law as

$$\mathbf{q} = -\mathbf{K} \cdot \nabla h. \tag{4.16}$$

Combination of Eq. (4.16) and the mass balance Eq. (4.15) yields

$$\frac{\partial (\varepsilon \rho)}{\partial t} - \nabla \cdot \rho (\mathbf{K} \cdot \nabla h) = 0. \tag{4.17}$$

This allows us to replace three unknowns (the components of $\mathbf{v}$) with one new one, namely, hydraulic head h. We are left with one equation and two unknowns (or four equations in five unknowns).

To further our cause, let us use the chain rule of differentiation to expand the left-hand side of Eq. (4.17):

$$\frac{\partial (\varepsilon \rho)}{\partial t} = \varepsilon \frac{\partial \rho}{\partial t} + \rho \frac{\partial \varepsilon}{\partial t}. \tag{4.18}$$

Now we introduce two additional constitutive relationships. It is known through experiments conducted on various kinds of soil that soil particles consolidate or become more compact when the water pressure is decreased. The relationship is normally given as

$$\frac{\partial \varepsilon}{\partial t} = \frac{d\varepsilon}{dp} \frac{\partial p}{\partial t}$$

$$= C_v \frac{\partial p}{\partial t},$$

where the parameter C_v is related to the classical *coefficient of consolidation* used in soil mechanics [2].

The second relationship we will use relates the fluid pressure to the compressibility of water, that is,

$$\beta \equiv \frac{1}{\rho} \frac{d\rho}{dp},$$

where β is the *compressibility of water*. By introducing the compressibility of soil and compressibility of water relationships into Eq. (4.18), we obtain

$$
\begin{aligned}
\frac{\partial (\varepsilon \rho)}{\partial t} &= \varepsilon \frac{\partial \rho}{\partial t} + \rho \frac{\partial \varepsilon}{\partial t} \\
&= \varepsilon \frac{d\rho}{dp} \frac{\partial p}{\partial t} + \rho \frac{d\varepsilon}{dp} \frac{\partial p}{\partial t} \\
&= \rho (\varepsilon \beta + C_v) \frac{\partial p}{\partial t}.
\end{aligned}
\tag{4.19}
$$

We showed in Section 2.3 that the hydraulic head can be related to the fluid pressure via the relationship

$$h = \int_{z_0}^{z} dz + \int_{P_{\text{atm}}}^{p} \frac{d\pi}{g\rho(\pi)}. \tag{4.20}$$

To realize a relationship between the time derivative of head and the time derivative of pressure, we differentiate Eq. (4.20) using Leibnitz's rule to obtain (see page 94)

$$\frac{\partial h}{\partial t} = \frac{1}{g\rho(p)} \frac{\partial p}{\partial t},$$

which upon rearrangement yields

$$\frac{\partial p}{\partial t} = g\rho(p) \frac{\partial h}{\partial t}. \tag{4.21}$$

Finally, we can combine Eqs. (4.17), (4.19), and (4.21) to obtain

$$g\rho^2 (\varepsilon \beta + C_v) \frac{\partial h}{\partial t} - \nabla \cdot \rho (\mathbf{K} \cdot \nabla h) = 0$$

or

$$S_s \frac{\partial h}{\partial t} - \nabla \cdot (\mathbf{K} \cdot \nabla h) = 0, \tag{4.22}$$

where $S_s \equiv g\rho (\varepsilon \beta + C_v)$ is the *specific storage*, defined as the volume of water released from a unit volume of porous medium due to a unit decrease in hydraulic head. To derive Eq. (4.22) from Eq. (4.21), we have assumed that the spatial gradient of ρ is negligible compared to other terms in the equation and the spatial gradient of ρ has been neglected. Because the specific storage coefficient is a function of p, and therefore h, Eq. (4.22) is nonlinear; that is, there is a product of a coefficient that depends on h and a derivative of h in the time derivative term. However, because this is a weak nonlinearity (S_s does not change very much when p changes), it is generally neglected; S_s is normally given as constant or a function only of space. The spatial gradient of ρ in $(\mathbf{q} \cdot \nabla \rho)$ is usually much smaller than the divergence of the flux $(\rho \nabla \cdot \mathbf{q})$, which justifies neglecting the former term.

The preceding development could also have been written using fluid pressure instead of hydraulic head as the primary variable. While that formulation tends to be somewhat more general, the head formulation was used because Eq. (4.22) is the form of the groundwater flow equation generally found in the literature. It is also usually easier to specify boundary conditions on head than on pressure (see Section 4.3.3). We will now examine what other information we will need to solve this equation for problems of practical importance.

4.3.2 Parameter Estimates

Equation (4.22) has two parameters that need to be defined—the hydraulic conductivity tensor $\mathbf{K}$ and the specific storage S_s. These parameters need to be defined everywhere within the domain to which the equation applies. Let us denote that domain by the symbol Ω. In the simplest case, these parameters are defined as constants over the entire region defined by Ω. Of course, natural porous media will have spatially varying values of $\mathbf{K}$ and S_s. The degree to which spatial variability is included in the system description depends on the length scale of the REV used to define the equations, as well as practical considerations such as the objectives of the analysis and the available data. In general, spatially variable parameters require the use of some form of interpolation to extend information from limited data points to the entire domain Ω. *Kriging* is the methodology most often used in groundwater hydrology to accomplish this interpolation (see Section 7.7.4 or [3] for a discussion of kriging). One advantage of the kriging approach is the ability to quantify the parameter estimation uncertainty due to interpolation, a feature that is very important in models that accommodate uncertainty.

4.3.3 Boundary Conditions

As is evident from Eq. (4.22), the groundwater flow equation has second-order spatial derivatives. Such second-order equations require that information be provided about the behavior of the dependent variable (in this case head), or its derivative, everywhere on the perimeter of the domain in which the governing equation is to be solved. These additional data are referred to as boundary conditions. In essence, the purpose of *boundary conditions* is to tell the governing equation that the world of interest from the model's point of view is limited. Moreover, it informs the equation about the behavior of the dependent variable as defined by the external world as it intersects with the domain of interest.

If the model is two dimensional, boundary conditions must be specified around the one-dimensional line perimeter. In the case of a three-dimensional model, boundary conditions must be specified along the two-dimensional surface that defines the boundary of the domain. If the domain is denoted by Ω, the boundary is often denoted by the symbol $\partial \Omega$ (see Figure 4.6).

In the case of groundwater flow, boundary conditions can have a number of different forms, including the three standard forms for second-order partial differential equations of this type. The forms are *Dirichlet,* also known as *fixed head; Neumann*, also known as *fixed flux*; and *Robbins,* also known as *induced flux.* Mathematically, the boundary

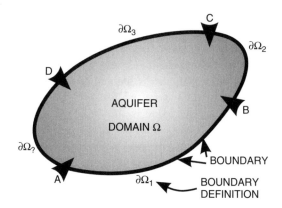

FIGURE 4.6. Diagrammatic representation of an areal two-dimensional aquifer model illustrating the requirement of specifying boundary conditions $\partial\Omega_i$ along the model perimeter. Each boundary segment is limited by the arrows and identified by $\partial\Omega_i$, where the subscript i indicates that any of the three types of boundaries can be defined. The segment $\partial\Omega_?$ indicates that this segment has yet to be defined.

conditions are stated as follows:

$$h(\mathbf{x}, t) = h_0(\mathbf{x}, t) \quad \mathbf{x} \in \partial\Omega_1 \quad Dirichlet, \tag{4.23}$$

where h_0 is the specified head along the boundary segment $\partial\Omega_1$ of the overall domain $\partial\Omega$;

$$\frac{\partial h(\mathbf{x}, t)}{\partial n} = \frac{\partial h(\mathbf{x}, t)}{\partial n}\bigg|_0 \quad \mathbf{x} \in \partial\Omega_2 \quad Neumann, \tag{4.24}$$

where $\partial h(\mathbf{x})/\partial n|_0$ is the specified outward normal gradient along the boundary segment $\partial\Omega_2$; and

$$\alpha h(\mathbf{x}, t) + \gamma\frac{\partial h(\mathbf{x}, t)}{\partial n} = G_0(\mathbf{x}, t) \quad \mathbf{x} \in \partial\Omega_3 \quad Robbins, \tag{4.25}$$

where α, γ, and G_0 are specified constants or functions along the boundary segment $\partial\Omega_3$. For problems in which all boundary conditions correspond to one of the three standard types, the sum of the above segments must equate to the whole, that is,

$$\partial\Omega = \partial\Omega_1 + \partial\Omega_2 + \partial\Omega_3. \tag{4.26}$$

Of course, there can be more than one segment $\partial\Omega_1$ and so on, or there might be a part of the boundary along which a specialized boundary condition is specified that does not correspond to one of the standard types (for an example see Section 4.4). This concept is shown on Figure 4.6, where one segment of the perimeter of the model is identified as $\partial\Omega_?$. The idea here is that $\partial\Omega_?$ could take on any of the three types of boundary specifications or perhaps a specialized condition. In addition, we have implicitly assumed the boundary location is known; for some special problems, such as so-called moving boundary problems, we might have to solve for the boundary location as part of the overall problem solution. In that case we need additional information to tell us something about the boundary movement.

Equation (4.23) is used to specify the head along the boundary and this can be a function of time if the time-dependent behavior is known. Tidal behavior of the water surface of an estuary for example, could be described by this kind of boundary specification.

Equation (4.24), when combined with the hydraulic conductivity or, in the case of a two-dimensional areal model, *transmissivity*, provides a statement of the *flow across the boundary*. In other words, in terms of flow per unit length of boundary we have

$$\mathbf{q}(\mathbf{x}) \cdot \mathbf{n} = -\mathbf{K} \cdot \nabla h(\mathbf{x}) \cdot \mathbf{n}, \tag{4.27}$$

where $\mathbf{q}(\mathbf{x}) \cdot \mathbf{n}$ is the specified normal flux across the boundary $\partial\Omega_2$.

The Robbins or third-type boundary condition specified in Eq. (4.25) is often used to describe what is called a *leakage condition* associated with a flux across the boundary that depends on the head values on either side of the boundary. It can also be used to represent a physical boundary at a long distance from a model boundary when it is used in a two- or three-dimensional model setting. The form that captures both of these conditions is

$$\mathbf{K} \cdot \nabla h(\mathbf{x}) \cdot \mathbf{n} = \kappa(h_0(\mathbf{x}) - h(\mathbf{x})), \tag{4.28}$$

where h_0 is the *head external to the model* and is assumed known. For example, h_0 might be the *elevation in a lake* above the top of the model in the case of a leakage situation. In this case the coefficient κ has the meaning of the resistivity to flow across the boundary. Alternatively, h_0 could be considered as the *head in the aquifer at some distance* from the boundary and the coefficient κ in this case would be used as a surrogate for the distance to the location of h_0 relative to $\partial\Omega_3$, that is, $\kappa \equiv K/L$, where L is the distance from the model boundary to the real boundary where h_0 is observed.

4.3.4 Initial Conditions

Equation (4.22) contains a first-order time derivative term. Thus it is necessary, from a mathematical perspective, to define the state of the system at the beginning of the analysis period. Because the dependent variable is the head, one is required to specify the head everywhere in Ω at the initial time t_0.

In general, the role of the initial condition in a groundwater flow problem is relatively minor. The reason is that for problems defined at a scale of practical importance, the head quickly adjusts to the boundary conditions. Because the system quickly adjusts from the initial state to that consistent with the boundary conditions, we can often ignore the time derivative term and solve the steady-state version of Eq. (4.22), which involves all terms except the time derivative.

4.3.5 Sources and Sinks

To derive the equation governing the three-dimensional transient evolution of hydraulic head, that is, Eq. (4.22), we have assumed no source or sink terms. This means that the only way to get fluid into or out of a volume of porous medium is via flow through the boundaries of the volume. Other mechanisms that allow for addition or subtraction of liquid water to or from a given volume of porous medium include phase change and external sources or sinks. An example of phase change is melting or freezing, which obviously changes the amount of liquid water in a volume. An example of an external source or sink is a well. Because the radius of a well, which is on the order of 10–20 cm, is very much smaller than the usual size of a domain, which might be on the order of kilometers, we often choose to represent the well as having a vanishingly small radius. That is, we represent the well as a line source or sink, typically with a specified flow rate along the length of the line.

Mathematically, we represent these sources by a general right-hand-side function added to Eq. (4.22), so that the general three-dimensional groundwater flow equation takes the form

$$S_s \frac{\partial h}{\partial t} - \nabla \cdot \mathbf{K} \cdot \nabla h = Q,$$

where the function Q represents all source or sink terms other than the divergence of the volumetric flux vector. Dimensions of this function are $[T^{-1}]$, which is interpreted as the volume of liquid water per volume of porous medium per time. For representation of a well, the function associated with a line source or sink involves Dirac delta functions. For practical purposes, we may define the Dirac delta function as a function that has a nonzero value only over an interval $[x_i - \Delta,\ x_i + \Delta]$, is restricted to having an integral equal to 1, and is defined by taking the limit as $\Delta \to 0$. This function is represented symbolically as $\delta(x - x_i)$, and it has dimension of $[L^{-1}]$. With this concept, a well is represented mathematically as

$$Q = q_w(z)\, \delta\left(x - x_i^w\right) \delta\left(y - y_i^w\right),$$

where $q_w(z)$ represents the flow rate to or from the well per unit length of the well, with associated dimension $[L^2/T]$ interpreted as volume of liquid water per time per unit length of well.

Finally, we note that when governing equations are written in reduced dimensionality domains, information that appears as boundary conditions in the three-dimensional equation is represented instead as apparent source or sink terms in the reduced dimensionality representation. We will see examples of this shortly, in Section 4.5, where we derive the governing equation applied to two-dimensional (x, y) domains.

4.4   THE FREE-SURFACE CONDITION

When analyzing water-table aquifers, we often take the water table as the upper boundary of our domain. We already know that, by definition, at the water table the water pressure is equal to atmospheric pressure, which we typically assign as a reference pressure (gauge pressure) of zero. This provides a first-type boundary condition. However, while we know this condition, the problem we face is that we usually do not know the location of the water table; indeed, the water-table location may be an important part of what we wish to predict. Because the water table represents a moving boundary, we need to develop an additional equation to define the movement of the boundary.

The water table is shown as the surface labeled B in Figure 4.7. Let the height of the water table be denoted by $z = b(x, y, t)$. Since z is a function of x, y, and t, the location of the water table can vary in (x, y) space and time. We can express this equivalently by construction of a function that by definition is equal to zero whenever $z = b$, which

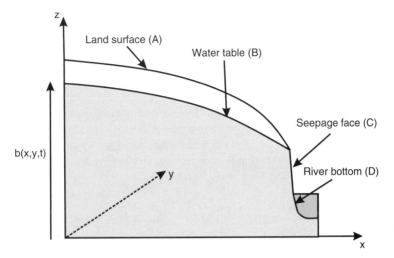

FIGURE 4.7. Definition sketch of a sharp interface representation of the water table.

we choose as the function $F(x, y, z, t) = z - b(x, y, t)$. This means that a point on the surface corresponding to the water table will always have a value of $F(x, y, z, t) = 0$. Keep in mind that the surface is a geometric entity and water can pass through it. It should not be identified with a particle of water. One can think of this surface as a piece of permeable cloth floating on the water table. Once a point is on the free surface, it is always on that free surface; that is, the surface is always defined by $F(x, y, z, t) = 0$.

On page 167 we introduced the concept of a substantial derivative. It described the time-dependent behavior of a fluid system given you were riding along at the same velocity as the fluid. In the case of the free surface, this is precisely what we need to describe its behavior, because we want to know what is going on around us if we are riding on a point located on the free surface (a location on our hypothetical piece of cloth). Recalling our earlier analogy of a person counting fish from a canoe floating on a stream to describe a substantial derivative, in this problem we are riding along with the free surface, perhaps counting soil grains.

Because the surface is always defined by $F(x, y, z, t) = 0$, if we remain on the surface, then

$$
\begin{aligned}
\frac{DF}{Dt} &= \frac{D}{Dt}\left(z - b\left(x, y, t\right)\right) \\
&= \frac{\partial}{\partial t}\left(z - b\left(x, y, t\right)\right) + \mathbf{v}_{\mathrm{wt}} \cdot \mathbf{\nabla}\left(z - b\left(x, y, t\right)\right) \\
&= -\frac{\partial b}{\partial t} - v_{\mathrm{wt}x}\frac{\partial b}{\partial x} - v_{\mathrm{wt}y}\frac{\partial b}{\partial y} + v_{\mathrm{wt}z}\frac{\partial z}{\partial z} \\
&= 0,
\end{aligned} \tag{4.29}
$$

where the subscript wt refers to water table.

We found in Chapter 2 that the head is given by

$$
h(\mathbf{x}, t) = h_p(\mathbf{x}, t) + \zeta(\mathbf{x}, t), \tag{4.30}
$$

where h is the total head, h_p is the pressure head, and ζ is the elevation head. The pressure head was defined to be

$$h_p = \frac{p(\mathbf{x}, t)}{\rho g}, \tag{4.31}$$

where p is the fluid pressure, ρ the fluid density, and g the gravitational acceleration. Since, by definition, the pressure at the free surface is zero, by virtue of Eq. (4.31) so also is the pressure head. Thus the head at the free surface becomes, using Eq. (4.30), $h = \zeta$. Because b in Eq. (4.29) is the equation of the free surface, one can write $b = \zeta = h$. Substitution of this expression for b in Eq. (4.29) yields

$$-\varepsilon_D \frac{\partial h}{\partial t} - \varepsilon_D v_{\text{wtx}} \frac{\partial h}{\partial x} - \varepsilon_D v_{\text{wty}} \frac{\partial h}{\partial y} + \varepsilon_D v_{\text{wtz}} = 0. \tag{4.32}$$

In this expression, the porosity ε_D is to be thought of as the drainable porosity, which we will consider shortly.

One can think about this condition from a physical perspective. If we could determine the velocity of the free surface, either explicitly or as a function of the state variable h, Eq. (4.32) would provide an auxiliary condition for the free-surface condition and allow us to determine the position of the water table. In the case where we have no infiltration to, or evaporation from, the water table, then the only way for the water table to move is for the water in the vicinity of the water table to move. In that case, the velocity of the water at the table corresponds to the velocity of the water table. Therefore we can replace the water-table velocity by the actual groundwater velocity in Eq. (4.32).

However, the velocity in the vicinity of the water table is modified by the fact that above the water table is the unsaturated zone. When water drains out of the pore space, to form unsaturated conditions, not all of the pore space is drained because some of the water remains behind as so-called residual saturation (see Sections 11.3 and 2.5). Therefore the relationship between specific discharge and actual water-table velocity is not based on total porosity but instead on the porosity that drains, which we have called *drainable porosity* above. Therefore $\mathbf{q} = \varepsilon_D \mathbf{v}$. We can use this, coupled with the Darcy equation, to write Eq. (4.32) as

$$-\varepsilon_D \frac{\partial h}{\partial t} + K \frac{\partial h}{\partial x} \frac{\partial h}{\partial x} + K \frac{\partial h}{\partial y} \frac{\partial h}{\partial y} - K \frac{\partial h}{\partial z} = 0. \tag{4.33}$$

If infiltration due to rainfall moving from the land surface downward through the unsaturated zone and to the water table is to be included, then the velocity of the water table is the sum of the local water velocity and the water being added to the domain. With infiltration given as a vertical flow with rate N (volume per area per time), Eq. (4.33) is modified by adding N to the right-hand side.

In Figure 4.7 there is an area forming the bank of the river that is identified as the *seepage face* and denoted with the letter C. While the pressure along this face is atmospheric, it is different from the water table because it is not a moving boundary, and therefore we do not need another equation to solve for the position of the boundary. Along this face one assumes the head is specified as a constant (type one) boundary condition.

However, since the pressure is prescribed as atmospheric, the head is given simply as (see Eq. (4.30))

$$h(\mathbf{x}, t) = \zeta(\mathbf{x}, t) \quad \mathbf{x} \in \partial\Omega_C.$$

Finally, along the boundary that represents the stream bed, identified as "River bottom" and given the letter designation D, the head is constant and equal to that in the stream. Thus anywhere along D, the constant head is equal to the elevation of the stream surface measured relative to the same reference plane used for the definition of other variables in this problem. Along this surface, the decrease in the pressure head as one moves vertically upward is exactly balanced by the increase in elevation.

The boundary conditions along the remaining sides are those normally defined, such as specified head, specified flux (second type), or a leakage condition (third type).

We have assumed in this section that the water table is a sharp interface. While this may be a reasonable approximation in some locations, it may not be in others. When a sharp interface assumption is inappropriate, that is, there is a significant capillary fringe, one must incorporate in the simulation the effects of capillarity. Because of the need for additional knowledge of the unsaturated zone and its associated physics in order to model the water table in the presence of significant capillary effects, we will postpone discussion of this approach until Chapter 11.

4.5   REDUCTION IN DIMENSIONALITY

Although, in general, the world is most accurately described in terms of three space coordinate directions, plus time if the problem is transient, it is often convenient to reduce the dimensionality. The motivation to do this stems from a desire to simplify mathematical calculations for cases where such simplification is appropriate. For example, many aquifers are very thin in the vertical dimension, on the order of tens of meters, relative to their areal extent, which is often many kilometers (see Figure 1.4). Therefore elimination of the vertical dimension may be reasonable. Computational considerations also motivate reduction in dimensionality. In general, the mathematical apparatus needed to solve problems is more complex when employed in higher dimensions. When numerical methods are used, the computational effort needed to solve a problem increases dramatically with each added space dimension. A two-dimensional problem, for example, generally requires much more than twice the computational effort needed to solve a one-dimensional representation of the same problem. Of course, there is no free lunch, and with a reduction in dimensionality there is normally a reduction in the accuracy with which a physical system is being portrayed. As we illustrate how one reduces the dimensionality of groundwater problems, we will point out how the procedure impacts the accuracy of the simulation.

4.5.1   Physical Dimensions of the Model

In general, one can assume that groundwater flow and transport analyses require a three-dimensional representation in space. In other words, groundwater flow is a three-dimensional process. In special circumstances, one can simplify the simulation to require only a two-dimensional simulation. For example, when studying salt-water intrusion in a coastal aquifer a cross-sectional model, involving one horizontal dimension and the vertical dimension, may, in some circumstances, be adequate. Similarly, when a vertically homogeneous aquifer is to be considered and any wells involved are nearly fully penetrating, a two-dimensional *horizontal (areal) model* may be sufficient.

In general, it is necessary to justify simplifying a three-dimensional world to two dimensions.[2] For example, in a cross-sectional model, one is assuming that *the behavior of the groundwater system within any cross section along a line perpendicular to the section being considered (i.e., perpendicular to the paper) is the same as that for the selected cross section.* A very common error in this regard is to assume that groundwater flow in response to multiple wells can be represented in a Cartesian (x, z) cross section. This is not possible because multiple wells generate radial flow patterns that cannot be represented, in general, in a Cartesian cross section. One can represent flow to a single well in two dimensions, but this requires the use of a *cylindrical (r, z) coordinate system*, not a Cartesian (x, z) coordinate system.[3]

The most common simplification of the three-dimensional world of groundwater flow is to average over the vertical dimension to generate a two-dimensional areal model. While this can be justified when flow is truly horizontal, one often hears the argument made that a two-dimensional model is desirable because too little is known about the hydrogeological properties in the vertical dimension to justify modeling it. This is an incorrect concept. Even when the aquifer is homogeneous vertically, flow in the third dimension may still be very important. The correct question to ask is: Can the flow behavior in the vertical dimension be neglected without compromising the effectiveness of the model?

4.5.2 Vertical Integration of the Flow Equation

To understand what is involved in reducing model dimensionality by disregarding the vertical dimension in modeling groundwater flow, one must realize what is happening from the mathematical–physical point of view. This is best achieved by formally developing the areal two-dimensional model from the more general three-dimensional model and explaining the inherent simplifications and assumptions. Let us begin with the flux form of the groundwater fluid mass-conservation equation

$$\nabla \cdot \mathbf{q} + S_s \frac{\partial h}{\partial t} + Q = 0, \tag{4.34}$$

where Q represents a source or a sink (see Section 4.3.5) and is positive for fluid discharge. Coupled to Eq. (4.34) is Darcy's law given by

$$\mathbf{q} = -\mathbf{K} \cdot \nabla h. \tag{4.35}$$

Consider the diagrammatic representation of a three-dimensional aquifer illustrated in Figure 4.8. The goal is to eliminate the vertical dimension while accounting for all physical processes inherent in the system. The mathematical approach we will take is

---

[2]In actual fact, the dimension that is eliminated is not totally disregarded. Formally one is integrating over the neglected dimension and the missing dimension is therefore being accommodated in this approximate sense.

[3]An exception to this statement is when a series of wells are located along a straight line such that an approximate line sink is created. This, of course, is not a likely scenario.

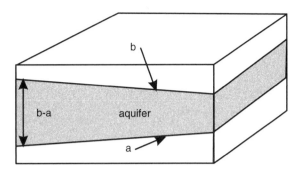

FIGURE 4.8. Diagrammatic representation of an aquifer system providing notation to be used for vertical integration.

vertical averaging. In other words, we will no longer represent vertical variability within the domain. Note that this may be seen as an extension of the volume averaging presented earlier in this chapter. In the case of volume averaging, we accepted that we cannot represent variability below the scale of the REV; in vertical averaging, we extend this to the entire vertical dimension.

In normal averaging, one creates a sum. In our approach, we will replace the summation with the more abstract concept of integration. The aquifer to be vertically integrated is illustrated in Figure 4.8. Performing a formal integration of Eq. (4.34) over the aquifer thickness $(b - a)$ one obtains

$$\int_{a(x,y,t)}^{b(x,y,t)} \left(\nabla \cdot \mathbf{q} + S_s \frac{\partial h}{\partial t} + Q \right) dz = 0, \tag{4.36}$$

where $b(x, y, t)$ is the *surface corresponding to the top of the aquifer* and $a(x, y, t)$ is the *lower boundary of the aquifer*. By applying *Leibnitz's rule* (introduced earlier, Section 2.5.3) for differentiation of an integral, one can show the following equivalence relationship:

$$\int_{a(x,y,t)}^{b(x,y,t)} \nabla \cdot \mathbf{q}(\mathbf{x}) \, dz = \nabla_{xy} \cdot \int_a^b \mathbf{q}(\mathbf{x}) \, dz + \mathbf{q}(b) \cdot \nabla b - \mathbf{q}(a) \cdot \nabla a, \tag{4.37}$$

where $\nabla_{xy}(\cdot)$ is defined as the two-dimensional gradient operator

$$\nabla_{xy}(\cdot) \equiv \frac{\partial(\cdot)}{\partial x}\mathbf{i} + \frac{\partial(\cdot)}{\partial y}\mathbf{j}.$$

Expansion of the coordinate vector in Eq. (4.36) yields

$$\int_a^b \left(\frac{\partial q_x}{\partial x} + \frac{\partial q_y}{\partial y} + \frac{\partial q_z}{\partial z} \right) dz = \frac{\partial}{\partial x} \int_a^b q_x \, dz - q_x|_b \frac{\partial b}{\partial x} + q_x|_a \frac{\partial a}{\partial x} \tag{4.38}$$

$$+ \frac{\partial}{\partial y} \int_a^b q_y \, dz - q_y|_b \frac{\partial b}{\partial y} + q_y|_a \frac{\partial a}{\partial y} + q_z|_a^b$$

or

$$\int_a^b \nabla \cdot \mathbf{q} \, dz = \nabla_{xy} \cdot \int_a^b \mathbf{q}_{xy} \, dz + \mathbf{q}|_b \cdot \nabla (z - b) - \mathbf{q}|_a \cdot \nabla (z - a). \tag{4.39}$$

The *time derivative term* is treated in a similar way; that is,

$$\overline{S_s} \int_a^b \frac{\partial h}{\partial t} \, dz = \overline{S_s} \frac{\partial}{\partial t} \int_a^b h \, dz - \overline{S_s} h|_b \frac{\partial b}{\partial t} + \overline{S_s} h|_a \frac{\partial a}{\partial t}, \tag{4.40}$$

where we are using an average value of S_s, that is, $\overline{S_s}$, so as to permit its removal from under the spatial integration.

Let us now leave the continuity equation momentarily while we examine the vertical integration of Darcy's law. From the general form of Darcy's law, Eq. (4.35), we obtain

$$\mathbf{q} = -\mathbf{K} \cdot \nabla h. \tag{4.41}$$

Employment of Leibnitz's rule and the assumption that a vertical averaged $\mathbf{K}$, namely $\overline{\mathbf{K}}$, can be defined, as we did in Eq. (4.37), results in

$$\int_a^b \mathbf{q}_{xy} \, dz = -\overline{\mathbf{K}} \cdot \int_a^b \nabla_{xy} h \, dz \tag{4.42}$$

$$= -\overline{\mathbf{K}} \cdot \left[\nabla_{xy} \int_a^b h \, dz - h|_b \nabla_{xy} b + h|_a \nabla_{xy} a \right], \tag{4.43}$$

where $\mathbf{k}$ is the unit vector in the z coordinate direction.

The combination of Eqs. (4.39), (4.40), (4.41), and (4.42) yields

$$-\nabla_{xy} \cdot \overline{\mathbf{K}} \cdot \left[\nabla_{xy} \int_a^b h \, dz - h|_b \nabla_{xy} b + h|_a \nabla_{xy} a \right] + \mathbf{q}|_b \cdot \nabla_{xy} (z - b) - \mathbf{q}|_a \cdot \nabla_{xy} (z - a)$$

$$+ \overline{S_s} \frac{\partial}{\partial t} \int_a^b h \, dz - \overline{S_s} h|_b \frac{\partial b}{\partial t} + \overline{S_s} h|_a \frac{\partial a}{\partial t} + \int_a^b Q dz = 0. \tag{4.44}$$

Equation (4.44) is equivalent to the groundwater flow equation, vertically integrated. We have made no additional assumptions to this point, so Eq. (4.44) retains all the physics inherent in the groundwater flow equation.

We now proceed to simplify Eq. (4.44) by eliminating the vertical dimension. In doing this, we introduce assumptions that reduce the generality of this equation and limit its utility to a subset of groundwater systems.

Our first step is the definition of the following vertical averages:

$$\overline{h} \equiv \frac{1}{l} \int_a^b h(\mathbf{x}) \, dz, \tag{4.45}$$

$$\overline{Q} \equiv \frac{1}{l} \int_a^b Q(\mathbf{x}) \, dz, \tag{4.46}$$

where $l(x, y, t) = b(x, y, t) - a(x, y, t)$.

To proceed, we now make the following rather profound assumption, namely, that $h|_a \simeq h|_b \simeq \overline{h}$. The physical interpretation of this assumption is that the head values at any (x, y) location are essentially the same as one moves vertically from the bottom $(z = a)$ to the top $(z = b)$ of the aquifer. The implication of this is that there is an insignificant vertical gradient in the aquifer, meaning that the dominant flow direction is horizontal. Therefore, *in the presence of significant vertical gradients, use of the areal two-dimensional form of the groundwater flow equation is not appropriate unless $h|_a$ and $h|_b$ are known. The equation is only appropriate for aquifers with essentially horizontal flow.*

Substitution of this assumption into Eq. (4.44) yields

$$-\nabla_{xy} \cdot \overline{\mathbf{K}} \cdot (\nabla_{xy} l\overline{h} - \overline{h}\nabla (b - a)) - \mathbf{q}_{xy}|_b \cdot \nabla_{xy} b + \mathbf{q}_{xy}|_a \cdot \nabla_{xy} a + q_z|_b - q_z|_a$$
$$+ \overline{S_s}\frac{\partial}{\partial t} l\overline{h} - \overline{S_s h}\frac{\partial}{\partial t}(b - a) + \overline{Ql} = 0, \tag{4.47}$$

which simplifies, upon expansion of the derivatives, to

$$\nabla_{xy} \cdot \mathbf{T}_{xy} \cdot \nabla_{xy}\overline{h} = S\frac{\partial \overline{h}}{\partial t} + \underbrace{\mathbf{q}|_b \cdot \nabla (z - b)}_{A} - \underbrace{\mathbf{q}|_a \cdot \nabla (z - a)}_{B} + \overline{Ql}, \tag{4.48}$$

where the *storage coefficient* is defined as $S \equiv \overline{S_s}l$ and the *transmissivity tensor* is defined as $\mathbf{T} \equiv l\overline{\mathbf{K}}$. The terms denoted by the letters A and B are the net flux through the surfaces $z = b (x, y, t)$ and $z = a (x, y, t)$, respectively.

While the new parameters $\mathbf{T}$ and S are derived directly from those defined for the three-dimensional system case, their physical meanings are quite different because they reflect the vertical averaging process. The transmissivity describes the flow through a unit width of the entire vertical thickness of the aquifer. Thus it has units of $[L^2/T]$ rather than $[L/T]$ as is the case for the hydraulic conductivity $\mathbf{K}$. Similarly, the storage coefficient S represents the amount of water released from storage over a unit area of aquifer over its entire vertical thickness in response to a unit drop in the average head $\overline{h}$ defined in Eq. (4.45).

Denoting the flux through the top of the aquifer as q_T and that through the bottom as q_B, and defining the average flux being added to the aquifer from other sources at a particular areal location by $q_{\text{ext}}(x, y)$, one obtains

$$\nabla_{xy} \cdot \mathbf{T} \cdot \nabla_{xy}\overline{h} = S\frac{\partial \overline{h}}{\partial t} + q_T + q_B + q_{\text{ext}}, \tag{4.49}$$

where $q_T \equiv \mathbf{q}|_b \cdot \nabla (z - b)$ and $q_B \equiv -\mathbf{q}|_a \cdot \nabla (z - a)$. For systems with wells, where the well is represented by a Dirac delta function (see Section 4.3.5), this term is

$$q_{\text{ext}} = \int_z q_z \delta (x - x_i) \delta (y - y_i) \, dz = \left(\int_z q(z) \, dz\right) \delta (x - x_i) \delta (y - y_i)$$
$$= Q_w \delta (x - x_i) \delta (y - y_i),$$

where Q_w is the total flow rate (volume per time) in the well. Equation (4.49) is the two-dimensional areal groundwater flow equation. It is the most commonly used formulation and is a reasonable compromise between the relatively simplistic one-dimensional formulation, which we consider in Section 4.7, and the computationally burdensome three-dimensional alternative.

4.5.3 The Free-Surface Condition in the Areal Model

The above analysis assumes a confined aquifer, that is, an aquifer wherein there exist relatively impermeable geologic units above and below the reservoir, such that the reservoir remains totally saturated at all times. Let us consider the case when the aquifer is unconfined, that is, the reservoir contains the water table (see Figure 4.9). We then use the analysis of Section 4.4 and write

$$\frac{DF}{Dt} = \frac{D}{Dt}(z-b) = \left(-\frac{\partial b}{\partial t} - \mathbf{v}_{xy} \cdot \nabla_{xy} b + v_z\right)\Bigg|_b = 0, \qquad (4.50)$$

which is identical to Eq. (4.29). Recall that at the water table we used the relationship

$$\mathbf{q} = \varepsilon_D \mathbf{v},$$

where ε_D is the *drainable porosity*. Given that $b = h$ at the water table, we can write

$$\mathbf{q}|_b \cdot \nabla(z-b) = \varepsilon_D \frac{dh|_{z=b}}{dt} - N.$$

This equation may be used to substitute for the top flux q_T. After multiplication of Eq. (4.50) by ε_D, substitution for q_T and subsequent subtraction of the result from Eq. (4.49) yields

$$\nabla_{xy} \cdot \overline{\mathbf{K}l}(h) \cdot \nabla \overline{h} = \overline{S_s l}(h) \frac{\partial \overline{h}}{\partial t} + \varepsilon_D \frac{\partial \overline{h}}{\partial t} + q_T + N + q_{\text{ext}}, \qquad (4.51)$$

where $q_T = -\mathbf{q}_{xy}|_b \cdot \nabla_{xy} b + \varepsilon \mathbf{v}_{xy} \cdot \nabla_{xy} b + (q_z - \theta v_z)|_b$ represents the net flow out of the aquifer across the water table. Note that the coefficients $\overline{\mathbf{K}l}$ and $\overline{S_s l}$ are now functions of the solution $\overline{h}$ because $l = (b - a) = (\overline{h} - a)$ and therefore the *partial differential equation is nonlinear*. The interesting element of this formulation is that there is no boundary condition per se on the top of the model, the vertical dimension having been integrated out. Rather, we have a source term q_T that represents the relative flux of fluid across the top of the water table. We also have an additional term involving the time derivative of $\overline{h}$ with coefficient ε_D. The drainable porosity is often replaced by the

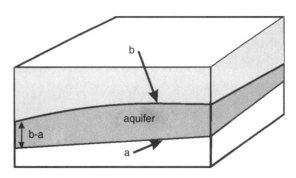

FIGURE 4.9. Diagrammatic representation of an unconfined aquifer system providing notation to be used for vertical integration.

equivalent term called the "specific yield," denoted by S_y, and defined as the volume of water that drains from a unit area of aquifer when the water table is lowered by a unit amount. Thus Eq. (4.51) may be written

$$\left(S + S_y\right) \frac{\partial \overline{h}}{\partial t} - \nabla_{xy} \cdot \mathbf{T} \cdot \nabla \overline{h} = +q_T + N + q_{\text{ext}}.$$

4.6 SALT-WATER INTRUSION

In some ways this section is out of place. We are talking about salt-water intrusion before we have considered dissolved mass transport. The rationale behind this lies in the fact that salt-water intrusion, as described in Section 1.8.1 can be described via a two-fluid analogy separated by a sharp interface. The interface between the salt-water body and the freshwater body is dispersive. However, in some instances the dispersive zone can be considered small relative to the salt-water–freshwater system and can be replaced by a sharp interface. Such a situation is shown diagrammatically in Figure 4.10. The zone of dispersion is outlined by the dashed lines. The approximate sharp interface is represented by the solid line. Given this approximation, the equations governing the salt-water interface behavior are represented using the following extension of the water-table formulation given above.

The sharp interface assumption applied to salt-water intrusion implies the existence of an interface separating two fluids, salt water and fresh water. The two fluids are essentially the same but for their densities. Salt water has a density about 2.5% larger than fresh water. Figure 4.10 provides a diagrammatic sketch that defines the physical concepts and notation we will need to obtain the governing equations of the sharp salt-water interface equations. In this system we assume the existence of two pseudofluids, the salt water and the fresh water.

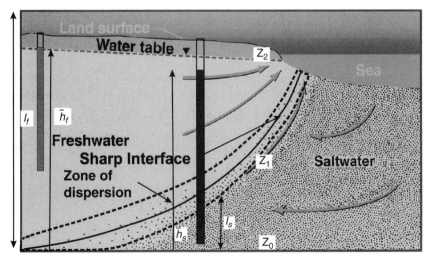

FIGURE 4.10. Sharp interface approximation (solid line) of a salt-water intrusion zone of dispersion defined by dashed lines.

The salt water and the fresh water do not behave independently. The movement of the fresh water affects the movement of the salt water and vice versa. The key that unlocks the way these two "fluids" interact is found at the interface. Along the interface the pressure exhibited by the two fluids is the same. Thus we can write the head equation for each fluid at an elevation z_1 on the interface as

$$h_f = \frac{p}{\rho_f g} + z_1,$$

$$h_s = \frac{p}{\rho_s g} + z_1,$$

which can be written after some algebra as

$$\frac{\rho_f h_s - \rho_s h_f}{\rho_s - \rho_f} = z_1. \tag{4.52}$$

Note that in a system at equilibrium, the elevation of the interface, z, can be determined from a knowledge of the constant heads in each fluid and their respective densities; for dynamic cases Eq. (4.52) still applies at the interface, but each fluid has spatially and temporally varying hydraulic heads.

From the free-surface condition applied to the sharp interface, namely, that a point on the interface will remain on the interface, we have the relationship

$$v_z|_{z_1} - \left(\mathbf{v}_{xy} \cdot \nabla z_1 + \frac{\partial z_1}{\partial t} \right) = 0. \tag{4.53}$$

If an assumption of essentially horizontal flow for the two fluids is appropriate, then we can apply the vertical averaging procedures of the last section to each of the two fluids. Application of the integration procedure to the freshwater layer yields

$$\nabla \cdot (\overline{\mathbf{q}}_{fxy} l_f) - \mathbf{q}_f|_{z_1} \cdot \nabla (z - z_1) + \mathbf{q}_f|_{z_2} \cdot \nabla (z - z2) + \overline{S_f} l_f \frac{\partial \overline{h}_f}{\partial t} = 0, \tag{4.54}$$

and application to the salt-water layer gives

$$\nabla \cdot (\overline{\mathbf{q}}_{sxy} l_s) + \mathbf{q}_s|_{z_1} \cdot \nabla (z - z_1) - \mathbf{q}_s|_{z_2} \cdot \nabla (z - z_0) + \overline{S_s} l_s \frac{\partial \overline{h}_s}{\partial t} = 0. \tag{4.55}$$

We define $\overline{h}_f$ as the average head in the fresh water, $\overline{h}_s$ as the average head in the salt water, $\overline{\mathbf{q}}_{fxy}$ as the average flux of fresh water, and $\overline{\mathbf{q}}_{sxy}$ as the average flux of salt water, with all averages defined over the vertical extent of each of the two fluids. We indicate the thickness of the freshwater layer as l_f and that of the salt-water layer as l_s (i.e., $l_f = z_1 - z_0$, $l_s = z_2 - z_1$).

Multiplication of Eqs. (4.53) and (4.50) by the porosity ε and addition of the result to Eq. (4.54) produces the following equation:

$$\nabla \cdot (\overline{\mathbf{q}}_{fxy} l_f) + q_{fz}|_{z_2} - (q_{fz} - \varepsilon v_z)|_{z_1} - (\mathbf{q}_{fxy} - \varepsilon \mathbf{v}_{fxy})|_{z_2} \cdot \nabla_{xy} z_2$$

$$+ (\mathbf{q}_{fxy} - \varepsilon \mathbf{v}_{fxy})|_{z_1} \cdot \nabla_{xy} z_1 - \varepsilon \frac{\partial z_1}{\partial t} + \varepsilon \frac{\partial z_2}{\partial t} + \overline{S_f} l_f \frac{\partial \overline{h}_f}{\partial t} = 0.$$

Similarly, by multiplying Eq. (4.53) by ε and subsequently subtracting the result from Eq. (4.55), we obtain

$$\nabla \cdot (\overline{\mathbf{q}}_{sxy} l_s) - q_{sz}|_{z_0} + \mathbf{q}_{sxy}|_{z_0} \cdot \nabla_{xy} z_0$$

$$+ \varepsilon \frac{\partial z_1}{\partial t} + \overline{S}_s l_s \frac{\partial \overline{h}_s}{\partial t} = 0.$$

We now differentiate Eq. (4.52) with respect to time:

$$\rho_s^* \frac{\partial h_s}{\partial t} - \rho_f^* \frac{\partial h_f}{\partial t} = \frac{\partial z_1}{\partial t},$$

where

$$\rho_f^* \equiv \frac{\rho_f}{\rho_s - \rho_f},$$

$$\rho_s^* \equiv \frac{\rho_s}{\rho_s - \rho_f}.$$

Substitution of Darcy's law and imposition of the requirement that the pressure on the water table is zero yields

$$\left(l_f \overline{S}_s + S_y + \varepsilon \rho_f^*\right) \frac{\partial \overline{h}_f}{\partial t} - \varepsilon \rho_s^* \frac{\partial \overline{h}_s}{\partial t} - \nabla_{xy} \cdot l_f \overline{\mathbf{K}_f} \cdot \nabla_{xy} \overline{h}_f + Q_f - R = 0, \qquad (4.56)$$

$$\left(l_s \overline{S}_s + \varepsilon \rho_f^*\right) \frac{\partial \overline{h}_s}{\partial t} - \varepsilon \rho_f^* \frac{\partial \overline{h}_f}{\partial t} - \nabla_{xy} \cdot l_s \overline{\mathbf{K}_s} \cdot \nabla_{xy} \overline{h}_s + Q_s = 0. \qquad (4.57)$$

In these equations we assume that (1) flow is essentially horizontal, (2) all fluxes from the surface are incorporated in the term R, (3) there are no fluxes from the base of the aquifer, (4) water does not move across the salt-water–freshwater interface, and (5) the subscripts s and f identify salt-water and freshwater related properties, respectively.

Examination of Eqs. (4.56) and (4.57) reveals that the coupling between the salt water and the fresh water is through both the time derivative (both $\partial \overline{h}_f / \partial t$ and $\partial \overline{h}_s / \partial t$ are found in each equation) and through l_f and l_s, which depend on $\overline{h}_s$ and $\overline{h}_f$ via the z_1 and z_2 values. It is important to observe that nowhere in these equations does the vertical velocity appear. It has been eliminated through the averaging process. Thus the vertical movement of the interface is dependent only on the horizontal velocity, which, in turn, depends on the horizontal hydraulic conductivity values. Therefore one should not expect the vertical velocity of the interface to be accurately represented in this formulation. We will see in Section 8.3 another way to describe mathematically salt-water intrusion that will circumvent this difficulty.

4.7 ONE-DIMENSIONAL FORMULATION

We have seen how one would formally reduce the dimensionality of a problem from three to two dimensions. It was pointed out that in the reduction of dimensionality, one also lost information, especially regarding the dimension over which the integration was being performed. In this brief section we will take the problem one step further and reduce the dimensionality from two to one.

Let us use as our "type equation"

$$\nabla \cdot \mathbf{q} + S_s \frac{\partial h}{\partial t} + Q = 0, \tag{4.58}$$

which describes groundwater flow in three areal space dimensions.

Equation (4.58) has already been integrated vertically over the z dimension in Section 4.5. Consider now integrating it instead in the x and y directions, so that we end up with an equation that has only z and t as independent variables. Let us denote the (x, y) planes that are within the overall three-dimensional domain by $\Omega_{x,y}(z)$. The new notation is found in Figure 4.11. Then we have

$$\int_{\Omega_{xy}(z)} \left(\nabla \cdot \mathbf{q}(x, y, z, t) + S_s \frac{\partial h(x, y, z, t)}{\partial t} + Q \right) dx\, dy = 0. \tag{4.59}$$

Expansion of the divergence provides three terms so we have, including the time term, a total of four terms to consider. We first consider the x and y derivatives,

$$\int_{\Omega_{xy}(z)} \nabla_{xy} \cdot \mathbf{q}(x, y, z, t)\, dx\, dy.$$

Using Gauss's theorem, we can rewrite this as

$$\int_{\Omega_{xy}(z)} \nabla_{xy} \cdot \mathbf{q}\, dx\, dy = \oint \mathbf{q} \cdot \mathbf{n}_{xy}\, ds. \tag{4.60}$$

The q_z component is more challenging. To address this term, we employ the two-dimensional form of Leibntiz's rule for differentiation of an integral. We obtain

$$\int_{\Omega_{xy}(z)} \frac{\partial q_z}{\partial z} dy\, dx = \frac{\partial}{\partial z} \int_{\Omega_{xy}(z)} q_z dx\, dy - \oint_{\partial \Omega_{xy}(z)} q_z \mathbf{k} \cdot \mathbf{n}_{xy}\, ds, \tag{4.61}$$

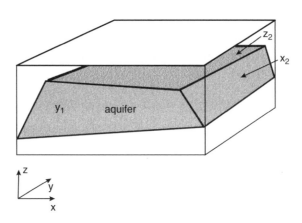

FIGURE 4.11. Definition sketch for integrating the groundwater flow equation over two space dimensions.

where as usual **k** is the unit vector in the z coordinate direction. The surface integral is taken over the sides of the deformed pyramid shown in Figure 4.11 for each z.

The time derivative employs Leibnitz's rule as well, but in a slightly different context. For the case of a fixed aquifer, the form of the rule is

$$\overline{S_s} \int_{\Omega_{xy}(z)} \frac{\partial h}{\partial t} dy \, dx = \overline{S_s} \frac{\partial}{\partial t} \int_{\Omega_{xy}(z)} h \, dx \, dy, \tag{4.62}$$

where, as earlier, we are assuming an average value of S_s so as to permit its removal from the spatial integration.

We now define average values of the flux q_z and the head h as

$$\overline{q_z}(z, t) = \frac{1}{A_{xy}(z)} \int_{\Omega_{xy}(z)} q_z(x, y, z, t) \, dx \, dy, \tag{4.63}$$

$$\overline{h}(z, t) = \frac{1}{A_{xy}(z)} \int_{\Omega_{xy}(z)} h(x, y, z, t) \, dx \, dy, \tag{4.64}$$

where $A_{xy}(z)$ is the area in the xy plane. We now reassemble the terms we considered separately in Eqs. (4.60)–(4.62) and introduce the averages defined in Eqs. (4.63) and (4.64):

$$\frac{\partial}{\partial x} \left(A_{xy} \overline{q_z}(z, t) \right) + \overline{S_s} \frac{\partial}{\partial t} \left(A_{xy} \overline{h}(z, t) \right) + \int_{\partial\Omega} \mathbf{q} \cdot \mathbf{n}_{xy} \, ds - \int_{\partial\Omega} q_z \mathbf{k} \cdot \mathbf{n} \, ds = 0. \tag{4.65}$$

Equation (4.65) is the general form of the one-space-dimensional groundwater flow equation written in terms of fluid fluxes. The first term describes the behavior of a horizontally averaged vertical flux, which we have denoted as $\overline{q_z}(z, t)$. Note that it is now a function only of the z and t coordinates, the x and y dependence having been removed via averaging (see Eqs. (4.63) and (4.64)). Similarly, the average head value $\overline{h}(z, t)$ is also a function only of the z and t coordinates. All information regarding the behavior of these variables on the interior of the pyramid in the horizontal plane has been lost through the averaging process.

Terms three and four in Eq. (4.65) account for the fluid moving into and out of the aquifer from the sides. The sides can be located anywhere; that is, you can make the horizontal extent of your model any size you wish, but whatever area on the xy plane you select, it is over that horizontal area that the average is being taken, and the associated boundary fluxes are with respect to the boundary of that area.

We now consider the averaging of Darcy's law. Performing integration over the x and y directions, we obtain

$$\int_{\Omega_{xy}(z)} q_z \, dx \, dy = -\overline{K_{zz}} \cdot \int_{\Omega_{xy}(z)} \frac{\partial h}{\partial z} \, dx \, dy$$

$$= -\overline{K_{zz}} \cdot \left(\frac{\partial}{\partial z} \int_{\Omega_{xy}(z)} h \, dx \, dy - \int_{\partial\Omega_{xy}(z)} h \mathbf{k} \cdot \mathbf{n} \, ds \right), \tag{4.66}$$

which using the averaging notation gives

$$\overline{q_z}(z, t) \times A_{xy}(z, t) = -\overline{K_{zz}} \cdot \left(\frac{\partial}{\partial z} \left(A_{xy}(z) \overline{h}(z, t) \right) - \int_{\partial\Omega_{xy}(z)} h \mathbf{k} \cdot \mathbf{n} \, ds \right),$$

which, when substituted into Eq. (4.65) yields

$$
-\frac{\partial}{\partial z}\overline{K_{zz}}\left(\frac{\partial}{\partial z}\left(A_{xy}\left(z\right)\overline{h}\left(z,t\right)\right) - \int_{\partial\Omega_{xy}(z)} h\mathbf{k}\cdot\mathbf{n}\,ds\right)
$$

$$
+\overline{S_s}\frac{\partial}{\partial t}\left(A_{xy}\overline{h}\left(z,t\right)\right) - \int_{\partial\Omega_{xy}(z)} q_z\mathbf{k}\cdot\mathbf{n}\,ds + \int_{\partial\Omega_{xy}(z)} \mathbf{q}\cdot\mathbf{n}_{xy}\,ds = 0. \qquad (4.67)
$$

If we assume the aquifer has vertical sides and that there is no external flow entering the aquifer along the sides, Eq. (4.67) reduces to the generally found form of the one-dimensional flow equation,

$$
-\frac{\partial}{\partial z}\overline{K_{zz}}\left(\frac{\partial}{\partial z}\left(\overline{h}\left(z,t\right)\right)\right) + \overline{S_s}\frac{\partial}{\partial t}\left(\overline{h}\left(z,t\right)\right) = 0. \qquad (4.68)
$$

More generally, we would have a nonzero right-hand side, which would correspond to source or sink terms representing flows through the lateral boundaries (as a function of z and t) and possibly other sources or sinks. To arrive at Eq. (4.68), a number of simplifying assumptions were made. Included in this list are:

1. The validity of the use of average parameter values in the directions over which averaging has been performed
2. An aquifer with vertical sides
3. No horizontal flow through the vertical aquifer sides
4. The validity of area-averaged values for the state variables, head, and fluid flux

While the above assumptions are not required when developing the one-dimensional flow equation in its most general form, those terms eliminated by assumptions numbered 1 through 3 would have to be evaluated if the general form of the one-dimensional equation were to be solved.

The strategy used above to generate two and finally one-space-dimensional forms of the flow equation from their fully three-dimensional counterparts can be used to further reduce the equation to zero-space-dimensions, and time. Since there are no new concepts involved and a zero-space-dimensional formulation is seldom of interest, we do not develop the time-only dependent flow equation here.

4.8 CYLINDRICAL COORDINATES

When we have problems involving a well, the flow in the vicinity of the well tends to be radial, either toward (pumping) or away from (injection) the well. Therefore we are motivated to write the governing equation in cylindrical coordinates. We begin by rewriting the flux balance equation, that is, Eq. (4.58):

$$
\nabla\cdot\mathbf{q} + S_s\frac{\partial h}{\partial t} + Q = 0. \qquad (4.69)
$$

The divergence operator presented in this expression has heretofore been considered as defined in the Cartesian coordinate system, that is, (x, y, z, t). However, it can also be

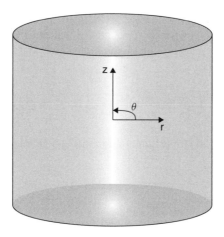

FIGURE 4.12. Definition sketch for cylindrical coordinates.

written in the cylindrical coordinate system defined on (r, θ, z, t), which is illustrated in Figure 4.12. The vertical dimension is denoted as z, the radial as r, the angle as θ, and the time as t. In this coordinate system, the gradient operator is defined as

$$\nabla\,(\cdot) \equiv \frac{\partial\,(\cdot)}{\partial r}\mathbf{r} + \frac{1}{r}\frac{\partial\,(\cdot)}{\partial \theta}\boldsymbol{\theta} + \frac{\partial\,(\cdot)}{\partial z}\mathbf{z} \tag{4.70}$$

and the vector $\mathbf{q}$ as

$$\mathbf{q} \equiv q_r\mathbf{r} + q_\theta\boldsymbol{\theta} + q_z\mathbf{z}, \tag{4.71}$$

where $\mathbf{r}$, $\boldsymbol{\theta}$, and $\mathbf{z}$ are the unit vectors in the r, θ, and z coordinate directions, respectively. To arrive at a groundwater flux equation in the cylindrical coordinate system, we begin with the substitution of Eq. (4.71) into Eq. (4.70) and obtain the expansion (the development of which is a little subtle[4])

$$\nabla \cdot \mathbf{q} = \left(\frac{\partial q_r}{\partial r} + \frac{q_r}{r}\right) + \left(\frac{1}{r}\frac{\partial q_\theta}{\partial \theta}\right) + \frac{\partial q_z}{\partial z}. \tag{4.72}$$

Substitution of this definition of the divergence into Eq. (4.69) yields

$$\left(\frac{\partial q_r}{\partial r} + \frac{q_r}{r}\right) + \left(\frac{1}{r}\frac{\partial q_\theta}{\partial \theta}\right) + \frac{\partial q_z}{\partial z} + S_s\frac{\partial h}{\partial t} + Q = 0. \tag{4.73}$$

The form of the groundwater flow equation shown in Eq. (4.73) is normally encountered only in problems involving radial flow to wells. In such circumstances, it is generally assumed that the flow is radially uniform; that is, the state variables $\mathbf{q}$ and h do not change with the angle θ. When one makes this assumption, the second term in Eq. (4.73) vanishes and we are left with

$$\left(\frac{\partial q_r}{\partial r} + \frac{q_r}{r}\right) + \frac{\partial q_z}{\partial z} + S_s\frac{\partial h}{\partial t} + Q = 0.$$

---

[4]Use the fact that $d\mathbf{r}/d\theta = \boldsymbol{\theta}$.

Early efforts to solve this equation for problems of practical importance reduced the dimensionality of the problem further through vertical integration, a process similar to that considered in Section 4.5.2. Using this approach, we first integrate over the vertical to obtain

$$\int_{a(r,t)}^{b(r,t)} \left(\left(\frac{\partial q_r}{\partial r} + \frac{q_r}{r} \right) + \frac{\partial q_z}{\partial z} + S_s \frac{\partial h}{\partial t} + Q \right) dz = 0.$$

Application of Leibnitz's rule for differentiation through an integral and direct integration of the term $\partial q_z / \partial z$ gives

$$\frac{\partial}{\partial r} \int_{a(r,t)}^{b(r,t)} q_r \, dz + \frac{1}{r} \int_{a(r,t)}^{b(r,t)} q_r \, dz - q_r|_b \frac{\partial b}{\partial r} + q_r|_a \frac{\partial a}{\partial r} + q_z|_b - q_z|_a$$

$$+ \overline{S_s} \frac{\partial}{\partial t} \int_{a(r,t)}^{b(r,t)} h \, dz - h|_b \frac{\partial b}{\partial t} + h|_a \frac{\partial a}{\partial t} + \int_{a(r,t)}^{b(r,t)} Q \, dz = 0. \tag{4.74}$$

We now proceed as we did in Section 4.5.2 and define the average quantities

$$\overline{q_r}(r,t) \equiv \frac{1}{l(r,t)} \int_{a(r,t)}^{b(r,t)} q_r(z,r,t) \, dz$$

and

$$\overline{h}(r,t) \equiv \frac{1}{l(r,t)} \int_{a(r,t)}^{b(r,t)} h(z,r,t) \, dz \tag{4.75}$$

and substitute them into Eq. (4.74) to give

$$\frac{\partial}{\partial r} (l(r,t) \overline{q_r}(r,t)) + \frac{1}{r} (l(r,t) \overline{q_r}(r,t)) - q_r|_b \frac{\partial b}{\partial r} + q_r|_a \frac{\partial a}{\partial r} + q_z|_b - q_z|_a$$

$$+ \overline{S_s} \frac{\partial}{\partial t} \left(l(r,t) \overline{h}(r,t) \right) - h|_b \frac{\partial b}{\partial t} + h|_a \frac{\partial a}{\partial t} + Q^* = 0,$$

where $Q^* \equiv \int_{a(r,t)}^{b(r,t)} Q \, dz$. We can simplify this expression by assuming essentially horizontal flow, such that $\overline{h}(r,t) = h|_b = h|_a$, to obtain

$$\frac{\partial}{\partial r} (l(r,t) \overline{q_r}(r,t)) + \frac{1}{r} (l(r,t) \overline{q_r}(r,t)) + \overline{S_s} l(r,t) \frac{\partial}{\partial t} \left(\overline{h}(r,t) \right) + q_i(r,t) + Q^* = 0, \tag{4.76}$$

where

$$q_i(r,t) = -q_r|_b \frac{\partial b}{\partial r} + q_r|_a \frac{\partial a}{\partial r} + q_z|_b - q_z|_a \tag{4.77}$$

is the net flux through the top and bottom surfaces of the aquifer. Notice that the possibility of a "cone-shaped" surface to the cylinder is taken into account through the second and third terms on the right-hand side of Eq. (4.77). The second term describes the horizontal flux through the sloping surface $\partial z_2 / \partial r$ and the fourth term the flux through the surface $\partial z_1 / \partial r$.

Let us now vertically average Darcy's law. Following a similar procedure to that presented above for the flux equation, and employing the average head specified earlier as Eq. (4.75), we obtain

$$\int_{a(r,t)}^{b(r,t)} q_r \, dz = l\,(r,t)\,\overline{q_r}\,(r,t) = -\overline{K_{rr}}\frac{\partial}{\partial r}\left(l\,(r,t)\,\overline{h}\,(r,t)\right) - \frac{\overline{K_{rr}}}{r}\left(l\,(r,t)\,\overline{h}\,(r,t)\right).$$

(4.78)

If we make the assumption that the top and bottom of the aquifer are horizontal, then Eq. (4.78) in combination with Eq. (4.76) becomes

$$\frac{\partial}{\partial r}\left(-\overline{K_{rr}}l\,(r,t)\frac{\partial}{\partial r}\left(\overline{h}\,(r,t)\right)\right) + \frac{1}{r}\left(-\overline{K_{rr}}l\,(r,t)\frac{\partial}{\partial r}\left(\overline{h}\,(r,t)\right)\right)$$

$$+ \overline{S_s}l\,(r,t)\frac{\partial}{\partial t}\left(\overline{h}\,(r,t)\right) + q_i\,(r,t) + Q^* = 0,$$

(4.79)

where $q_i\,(r,t)$ now contains only the vertical flux. Equation (4.79) can be written in a more traditional form as

$$T\left(\frac{\partial^2}{\partial r^2}\overline{h}\,(r,t) + \frac{1}{r}\frac{\partial}{\partial r}\overline{h}\,(r,t)\right) - S\frac{\partial}{\partial t}\left(\overline{h}\,(r,t)\right) + q_i\,(r,t) + Q^* = 0,$$

(4.80)

where the transmissivity $T \equiv \overline{K_{rr}}l\,(r)$ isassumed to be a scalar constant and $S \equiv \overline{S_s}l\,(r)$ is once again the storage coefficient. This is the equation that is typically used in classical well hydraulics.

4.9 SUMMARY

The goal of this chapter is to provide insight into the concepts that furnish the foundation supporting the groundwater flow equations generally used in practice. The point of departure was development of the equation for fluid flow defined at the microscopic level of observation. The resulting point equation was then extended to the porous medium, which consists of both fluid and solid phases—the grains constituting the solid phase and air, water, and nonaqueous phase fluids constituting the fluid phase. The general porous medium equations were then combined with constitutive (experimental) equations to provide an expression describing groundwater flow. To address specific problems, it was found necessary to augment the groundwater flow equation with boundary and initial conditions. Three traditional boundary conditions were discussed that accommodate the majority of practical problems encountered in the field. However, special cases, such as description of the behavior of the water table, required additional discussion.

Having developed the general equations descriptive of groundwater flow, we turned our attention to simplification of these equations, which may be applicable in selected situations. The concept of reducing dimensionality through formal integration over selected dimensions was described. The practical impact of the assumptions inherent in using equations of reduced dimensionality was thereby revealed.

4.10 PROBLEMS

4.1. Derive Eq. (4.5) using a "box balance" method. That is, write a mass balance with respect to a simple box of length δx by δy by δz, which has its centroid located at point $\mathbf{x} = (x, y, z)$. For this box, relate the change of mass within the volume, over a time increment δt, to the net flux entering the box through its six sides. Then take the limit as the increments in space and time go to zero.

4.2. Consider a two-dimensional porous medium composed of solids that are squares of length ℓ. Let the centers of each square be located on a regular grid, with separation distance between centers being 2ℓ. Let the porosity of this material be defined using a square averaging area of length $\mathcal{L}$ on each side. Choose one or two points $\mathbf{x}$ as the center of an averaging volume, and define porosity at that point $\mathbf{x}$ as a function of averaging length $\mathcal{L}$. Plot this function and show the existence of a representative elementary volume (REV).

4.3. Derive Eq. (4.10) from the equation immediately preceding it, using the definition of the mass average operator.

4.4. Consider the so-called Heaviside step function, which is defined as follows:

$$H(x - x^*) = \begin{cases} 0, & x < x^* \\ 1, & x > x^*. \end{cases}$$

This function has the property that it is zero to the left of the point x^*, and one to the right of x^*. Given the properties of the Dirac delta function, as described in Section 4.3.5, show that the Dirac delta function may be viewed as the derivative of the Heaviside step function. That is,

$$\delta(x - x^*) = \frac{d}{dx} H(x - x^*) \quad \text{and} \quad H(x - x^*) = \int_{-\infty}^{x} \delta(x' - x^*) \, dx'.$$

4.5. Explain why the vertically integrated groundwater flow equation, Eq. (4.49), has the terms q_T and q_B, while the original three-dimensional groundwater flow equation, Eq. (4.34), does not.

4.6. A major assumption is made when deriving the vertically integrated groundwater flow equation, Eq. (4.48). That assumption is stated in the text as $h|_a \simeq h|_b \simeq \overline{h}$. Write the analogous vertically integrated equation if this assumption is *not* made. Comment on the additional terms that appear in the equation.

4.7. Derive Eqs. (4.56) and (4.57) from Eqs. (4.54) and (4.55), following the procedures presented in Section 4.6, but filling in all of the details of the derivation.

4.8. Consider a porous media system that can be treated as a one-dimensional, homogeneous formation that is at steady state. Let the domain of the system be defined by $0 < x < L$. This formation has a sink term that can be represented as a Dirac delta function, applied at the midpoint of the domain. No other sources or sink terms exist for this one-dimensional domain. (a) Show that the governing equation is given by

$$K \frac{d^2 h}{dx^2} = Q \delta \left(x - \frac{L}{2} \right), \quad 0 < x < L,$$

where h is interpreted as the hydraulic head averaged over the y and z directions, K is the hydraulic conductivity, and Q represents the strength of the sink term. (b) Give the dimensions of Q. (c) Assume the boundary conditions are given by $h(0) = h(L) = 0$. Show that the solution to this equation is given by

$$h(x) = \begin{cases} \dfrac{-Q}{2K}x, & 0 < x < \dfrac{L}{2} \\ \dfrac{Q}{2K}(x - L), & \dfrac{L}{2} < x < L. \end{cases}$$

Note that, in this solution, the head $h(x)$ is continuous but the derivative dh/dx is not continuous, nor is the Darcy flux q. The discontinuity in dh/dx occurs at the point of action of the Dirac delta function ($x = L/2$), and the size of the discontinuity is given by the factor Q/K.

4.9. Derive Eq. (4.73) using the "box balance" approach described in Problem 4.1, now applied in the cylindrical coordinate system (r, θ, z).

BIBLIOGRAPHY

[1] R. B. Bird, W. E. Stewart, and E. N. Lightfoot, *Transport Phenomena*, John Wiley & Sons, Hoboken, NJ, 1960.

[2] K. Terzaghi and R. B. Peck, *Soil Mechanics in Engineering Practice*, John Wiley & Sons, Hoboken, NJ, 1964.

[3] A. G. Journel, Nonparametric estimation of spatial distributions, *Math. Geol.* **15**(3): 445, 1983.

CHAPTER 5

ANALYTICAL SOLUTIONS FOR FLOW PROBLEMS

We write governing equations to describe groundwater flow systems because solutions to those equations tell us how groundwater systems behave. That is, if we solve the groundwater flow equation, for example, Eq. (4.17), for the hydraulic head $h(x, y, z, t)$, we can predict the behavior of the system at any point in space (x, y, z), and at any time t. Differentiation of h and subsequent substitution into Darcy's law allows us to calculate the flow rate or, in combination with the porosity, the velocity of the flow field. This can tell us about how much water we can extract for a water supply, or about how contaminants will move within this flow system. When solving a governing differential equation, if the equation is sufficiently constrained (or sufficiently "simple"), we may be able to derive a closed-form analytical solution. In such a case we simply write the solution and use that solution for practical calculations. If the system is too complex, such that we are not able to derive an analytical solution, then we usually resort to methods that provide approximate solutions. In the groundwater area, these approximate solutions are usually based on finite-difference or finite-element methods. In this chapter, we examine a set of practical analytical solutions for groundwater flow problems.

Recall that when writing governing equations for groundwater flow systems, the result is usually a partial differential equation that has as independent variables one, two, or three spatial coordinates and time. For any such equation, the domain over which the equation applies must be defined, and boundary and initial conditions must be specified. Because the groundwater flow equation involves second derivatives in space, the requirements for boundary conditions are that the location of the boundary needs to be specified, and one boundary equation must be given at each point along the boundary (see Section 4.3.3). For transient problems, an initial condition must be specified for all points within the domain. Of course, steady-state problems do not involve changes in time and therefore do not require an initial condition. As a general rule, analytical solutions can only be derived for systems with boundaries that align with the coordinate axes, and for equations

Subsurface Hydrology   By George F. Pinder and Michael A. Celia
Copyright © 2006 John Wiley & Sons, Inc.

that have constant coefficients. While there are exceptions, these two conditions often serve as useful guidelines when thinking about analytical solution options.

In this chapter, we consider a sequence of analytical solutions for problems involving groundwater flow. We begin with relatively simple systems in one spatial dimension and use these solutions to demonstrate certain principles of solution behavior and to introduce certain approximations that allow for derivation of analytical solutions. In later chapters we will consider specific analytical solutions that allow estimation of hydraulic parameters (Chapter 6) and we will also briefly consider analytical solutions associated with contaminant transport (Section 8.3.2).

5.1 ONE-DIMENSIONAL FLOW PROBLEMS

In this section we develop several analytical solutions for problems defined in one spatial dimension. We use these to highlight the importance of choice of domain, including consideration of finite versus semi-infinite domains, as well as choice of boundary conditions. We also introduce the so-called Dupuit assumption for flow in water-table aquifers, introduce initial ideas about recharge to the water table and leakage through aquitards, and begin discussion of radial flow to pumping wells. These ideas are continued in the chapter on well hydraulics (Chapter 6), where additional analytical solutions are presented in the context of parameter estimation and well testing.

5.1.1 Darcy Column Experiments

To begin, consider the simplest flow equation: steady-state, one-dimensional flow in a homogeneous porous medium of finite length, with specified head boundary conditions on each end of the domain. The governing differential equation then derives from Eq. (4.68), which for a homogeneous medium with no lateral fluxes reduces to

$$-K \frac{\partial^2 h}{\partial x^2} = 0, \quad 0 < x < l,$$
$$h(0, t) = h_L(t),$$
$$h(l, t) = h_R(t).$$
(5.1)

In this equation, h_L and h_R are head values at the left and right boundaries, respectively. If these values do not change in time, then the solution h is a function only of x, and the partial derivative in Eq. (5.1) can be written as a total derivative. The equation is then an ordinary differential equation. Otherwise, it remains formally a partial differential equation because h is a function of both x and t. In either case, the solution is simply a straight line in space, connecting the two boundary values,

$$h(x, t) = h_L(t) + \frac{(h_R(t) - h_L(t))}{l} x.$$
(5.2)

Notice that this equation, with time-invariant boundary conditions, corresponds to the experiments of Darcy, where one-dimensional flow in a column of finite length was driven by fixed head values at the top and bottom boundaries. In fact, knowledge of this analytical solution allows us to perform experiments for the purpose of *parameter estimation.* In this case, performance of a column test allows the hydraulic conductivity of the material to be determined. Given the solution for head, we may differentiate it and insert it into Darcy's law, where, upon measurement of the flow rate, the hydraulic conductivity may be determined from the following (rearranged) form of Darcy's law:

$$K = \frac{Q}{A\frac{\partial h}{\partial x}} = \frac{Ql}{A\,(h_R - h_L)}. \tag{5.3}$$

It turns out that the most frequent use of analytical solutions for groundwater flow problems is for parameter estimation. The entire field of well testing is based on different analytical solutions for flow to wells. As noted earlier, we devote all of Chapter 6 to that topic.

While the steady-state solution associated with the Darcy experiments is useful, we might consider how that steady state is reached. In particular, consider a soil column in which we initially have no flow (therefore $\nabla h = 0$, meaning h is constant in space). Then at some time we impose boundary conditions that induce a flow through the column. If we wish to describe the transient response of this system to the imposed boundary conditions, we must solve the transient version of the flow equation. Assuming spatial homogeneity of parameters, a one-dimensional column with no-flow boundaries on the lateral sides, and imposed head boundary conditions on the two ends of the column, the governing system of equations takes the following form:

$$S_s \frac{\partial h}{\partial t} - K \frac{\partial^2 h}{\partial x^2} = 0, \quad 0 < x < l, t > 0,$$
$$h(0, t) = h_L(t),$$
$$h(l, t) = h_R(t), \tag{5.4}$$
$$h(x, 0) = h_{\text{init}}(x).$$

Consider a case of transient flow driven by an instantaneous change of head at one of the boundaries, which serves to disturb the initial steady state of the column. If the initial condition is given by $h = 0$, we can change one of the boundary conditions at time $t = 0$ to induce flow. In this instance, the boundary and initial conditions are: $h(0, t) = h_L = \text{constant}(h_L > 0), h(L, t) = h_R = 0$, and $h(x, 0) = h_{\text{init}} = 0$. This equation is sufficiently simple to solve, for example, by the method of separation of variables. The solution is given by

$$h(x, t) = \frac{2}{\pi} \sum_{n=1}^{+\infty} \frac{h_L}{n} \sin\left(\frac{n\pi x}{l}\right) \left[1 - \exp\left(-\frac{kn^2\pi^2 t}{S_s l^2}\right)\right]. \tag{5.5}$$

Figure 5.1 shows hydraulic head as a function of spatial location for three different times, using the parameters $h_L = 1.0$, $h_R = h_{\text{init}} = 0$, $l = 1$, and $K/S_s = 0.1$. The first time corresponds to an "early time," where the pressure front has begun to move into the domain but remains far from the right boundary. The second time, which we will call an "intermediate time," shows the influence of both boundary conditions on the solution, but the solution is still changing in time. And the third time we call "late time," where we have essentially reached steady state. For this problem, the steady-state solution is just a straight line in space connecting the two boundary values.

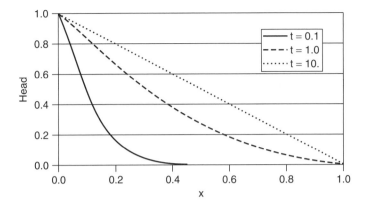

FIGURE 5.1. Transient solutions evaluated at different times.

For the late-time solution, we have a steady-state situation that reduces the governing equation to a simple ordinary differential equation (because at steady state $\partial h/\partial t = 0$). For the early-time solution, we may also use a simplification that involves the observation that the right boundary condition does not have any significant influence on the solution. For such cases, we often treat the domain as though the right boundary were infinitely far away, and the domain is therefore "semi-infinite," meaning we have one boundary identified, but the second one is very far away and does not influence the solution. In this case the analytical solution usually becomes simpler. In the case of a semi-infinite approximation to the domain, the governing equation takes the following form:

$$S_s \frac{\partial h}{\partial t} - K \frac{\partial^2 h}{\partial x^2} = 0, \quad 0 < x < +\infty, \ t > 0,$$

$$h(0, t) = h_L,$$

$$\lim_{x \to +\infty} h(x, t) = h_{\text{init}}, \tag{5.6}$$

$$\lim_{x \to +\infty} \frac{\partial h}{\partial x} = 0,$$

$$h(x, 0) = h_{\text{init}}.$$

Again we can derive an analytical solution for this case. For the specific example we are considering, we find that the solution for head propagation into a semi-infinite domain is given by

$$h(x, t) - h_{\text{init}} = (h_L - h_{\text{init}}) \ \text{erfc}\left(\frac{x}{\sqrt{4(K/S_s)t}}\right), \tag{5.7}$$

where erfc denotes the *complementary error function*, defined by

$$\text{erfc}\,(x) = \frac{2}{\sqrt{\pi}} \int_x^{+\infty} e^{-z^2} \, dz. \tag{5.8}$$

In Figures 5.2 and 5.3, the complementary error function solution associated with the semi-infinite domain approximation is plotted and compared to the full solution, Eq. (5.5).

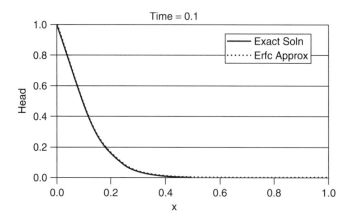

FIGURE 5.2. Comparison of series solution ("exact") to the error function solution, for an early time.

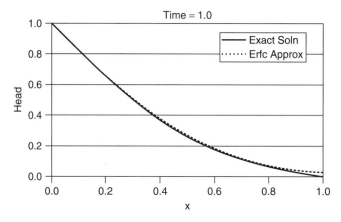

FIGURE 5.3. Comparison of series solution ("exact") to the error function solution, for an intermediate time.

Notice that at early times (Figure 5.2), the solutions are virtually identical. For the intermediate time of $t = 1.0$ (Figure 5.3), the solutions begin to differ, due to the influence of the downstream (right side) boundary condition. The solutions become progressively more different as time increases. A practical strategy for when the semi-infinite solution matches the full solution may be taken as $t \leq 0.1 \left(l^2/(K/S_s) \right)$, which may be taken as the limit for "early time."

This example also points out the importance of dimensionless groups. The time limit for "early time" only makes sense if it is written in terms of the parameters and associated units used to define the problem. For the groundwater flow equation, we see that the natural dimensionless group that arises is given by $l^2 S_s/Kt$, where l represents length and t time. This grouping appears in the analytical solutions above, and a modified version of this grouping will play a central role in the parameter estimation methods described in Chapter 6.

5.1.2 One-Dimensional Regional Flow

As a second example, consider flow in a water-table aquifer, subject to recharge from above. We will consider a two-dimensional vertical cross section and apply vertical averaging (Section 4.5.2) to derive a one-dimensional governing equation. A schematic of the system is shown in Figure 5.4, which shows a vertical cross section with independent variables x and z. Boundaries exist at $x = 0$ and $x = l$, and for simplicity we assume the aquifer is underlain by an impermeable formation (at $z = 0$). The top boundary corresponds to the water table, whose location needs to be determined as part of the solution. The boundary condition appropriate for the water table has been presented in Section 4.4. Because of the complexity of that boundary condition, we will seek simplifications that allow an analytical solution to be derived.

Assume the infiltration rate is known and is taken to be a constant, in both space and time, that corresponds to the long-term average infiltration rate (e.g., based on yearly average precipitation). While this system exhibits multidimensional flows, we often use vertical averaging to replace the two-dimensional governing equation with a one-dimensional equation that accommodates the water table by appropriate introduction of the specific yield, modification of the transmissivity to include the saturated thickness, and inclusion of recharge as a source term in the governing equation (see Eq. (4.68)). Under the assumptions of essentially horizontal flow and steady state, the governing equation takes the form

$$-\frac{d}{dx}\left(\overline{T}\frac{d\overline{h}}{dx}\right) = -\frac{d}{dx}\left(\overline{Kh}\frac{d\overline{h}}{dx}\right) = N, \quad 0 < x < l,$$

$$\overline{h}(0, t) = h_L, \tag{5.9}$$

$$\overline{h}(l, t) = h_R.$$

In this equation, the overbar once again indicates a vertically averaged quantity, the recharge rate is given by N, and the left and right boundary conditions are taken to be fixed head values that are constant in time. Notice that the transmissivity is a function

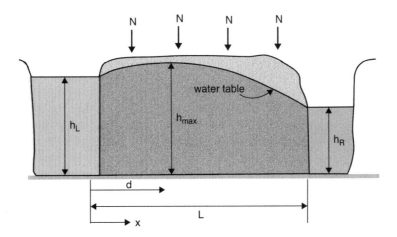

FIGURE 5.4. Diagrammatic representation of idealized aquifer system (adapted from [1]).

of the hydraulic head, because the aquifer thickness depends on the location of the water table, and that location, in turn, depends on the (vertically integrated) hydraulic head. This dependence of the coefficient (transmissivity) on the dependent variable (head) makes the equation nonlinear. While most nonlinear equations cannot be solved analytically, this one can be solved fairly easily. To do so, observe that the left-hand side of Eq. (5.9) may be rewritten (dropping the overbars for simplicity and assuming K is constant) as

$$-K\frac{d}{dx}\left(h\frac{dh}{dx}\right) = -\frac{K}{2}\frac{d^2h^2}{dx^2}.$$

Therefore the governing equation tells us that the square of hydraulic head has a constant second derivative in space, proportional to the infiltration rate N. Therefore the solution for $h^2(x)$ is a quadratic polynomial in x. The form of the solution is easily determined and is given by

$$h^2(x) = \frac{N}{K}x\,(l-x) + \left(h_R^2 - h_L^2\right)\frac{x}{l} + h_L^2. \tag{5.10}$$

As an example, consider a long island (similar to the actual case on Long Island, New York) that is bounded on the left and right by conditions of $h_L = h_R = B$, where B denotes distance between the bottom of the aquifer and sea level. The solution for $h(x)$ then takes the form

$$h^2(x) = \frac{N}{K}x\,(l-x) + B^2$$

or

$$h(x) = \sqrt{\frac{N}{K}x\,(l-x) + B^2}. \tag{5.11}$$

We see that, without recharge, there is no driving force for flow and the solution is simply $h(x) = B$. With recharge, the solution may be rewritten

$$h^2(x) - B^2 = [h(x) - B][h(x) + B] = \frac{N}{K}x\,(l-x). \tag{5.12}$$

From this solution we make two observations. The first is that the solution is symmetric about the midpoint of the domain ($x = l/2$), which is where the maximum height of the water table occurs. This is because the two boundary conditions have the same value. So water infiltrates, joins the flow system, and flows outward (horizontally) from the middle of the domain to the left and right boundaries. The second observation is that, when the increase in head h above B is small relative to thickness B, then the nonlinearity in the problem is not significant and the transmissivity may be reasonably approximated by $T \approx KB$. This may be argued by representing the head as $h(x) = B + \epsilon(x)B$. So when $\epsilon \ll 1$, head h does not differ much from thickness B. In that case, terms involving ϵ^2 can be ignored because they are very small. Substitution of this representation of h into the solution yields

$$[h(x) - B][h(x) + B] = [\epsilon B][2B + \epsilon B] = 2\epsilon B^2 + \epsilon^2 B^2 \approx 2\epsilon B^2$$
$$= 2B\,(\epsilon B) = 2B\,[h(x) - B] = \frac{N}{K}x\,(l-x), \tag{5.13}$$

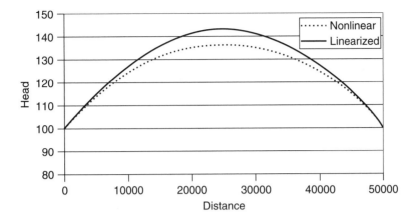

FIGURE 5.5. Water table location based on linear and nonlinear solutions.

which may be solved for $h(x)$ as

$$h(x) = B + \frac{N}{2KB} x \, (l - x).$$ (5.14)

Figure 5.5 shows both the nonlinear solution (Eq. (5.12)) and the linearized solution (Eq. (5.14)) for the case of an aquifer that is 100 ft thick, extends over an island of length 10,000 ft, has permeability of 300 ft/day, and has a recharge rate of 1.5 ft/yr. The solution that includes the transmissivity nonlinearity shows a somewhat deeper water table as compared to the linearized version that uses a constant transmissivity (the curious reader should offer an explanation of why this happens). However, as must be the case for this problem, the fluxes are the same through the aquifer; for example, the flux out the boundaries for each case is given by $K(dh/dx) = \pm Nl/2B$. Again the interested reader may easily verify this via differentiation of the solutions for hydraulic head. Finally, we note that the dimensionless grouping that is characteristic of this equation is given by Nl^2/KB^2. When this ratio is small, the linear approximation will be close to the nonlinear result.

The analytical treatment presented here is based on the application of vertical averaging and its associated assumptions. The main assumption that allows vertical integration to produce a simplified governing equation is that the flow is essentially horizontal. This means that equipotential lines are assumed to be essentially vertical, and that the hydraulic head at any point (x, z) in the domain may be represented well by the vertically averaged head, $h(x, z) \approx \overline{h}(x)$. When the bottom of the aquifer is assigned the elevation $z = 0$, the location of the water table will then be given by the value of the hydraulic head at any x location, because by definition the pressure at the water table is atmospheric (taken as the datum $p = 0$): that is,

$$z_{\text{WT}} = \frac{p_{\text{WT}}}{\rho g} + h_{\text{WT}} = 0 + h_{\text{WT}} = \overline{h}(x),$$

where z_{WT} denotes the vertical location of the water table. Because all flow is assumed to be horizontal, the flow rate is proportional to the horizontal gradient of $\overline{h}$. Since the water-table elevation is equal to $\bar{h}$, the flow rate is then given equivalently by the gradient of the water-table elevation.

The use of vertical averaging in the case of water-table aquifers, given the constraints discussed in the previous paragraph, is often referred to as the *Dupuit approximation*, after the pioneering work of Dupuit (1863) [2] in approximating flows in free-surface systems. When the Dupuit approximation is stated, it is usually associated with the following two assumptions: (1) flow is horizontal, and (2) the head gradient is constant with depth and given by the slope of the free surface, which in this case is the water table. The reader will note that the first of these assumptions is invoked during the derivation of vertically averaged governing equations, because without that approximation the averaged equations contain many additional terms that are not easily dealt with (see Section 4.7). Once the assumption of horizontal flow is made, then the second Dupuit assumption follows as a consequence, because vertical equipotentials imply that head does not change as a function of depth, and therefore horizontal gradients cannot change with depth.

If the Dupuit assumptions (or the application of vertical averaging—they are essentially equivalent statements) are not valid, then the problem must be solved without use of vertical averaging. That more complex case involves a multidimensional partial differential equation. The equation itself is linear (simply Laplace's equation for the case of homogeneous and isotropic steady-state flow), but the location of the free-surface boundary is unknown and needs to be determined. Moreover, the boundary conditions that apply at the water table involve nonlinearities. In general, this equation cannot be solved analytically.

Overall, the flow system in the vicinity of the water table is complex, involving the transition between saturated and unsaturated porous media. A full treatment of the system requires a model of combined saturated and unsaturated flow. We revisit this idea when we discuss unsaturated flow systems in Chapter 11.

5.1.3 Flow in Radial Coordinates

As a final example of analytical solutions in one spatial dimension, consider the case of radial flow to a well in a homogeneous, confined aquifer. The domain of the porous medium begins at the well radius, denoted by r_w, and extends out to an outer radius denoted by r_{outer}. Radial symmetry is assumed, so there is no variation in head with angular location θ, and vertical averaging is applied. The thickness of the aquifer is denoted by B. Then the governing flow equation, written in radial coordinates and under steady-state conditions, takes the form (see Eq. (4.80))

$$-T\frac{\partial}{\partial r}\left(r\frac{\partial h}{\partial r}\right) = 0, \quad r_w < r < r_{outer}, \tag{5.15}$$

where the overbar to denote vertical averaging again has been dropped for clarity of presentation. To solve this equation, boundary conditions must be specified at the inner and outer boundaries. For the case of fixed head conditions, with

$$h(r_w) = h_w,$$

$$h(r_{outer)} = h_{outer},$$

the solution is a logarithm with the following form:

$$h(r) = h_w + (h_{outer} - h_w) \frac{\ln\left(\dfrac{r}{r_w}\right)}{\ln\left(\dfrac{r_{outer}}{r_w}\right)}. \tag{5.16}$$

If instead the flow rate at the well is given, then the inner boundary condition is a flux condition, which may be written

$$\left(KB(2\pi r)\frac{\partial h}{\partial r}\right)_{r=r_w} = Q_w.$$

Then the solution takes the form

$$h(r) = h_{outer} - \frac{Q_w}{2\pi T} \ln \frac{r_{outer}}{r}. \tag{5.17}$$

This equation is often referred to as the *Thiem equation*, after A. Thiem, who presented this equation in 1906 [3].

Notice that, for both solutions, the total (θ integrated) flow rate in the radial direction crossing a cylinder of radius r is the same for all values of r within the domain. In the case of two fixed head conditions, the total flow rate is given by

$$2\pi T \frac{h_{outer} - h_w}{\ln\left(\dfrac{r_{outer}}{r_w}\right)}.$$

In the case of a given flow rate in the well, the total flow rate toward the well is given for any radius r by Q_w. This is consistent with the governing equation, which states that the total flow in the radial direction (equal to $2\pi T r\, \partial h/\partial r$) does not change with a change in the radial coordinate; that is,

$$\frac{\partial}{\partial r}\left(2\pi T r \frac{\partial h}{\partial r}\right) = 0.$$

This is also consistent with simple physical reasoning, in that given there are no sources or sinks of water within the domain, any inflow or outflow can only be supplied through the boundaries. Therefore within the domain, at steady state, the total flow rate must be constant for any radius r.

Before leaving this example, we might consider what a source or sink of water might look like in this radial case, and what the implications are. One possibility is to have recharge, as in the previous example. But in the case of a confined aquifer, we often have fluid flowing into (or sometimes out of) an aquifer via vertical leakage from an adjacent aquifer through the adjoining aquitard that separates the two aquifers. For reasons that we discuss below (the "tangent law"), we often treat flow in the aquitard as essentially vertical, while that in the aquifer is treated as essentially horizontal. If we assume water leaks through the aquitard due to decreases in head in the aquifer caused by pumping of a well in that aquifer, then under an assumption of essentially constant head in the aquifer above the aquitard and steady-state flow in both the aquitard and the aquifer, we may write the equation for flow in the aquitard as

$$-\widetilde{K}\frac{\partial^2 \widetilde{h}}{\partial z^2} = 0, \quad B < z < B + \widetilde{B},$$

$$\widetilde{h}(B) = h(r), \qquad\qquad (5.18)$$

$$\widetilde{h}(B + \widetilde{B}) = h_{\text{top}},$$

where the hydraulic head in the aquitard is denoted by $\widetilde{h}(r, z)$, the hydraulic conductivity in the aquitard is denoted by $\widetilde{K}$, and the thickness of the aquitard is $\widetilde{B}$. Note that the radial dependence comes from the bottom boundary condition, which serves to couple the flux in the aquitard to the head in the lower aquifer ($h(r)$). No derivatives of $\widetilde{h}$ with respect to r appear in the equation because of the assumption of essentially vertical flow in the aquitard. Solution of this equation is simple and is given by

$$\widetilde{h}(r, z) = h(r) + \left(h_{\text{top}} - h(r)\right) \frac{z - B}{\widetilde{B}}. \qquad\qquad (5.19)$$

Differentiation of the equation gives the volumetric flux through the aquitard:

$$\widetilde{q}_z = -\frac{\widetilde{K}}{\widetilde{B}}\left(h_{\text{top}} - h(r)\right). \qquad\qquad (5.20)$$

Because the amount of water leaving the aquitard at its bottom boundary is the same as the amount of water entering the lower aquifer through its top boundary, the flow into the aquifer (which is a source term for that aquifer) must be given by Eq. (5.20). This must appear in the governing equation for the aquifer (see Section 4.8), such that the governing equation for the aquifer becomes

$$-T\frac{\partial}{\partial r}\left(r\frac{\partial h}{\partial r}\right) - \frac{\widetilde{K}}{\widetilde{B}}\left(h_{\text{top}} - h\right) = 0, \quad r_w < r < r_{\text{outer}}. \qquad\qquad (5.21)$$

While this equation is more complicated than other equations we have solved so far, it turns out that an analytical solution may be derived for $h(r)$ involving specific infinite series that occur often in mathematical physics and are given the name *Bessel functions*, after the mathematician Bessel.

One interesting observation is that existence of the internal source term (leakage through the aquitard) means that the importance of the outer boundary decreases the further away it is, until in the limit of a semi-infinite domain, all of the water supplying the well comes from leakage. Therefore physically meaningful analytical solutions may be derived on semi-infinite domains when an internal leakage (source) term is present. This is not true in the absence of such a source term, because all of the supply must come through the boundary, and for semi-infinite domains this leads to heads that are unbounded (meaning they go to negative infinity for any finite head at the outer boundary) and therefore have no practical meaning. We return to this idea in the next chapter when discussing pumping tests.

In the semi-infinite case, the governing equation and appropriate boundary conditions may be written

$$\frac{d}{dr}\left(r\frac{dh}{dr}\right) + \frac{\widetilde{K}}{KB\widetilde{B}}\left(h_{\text{top}} - h\right) = \frac{d}{dr}\left(r\frac{dh}{dr}\right) + \frac{\left(h_{\text{top}} - h\right)}{\lambda^2} = 0, \quad r_w < r < \infty,$$

$$\left(KB(2\pi r)\frac{dh}{dr}\right)_{r=r_w} = Q_w,$$

$$\lim_{r \to \infty} h(r) = h_{\text{outer}},$$

$$\lim_{r \to \infty}\frac{dh}{dr}(r) = 0.$$

(5.22)

In this equation, the length scale $\lambda \equiv \sqrt{KB\widetilde{B}/\widetilde{K}}$ has been introduced. This is a characteristic length scale that is usually called the *leakage factor*. The governing equation is a second-order ordinary differential equation whose general solution is a linear combination of the Bessel functions I_0 and K_0, of the form $h(r) = C_1 I_0(r/\lambda) + C_2 K_0(r/\lambda)$. These Bessel functions are specific infinite series that go by the name *modified Bessel function of the first kind, order zero (I_0)* and *modified Bessel function of the second kind, order zero (K_0)*. Notice that the solution is a function of the ratio r/λ. For the case of a semi-infinite domain, the solution simplifies to

$$h_{\text{outer}} - h(r) = \frac{Q_w}{2\pi T}\frac{K_0(r/\lambda)}{(r_w/\lambda)K_1(r_w/\lambda)},$$

(5.23)

where K_1 is a *modified Bessel function of the second kind, order one*. Typically, we have $r_w/\lambda \ll 1$. In that case, the behavior of the Bessel function K_1 in the limit of a small argument is such that

$$\frac{r_w}{\lambda}K_1(r_w/\lambda) \approx 1 \quad \text{for } r_w/\lambda \ll 1.$$

Therefore the solution simplifies to the following form:

$$h_{\text{outer}} - h(r) = \frac{Q_w}{2\pi T}K_0(r/\lambda)$$

For any given radius r, one can imagine a cylinder centered at $r = 0$ and having radius r and thickness covering the thickness of the aquifer. Flow to the well can be partitioned into that coming from leakage through the aquitard in the area along the top of the cylinder, and that coming from regions in the aquifer beyond the radius r. The sum of these two contributions must equal the flow entering the well, Q_w. As r increases from $r = r_w$ to larger values, the fraction of Q_w supplied by leakage through the aquitard increases from zero (when $r = r_w$) to a limit of one as r becomes very large.

Figure 5.6 shows the fraction of total flow Q_w that is supplied from the aquifer beyond radius r, as a function of r. The plotted fraction begins at one when $r = r_w$ and decreases according to the Bessel function solution. Once the radial distance has reached $r = 4\lambda$, the amount supplied from leakage is about 95%, and only 5% comes from the aquifer

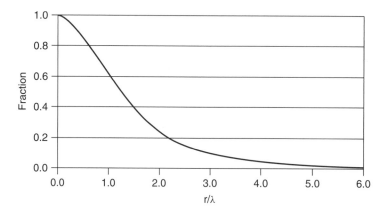

FIGURE 5.6. Fraction of total discharge supplied by aquifer beyond radius r/λ.

beyond this radial distance. We sometimes define a *radius of influence* of a well as the radial distance within which most of the flow is supplied by leakage from within that radius, with only a very small fraction coming from further out in the formation. That fraction is often taken as 95%, although the choice is somewhat arbitrary. Therefore, for a well pumping from a leaky aquifer in which the leakage comes from an adjacent aquifer with constant head, the radius of influence is approximated by four times the leakage factor. Notice that the case of a totally confined aquifer, with aquicludes above and below, leads to a leakage factor $\lambda \to \infty$. As such, there is no finite radius of influence, meaning that the hydraulic head solution is not well defined for this case of a semi-infinite domain. This is consistent with the observation that the Thiem solution becomes unbounded if the outer boundary is extended toward infinity.

5.2 TWO-DIMENSIONAL FLOW PROBLEMS

In this section we consider analytical solutions in two space dimensions. While analytical mathematics permit us to write quite general solutions for multidimensional, transient equations, we focus on steady-state systems in two spatial dimensions because these allow several important concepts to be developed. We begin with a discussion of *flow lines* and *equipotential lines* and present conditions under which those sets of lines are orthogonal to one another. This is used as a basis for construction of graphical solutions. We then consider a specific analytical solution to a two-dimensional problem defined on a vertical cross section corresponding to a hill slope, identifying recharge and discharge patterns and developing a simple but fairly general view of subsurface flow dynamics and their relationship to surface hydrology. Finally, we consider extensions to more complex systems that include certain kinds of heterogeneity.

5.2.1 Graphical Solutions

Graphical solution methods are based on the observation that flow lines are perpendicular to lines of equal potential. We begin by exploring this in some detail. First, let us define a *streamline* as a curve in space that is everywhere tangent to the volumetric flux vector **q**. In a steady-state system, a streamline traces the trajectory of a particle moving in the flow field. That flow field is driven by gradients in hydraulic head as given by Darcy's law, which states that the volumetric flux vector is related to the head gradient vector by a linear transformation that is defined by the hydraulic conductivity. When the material is isotropic, the hydraulic conductivity is a single number (at any point in space) and the mapping between $\mathbf{q}$ and $-\nabla h$ is a simple scalar mapping (we include the negative sign because the direction of $\mathbf{q}$ is in the direction of $-\nabla h$). So in isotropic media the flow vector is in the same direction as the negative of the head gradient.

When the material is anisotropic, the situation is not so simple, because the mapping between $\mathbf{q}$ and $-\nabla h$ involves a matrix of hydraulic conductivity values, and therefore the vectors $\mathbf{q}$ and $-\nabla h$ are generally not collinear (see the section "Anisotropy and Coordinate Maps" beginning on page 213). We will deal with the case of anisotropy shortly.

For now, let us assume isotropy, so that we are assured that $\mathbf{q}$ and $-\nabla h$ both have the same direction. Therefore we can say that streamlines follow the direction of $-\nabla h$ because this direction is the same as that of $\mathbf{q}$, which serves to define the streamlines. Next, we define *equipotentials* to be lines along which the potential, or hydraulic head, is constant. Recall from analytical geometry that, by definition, the gradient of any function has a direction that is perpendicular to lines of constant values of that function. Therefore the gradient of hydraulic head must be perpendicular to the lines of constant hydraulic head, which means that *streamlines must be perpendicular to equipotential lines in isotropic media*.

We use this observation to advantage because often we can estimate equipotential lines based on measurements. From these we can draw flow lines based on the orthogonality relationship between equipotential lines and flow lines. We present three examples next to demonstrate the use of this simple graphical approach.

Classical Flow Nets   In the field of soil mechanics and foundation engineering, one is often interested in flow through or around engineered structures that involve the subsurface and groundwater. This usually involves introduction of a low-permeability object into the subsurface, or perhaps construction of dams aboveground that create new porous media (e.g., an earthen dam) with associated flow possibilities through that object.

As an example, consider the situation shown in Figure 5.7, where a sheet of impermeable material is introduced into the ground that serves as a dam. Water pools on the left side of the dam ("behind" the dam) and flows away on the right side. We may represent the flow in this system by use of the orthogonality principle, given that we can identify appropriate boundaries along which either hydraulic head is constant (an equipotential line) or there is no flow in the direction normal to the boundary (in which case the boundary corresponds to a streamline). In Figure 5.8, the lines labeled A–B and E–F–G–H–I–J are lines across which no flow occurs. Because the flux vector $\mathbf{q}$ has no component in the direction normal to the line, the only nonzero component can be in the direction tangent to the line. Therefore, by definition of a streamline, the lines A–B and E–F–G–H–I–J are streamlines. The lines labeled D–F and I–C are lines of equal hydraulic head, with the head along D–F equal to H_L and the head along I–C equal to H_R. So any representation of the flow system through the porous medium must include equipotentials along D–F and I–C, and streamlines along A–B and E–F–G–H–I–J. With this information, and the requirement that the set of streamlines and equipotentials must be mutually orthogonal, we can construct a set of such lines. One representation is shown in Figure 5.9.

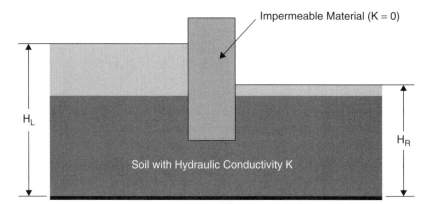

FIGURE 5.7. Schematic of dam structure.

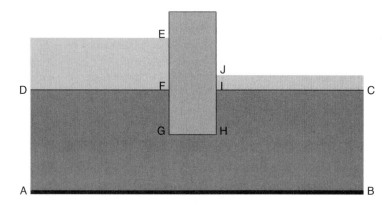

FIGURE 5.8. Schematic of dam with boundary lines labeled.

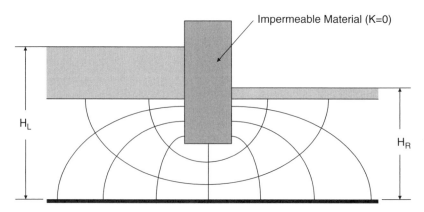

FIGURE 5.9. Streamlines and equipotential lines for the dam problem.

While the number of equipotentials and streamlines, and their spacing, may be chosen arbitrarily (restricted only by the orthogonality requirement), it is often useful to follow a more systematic approach to construction of these lines. In particular, we often try to draw lines so that the flow that occurs between any two adjacent streamlines is the same, and the head drop between any two adjacent equipotential lines is the same.

Regional Groundwater Flow: Hydrologic Maps   A hydrologic map refers to a graphical representation of groundwater flow based on equipotential lines and streamlines applied to an aquifer within which vertical averaging has been applied (see Section 4.5.2 for a further discussion of this topic). Therefore the map represents the hydraulic head and volumetric flux in the (x, y) areal plane wherein the flow within the aquifer is assumed to be essentially horizontal. For homogeneous, isotropic materials, the *principle of orthogonality* may be applied. Therefore streamlines may be sketched based on knowledge of equipotential lines. The practical application of this kind of map rests on the observation that measurements of water levels in wells are direct measures of hydraulic head over the length of the well screen. So for wells screened within the same aquifer, water level measurements provide point measures of hydraulic head. Contour maps based on these well measurements then provide the equipotential lines, because by definition a contour line is a line along which the value of the variable being contoured is constant. When the variable is hydraulic head, the contours are equipotential lines. Once these lines are drawn, streamlines may be constructed using the orthogonality principle. An example of such a map is provided in Figure 5.10, where water level measurements in the vicinity of two pumping wells are contoured, and from those contours flow directions are inferred and drawn.

Anisotropy and Coordinate Maps   Up to this point, we have restricted our analysis to isotropic systems, in which hydraulic conductivity is the same in all directions. The reason for this, as argued earlier, is that the orthogonality principle does not hold in general for anisotropic systems. We may observe this using a simple calculation. Assume we have a two-dimensional (x, y) system, and the principal directions of the hydraulic conductivity align with the coordinate axes, so that in the (x, y) coordinate system the hydraulic conductivity matrix is given by

$$\mathbf{K} = \begin{bmatrix} K_x & 0 \\ 0 & K_y \end{bmatrix}. \tag{5.24}$$

Now consider the resulting direction of flow when the (negative of the) hydraulic head gradient is aligned at an angle α from the x-axis (see Figure 5.11). We may represent the two components of the gradient vector as

$$\frac{\partial h}{\partial x} = |\nabla h| \cos \alpha \quad \text{and} \quad \frac{\partial h}{\partial y} = |\nabla h| \sin \alpha, \tag{5.25}$$

where $|\nabla h|$ denotes the magnitude of the vector ∇h. Then we have the following equations governing the flow, derived directly from Darcy's law:

$$q_x = -K_x \frac{\partial h}{\partial x} = -K_x |\nabla h| \cos \alpha,$$

$$q_y = -K_y \frac{\partial h}{\partial y} = -K_y |\nabla h| \sin \alpha. \tag{5.26}$$

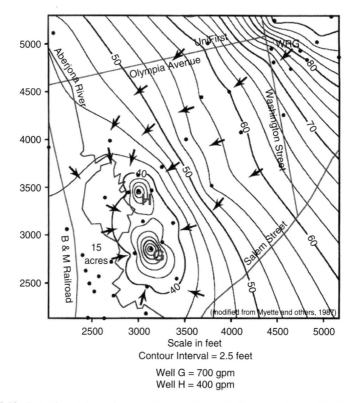

FIGURE 5.10. Potentiometric surface on January 3, 1986 after pumping wells G and H for 30 days (from [4]).

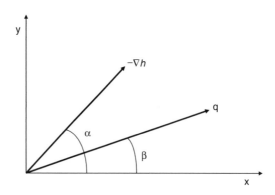

FIGURE 5.11. Alignment of flow and head-gradient vectors.

The angle at which water flows is then given by β, which is defined by

$$\tan \beta = \frac{q_y}{q_x} = \frac{K_y \, \sin \, \alpha}{K_x \, \cos \, \alpha}. \tag{5.27}$$

Clearly, when the material is isotropic, $K_x = K_y$ and therefore $\tan \, \beta = \tan \, \alpha$, which means that $\alpha = \beta$. However, when $K_x \neq K_y$, the flow vector does not align with the head gradient vector except when the head gradient happens to align with the principal directions of the hydraulic conductivity.

It turns out that for anisotropic systems we can define a new coordinate system in which the flow vector and the head gradient align. In this new coordinate system, the orthogonality principle between equipotentials and streamlines will apply, and therefore graphical methods based on orthogonality between equipotentials and streamlines can be used. However, the mapping between coordinate systems complicates the actual implementation of this idea.

Let the new coordinate system be given by the symbols (x', y'). Then the appropriate definition of the new coordinate system is given by a simple coordinate map that basically involves coordinate "stretching" and "shrinking" defined by the following relations:

$$x' = \sqrt{\frac{K_y}{K_x}}\, x \quad \text{and} \quad y' = \sqrt{\frac{K_x}{K_y}}\, y. \tag{5.28}$$

We observe that the flux vector components may now be written

$$q_x = -K_x \frac{\partial h}{\partial x} = -K_x \frac{\partial h}{\partial x'} \frac{\partial x'}{\partial x} = -K_x \sqrt{\frac{K_y}{K_x}} \frac{\partial h}{\partial x'} = -\sqrt{K_x K_y} \frac{\partial h}{\partial x'},$$

$$q_y = -K_y \frac{\partial h}{\partial y} = -K_y \frac{\partial h}{\partial y'} \frac{\partial y'}{\partial y} = -\sqrt{K_x K_y} \frac{\partial h}{\partial y'}. \tag{5.29}$$

Therefore calculation of the direction of flow shows that flow is in the direction of the head gradient *defined in the new coordinate system*,

$$\tan \beta = \frac{-\sqrt{K_x K_y} \partial h/\partial y'}{-\sqrt{K_x K_y} \partial h/\partial x'} = \frac{\partial h/\partial y'}{\partial h/\partial x'} = \tan \alpha. \tag{5.30}$$

So in the transformed space, the directions of flow may be determined from the usual flow net analysis. The entire flow system defined in the original, or physical, coordinates may then be determined by transformation back to the original (physical) coordinate system defined by (x, y).

Material Interfaces and the Tangent Law Heterogeneity of material properties such as hydraulic conductivity means that the value of the property changes with spatial location. A special kind of heterogeneity occurs when the value of the property is constant within a region of the domain, then it abruptly changes value when crossing a particular line (in two dimensions) or surface (in three dimensions). After crossing the line or surface, the property maintains a different constant value in this new subregion of space. The line or surface across which the property changes is referred to as a material interface. An example of such a situation is shown in Figure 5.12, where a body of homogeneous material has a circular inclusion within it that has different material properties.

In a saturated groundwater system, movement across a material interface must obey two fundamental rules: (1) continuity of hydraulic head and (2) continuity of normal flux. The term continuity has the usual mathematical meaning; that is, a continuous variable has the same value at any point when that point is approached from any direction.

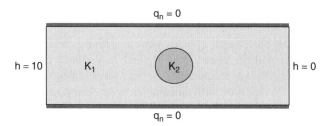

FIGURE 5.12. Inclusion of different material within a homogeneous medium.

The first of these rules follows from the fact that a discontinuous hydraulic head implies an infinite flux for any finite hydraulic conductivity (due to Darcy's law), which is physically impossible. Therefore discontinuities in hydraulic head are impossible in any media with nonzero permeability.

The second rule follows from simple mass balance arguments: any mass entering the interface must emerge on the other side, because the interface cannot store mass (it has zero volume).

Based on the two rules that apply to material interfaces, we can derive a specific result that indicates how streamlines behave when crossing a material interface. Based on this result, we may construct flow nets and hydrologic maps when the domain of interest is composed of regions of homogeneous materials, each separated from its neighbor by an abrupt (material) interface.

Consider, for example, the material interface identified by the line A–A' in Figure 5.13. On one side of the interface is material 1, with hydraulic head h_1, hydraulic conductivity K_1, and volumetric flux vector $\mathbf{q}_1$; and on the other side is material 2 with h_2, K_2, and $\mathbf{q}_2$. Let the direction normal to the interface be denoted by n and the direction along the interface as s. The vectors $\mathbf{q}_1$ and $\mathbf{q}_2$ are shown in Figure 5.13, as are the components of those vectors in the (s, n) directions. The angle that the two vectors make with the normal direction are denoted by β_1 and β_2. Continuity of normal flux across the interface requires $q_{n1} = q_{n2}$, where subscript n denotes the component in the normal direction. Next, observe that continuity of hydraulic head implies that the derivative of h along the s-direction is the same on both sides of the interface, so that $\partial h_1/\partial s = \partial h_2/\partial s$

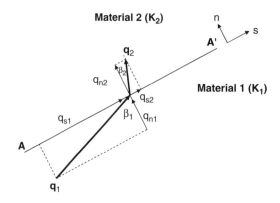

FIGURE 5.13. Flow vectors at a material interface.

along the interface. From Darcy's law,

$$q_{s1} = -K_1 \frac{\partial h_1}{\partial s} \quad \text{and} \quad q_{s2} = -K_2 \frac{\partial h_2}{\partial s}. \tag{5.31}$$

Therefore

$$\frac{q_{s2}}{q_{s1}} = \frac{K_2}{K_1}. \tag{5.32}$$

We now relate the angles β_1 and β_2 using the definition of the tangent of an angle:

$$\frac{\tan \beta_2}{\tan \beta_1} = \frac{q_{s1}/q_{n1}}{q_{s2}/q_{n1}} = \frac{q_{s2}}{q_{s1}} = \frac{K_2}{K_1}. \tag{5.33}$$

This last equation says that the change in angle of the flow vectors across the material interface is governed by the permeability contrast across the interface. In particular, we see that if $K_1 \gg K_2$, then $\tan \beta_1 \gg \tan \beta_2$. Therefore the vector in the more permeable material has a direction close to parallel to the interface, while that in the less permeable material is close to the normal direction. We refer to Eq. (5.33) as the tangent law.

We highlight two consequences of the tangent law. First, it provides a methodology for construction of flow nets in heterogeneous materials where the heterogeneity involves simple homogeneous regions separated by material interfaces. Figure 5.14 shows an example of a flow net constructed using the tangent law as a guide to match the results for the two homogeneous regions that comprise the overall domain. The second consequence involved flows in layered aquifer–aquitard systems. Because by definition aquifers have permeability much greater than the permeability in an aquitard, flow lines in the vicinity of

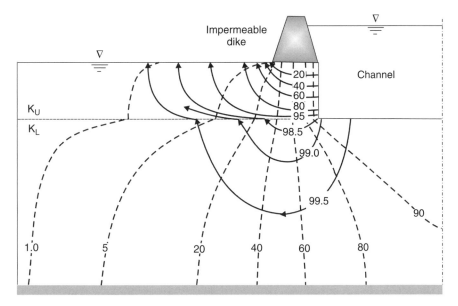

FIGURE 5.14. Flow nets for seepage from the upstream side of a dam to the downstream. The hydraulic conductivity ratio between the upper and lower formations is $K_U/K_L = 50$ and the anisotropy ratio of horizontal to vertical is 10 (adapted from [5]).

the aquifer–aquitard boundary must be close to parallel to the interface in the aquifer and close to the normal direction in the aquitard. For layers that are essentially horizontal, this means that flow in the aquifer will be essentially horizontal (at least in the vicinity of the boundary) and flow in the aquitard will be essentially vertical. Therefore the tangent law supports the assumption of essentially horizontal flow in aquifers and essentially vertical flow in aquitards. These are common assumptions used to simplify the mathematical description of groundwater flow systems.

5.2.2   Analytical Solutions in Two Dimensions

Analytical solutions for groundwater flow problems in more than one dimension become more complicated because the governing equation is a partial differential equation rather than an ordinary differential equation. Ordinary differential equations typically have finite-dimensional solution spaces, meaning that the solution may be written as a linear combination of a finite number of fundamental solutions (the Bessel functions in Eq. (5.23) are an example).

Partial differential equations usually have infinite-dimensional solution spaces, meaning that solutions will involve infinite series and/or various kinds of integrals. The infinite series in Eq. (5.5) is an example of a solution to a partial differential equation—in that case the transient one-dimensional groundwater flow equation.

In general, analytical solutions can only be derived for linear equations with constant coefficients defined on domains whose boundaries are parallel to the coordinate axes. In this section we will focus on one specific solution for steady-state groundwater flow in two spatial dimensions. That case involves groundwater flow in small drainage basins and is based on the classic work of Toth [6].

Groundwater Flow in Small Drainage Basins   In his 1963 paper, Joseph Toth used an analytical solution to the two-dimensional groundwater flow equation for homogeneous and isotropic aquifers to study recharge, discharge, and subsurface flow patterns in small drainage basins. The basins are characterized by surface features, particularly streams, and by topographic highs that correspond to water divides.

Toth defines a small drainage basin as follows [6, pp. 4796–4797]:

> an area bounded by a topographic high, its lowest parts being occupied by an impounded body of surface water or by the outlet of a relatively low order stream having similar physiographic conditions over the whole of its surface. The upper limit for such basins is usually several hundred square miles.

Of course, a river basin, or drainage basin, is inherently three dimensional. The two-dimensional analysis involves a cross section of the basin that is orthogonal to the basin axis. Toth argues that this is a reasonable assumption as follows [7, p. 4797]:

> The assumption that the problem can be treated as a two-dimensional one is supported by the recognition that in most small basins the slopes of the valley flanks greatly exceed the longitudinal slopes of the valley floors. This difference in slopes causes the longitudinal component of the flow to become negligible compared to the lateral component.

In an earlier paper, Toth had expanded on this assumption with the following additional commentary [7, p. 4379]:

The deeper below the surface, however, that the point is located at which **q** ... is investigated, the less valid is the assumption. And if the flow lines are extended sufficiently (in the case where no impermeable boundary intercepts the downward motion) they will finally merge with the flow lines of deeper flow systems following the path prescribed by the regional land surface.

The system being modeled is illustrated in Figure 5.15, which is a reproduction of Toth's original Figure 1 [6, p. 4796] [6]. The valley bottom ($x = 0$) and the water divide ($x = S$) are taken as lines of symmetry, meaning that no flow crosses those lines in the normal direction. Therefore along those lines $\partial h / \partial x = 0$. Along the bottom ($z = 0$), an impermeable boundary is assumed, so that $\partial h / \partial z = 0$. The top boundary of the saturated zone is the water table. Toth prescribes the location of the water table by noting that observations indicate a "close correlation of the piezometric surface with the topography" [7, p. 4376]. This is repeated somewhat more expansively in 1963 with the comment that "It has been observed in Alberta ... as well as elsewhere ... that the water table is generally similar in form to the land surface" [6, p. 4797]). Therefore representation of the land surface is taken as a good representation of the water-table location.

The water-table location is represented by a straight line that connects the elevation of the valley bottom to the elevation of the water divide. Superimposed on this sloping line is a sinusoidal function (see Figure 5.15) meant to represent spatial variability of topography within the basin ("the highs and lows [of the sinusoidal function] ... are thought to be representative of the hills and depressions of the natural land surface" [6, p. 4797]).

While these boundary conditions, coupled with the groundwater flow equation for steady-state flow in a homogeneous and isotropic aquifer, provide a complete mathematical description of the problem, the equations cannot be solved analytically because the location of the water-table is not parallel to a coordinate axis. To obtain a solution, Toth applies the water table condition along the line $z = z_0$, which is parallel to the x-axis. He comments that "this restricts the validity of the numerical results to slopes of about $3°$ or less" [6, p. 4797]. The mathematical statement of the problem solved by Toth is

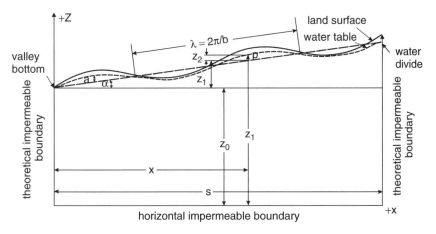

FIGURE 5.15. Schematic of domain and boundary conditions used by Toth [6].

therefore given by the following:

$$\frac{\partial^2 h}{\partial x^2} + \frac{\partial^2 h}{\partial z^2} = 0, \quad 0 < x < S, \ 0 < z < z_0,$$

$$\frac{\partial h}{\partial x}(0, z) = 0,$$

$$\frac{\partial h}{\partial x}(S, z) = 0, \tag{5.34}$$

$$\frac{\partial h}{\partial z}(x, 0) = 0,$$

$$h(x, z_0) = z_0 + c'x + a' \, \sin \, (b'x).$$

This equation may be solved analytically, yielding an infinite-series solution involving trigonometric functions. This solution may then be evaluated, manipulated, and differentiated, and the results displayed graphically to illustrate how these kinds of systems behave.

The classic result that is derived directly from this solution is the diagram shown in Figure 5.16 (Toth's original Figure 3). From this figure we make the following observations. First, notice that the solution is represented by flow lines and that the flow lines are only drawn within the domain where the solution is actually obtained, that is, for $z \leq z_0$.

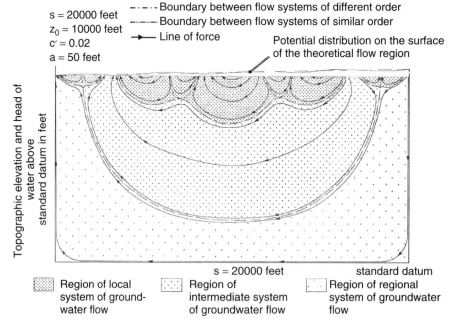

FIGURE 5.16. Classic figure from Toth [6] showing local, intermediate, and regional flow systems.

Second, we may use this figure to identify recharge areas and discharge areas. A *recharge area* is an area along the top of the domain (for a two-dimensional cross section, it is actually a *recharge line*, because the "area" involves the y-direction, which is ignored in the two-dimensional cross section) along which hydraulic head decreases with depth, implying that water flows downward into the domain, across the upper boundary. Conversely, a *discharge area* is one for which the head increases with depth, so that water is flowing upward and out of the domain across the top boundary. Recharge and discharge areas are separated (in the xy plane) by *hinge lines* (a hinge line appears on the two-dimensional cross section as *hinge points* along the top boundary).

The third observation is that Toth has identified three general zones on the figure, which are denoted as local flow systems, intermediate flow systems, and regional flow systems. A *local flow system* has its recharge area along a local (sinusoidal) topographic high and its discharge area along the adjacent topographic low. An *intermediate flow system* has recharge and discharge areas that are separated by one or more other topographic highs and lows but does not span the entire hill slope length. A *regional flow system* is a system whose recharge area includes the water divide and discharge area includes the valley bottom.

Some of the general observations that follow from the range of problems that were solved include the following: (1) for cases of higher amplitude (a) and smaller depth (z_0), local systems tend to dominate; (2) where local relief is negligible, only regional systems develop; (3) highlands are recharge areas while lowlands are discharge areas; and (4) because of the local flow systems, recharge and discharge areas alternate across the valley, and complicated flow patterns can mean that water samples taken nearby one another have very different recharge locations and therefore different travel paths and times. Note that for most common topographic configurations, hinge lines lie closer to the valley bottom than to ridges, so that on an areal (x, y) map discharge areas commonly cover 5–30% of the surface area of the watershed.

5.3 SUMMARY

In this chapter, we have begun to examine how groundwater flow equations can be solved by focusing on analytical solutions for the governing differential equations. Analytical solutions require the system to be fairly simple, which usually restricts us to systems that are homogeneous, and whose boundaries align with the coordinate axes. Analytical solutions are usually easy to compute and often provide important practical insights into system behavior. They also provide the basis for parameter estimation, as we have seen to some extent in this chapter, and will see again in the next chapter. In addition to explicit analytical expressions for the solution to the flow equation, an understanding of simple graphical relationships between flow lines and equipotential lines allows for graphical analyses that also provide significant insights into overall behavior of the system. In addition, the tangent law, which describes the flow direction at material interfaces, has important consequences for groundwater flow systems. While numerical simulations have largely replaced traditional analytical methods for analysis of groundwater problems, the simplicity and insights offered by analytical solutions make them valuable analysis tools, independent of the amount of computational power that might be available.

5.4 PROBLEMS

5.1. Calculate the solutions given in Eqs. (5.5) and (5.7) for a series of time values, and plot the results as is done in Figure 5.2. Observe how the solutions overlap for early times and then diverge for later times. Use these results to propose and justify a criterion for when the two solutions can be used interchangeably.

5.2. The Bessel function $K_0(x)$ has the property that

$$\frac{d}{dx} K_0(x) = x K_1(x).$$

With this relationship, show that the solution given in Eq. (5.23) satisfies the boundary condition at the well. Also, identify the property of the Bessel function that shows that the outer boundary condition (as $r \to \infty$) is also satisfied.

5.3. The Thiem equation gives a solution for hydraulic head that is a logarithmic function of radial distance r. It applies to a radial flow system that is finite in extent. Show that the analogous one-dimensional system in Cartesian coordinates (e.g., with independent variable x) over a finite domain gives a solution for hydraulic head that is linear in x, as opposed to logarithmic. Relate this linear solution to the Darcy experiments. Then provide a physical explanation for why the solution in Cartesian coordinates is linear while that in radial coordinates is logarithmic.

5.4. Consider the Thiem equation. If the solution were plotted in the (x, y) coordinate system, and both equipotential lines and flow lines were drawn, explain how the resulting figure would look. Sketch an example of equipotential and flow lines for the Thiem equation.

5.5. Evaluate Eq. (5.23) and use the results to create Figure 5.6. Identify values of r/λ for which the fraction of flow from leakage is 95% and 99%.

5.6. Draw flow lines and equipotential lines for the system shown in Figure 5.12. Assume first that $K_2 = 0.1K_1$, and then that $K_2 = 10K_1$.

5.7. Develop a criterion that will guarantee that the maximum difference between the nonlinear solution of Eq. (5.12) and the linearized approximation given by Eq. (5.14) is not more than 10%.

BIBLIOGRAPHY

[1] C. W. Fetter, Jr. *Applied Hydrogeology*, Charles E. Merrill Publishing, Columbus, OH, 1980, p. 488.

[2] J. Dupuit, *Étude théorique et pratique sur le mouvement des eaux dans les canaux découverts et à travers les terrains perméables*, 2nd ed., Dunod, Paris, 1863.

[3] G. Thiem, *Hydrologische Methoden*, Vol. 56. J. M. Gebhart, Leipzig, 1906.

[4] C. F. Myette, J. C. Olimpio, and D. G. Johnson, *Area of Influence and Zone of Contribution to Superfund-Site Wells G and H, Woburn, Massachusetts*, U.S. Geological Survey Water Resources Investigations Report 87-4100, 1987.

[5] D. K. Todd and J. Bear, Seepage through layered anisotropic porous media, *J. Hydraulics Div. Am. Soc. Civil Eng.* **87**(HY3): 31–57, 1961.

[6] J. Toth, A theoretical analysis of groundwater flow in small drainage basins, *J. Geophys. Res.*, **68**(16): 4795–4387, 1963.

[7] J. Toth, A theory of groundwater motion in small drainage basins in central Alberta, Canada, *J. Geophys. Res.*, **67**(11): 4375–4387, 1962.

CHAPTER 6

WELL HYDRAULICS

In Chapter 4 we developed the equations describing groundwater flow. In the course of this development various physical parameters appeared as coefficients in these equations. To solve the equations, these coefficients must be known. Since they are unique to each physical system, these coefficients must be determined for each system of interest. The goal of this chapter is to explore two approaches as to how this determination is made using the theory of well hydraulics.

In Section 1.5 we saw that the hydraulic conductivity of a saturated soil sample could be determined using a properly instrumented soil column called a permeameter. However, as this approach examines and evaluates only a small soil sample, it does not, in general, provide adequate insight into regional hydraulic conductivity. In other words, the hydraulic conductivity in the field varies from point to point, and by taking selected core samples for laboratory analysis one may not generate parameter values representative, in a regional sense, of the heterogeneous soil from which they were selected.

By way of analogy, consider the determination of grain size. If one chose to examine ten soil grains taken at random from a sample, the individual grain sizes may or may not be similar to the average obtained in the event that all grains in the sample were measured. What is needed is a technique that measures the average grain size for the soil sample directly. Returning once again to hydraulic conductivity, what we need is a technique that measures this parameter directly in the field and is representative of a relatively large soil volume. Two such field scale approaches are the *slug test* and the *pumping test.*

6.1 THE SLUG TEST

The slug test is an approach to the measurement of hydraulic conductivity in the field using information obtained from a well. The concept is best understood with the help

Subsurface Hydrology   By George F. Pinder and Michael A. Celia
Copyright © 2006 John Wiley & Sons, Inc.

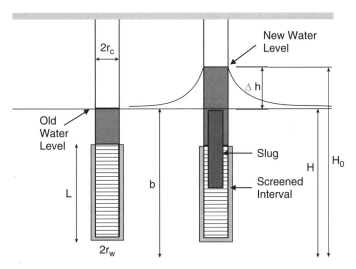

FIGURE 6.1. Definition sketch for a slug test. The two water levels indicated denote before and after the slug is placed in the well, that is, $\Delta h = H_0 - H$.

of Figure 6.1. However, before proceeding further, it is important to introduce or, perhaps, reintroduce some definitions. A *confined aquifer* is one that has low-permeability formations, or *confining beds*, bounding it from both above and below such that during the pumping process the aquifer under consideration will remain saturated. A *leaky aquifer* is similar to a confined aquifer but for the ability of the aquifer to exchange significant quantities of water with the "confining beds." An aquifer is in steady state when the changes in water level with time are small and within a predefined tolerance. In other words, the transient behavior of the steady-state aquifer is no longer considered of practical importance. A *fully penetrating well* is one that extends vertically through, and receives water from, the entire thickness of the aquifer. A *partially penetrating well,* on the other hand, extends vertically through and accesses only a limited portion of the aquifer. An *infinite aquifer* is one that has dimensions sufficiently large that its boundaries in the (x, y) plane are unaffected by the pumping strategy under consideration.

In Figure 6.1 we illustrate the water levels in a well before and after the introduction of a solid (as opposed to liquid) *slug*. The solid slug, which can be any cylindrical object of suitable size that will sink through the water column, is placed quickly in the well. The resulting volume displacement of water is equal to the volume of the slug. The displaced water rises up the well from the original water-level elevation in the well, H, to a new level, H_0. Over a period of time the water level decays back to the original level H. The reason the water level decays is that the water seeps into the formation along the length of the well screen. The higher the hydraulic conductivity of the aquifer, the more quickly the water level decays.

In conducting this experiment in the field, one should recognize some practical limitations:

- If the slug is introduced too quickly, the water level in the well may begin to oscillate, making analysis of the data difficult.

- If the formation is very permeable, a significant volume of water may be entering the formation as the slug is introduced. Thus the water level observed does not necessarily represent the volume of the slug. This provides an inaccurate reading for the water level H_0.

- If the formation has a very low hydraulic conductivity, the water level in the well may decay very slowly. In this situation the slug test may take several hours to complete.

- Since the rate of flow into the aquifer is proportional to the initial height Δh in Figure 6.1, it may be helpful in low-permeability formations to add water to the well to raise the elevation rather than introduce a slug. This protocol will allow a larger increment of head Δh, with the consequent effect of a more rapid initial water-level decline. In very low-permeability formations, practical limitations may require that only early data are obtainable, in which case the larger increment may provide changes in water levels that are larger and therefore more accurately analyzed.

- The annulus between the well casing and the surrounding soil must be sealed above the screen. If this is not done, the water moves upward along the interface between the casing and the soil rather than into the aquifer adjacent to the screen. Such a "short circuit" yields inaccurate analyses.

With these qualifications in hand, let us now assume that we have conducted a slug test in the field and would like to analyze the results.

6.1.1 Hvorslev Method

To begin assume the flow of water from a well is proportional to (1) the height of the slug-induced excess water level in the well relative to the water level in the soil outside the well and (2) the hydraulic conductivity in the radial direction from the well screen. We will denote this hydraulic conductivity value in the radial r direction as K_{rr} and the excess hydraulic head in the well as $(H_0 - H)$; that is, we assume

$$Q = F K_{rr} (H_0 - H), \tag{6.1}$$

where F is the proportionality factor and is assumed to be dependent on the geometry of the well screen. Note that at $t = 0$ we have as initial conditions (1) that the head in the well is H_0 and (2) the head in the aquifer immediately adjacent to the well is H.

Now consider what is actually happening in the well bore as the water is exiting the screen due to the slug-induced change in water level in the well bore. The volume of water in the well attributable to the slug at any time t is $\pi r_c^2 (h - H)$, where r_c is the radius of the well casing and h is the head in the well. Because the rate of change of the volume of water in the well must be equal to that leaving the well through the screen, the following relationship holds:

$$\pi r_c^2 \frac{d(h-H)}{dt} = -Q. \tag{6.2}$$

Because H is the water level external to the well and is assumed constant over the period of the test, substitution of Eq. (6.1) into Eq. (6.2) yields

$$\frac{dh}{dt} = -\frac{F K_{rr} (h - H)}{\pi r_c^2}. \tag{6.3}$$

One now defines a time lag, t_l, as the time required for the excess head Δh to dissipate if one were to assume that the initial flow rate $Q_0 = F K_r (H_0 - H)$ were maintained.[1] It is formally defined as

$$t_l = \frac{V_{\text{well}}}{Q_0} = \frac{\pi r_c^2 (H_0 - H)}{F K_{rr} (H_0 - H)} = \frac{\pi r_c^2}{F K_{rr}}, \tag{6.4}$$

where V_{well} is the volume of water displaced by the slug or added to the well in the event a slug of water rather than a solid cylinder were to be used.

Let us now substitute Eq. (6.4) into Eq. (6.3) to give

$$\frac{dh}{dt} = -\frac{h - H}{t_l} \tag{6.5}$$

or, rearranging,

$$dt = -t_l \frac{dh}{h - H}, \tag{6.6}$$

which upon integration yields[2]

$$t = -t_l \, ln \, (h - H) + C. \tag{6.7}$$

The constant of integration C is evaluated through substitution of the initial condition

$$h(t = t_0 = 0) = H_0 \tag{6.8}$$

into Eq. (6.7). One obtains upon evaluation of C and introduction of the result into Eq. (6.7) the expression

$$t_l = \frac{-t}{\ln \left(\dfrac{h - H}{H_0 - H} \right)}, \tag{6.9}$$

which we can combine with Eq. (6.4) to obtain

$$\frac{\pi r_c^2}{F K_{rr}} = \frac{-t}{\ln \left(\dfrac{h - H}{H_0 - H} \right)} \tag{6.10}$$

or

$$\frac{F K_{rr}}{\pi r_c^2} = \frac{\ln \left(\dfrac{h - H}{H_0 - H} \right)}{-t}. \tag{6.11}$$

---

[1] Of course, due to the fact that the water level is dropping, this situation would not, in fact, be realized. The outward flow would actually decrease as the water level in the well decreased.

[2] Note that $\ln x = 2.30258 \log_{10} x$ and $\log_{10} x = 0.43429 \ln x$.

From Eq. (6.11) we see that one can obtain the hydraulic conductivity by plotting the log of the ratio $(h - H)/(H_0 - H)$ against time and then taking the slope of the curve. More specifically, given the factors F and r_c^2 are known, the calculation of K_{rr} can be obtained from Eq. (6.11). The factor F was presented for a number of different borehole configurations in the original work of Hvorslev [1]. Possibly the most generally applicable of these leads to the following formula[3]:

$$\ln(H_0 - H) - \ln(h - H) = \frac{2K_{rr}Lt}{r_c^2 \ln\left(\frac{L}{2r_{we}} + \sqrt{1 + \left(\frac{L}{2r_{we}}\right)^2}\right)}, \qquad (6.12)$$

where the effective well radius r_{we} is given by

$$r_{we} = r_w\sqrt{\frac{K_{zz}}{K_{rr}}}.$$

We can rewrite Eq. (6.12) to obtain an explicit relationship for K_{rr}:

$$K_{rr} = \frac{r_c^2 \ln\left(\frac{L}{2r_{we}} + \sqrt{1 + \left(\frac{L}{2r_{we}}\right)^2}\right)}{2L} \times \frac{\ln(H_0 - H) - \ln(h - H)}{t}. \qquad (6.13)$$

Let us now consider an example. In Figure 6.2 we find the necessary information to analyze our problem. We observe that the well bore radius is 0.125 m and the casing radius is 0.064 m. The screen length is 1.52 m and the aquifer thickness is 36.89 m; this satisfies the criteria for a partially penetrating well since the screen does not extend the full thickness of the aquifer [4]. The initial displacement $H_0 - H$ is 0.38 m. The number of values of Δh measured over a period of time of 1 hour is 44. When these values are plotted on a graph that is logarithmic in normalized Δh (displacement normalized with respect to the initial displacement) and linear in time, the result is found in Figure 6.3. By using an automatic fitting routine, the best-fitting linear regression curve has been obtained. Introducing this slope value and the aquifer and well parameters shown in Figure 6.2 into Eq. (6.13), we obtain the results presented in Figure 6.3, namely, a K_{rr} value of 16.57 m/day.

A modification of the original concept of Hvorslev to address issues of anisotropy was proposed by Zlotnik [5] but is not discussed here.

6.1.2   Cooper–Bredehoeft–Papadopulos Approach

An alternative slug test analysis approach developed by Cooper et al. [6] is based on Eq. (4.80), which we rewrite here for convenience as

$$T\left(\frac{\partial^2}{\partial r^2}\overline{h}(r, t) + \frac{1}{r}\frac{\partial}{\partial r}\overline{h}(r, t)\right) - S\frac{\partial}{\partial t}\left(\overline{h}(r, t)\right) + q_i(r, t) + Q^* = 0, \qquad (6.14)$$

---

[3]Note that the r_c and r_w are shown in Figure 6.1.

AQUIFER DATA

Saturated Thickness: 47.87 m
Anisotropy Ratio (Kz/Kr): 1.

SLUG TEST WELL DATA

Test Well: : Well 3

X Location: 0. m
Y Location: 0. m

Initial Displacement: 0.38 m
Static Water Column Height: 36.89 m
Casing Radius: 0.064 m
Wellbore Radius: 0.125 m
Well Skin Radius: 0.125 m
Screen Length: 1.52 m
Total Well Penetration Depth: 36.89 m

No. of Observations: 44

Observation Data					
Time (sec)	Displacement (m)	Time (sec)	Displacement (m)	Time (sec)	Displacement (m)
0.1	0.369	2.	0.343	11.3	0.189
0.2	0.388	2.3	0.336	12.6	0.175
0.3	0.377	2.6	0.329	14.2	0.16
0.4	0.388	2.9	0.322	15.9	0.142
0.5	0.365	3.2	0.314	17.8	0.125
0.6	0.377	3.6	0.311	20.	0.109

FIGURE 6.2. Slug-test data used in the example problem for Hvorslev's method (from [2] with data from [3]).

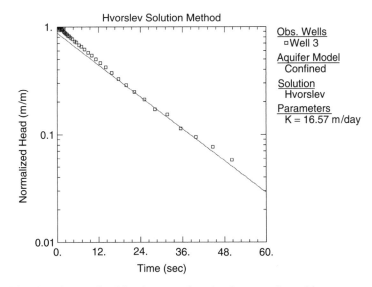

FIGURE 6.3. Plot of normalized head versus time for the example problem presented, and the best linear fit line (from [2] with data from [3]).

where $\overline{h}$ is once again the hydraulic head, T is transmissivity, S is the storage coefficient, q_i is vertical leakage into the aquifer, and Q^* is the total well discharge (the discharge integrated over the vertical dimension). Recall that this equation represents radial flow in an aquifer. It has been assumed that the flow is radially symmetric; that is, the flow is assumed to be independent of the angle θ in the cylindrical coordinate system. In addition, the resulting simplified equation (no dependence on θ) has been integrated vertically such that the parameters T and S and the state variable $\bar{h}$ are by the process of vertical integration required to be vertical averages. More specifically, if we use the bar overscore to indicate an average quantity, we have

$$\bar{h}\,(r, t) \equiv \frac{1}{l} \int_{z_1}^{z_2} h(r, t)\, dz,$$

$$T \equiv \overline{K_{rr}}\, l, \tag{6.15}$$

$$S \equiv \overline{S_s}\, l,$$

where $\overline{K_{rr}}$ is the vertical average hydraulic conductivity in the radial direction, $\overline{S_s}$ is the specific storage coefficient, and l is the thickness of the aquifer. If, for clarity in presentation, we drop the overbar notation, assume that there is no pumping well or leakage, and also that the aquifer is of uniform thickness, we can rewrite Eq. (6.14) as

$$\frac{\partial^2}{\partial r^2} h\,(r, t) + \frac{1}{r}\frac{\partial}{\partial r} h\,(r, t) - \frac{S}{T}\frac{\partial}{\partial t} h\,(r, t) = 0. \tag{6.16}$$

Cooper et al. [6] formulated the slug test problem in mathematical terms by imposing suitable initial and boundary conditions on Eq. (6.16). At the *face of the well*, that is, at the outer edge of the screen, they assumed that the head is equal to the head in the well at any time t: that is,

$$h\,(r_w, t) = H\,(t), \quad t > 0.$$

They also presumed an aquifer of infinite extent in the sense that the aquifer is so large that the boundary is never impacted by the test. One can state this mathematically as

$$\lim_{r \to \infty} h\,(r, t) = 0, \quad t > 0.$$

Conservation of mass between the well and the aquifer is written

$$2\pi r_w T \frac{\partial h\,(r_w, t)}{\partial r} = \pi r_w^2 \frac{\partial H\,(t)}{\partial t}, \quad t > 0,$$

where the left-hand side describes the flow out of the well and the right-hand side describes the change in volume of excess fluid in the well bore. For convenience the initial head is taken as zero everywhere:

$$h\,(r, 0) = 0, \quad r > r_w.$$

Finally, the initial excess head in the well is determined from the volume of the slug, realized through the introduction of either excess water or a metal cylinder as

$$H\,(0) = \frac{V}{\pi r_w^2}. \tag{6.17}$$

The solution to this equation for the head inside the well is

$$\frac{H}{H_0} = F(\alpha, \beta),$$ (6.18)

where

$$F(\alpha, \beta) = \frac{8u^2}{\alpha} \int_0^\infty \exp\left(-\frac{\beta u^2}{\alpha}\right) \frac{1}{u\,\Delta(u)} du.$$ (6.19)

The variables appearing in Eq. (6.19) are as follows:

$$\alpha = \frac{r_w^2 S}{r_c^2},$$ (6.20a)

$$\beta = \frac{Tt}{r_c^2},$$ (6.20b)

and

$$\Delta(u) = \left[u J_0(u) - 2\alpha J_1(u)^2\right] + \left[u Y_0(u) - 2\alpha Y_1(u)\right]^2,$$ (6.20c)

where J_0 and Y_0 are zero-order and first-order *Bessel functions* of the first kind and second kind, respectively [7]. Because the Bessel functions are of importance only for the evaluation of this solution and the values of these functions are tabulated in the literature (see [8]), we will not discuss them further here.

The approach that the authors proposed for the use of the solution given by Eq. (6.18) involves a *curve-matching procedure* (this procedure is described in more detail in Section 6.2.2). The first step is to obtain a plot of the values of $H(t)/H_0$ versus the logarithm of Tt/r_c^2 for selected values of α as determined from Eq. (6.18). The values of $H(t)/H_0$ should be plotted along the ordinate (vertical axis) of the graph. The required "type curves" are available in the literature so it is not essential that they be computed.

The next step is to determine the value of H_0. Two methods can be used to make this determination. If the value of H_0 was measured, the measured value can be used directly. If it was not, but the volume of the slug is known, the value of H_0 can be calculated from Eq. (6.17).

The next step is to plot $H(t)/H_0$ versus the logarithm of t. Once again, $H(t)/H_0$ should be plotted along the ordinate of the graph.

We now have two curves that purportedly represent the same data, one is the plot of the field data and the other is one of the family of plots that were created from Eq. (6.18). If one can determine the type curve that is, in some sense, the closest to that of the field data curve, then it is possible to determine the field parameters. To do this one first overlays the curve of the data on the family of type curves (it is convenient to have the data plotted on a transparency or to use a light table to do this). Next, one shifts the data plot, keeping the ordinates and the abscissas parallel, until the data "fit" one of the type curves. The α value associated with this best-fitting curve is recorded.

At this point, the two sets of values on the abscissas satisfy the data and the selected type curve. Thus one can select an arbitrary point along the abscissa of the data plot and determine the corresponding value on the abscissa of the selected type curve. A corresponding value of t and $\beta = Tt/r_c^2$ is thereby selected. Given the known values of

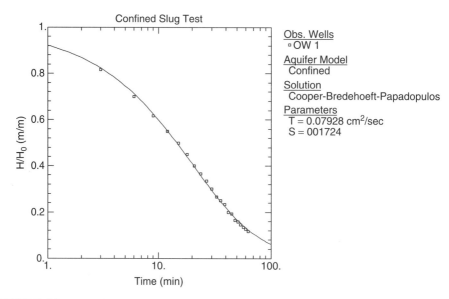

FIGURE 6.4. Plot of data and best-fitting type curve for analysis via the Cooper–Bredehoeft–Papadopulos [6] curve-fitting approach (from [2] using data from [6]).

t, Tt/r_c^2, and r_c, one can calculate a unique value for T. Given the known value of α and Eq. (6.20a), one can calculate S, the storage coefficient.

Using currently available software (e.g., see the AQTESOLVE web site at http://www.aqtesolv.com/), the curve matching is done automatically via optimization techniques. The results of such an automatic calculation are presented in Figure 6.4.

In comparing the results of the Hvorslev and Cooper–Bredehoeft–Papadopulos approaches, one must convert the transmissivity value to hydraulic conductivity or vice versa. Since the aquifer is 100 m thick, we can obtain the hydraulic conductivity from the transmissivity from

$$K = \frac{T}{b} = \frac{8.95 \times 10^{-2} \text{ cm}^2/\text{s}}{1.00 \times 10^4 \text{ cm}} = 8.95 \times 10^{-6} \text{ cm/s},$$

which compares extraordinarily well with the hydraulic conductivity value of $K = 8.94 \times 10^{-6}$ cm/s obtained using Hvorslev's method.[4] While the storage coefficient is obtained from this analysis, it has, by the nature of the matching procedure, less accuracy than does the transmissivity.

6.2 PUMPING TESTS

Pumping tests involve pumping a well at a prespecified rate of discharge and observing the resultant change in water level in the pumping well and, in general, nearby observation wells. The approach taken to analyze a pumping test may involve the use of linear

---

[4]We use K to denote K_{rr} to simplify the notation, given there is little likelihood of confusion.

relationships such as employed in the Hvorslev approach (see Section 6.1.1) or curve-matching techniques similar to the approach used in the Cooper–Bredehoeft–Papadopulos method (see Section 6.1.2).

An enormous literature has developed dedicated to the analysis of pumping tests for groundwater environments of varying degrees of complexity. An especially comprehensive treatise detailing much of this work was written by Batu [7]. Another helpful practical reference is Kruseman and Ridder [9].

6.2.1 Thiem Method

In general, we will proceed in our consideration of pumping test methods from the simple to the more complex. In so doing, we will usually be considering scientific contributions made in approximate chronological order. The simplest and one of the earliest methods of well-test analysis was introduced by Thiem in 1906 [10] and is known as the *Thiem method*. The Thiem solution was discussed in a different context in Section 5.1.3 .

Assuming a confined aquifer and steady-state flow, Thiem proposed a method to extract the aquifer transmissivity using a pumping well and two observation wells (or piezometers). The point of departure for this formulation is the steady-state fluid balance applicable to flow to a well.

Given radial flow to a well, the flow entering the well, Q^*, is the same as that crossing the vertical surface of a cylinder that fully penetrates the aquifer and has the well as its central axis of symmetry. Thus one can write

$$Q^* = -K_{rr} \frac{dh}{dr} 2\pi r l, \qquad (6.21)$$

where l is the aquifer thickness, r is the distance from the axis of symmetry, and h is defined in Eq. (6.15).

Let us assume that there is a radius r_e sufficiently distant from the well that one can assume, for all practical purposes, that the head in the aquifer is undisturbed by the pumping (radius of influence), and let us call that head value H_e. We can now integrate Eq. (6.21) to give

$$\int_{\overline{h}}^{H_e} dh = \frac{Q^*}{2\pi K_{rr} l} \int_r^{r_e} \frac{dr}{r},$$

which yields

$$H_e - h = \frac{Q^*}{2\pi K_{rr} l} \ln \frac{r_e}{r}. \qquad (6.22)$$

Eq. (6.22) is generally referred to as the *Thiem equation*.

If we assume that there are two wells at different radii from the pumping well for which water-level data are available (since we are assuming radial symmetry, the wells need not be at a constant angle θ), Eq. (6.22) can be written

$$H_2 - H_1 = \frac{Q^*}{2\pi K_{rr} l} \ln \frac{r_2}{r_1},$$

from which one can obtain $K_{rr}l$ or T from

$$K_{rr}l = T = \frac{Q^*}{2\pi (H_2 - H_1)} \ln\frac{r_2}{r_1}. \tag{6.23}$$

From Eq. (6.23) one observes that if Q^* and the two head measurements are known, $K_{rr}l$ can be calculated, or if $K_{rr}l$ is known, then Q^* can, in concept, be determined.

6.2.2   Theis Method

Theis [11], drawing on the analogy between heat flow and groundwater flow (e.g., temperature is analogous to head), developed a method for determining both the transmissivity and the storage coefficient for fully penetrating aquifers. The method requires use of a pumping well and one or more observation wells. Unlike the Thiem approach, the Theis method requires that water-level changes be recorded from the onset of the pumping. Since, given the pumping is maintained constant, the change in head tends to decline exponentially with time, the head measurements are initially made very frequently and then less often as the test proceeds. Determination of the transmissivity and storage coefficient is made via a curve-matching approach similar to that used in the Cooper–Bredehoeft–Papadopulos slug test method described in Section 6.1.2.

The point of departure for the development of this method is the form of the groundwater flow equation found in Eq. (6.16) and the associated well configuration shown in Figure 6.5. The center of the pumping well is defined as $r = 0$. One or more wells are located at different radii from the pumping well, in the case of this figure at r_1 and r_2.

$$\frac{\partial^2}{\partial r^2}h(r,t) + \frac{1}{r}\frac{\partial}{\partial r}h(r,t) - \frac{S}{T}\frac{\partial}{\partial t}(h(r,t)) = 0. \tag{6.24}$$

Rewrite Eq. (6.24) in terms of drawdown $s(r,t)$ using the relationship

$$s \equiv H_0 - h(r,t),$$

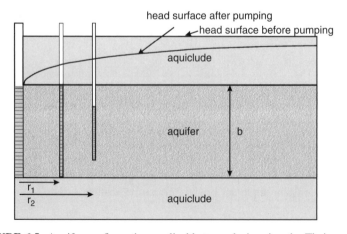

FIGURE 6.5. Aquifer configuration applicable to analysis using the Theis method.

where H_0 is the initial head value assumed constant in space. Then Eq. (6.24) becomes

$$\frac{\partial^2}{\partial r^2} s\,(r, t) + \frac{1}{r} \frac{\partial}{\partial r} s\,(r, t) - \frac{S}{T} \frac{\partial}{\partial t} \left(s\,(r, t) \right) = 0.$$

If we assume that the pumping rate is Q^* from the time $t = 0$ to $t = t$, and that the aquifer is infinite and of uniform thickness, the boundary and initial conditions are

$$s\,(r, 0) = 0,$$

$$s\,(\infty, t) = 0,$$

$$Q = 0, \quad t < 0,$$

$$Q = Q^* = \text{constant}, \quad t \geq 0,$$

$$\lim_{r \to 0} \left(r \frac{\partial s}{\partial r} \right) = \frac{-Q^*}{2\pi T}, \quad t \geq 0.$$

The solution to this set of equations [12] is

$$s\,(r, t) \equiv H_0 - \overline{h}\,(r, t) = \frac{Q^*}{4\pi T} \int_{y=u}^{\infty} \frac{e^{-y}}{y}\,dy, \tag{6.25}$$

where H_0 is the head in the aquifer at $t = 0$ and

$$u = \frac{r^2 S}{4T t}. \tag{6.26}$$

Eq. (6.25) is normally rewritten in the equivalent form

$$s\,(r, t) = \frac{Q^*}{4\pi T} W(u), \tag{6.27}$$

where

$$W\,(u) = \int_{y=u}^{\infty} \frac{e^{-y}}{y}\,dy = -Ei\,(-u),$$

and $W(u)$ is called the *well function* and $-Ei\,(-u)$ is called, in the mathematical literature, the *exponential integral*. To use the information found in Eq. (6.27), a plot is first made of $\log u$ versus[5] $\log W\,(u)$ (or alternatively $\log\,(1/u)$ versus $\log W\,(u)$). The solid heavy line in Figure 6.6 is such a plot and the resulting curve is called a *type curve*. From the information obtained from the pumping test, the logarithm of the observed drawdown $s\,(r, t)$ is plotted against the logarithm of the ratio of the elapsed time to the square of the radial distance r_{ow} of the observation well from the pumping well, that is, t/r_{ow}^2. Such a plot of data is shown by the crosses ($\times$'s) on Figure 6.6. The dashed lines and italic lettering are associated with this data plot. Keeping the two sets of horizontal and vertical axes parallel, the two plots are adjusted until the data points match the type curve optimally. One can then argue that we have solved two equations in two unknowns; the

---

[5]By the notation $\log\,(\cdot)$ we infer $\log_{10}\,(\cdot)$.

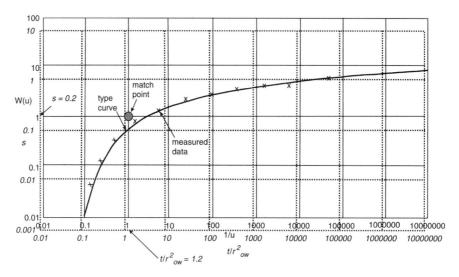

FIGURE 6.6. Illustration of matching concept using the Theis type curve. The solid lines and normal lettering denote the type curve coordinates and the dashed lines and italic lettering denote data and its coordinates.

equations are those generated by the data and by the theoretical curve and the unknowns are the storage coefficient S and the transmissivity T. Since both equations are valid anywhere within their span, we can select an arbitrary point in the overlapping portion of the graphs and argue the equivalence of the values on the ordinate and the abscissa. A convenient point to select is $W(u) = 1/u = 1$, which we call the *match point*. Let us assume that, as shown on Figure 6.6, with a pumping rate Q of 1.0 m^3/min the value of t/r_{ow}^2 that matches $1/u = 1$ is (60 min)/(7m)^2 and the corresponding value of the drawdown that matches $W(u) = 1$ is $s(r_{ow}, 60 \text{ min}) = 0.2$. From Eq. (6.27) we have

$$0.2 = \frac{Q}{4\pi T} \ (1),$$

from which we can determine

$$T = \frac{1 \text{ m}^3/\text{min}}{4\pi 0.2 \text{ m}} = 0.40 \text{ m}^2/\text{min}.$$

From Eq. (6.26) we have

$$\frac{1}{u} = 1 = \frac{4Tt}{r_{ow}^2 S} = \frac{4 \times 0.40 \text{ m}^2/\text{min} \times 60 \text{ min}}{7 \text{ m} \times 7 \text{ m} \times S}$$

or

$$S = \frac{u4Tt}{r_{ow}^2} = \frac{4 \times 0.40 \text{ m}^2/\text{min} \times 60 \text{ min}}{7 \text{ m} \times 7 \text{ m}} = 2.0.$$

Alternatively, one could replace the curve-matching strategy by using an automatic, optimization-based, optimal fit approach. The value of S and T would be determined by choosing the values of these parameters that provided the best fit of the data to the type curve. In this approach values of $1/u$ and $W(u)$ would be created using the s and t data value pairs. A measure of how well these values fit the type curve using the selected values of S and T would be determined. The procedure would be repeated for new values of S and T until the best fit of the data-based values of $1/u$ and $W(u)$ were realized. The values of S and T that resulted from this process would be deemed the best estimate of these parameters.

An example of the curve-fitting procedure is shown next. In Figure 6.7 the required information regarding the observation well characteristics and observed drawdown (displacement) as a function of time is provided. Figure 6.8 illustrates the best-fitting curve to the data and the resultant parameter estimates. In interpreting this curve (see Figure 6.9), it is helpful to realize that the coordinate axes shown apply only to the observed data, and the coordinate system in which the Theis curve is defined is not shown.

6.2.3 Cooper–Jacob Method

The Cooper–Jacob method [13] is a very useful variant on the Theis curve approach. The idea stems from the series expansion of $W(u)$, which is

$$W(u) = -0.5772 - \ln(u) + u - \frac{u^2}{2.2!} + \frac{u^3}{3.3!} - \frac{u^4}{4.4!} + \cdots, \tag{6.28}$$

where, as earlier, $u = r^2 S / 4Tt$.

Note that in Eq. (6.28), u appears as an increasing integer power after the third term. Thus, if u is small compared to unity, u gets increasingly smaller in those terms in the series beyond the third. Moreover, the $\ln(u)$ term is large relative to u for small values of u. Let us assume that we select our time values t large enough or r, the distance of the observation point from the pumping well, small enough, to assure that u is smaller than 0.01. Under these circumstances the first two terms in Eq. (6.28) dominate and we

OBSERVATION WELL DATA

No. of observation wells: 1

Observation Well No. 1:OW 1

X Location: 100. ft
Y Location: 0. ft

Radial distance from PW 1: 100. ft

Fully Penetrating Well

No. of Observations: 10

Observation Data

Time (min)	Displacement (ft)	Time (min)	Displacement (ft)	Time (min)	Displacement (ft)
0.5	0.0489	50.	3.355	5000.	7.94
1.	0.2194	100.	4.038	10000.	8.633
5.	1.223	500.	5.639		
10.	1.823	1000.	6.332		

FIGURE 6.7. Input data for the automatic determination of S and T using a least-squares fitting procedure (from [2]).

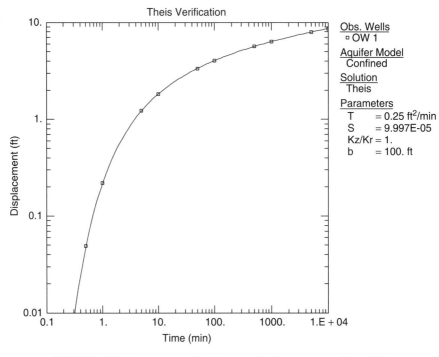

FIGURE 6.8. Least-squares fit of data to Theis type curve (from [2]).

AUTOMATIC ESTIMATION RESULTS

Estimated Parameters

Parameter	Estimate	Std. Error	
T	0.25	9.145E-06	ft^2/min
S	9.997E-05	1.738E-08	
Kz/Kr	1.	not estimated	
b	100.	not estimated	ft

K = T/b = 0.0025 ft/min

Parameter Correlations

	T	S
T	1.00	-0.86
S	-0.86	1.00

Residual Statistics

for weighted residuals

Sum of Squares. . . 4.799E-07 ft^2
Variance. . . . . 5.998E-08 ft^2
Std. Deviation. . . . . 0.0002449 ft
Mean. . . . . -9.196E-06 ft
No. of Residuals. . . 10
No. of Estimates. . . 2

FIGURE 6.9. Statistics descriptive of the curve-fitting procedure illustrated in Figure 6.8 (from [2]).

can write Eq. (6.27) as

$$s\left(r, t\right) \approx \frac{Q}{4\pi T}\left(-0.5772 - \ln\left(u\right)\right). \qquad (6.29)$$

Expansion of this expression yields

$$s\left(r, t\right) \approx -\frac{Q}{4\pi T}0.5772 - \frac{Q}{4\pi T}\ln\left(u\right), \qquad (6.30)$$

which is the expression of a straight line relating s and $\ln\left(u\right)$. Thus, if one plots s versus $\ln(u)$, the slope of the line is $Q/4\pi T$ from which one can extract T, given Q.

To obtain the value of S, one determines the intercept between Eq. (6.30) and the time axis, that is, where $s = 0$. Let us call that time t_0. At this point

$$0 = -\frac{Q}{4\pi T}0.5772 - \frac{Q}{4\pi T}\ln\left(u\right).$$

Rewrite the equation as

$$0.5772 = \ln\left(\frac{1}{u}\right);$$

substituting for u, we obtain

$$0.5772 = \ln\left(\frac{4T t_0}{r^2 S}\right).$$

We can take the antilog to obtain

$$\exp\left(0.5772\right) = \frac{4T t_0}{r^2 S}$$

or

$$S = \left(\frac{2.25 T t_0}{r^2}\right).$$

An alternative strategy exists for determining S that does not require interpolation. This alternative approach is described in Lohman [14] but will not be discussed here.

In common practice, $\log(\cdot)$ is more frequently used than the natural logarithm $\ln(\cdot)$ such that the appropriate equation becomes

$$s = \frac{2.30 Q}{4\pi T}\log\left(\frac{2.25 T t}{r^2 S}\right). \qquad (6.31)$$

The procedure used to obtain T and S from Eq. (6.31) is the same as that outlined above for the case of the natural logarithm.

In Figure 6.10 we present the least-squares fit of the same drawdown and time data used in Figure 6.8. The procedure outlined above was used to determine the optimal T and S parameters employing the Cooper–Jacob method. A comparison of the values obtained using these two approaches reveals that they are identical to four significant figures.

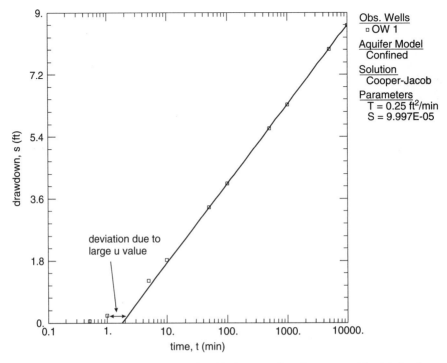

FIGURE 6.10. The best-fitting linear relationship between drawdown, s, and $\log(u)$ for the same example provided in Figure 6.8. The figure is an adaptation of one provided in Duffield [2].

6.2.4 Papadopulos–Cooper Method

Papadopulos and Cooper [15] extended the work of Theis to include the effects of a finite-diameter well. The solution generated by Theis had assumed a well of very small diameter such that storage in the well could be neglected. Figure 6.11 illustrates the aquifer characteristics applicable to analysis by their method.

The equation governing the time-dependent head behavior in this system is given by Eq. (6.24). The initial conditions are $s(r, 0) = 0$, $r \geq r_w$ in the aquifer and $s_w(0) = 0$ in the well casing. The boundary conditions are

$$s(r_w, t) = s_w(t),\tag{6.32}$$

$$\lim_{r \to \infty} s(r, t) = 0,\tag{6.33}$$

and

$$2\pi r_w T \frac{\partial s(r_w, t)}{\partial r} - \pi r_c^2 \frac{\partial s_w(t)}{\partial t} = -Q, \quad t \geq 0,\tag{6.34}$$

where $s_w(t)$ is the drawdown in the well. Eq. (6.32) states that the drawdown at the well screen is equal to that in the well, or in some instances the casing, which may have a different diameter than the well screen. Eq. (6.33) states that there will be no drawdown at the aquifer boundaries, which are located at infinity. The final boundary condition,

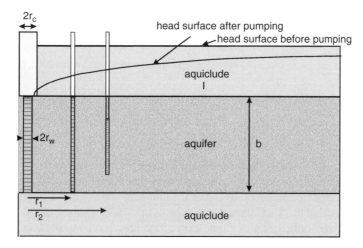

FIGURE 6.11. Aquifer system applicable to analysis by the Papadopoulos–Cooper [15] finite-diameter well method. The diameter of the well casing is $2r_c$ and the diameter of the screen is $2r_w$.

Eq. (6.34), says that flow from the well, Q, is balanced by the flow of water into the well plus the rate of change of storage in the well.

The solution to this set of equations is given by

$$s(r, t) = \frac{Q}{4\pi T} F(u, \alpha, \rho),$$ (6.35)

where

$$F(u, \alpha, \rho) = \frac{8\alpha}{\pi} \int_0^\infty \frac{C(\beta)}{D(\beta)\beta^2} d\beta$$

and

$$C(\beta) = \left[1 - \exp\left(-\beta^2 \frac{\rho^2}{4u}\right)\right] [J_0(\beta\rho) A(\beta) - Y_0(\beta\rho) B(\beta)],$$

$$A(\beta) = \beta Y_0(\beta) - 2\alpha Y_1(\beta),$$

$$B(\beta) = \beta J_0(\beta) - 2\alpha J_1(\beta),$$

$$D(\beta) = [A(\beta)]^2 + [B(\beta)]^2,$$

$$\alpha = \frac{r_w^2 S}{r_c^2},$$

and

$$\rho = \frac{r}{r_w}.$$ (6.36)

The functions J_0, Y_0, and Y_1 are zero-order and first-order Bessel functions of the first and second kind, respectively.

Drawdown in the pumped well can be obtained directly from Eq. (6.35). Substitution of $r = r_w$ in Eq. (6.36) yields $\rho = 1$ and from Eq. (6.35) we obtain

$$s_w(t) = \frac{Q}{2\pi T} F(u_w, \alpha), \qquad (6.37)$$

where

$$u_w = \frac{r_w^2 S}{4Tt}. \qquad (6.38)$$

The Papadopulos–Cooper method of aquifer analysis uses observations made in the pumping well. In this sense it is different from the Theis approach, which depends primarily on observation wells exclusive of the pumping well. A curve-matching strategy similar to that used in the Theis approach can be used to determine the aquifer coefficients T and S. However, rather than one type curve, one now encounters a family of curves, each generated using a different value of the parameter $\alpha = r_w^2 S / r_c^2$. On the vertical axis of the type curve plot is represented the log of $F(u_w, \alpha)$ and on the horizontal we plot the log of $1/u_w$. The corresponding data plot represents s_w on the vertical axis and the log of t on the horizontal. As in the Theis approach, one shifts the data plot, keeping the corresponding horizontal and vertical axes parallel, until the data overlies one of the type curves optimally. A match point is selected and the corresponding values of $F(u_w, \alpha)$ and s_w are determined. Given this information and a knowledge of the pumping rate, Eq. (6.37) is used to obtain T. The corresponding values of u_w and t are then used to obtain the value of S from Eq. (6.38).

It is also possible to use optimization software to obtain the best-fitting parameters. A solution obtained using this approach is presented in Figure 6.12. Notice the comparison with the Theis curve. Initially there is more drawdown using the small-diameter well assumption. During this period water in the well is being depleted. Eventually this source of water is exhausted and the two type curves become coincident in later time.

6.2.5 Hantush Leaky Aquitard Method

Hantush [17] extended the work of Theis to include the possibility that the upper boundary of the aquifer could consist of an aquitard, that is, a relatively low-permeability unit, but one that nevertheless is capable of providing significant quantities of water through vertical leakage. Figure 6.13 illustrates the type of system amenable to analysis using this approach. The aquifer is bounded from above by an *aquitard* (aquitard A), which is a geological unit considered to be of low permeability relative to the aquifer but capable of providing water to the aquifer via *vertical leakage*. The base of the aquifer is also bounded by an aquitard (aquitard B). Below the aquitard is an almost impermeable aquiclude. An important assumption in this regard is that the water in the aquitards moves only vertically; there is no horizontal component. Above the aquitard is an unpumped aquifer, which maintains a constant head during the pumping test. As in earlier methods, the aquifer is assumed to be of uniform thickness, to have infinite areal extent, and to be homogeneous.

The groundwater flow equation that describes this system is a variant on Eq. (6.14):

$$\frac{\partial^2}{\partial r^2}\overline{h}(r, t) + \frac{1}{r}\frac{\partial}{\partial r}\overline{h}(r, t) - \frac{S}{T}\frac{\partial}{\partial t}\left(\overline{h}(r, t)\right) + \frac{q_z' - q_z''}{T} = 0, \qquad (6.39)$$

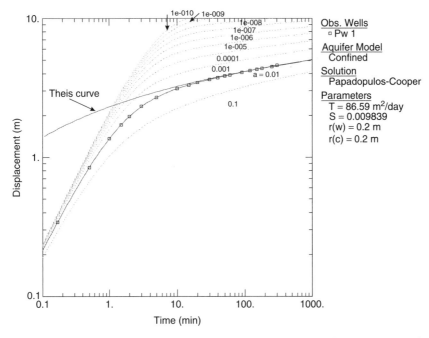

FIGURE 6.12. Optimal match of field data to type curves generated using the Papadopoulos–Cooper large-diameter well method (from [2] using data from [16]).

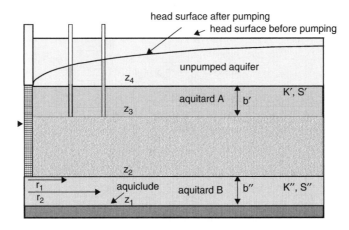

FIGURE 6.13. Definition sketch of aquifer system addressed by Hantush leaky aquitard solution.

where q_z' and q_z'' are the vertical fluxes in aquitard A and aquitard B, respectively. They provide a source of water to the aquifer through vertical drainage. It is convenient to express q_z' and q_z'' in terms of Darcy's law, remove the overbars for clarity of presentation, and then substitute the result into Eq. (6.39) to obtain

$$\frac{\partial^2}{\partial r^2} h\,(r, t) + \frac{1}{r} \frac{\partial}{\partial r} h\,(r, t) - \frac{S}{T} \frac{\partial}{\partial t}\,(h\,(r, t)) + \frac{K'}{T} \frac{\partial h_A}{\partial z} - \frac{K''}{T} \frac{\partial h_B}{\partial z} = 0. \qquad (6.40)$$

The one-dimensional form of the groundwater flow equation can be used to describe the transient head distribution in the aquitards, that is, Eq. (4.68), rewritten here using appropriate notation:

$$-\frac{\partial}{\partial z}\left(\frac{\partial}{\partial z}\left(h_A\left(z,t\right)\right)\right) + \frac{S'}{K'b'}\frac{\partial}{\partial t}\left(h_A\left(z,t\right)\right) = 0 \qquad (6.41)$$

and

$$-\frac{\partial}{\partial z}\left(\frac{\partial}{\partial z}\left(h_B\left(z,t\right)\right)\right) + \frac{S''}{K''b''}\frac{\partial}{\partial t}\left(h_B\left(z,t\right)\right) = 0. \qquad (6.42)$$

The solution of this set of equations requires specification of initial and boundary values for each of the state variables h, h_A, and h_B. To simplify notation and to be consistent with the general literature, we will substitute drawdown s for the head h in Eq. (6.40), (6.41), and (6.42) according to the relationships $s = H - h$ and $s_n = H_n - h_n$, where $n = A$, B, and H and H_n are the initial head values in the system. Having made this transformation, the auxiliary conditions (boundary conditions and initial conditions) can be stated as follows: for the upper aquitard the initial condition is

$$s_A\left(r,z,0\right) = 0$$

and the boundary condition at the top of the upper aquitard is

$$s_A\left(r,z_4,t\right) = 0$$

and that at the bottom is

$$s_A\left(r,z_3,t\right) = s\left(r,t\right),$$

where z_1, z_2, z_3, and z_4 are defined in Figure 6.13.

For the aquifer the initial condition is given as

$$s\left(r,0\right) = 0$$

and the boundary condition at $r \to \infty$ is

$$\lim_{r\to\infty} s\left(r,t\right) = 0.$$

The boundary condition at the infinitely small well bore is based on the rate of discharge and Darcy's law and is given by

$$\lim_{r\to 0}\left(r\frac{\partial s}{\partial r}\right) = -\frac{Q}{2\pi T}.$$

To understand the meaning of this relationship, it is convenient to cross-multiply the r and T such that the Darcy flow to the well is balanced by the flow through the well perimeter, which has a circumference of $2\pi r$.

The initial condition in the lower aquitard is given by

$$s_B (r, z, 0) = 0.$$

The boundary condition on the top of the bottom aquitard (i.e., is $z = z_2$) is

$$s_B (r, z_{2,}t) = s (r, t)$$

and on the bottom of the bottom aquitard (i.e., $z = z_1$) is

$$\frac{\partial s_B (r, z_1, t)}{\partial z} = 0.$$

In practice a restricted form of the general solution to the above defined problem is used. A small time solution was suggested by Hantush and is discussed in Batu [7]. The necessary conditions for the application of this solution are

$$\frac{b'S'}{K'} \geq 10r \tag{6.43}$$

and

$$\frac{b''S''}{K''} \geq 10r. \tag{6.44}$$

The solution form is

$$s (r, t) = \frac{Q}{4\pi T} H (u, \beta),$$

where

$$H (u, \beta) = \int_u^\infty \frac{e^{-y}}{y} \mathrm{erfc} \left\{ \frac{\beta u^{1/2}}{[y (y - u)]^{1/2}} \right\} dy, \tag{6.45}$$

$$u = \frac{r^2 S}{4Tt}, \tag{6.46}$$

and

$$\beta = \frac{r}{4} \left[\left(\frac{K'S'}{b'TS}\right)^{1/2} + \left(\frac{K''S''}{b''TS}\right)^{1/2} \right]. \tag{6.47}$$

Eq. (6.45) represents a family of curves, each curve corresponding to one value of the parameter β. A curve-matching procedure can be used that is analogous to that presented previously for the Theis method. However, in this case one determines not only values for $H (u, \beta)$ and u but also a value of β for each observation well. From the corresponding values of $H (u, \beta)$ and $s (r, t)$, compute the value of T, given Q is known and using a convenient match point. Then use the computed T value in combination with Eq. (6.46) and corresponding values of $1/u$ and t to obtain S. If there are two observation wells, and therefore two values of β, Eq. (6.47) can be written for each value of β. The resulting set of equations can be used to solve for the products $K'S'$ and $K''S''$, all other parameters in the equations being known from either field measurements or prior calculation. If the specific storage can be estimated from other information, the hydraulic conductivity can be determined. Given the estimated values of $K'S'$ and $K''S''$, the criteria presented in

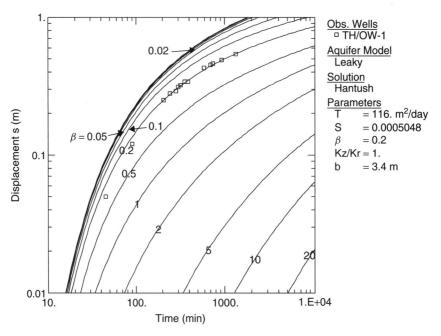

FIGURE 6.14. Determination of aquifer properties using the Hantush leaky aquifer with storage solution (from [2], with data from [7]).

Eq. (6.43) and (6.44) can be tested for compliance. If these conditions are not met, an alterative formulation developed by Hantush for large times can be employed (see [7]).

An example calculation taken from Duffield [2] is presented in Figure 6.14. In this example there is but one observation well and thus only one value of β can be determined. The heavy curve labeled 0.2 is the one selected as optimal by the automatic curve-matching algorithm.

6.2.6 Neuman Method for Unconfined Aquifers

While the Theis method (see Section 6.2.2) and the Cooper–Jacob method (see Section 6.2.3) could be adapted to accommodate unconfined aquifers, Neuman [18] extended their work by explicitly addressing the analysis of unconfined aquifers. The system being considered is shown in Figure 6.15. The pumping well is partially penetrating and the water-table surface responds to pumping. An observation well is located within the zone of influence of the pumping well.

The groundwater flow equation applicable to this system is the primitive (nonaveraged) form of Eq. (6.48):

$$\frac{\partial}{\partial r}\left(K_{rr}\frac{\partial}{\partial r}\left(s\left(r, z, t\right)\right)\right) + \frac{1}{r}\left(K_{rr}\frac{\partial}{\partial r}\left(s\left(r, z, t\right)\right)\right)$$

$$+ \frac{\partial}{\partial z}\left(K_{zz}\frac{\partial}{\partial z}\left(s\left(r, z, t\right)\right)\right) - S_s\frac{\partial}{\partial t}\left(s\left(r, z, t\right)\right) = 0. \tag{6.48}$$

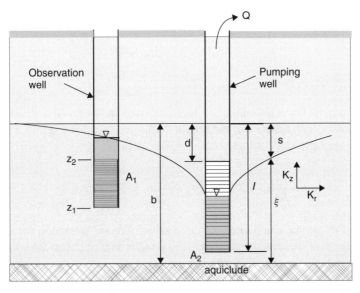

FIGURE 6.15. Definition sketch for physical system addressed by the Neuman method for unconfined aquifers.

The initial conditions applicable to this problem [7] are

$$s(r, z, 0) = 0,$$

which states that initially there is no drawdown, and

$$\xi(r, 0) = b,$$

which says that the water table is initially at b. The boundary condition at the infinite boundary is

$$\lim_{r \to \infty} s(r, z, t) = 0,$$

which states that at the furthest extent of the aquifer (or at least at the radius of influence), there is no drawdown. At the well, the condition for the saturated segment of the well that is identified with the impermeable bottom of the casing (A_2 in Figure 6.15) is

$$\frac{\partial}{\partial z}(s(r, 0, t)) = 0.$$

On the other hand, the condition over the screened segment of the well is that what is leaving the well comes through the saturated segment of the screen:

$$\lim_{r \to 0} \int_{b-l}^{\min(b-d, \xi)} r \frac{\partial s}{\partial r} dz = -\frac{Q}{2\pi K_{rr}}. \tag{6.49}$$

If the water level is below the top of the screen, the upper limit of integration in Eq. (6.49) is ξ. For those segments of the well for which there is no screen, the boundary condition is

$$\frac{\partial s\,(0, z, t)}{\partial z} = 0, \quad 0 \leq z \leq b - l, \quad b - d \leq z \leq d.$$

The free-surface boundary condition was discussed in Section 4.4. Neglecting infiltration, an equivalent expression written in terms of drawdown $s\,(r, z, t)$ and using cylindrical coordinates (see Section 4.8) is given as [7]

$$K_{rr} \frac{\partial s}{\partial r} n_r + K_{zz} \frac{\partial s}{\partial z} n_z - S_y \frac{\partial \xi}{\partial t} n_z = 0 \quad \text{at } x = \xi, \tag{6.50}$$

where n_r and n_z are the components of the outward directed normal to the free surface, ξ is the elevation of the free surface above the same reference datum for which the head is defined, and S_y is the specific yield.

The flow equation is written for the purposes of this analysis as

$$\frac{\partial^2 s}{\partial r^2} + \frac{1}{r} \frac{\partial s}{\partial r} + K_D \frac{\partial^2 s}{\partial z^2} - \frac{1}{\alpha_s} \frac{\partial s}{\partial t} = 0, \quad 0 < z < b,$$

where $K_D \equiv K_z / K_r$ and $\alpha_s \equiv K_r / S_s$.

If one assumes that the drawdown generated in the course of the pumping test is small relative to the saturated thickness of the aquifer, that is, $s \ll \xi$, one can linearize the boundary condition Eq. (6.50), which results in a problem defined by the following boundary conditions:

$$\lim_{r \to 0} \int_0^b r \frac{\partial s}{\partial r} ds = -\frac{Q}{2\pi K_{rr}},$$

$$\frac{\partial s\,(r, b, t)}{\partial z} = -\frac{1}{\alpha_y} \frac{\partial s\,(r, b, t)}{\partial t},$$

where $\alpha_y = K_z / S_y$. A further assumption that was made is that the flux along the screened portion of the well is uniform, giving rise to the condition

$$\lim_{r \to 0} \left(r \frac{\partial s}{\partial r} \right) = -\frac{Q}{2\pi K_{rr}\,(l - d)}, \quad b - l < z < b - d.$$

The solution to this set of equations is

$$s\,(r, z, t) = \frac{Q}{4\pi T} \int_0^\infty 4y J_0 \left(y\beta^{1/2} \right) \left[u_0\,(y) + \sum_{n=1}^\infty u_n\,(y) \right] dy$$

$$= \frac{Q}{4\pi T} s_D,$$

where

$$u_0(y) = \frac{\left\{1 - \exp\left[-t_s\beta\left(y^2 - \gamma_0^2\right)\right]\right\}\cosh(\gamma_0 z_D)}{\left[y^2 + (1+\sigma)\gamma_0^2 - \left(y^2 - \gamma_0^2\right)^2/\sigma\right]\cosh(\gamma_0)}$$

$$\cdot \frac{\sinh\left[\gamma_0(1-d_D)\right] - \sinh\left[\gamma_0(1-l_D)\right]}{(l_D - d_D)\sinh(\gamma_0)}$$

and

$$u_n(y) = \frac{\left\{1 - \exp\left[-t_s\beta\left(y^2 + \gamma_n^2\right)\right]\right\}\cos(\gamma_n z_D)}{\left[y^2 - (1+\sigma)\gamma_n^2 - \left(y^2 + \gamma_n^2\right)^2/\sigma\right]\cos(\gamma_n)}$$

$$\cdot \frac{\sin\left[\gamma_n(1-d_D)\right] - \sin\left[\gamma_n(1-l_D)\right]}{(l_D - d_D)\sin(\gamma_n)},$$

where J_0 is the Bessel function of the first kind of zero order and

$$\beta = K_D\left(\frac{r}{b}\right)^2 = \frac{K_{zz}}{K_{rr}}\left(\frac{r}{b}\right)^2,$$

$$t_s = \frac{Tt}{S_s b r^2} = \frac{Tt}{S r^2},$$

$$d_D = \frac{d}{b}, l_D = \frac{1}{b}, z_D = \frac{z}{b},$$

$$\sigma = \frac{S}{S_y} = \frac{t_y}{t_s},$$

$$t_y = \frac{Tt}{S_y r^2}.$$

The terms γ_0 and γ_n are given by the roots of the equations

$$\sigma\gamma_0\sinh(\gamma_0) - \left(y^2 - \gamma_0^2\right)\cosh(\gamma_0) = 0, \quad \gamma_0^2 < y^2$$

and

$$\sigma\gamma_n\sin(\gamma_n) + \left(y^2 + \gamma_n^2\right)\cos(\gamma_n) = 0,$$

where

$$(2n - 1)\frac{\pi}{2} < \gamma_n < n\pi, \quad n \geq 1.$$

The type curves that can be generated from the Neuman solution are shown along with sample data in Figure 6.16.

While we have focused on the Neuman method for unconfined aquifers, there are other methods that are applicable to unconfined aquifers as well, for example, the work of Streltsova [19] and more recently that of Moench [20]. We suggest the interested reader review these publications.

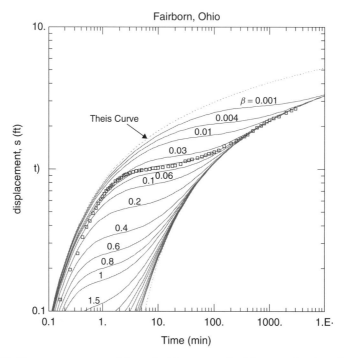

FIGURE 6.16. Type curves generated from the Neuman method for unconfined aquifers. The data are from a pump test conducted by Walton [21] and reported by Lohman [22]. The fitted curves were provided by Duffield [2].

6.2.7 Theis Recovery Method

To this point we have focused on using the response generated in an aquifer due to a prespecified pumping protocol. More specifically, we have mathematically derived drawdown at one or more wells in response to a pumping well to create type curves that describe the behavior of related idealized aquifers of varying but related properties and characteristics.

However, at the end of any pumping test, the wells are turned off and the aquifer recovers to its prepumping state. The recovery stage during which the water levels return to their prepumping conditions (assuming there have been no other changes in the system during the pumping test save the pumping itself) is a period during which the water-level changes in the aquifer can be used to advantage. The difference between the drawdown and the initial state of the system prior to initiating the pumping test is called the *residual drawdown*. During the recovery period, the residual drawdown is measured and recorded over time much as was the drawdown when the pumping test began. During the initial stages of the recovery, measurements of residual drawdown are often made, often decreasing in frequency logarithmically with time.

The most conceptually convenient way to analyze a recovery test is to think of the pumping rate as constant over the entire measurement period but with the introduction of a counterintuitive twist. At the point in time when the well stops pumping, let us call it t_{off}, introduce a new imaginary well injecting water at the same rate as the withdrawal well is pumping. The sum of these two pumping rates is zero from the point in time when the injection well is turned on (the production well is shut down) so, mathematically, this is equivalent to turning off the pumping well.

To analyze this mathematically, one can introduce a new initial time, t', which is zero at the point when the pumping well is turned off. Thus the amount of time the new well pumps will be represented by the elapsed time relative to the new reference time t'. Figure 6.17 illustrates the concept. The displacement (drawdown s) is shown in this figure as the solid line in the lower frame. As the pumping period increases, so also does the drawdown in this infinitely large aquifer. At time t_{off} the well is turned off, which is represented in this figure by the activation of the recharge well, denoted here as

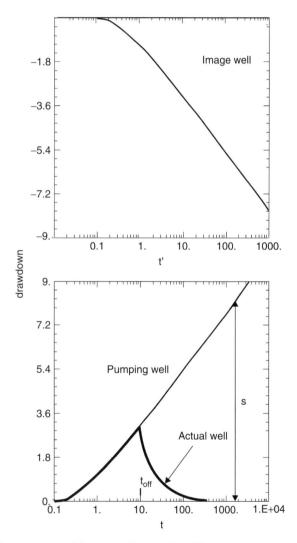

FIGURE 6.17. Curves generated from pumping test data illustrating the concept of recovery. The solid line in the lower frame is the pumping well, the line in the upper frame is the image well (recharge), and the bold line is the sum of the two effects, yielding the curve to be expected in the field.

the image well (note that the horizontal axes in the upper and lower frames are offset), and represented by the solid line in the upper frame. The sum of these two effects is represented by the bold curve, which represents the displacement one would expect to see in the observed well at this site. The concept of adding the solutions to these two linear equations is known as superposition. We will learn more about this shortly.

We now examine how one uses this concept of recovery in the determination of aquifer coefficients. The *residual drawdown* we define as s'. Using the Theis model, one can express s' as

$$s' = \frac{Q}{4\pi T} \left[W(u) - W(u') \right], \tag{6.51}$$

where

$$u = \frac{r^2 S}{4Tt}, \quad \text{and} \quad u' = \frac{r^2 S}{4Tt'}.$$

The term $(Q/4\pi T) W(u)$ describes the drawdown due to the pumping well and $(Q/4\pi T) W(u')$ describes the effect of the image well. The sum of the two terms describes the behavior of the recovery well (the one we actually measure in the field).

We now assume that u and u' are sufficiently small that we can use the linear approximation presented in Eq. (6.29) such that

$$s' = \frac{Q}{4\pi T} \left(\ln \frac{Tt}{r^2 S} - \ln \frac{Tt'}{r^2 S} \right) \tag{6.52}$$

or, using the rules of subtraction of logarithms,

$$s' = \frac{Q}{4\pi T} \ln \left(\frac{t}{t'} \right).$$

In terms of $\log(\cdot)$ this expression reads

$$s' = \frac{2.303 Q}{4\pi T} \log \left(\frac{t}{t'} \right). \tag{6.53}$$

To apply Eq. (6.53) to the determination of the transmissivity T, one plots the residual drawdown s' against the logarithm of the ratio of the elapsed real time t to the time since pumping stopped, t'. The slope of the resulting line gives the value of $2.303 Q/4\pi T$, which can be used to determine the transmissivity given the pumping rate Q. An example of this kind of calculation is found in Figure 6.18.

An estimate of the ratio of the storage coefficient calculated during pumping S to that obtained during recovery S' can be estimated. Ideally, it should be unity since this is a constant aquifer parameter. To see how this is calculated, return to Eq. (6.52) rewritten to identify the two variants on the storage coefficient as

$$s' = \frac{Q}{4\pi T} \left(\ln \frac{Tt}{r^2 S} - \ln \frac{Tt'}{r^2 S'} \right), \tag{6.54}$$

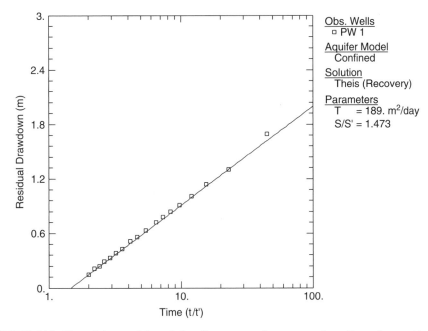

FIGURE 6.18. Plot of data and best-fitting linear curve for recover data. Example provided by Duffield [2] base on data from Batu [7].

where the s' curve encounters the time axis, $s' = 0$, and we have

$$0 = \frac{Q}{4\pi T} \left(\ln\frac{Tt}{r^2 S} - \ln\frac{Tt'}{r^2 S'} \right)_{s'=0},$$

which can be written

$$0 = \frac{Q}{4\pi T} \left[\ln\left(\frac{t}{t'}\right)\bigg|_{s'=0} - \ln\left(\frac{S}{S'}\right) \right].$$

Since $Q/4\pi T \neq 0$, then

$$\ln\left(\frac{t}{t'}\right)\bigg|_{s'=0} = \ln\left(\frac{S}{S'}\right)$$

and therefore

$$\left(\frac{t}{t'}\right)\bigg|_{s'=0} = \frac{S}{S'}.$$

Because at long time (small s') the ratio t/t' should approach unity (indicating S to be the same as S', as it should be), the deviation of their ratio from unity indicates the degree of influence of unaccounted-for factors such as boundaries (e.g., nearby lakes or streams) on the calculations.

An alternate strategy can be used that allows for the direct determination of S. We begin by rewriting Eq. (6.51):

$$s' = \frac{Q}{4\pi T} \left[W(u) - W(u') \right],$$ (6.55)

where t is the time since pumping started and t' is the time since pumping stopped,

$$u = \frac{r^2 S}{4Tt}, \quad \text{and} \quad u' = \frac{r^2 S}{4Tt'}.$$

Goode [23] proposed a new set of dimensionless variables that permits the determination of the storage coefficient. He defined dimensionless drawdown s_D as $s_D \equiv 4\pi T s/Q$, dimensionless time t_D as $t_D \equiv Tt/r^2 S = 1/4u$, and dimensionless time since pumping started as $t_{pD} = Tt_p/r^2 S$, where t_p is the duration of pumping. Using these definitions, Eq. (6.55) becomes

$$\frac{s'_D Q}{4\pi T} = \frac{Q}{4\pi T} \left[W\left(\frac{r^2 S}{4T}\right) \times \frac{T}{r^2 S t_D} - W\left(\frac{r^2 S}{4T} \times \frac{T}{r^2 S \left(t_D - t_{pD}\right)}\right) \right]$$

or, dividing both sides of this equation by $Q/4\pi T$,

$$s'_D = W\left(\frac{1}{4t_D}\right) - W\left(\frac{1}{4\left(t_D - t_{pD}\right)}\right).$$

The type curves generated using this equation are found in Figure 6.19. The ordinate is equivalent to s'_D. The abscissa is the normalized time defined as $t_n \equiv (t - t_p)/t_p = (t_D - t_{pD})/t_{pD}$.

To use this type curve, one plots the normalized time on the abscissa versus the observed drawdown. Using a curve-matching technique, one first shifts the data plot vertically, with the time axis common to both plots, until the observed curve matches one of those represented by the dimensionless pumping period t_{pD}. Next, one selects a convenient match point on the overlapping portion of the two graphs. Let us assume we select the point $s'_D = 1$, $t_n = 0.01$. Since the time axis is not shifted, the value of t_n is the same for both graphs, namely, 0.1. Let us assume that the value of the drawdown corresponding to the dimensionless drawdown $s'_D = 1$ is C_1 and the best matching member of the family illustrated in Figure 6.19 is $t_{pD} = 1.0$. From the definition

$$s'_D \equiv \frac{4\pi T s'}{Q},$$

we obtain

$$T = \frac{Q s'_D}{4\pi s'} = \frac{Q}{4\pi C_1}.$$ (6.56)

For convenience let T calculated from Eq. (6.56) be defined as C_2. From

$$t_{pD} = \frac{Tt_p}{r^2 S},$$

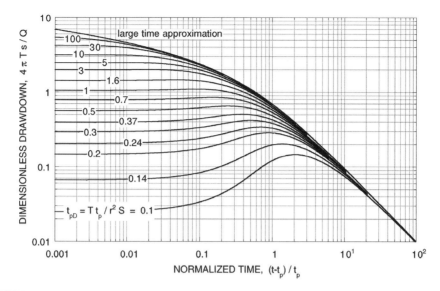

FIGURE 6.19. Type curves for the determination of the transmissivity and storage coefficient using recovery data (from [23]).

we have

$$S = \frac{Tt_p}{t_{pD}r^2} = \frac{C_2}{r^2} \times \frac{t_p}{t_{pD}}.$$

Recall that $t_{pD} = 1.0$, and that t_p is the duration of pumping and is known. Let us assume $t_p = C_3$. Then we can write

$$S = \frac{C_2 \times C_3}{r^2 \times 1.0}.$$

The radius r is normally that distance from the pumping well to the observation well where the head measurements are being made. Goode [23] describes how one can plot multiple well information and also how to scale the head values so they can be conveniently plotted using the same scale as used for the dimensionless drawdown s_D.

6.2.8 Image Well Theory

The analytical solutions that have been heretofore presented assume that the boundary of the aquifer is at an infinite distance from the pumping well. However, in many, if not most, instances physical barriers are located within the radius of a circle wherein significant drawdown is observed during a pumping test (in some sense the radius of influence). When this occurs, the barrier influences the drawdown attributable to the pumping well, and this effect should be taken into account in the parameter estimation. The impact of impermeable and constant head boundaries are addressed using image well theory. The approach is similar to that used in Section 6.2.7 for the analysis of recovery data.

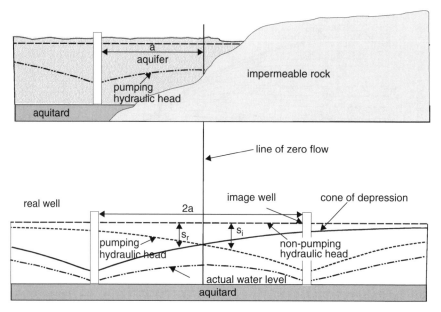

FIGURE 6.20. Illustration of the conceptual model employed in the analysis of drawdown near an impermeable boundary (adapted from [22]).

Consider the problem illustrated in the upper panel in Figure 6.20. In this cross section a pumping test is being conducted near an impermeable rock boundary. The presence of this boundary impacts the drawdown curve as seen by the asymmetry in the cone of depression generated by the pumping well.

The analytical solution for describing drawdown in an infinite aquifer can be used to accommodate this physical situation. The approach is best understood by viewing the lower panel of this figure. At some distance a from the well, it is assumed an impermeable boundary exists (denoted as the "line of zero flow" in Figure 6.20). By its nature, this location is, approximate since sharp boundaries are seldom seen in nature.

At a distance a from the line of zero flow denoted in this figure, an imaginary well (*image well*) is placed as shown on the lower panel. The distance from the *real well* to the image well is therefore $2a$. The real well and the image well are assumed to pump at the same rate Q, held constant for the duration of the test. As the figure illustrates, each pumping well generates a symmetric cone of depression. The line defining the cone created by the real well is solid, and that by the image well is dashed. If the solutions (drawdown surfaces) are summed, the resultant drawdown is indicated by the dash-dot curve. That portion of this curve located to the left of the line of zero flow mimics that shown as the pumping water level in the upper panel. Note that it is the nature of the equations that the slope of the water potential at the line of zero flow is indeed zero. Thus the formulation provides the desired no-flow effect at the impermeable rock wall.

One approach to utilizing the image well concept is to first write the drawdown at any point $s(r_o, t) \equiv s_o$ (the subscript o denotes the observation well) as the algebraic sum of the drawdown from the real and image wells, that is,

$$s_o = s_i + s_r \tag{6.57}$$

as shown in the lower panel of Figure 6.20. Using the Theis approach, Eq. (6.57) gives (see [24])

$$s_o = \frac{Q}{4\pi T} \left[W\left(u_p\right) + W\left(u_i\right) \right],$$

where

$$u_p = \frac{r_p^2 S}{4Tt} \tag{6.58}$$

and

$$u_i = \frac{r_i^2 S}{4Tt}. \tag{6.59}$$

The variable r_p is the distance from the pumping well to the observation well and r_i is the distance of the image well from the observation well. Combining Eq. (6.58) and (6.59), we obtain the relationship

$$u_i = \left(\frac{r_i}{r_p}\right)^2 u_p,$$

which can be simplified to $u_i = (\kappa)^2 u_p$, where $\kappa \equiv r_i/r_p$. A family of type curves can be created for different values of κ. The curve-matching strategy used earlier in the discussion of the Theis approach (see Section 6.2.2) can now be used to determine the transmissivity T and storage coefficient S.

A similar strategy as that presented for impermeable boundaries can be used to accommodate constant-head boundaries. Figure 6.21 illustrates the physical system to be considered. A fully penetrating stream with a constant head bounds the pumping test area. To achieve a constant head at a distance b from the pumping well, one places a recharge well injecting at the rate Q at a distance $2b$ from the pumping well. The resulting *cone of impression* exactly balances the *cone of depression* at the zero drawdown boundary. Thus the imposition of an image well at the appropriate location results in a hydrodynamic system equivalent to that created by a constant head at a distance b from the pumping well. The analysis of this physical system is analogous to that described for the constant-head boundary.

The systems we have addressed above are relatively simple. Only one boundary is addressed. Consider now the situation where multiple boundaries must be accommodated. A two-boundary example is presented in Figure 6.22.

In this situation, the recharging image well b on the right-hand side of the constant-head boundary balances the pumping well a on the left-hand side. Similarly, the discharging image well d balances well a, to create an impermeable boundary. However, this is not the end of the story. One must extend the boundaries because wells d and b must also be balanced. This is the role played by well c. Well c balances both wells d and b.

In Figure 6.23 an aquifer bounded to the east by a constant-head boundary is depicted. A pumping well is located at (0, 0). The drawdown curve and data applicable to this problem are found in Figure 6.24. Note that the drawdown does not increase indefinitely as we have seen earlier, but rather becomes constant at long times. The reason for this

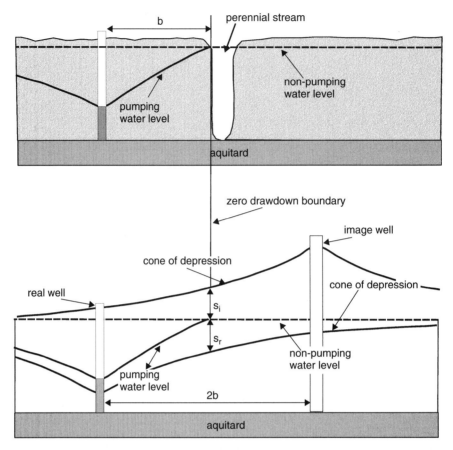

FIGURE 6.21. Illustration of the conceptual model employed in the analysis of drawdown near a constant-head boundary (adapted from [22]).

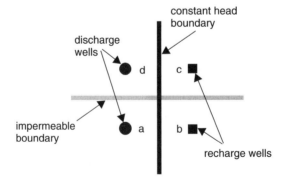

FIGURE 6.22. Diagrammatic sketch of a two-boundary example. The light grey is the impermeable boundary and the black is the constant-head boundary. The squares are recharge wells and the circles are discharge wells.

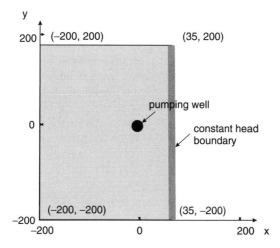

FIGURE 6.23. Definition for a constant-head boundary condition.

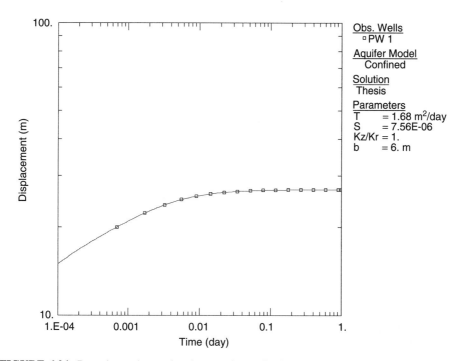

FIGURE 6.24. Drawdown observed at the pumping well when a constant-head boundary is located to the east (from [25]).

is the equilibrium established between the demand of the pumping well for fluid mass and the ability of the constant-head boundary to provide it.

If one has two parallel boundaries, the problem becomes even more complex and an infinite series of wells can result. The logic presented above for one boundary holds for two, except for the requirement of now summing over four wells.

6.3 SUMMARY

In this chapter we have introduced the concept of well hydraulics. We say "introduced" because this is a mature and broad topic with an enormous literature. We covered only three topics—(1) slug test, (2) pumping tests, and (3) image well theory. Even these topics were treated in a cursory fashion due to the scope of this book. We first defined the physical problem, followed by its mathematical description, and finally by its utility in the field. Where possible, we showed examples of how the technology could be used. The reader interested in investigating this topic further is referred to the treatise by Batu [7].

6.4 PROBLEMS

6.1. According to the slug test method of Hvorslev, the flow rate out of the well and into the formation is given by

$$Q = F K_{rr}(h - H),$$

where h is interpreted as the hydraulic head at the well, and H is the far-field value of head, assumed to be constant. (a) Write the expression for vertically averaged flow Q, in the radial direction, using Darcy's law as a basis. (b) Relate the Darcy expression from part (a) to the governing equation used in the Cooper–Bredehoeft–Papadopoulos method, given by Eq. (6.16). (c) Explain how the Hvorslev and the Darcy expressions for flow Q can be related to one another.

6.2. Use the data in Figure 6.2 to produce the graph shown in Figure 6.3, and show in detail how the hydraulic conductivity is estimated from that graph.

6.3. Evaluate and plot the well function $W(u)$ given in Eq. (6.27) and the following equation. Plot W versus u as well as W versus $1/u$. These plots should be produced on log–log paper, so that the actual plots are log W versus log u and log W versus log$(1/u)$.

6.4. Replot W versus u, but now display the plot on semilog paper: that is, plot W versus log u. On this plot, identify the values of u for which the curve plots as a straight line. Relate this to the region of the graph where the Cooper–Jacob approximation is applicable.

6.5. Use the data of Figure 6.7, together with the graph from Problem 6.3, to determine the transmissivity and storage coefficients associated with these data, using the Theis type-curve matching procedure.

6.6. Consider a system in which a well is pumped at a constant rate, beginning at $t = 0$. The pumping takes place in a homogeneous, horizontal, confined aquifer. Using the results of Problem 6.4, show that for this system the region within which the Cooper–Jacob approximation is valid expands radially in proportion to $\sqrt{t}$; that is, $r_{cj} = C\sqrt{t}$, where r_{cj} denotes the maximum radial extent away from the pumping well for which the Cooper–Jacob approximation is valid. Note that within this region, the solution has a logarithmic shape (see the Cooper–Jacob approximation equation), which corresponds to the solution to the steady-state flow problem (feel free to prove this). Note that this logarithmic solution can be seen as a transient version of the Thiem equation, where the transient part comes from the time-dependent location of the outer boundary (feel free to *think* about this and to formulate the generalized version of the Thiem solution).

6.7. Calculate values of u associated with the data points plotted in Figure 6.10. Identify those data points that should be amenable to the Cooper–Jacob approximation. Finally, explain why the Cooper–Jacob approximation is valid for *long times*, given a fixed value of distance r between the pumping and observation wells.

6.8. Explain why the leaky aquifer equation, Eq. (6.39), has a well-defined steady-state solution for all values of $K' \neq 0$, while the Theis equation (which is Eq. (6.39) with $K' = 0$) does not have a bounded steady-state solution. Explain how this observation relates to the equations and observations associated with Figure 5.6 in Chapter 5.

6.9. Consider an aquifer in which a well begins to pump, at constant rate, at time $t = 0$. At a distance b is a fixed-head boundary, like that shown in Figure 6.21. Assume that a second boundary, this one a no-flow boundary, exists at a distance b' from the pumping well, in the direction opposite to the fixed-head boundary. Therefore the pumping well experiences two parallel boundaries, on opposite sides, one that is no-flow and the other that is fixed-head, and the relevant distances are b and b'. Explain how image well theory can be used to solve this problem for the drawdowns in the aquifer between the two boundaries, and write the mathematical expression for drawdown at any point between the two boundaries.

BIBLIOGRAPHY

[1] M. J. Hvorslev, *Time Lag and Soil Permeability in Ground-Water Observations,* Bull. No. 36, Waterways Experimental Station, Corps of Engineers, U.S. Army, Vicksburg, MS, 1951, p. 50.

[2] G. M. Duffield, *AQTESOLV*, HydroSOLVE Inc., 2000.

[3] J. J. Butler, *The Design, Performance, and Analysis of Slug Tests*, Lewis Publishers, Washington, DC, 1998.

[4] H. Bouwer and R. C. Rice, A slug test method for determining hydraulic conductivity of unconfined aquifers with completely or partially penetrating wells, *Water Resour. Res.* **12**(3):423, 1976.

[5] V. Zlotnik, Interpretation of slug and packer tests in anisotropic aquifers, *Ground Water* **32**(5):761, 1994.

[6] H. H. Cooper, J. D. Bredehoeft, and S. S. Papadopulos, Response of a finite-diameter well to an instantaneous charge of water, *Water Resour. Res.* **3**(1):263 1967.

[7] V. Batu, *Aquifer Hydraulics*, John Wiley & Sons, Hoboken, NJ, 1998.

[8] S. S. Papadopulos, J. D. Bredehoeft, and H. H. Cooper, Jr., On the analysis of "slug test" data, *Water Resour. Res.* **9**(4):1087, 1973.

[9] G. P. Kruseman and N. A. de Ridder, *Analysis and Evaluation of Pumping Test Data*, 2nd ed. Publication 47, International Institute for Land Reclamation and Improvement, Wageningen, The Netherlands, 1991.

[10] A. Thiem, *Hydrologische Methoden*, Gebhardt, Leipzig, 1906.

[11] C. V. Theis, The relation between the lowering of the piezometric surface and the rate and duration of discharge of a well using ground-water storage, *Trans. Am. Geophys. Union* **16**:519, 1935.

[12] H. S. Carslaw and J. C. Jaeger, *Conduction of Heat in Solids*, 2nd ed., Oxford University Press, New York, 1959.

[13] H. H. Cooper and C. E. Jacob, A generalized graphical method for evaluating formation constants and summarizing well field history, *Trans, Am. Geophys. Union* **27**:526, 1946.

[14] S. W. Lohman, Method for determination of the coefficient of storage from straight-line plots without extrapolation, in R. Bentall, *Shortcuts and Special Problems in Aquifer Tests*, U.S. Geological Survey Water-Supply Paper 1545-C, p. C33, 1963.

[15] S. S. Papadopulos and H. H. Cooper, Drawdown in a well of large diameter, *Water Resour. Res.* **3**:241, 1967.

[16] R. S. Wikramaratna, A new type curve method for the analysis of pumping tests in large-diameter wells, *Water Resour. Res.* **21**(2):262, 1985.

[17] M. S. Hantush, Modification of the theory of leaky aquifers, *J. Geophys. Res.* **65**(11):3713, 1960.

[18] S. P. Neuman, Effect of partial penetration on flow in unconfined aquifers considering delayed gravity response, *Water Resour. Res.* **10**(2):303, 1974.

[19] T. D. Streltsova, Drawdown in compressible unconfined aquifer, *J. Hydrau. Div. Proc. Am. Soc. Civil Eng.* **100**(HY11):1601, 1974.

[20] A. F. Moench, Flow to a well of finite diameter in a homogeneous, anisotropic water table aquifer, *Water Resour. Res.* **33**(6):1397, 1997.

[21] W. C. Walton, Application and limitations of methods used to analyze pumping test data, *Water Well J. Pt. 1* **14**(2) and **14**(3), selected pages, 1960.

[22] S. W. Lohman, *Ground-Water Hydraulics*, U.S. Geological Survey Prof. Paper 708, 1979.

[23] D. J. Goode, Composite recovery type curves in normalized time from Theis' exact solution, *Ground Water* **35**(4):672, 1997.

[24] R. W. Stallman, Type curves for the solution of single boundary problems, in R. Bentall, *Shortcuts and Special Problems in Aquifer Tests,* U.S. Geological Survey Water-Supply Paper 1545-C, p. C45, 1963.

[25] G. M. Duffield, personal communication, 2002.

CHAPTER 7

NUMERICAL SOLUTIONS OF THE GROUNDWATER FLOW EQUATION

In this chapter we consider the use of numerical methods for the solution of the groundwater flow equations. We begin with a brief introduction to numerical methods as they are commonly used in groundwater flow modeling. Next, we illustrate their application in a regional setting. Insight into the accommodation of probabilistically defined parameters, such as hydraulic conductivity, follows.

7.1 INTRODUCTION TO NUMERICAL METHOD

The literature regarding numerical methods as applied to subsurface flow and transport is extensive and we will only provide a brief overview of the topic in this chapter. While there are several numerical methods currently used to solve the groundwater flow equations, they all have their roots in polynomial approximation theory. From this foundation it is possible to formulate, among others, the commonly employed finite-difference, finite-element, and finite-volume methods.

Initially we will use Eq. (4.68) as an example in our development. Extension of this one-space-dimensional equation to multiple space dimensions is conceptually straightforward. Removing the overbars from the averaged variables and the subscripts on K_{zz} in Eq. (4.68) to simplify notation, we have

$$\frac{\partial}{\partial z} K \left(\frac{\partial}{\partial z} \left(h\left(z, t \right) \right) \right) - S_s \frac{\partial}{\partial t} \left(h\left(z, t \right) \right) = 0, \quad z \in [0, L], \ t \in [0, \infty), \tag{7.1}$$

where $h\left(z, t \right)$ is the hydraulic head, K is the hydraulic conductivity, S_s is the storativity, and z and t are the space and time coordinates, respectively.

Subsurface Hydrology   By George F. Pinder and Michael A. Celia
Copyright © 2006 John Wiley & Sons, Inc.

Because Eq. (7.1) contains derivatives in both space and time, we must define both *initial* and *boundary conditions*. The occurrence of a second derivative in space demands two boundary conditions be specified. Candidates are:

$$h(z_i, t) = f_1(t), \quad z_i \in \{0, L\}, \ t \in [0, \infty); \tag{7.2}$$

$$\frac{\partial h(z_i, t)}{\partial z} = f_2(t), \quad z_i \in \{0, L\}, \ t \in [0, \infty); \tag{7.3}$$

$$\frac{\partial h(z_i, t)}{\partial z} + \alpha h(z_i, t) = f_3(t), \quad z_i \in \{0, L\}, \ t \in [0, \infty). \tag{7.4}$$

Conditions of the form found in Eq. (7.2) are called *type one*, *Dirichlet*, or *specified head*. Conditions such as found in Eq. (7.3) are known as *type two*, *Neumann*, or *specified flux*; and those of the form provided by Eq. (7.4) are denoted as *type three*, *Robbins*, or *specified leakage*. Of the choices in Eqs. (7.2) through (7.4), two must be applied, one at $z = 0$ and one at $z = L$.

The time derivative in Eq. (7.1) requires that we specify an initial condition, which takes the form[1]

$$h(0, z) = f_4(z), \quad z \in [0, L],$$

which states that within the domain of interest, $z \in [0, L]$, at time $t = 0$, the head must be specified.

Now that we have formulated a properly posed initial boundary value problem, let us proceed to solve it using a numerical approach.

7.2 POLYNOMIAL APPROXIMATION THEORY

The point of departure for this discussion is the definition of a polynomial. A polynomial is a mathematical expression that consists of a sum of powers in one or more variables each multiplied by a coefficient. For example, a polynomial in one variable would appear as

$$P(x) = a_0 + a_1 x + a_2 x^2 + \cdots + a_n x^n, \tag{7.5}$$

where a_i, $i = 1, \ldots, n$, are constant coefficients. A polynomial in two variables would be

$$P(x, y) = a_{00} + a_{01} y + a_{10} x + a_{11} xy + a_{12} xy^2 + a_{21} x^2 y + a_{22} x^2 y^2 + \cdots + a_{mn} x^m y^n. \tag{7.6}$$

An important theorem for our discussion, known as the *Weirstrass approximation theorem*, says that any continuous function defined on a closed and bounded interval can be uniformly approximated to any desired accuracy on the specified interval by polynomials. It is written formally as the following theorem.

---

[1]In this context, the notation $[a, b]$ indicates the set of values that lie between a and b inclusive of a and b.

Theorem 7.1   *If $f(x)$ is a continuous real-valued function on an interval $[a, b]$ and there is given an $\varepsilon > 0$, then there exists a polynomial P on $[a, b]$ such that*

$$|f(x) - P(x)| < \varepsilon \tag{7.7}$$

for all $x \in [a, b]$.

In our case, the function $f(x, y)$ we wish to approximate is the head $h(x, t)$.

In our work we will consider a special form of polynomial called a *Lagrange polynomial*. Of particular interest will be first, second, and third degree Lagrange polynomials. Linear Lagrange polynomials for location x_0, (i.e., ℓ_0) and x_1 (i.e., ℓ_1) are seen in Figure 7.1. The nodes are located at the points where one of the polynomials has a value of unity. Thus the node at x_0 is, by convention, identified with the polynomial ℓ_0. The equations that describe linear Lagrange polynomials are

$$\ell_0 = \frac{x_1 - x}{x_1 - x_0}, \tag{7.8}$$

$$\ell_1 = \frac{x - x_0}{x_1 - x_0}. \tag{7.9}$$

The quadratic Lagrange polynomial has the form

$$\ell_0^2 = \frac{(x - x_1)(x - x_2)}{(x_0 - x_1)(x_0 - x_2)}, \tag{7.10}$$

$$\ell_1^2 = \frac{(x - x_0)(x - x_2)}{(x_1 - x_0)(x_1 - x_2)}, \tag{7.11}$$

$$\ell_2^2 = \frac{(x - x_0)(x - x_1)}{(x_2 - x_0)(x_2 - x_1)} \tag{7.12}$$

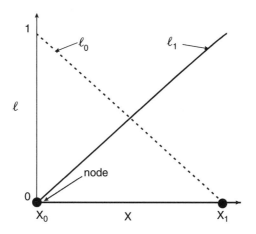

FIGURE 7.1. Linear Lagrange polynomials.

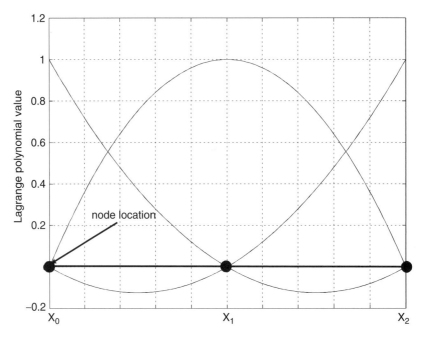

FIGURE 7.2. Quadratic Lagrange polynomial.

and has the form found in Figure 7.2. The general form of the Lagrange polynomial is given by

$$\ell_j^n = \frac{(x - x_0) \cdots (x - x_{j-1})(x - x_{j+1}) \cdots (x - x_n)}{(x_j - x_0) \cdots (x_j - x_{j-1})(x_j - x_{j+1}) \cdots (x_j - x_n)} \tag{7.13}$$

or, using a more condensed notation,

$$\ell_j^n = \prod_{\substack{i=0 \\ i \neq j}}^{n} \frac{x - x_i}{x_j - x_i}. \tag{7.14}$$

Now we will show how one uses Lagrange polynomials in approximating a function $f(x)$. We know that for a general form of $f(x)$, such as we would encounter when $f(x) = h(x)$, we would potentially require a very high degree polynomial to represent $f(x)$. Thus, if we select, for example, a quadratic polynomial such as provided in Eqs. (7.10) through (7.12), we would expect to encounter an error in the approximation. Having this in mind, we write

$$f(x) = \hat{f}(x) + E(x), \tag{7.15}$$

where $\hat{f}(x)$ is the quadratic Lagrange polynomial approximation given by

$$\hat{f}(x) = \sum_{j=0}^{n} \ell_j^n(x) f(x_j). \tag{7.16}$$

In this expression, $f(x_j)$ is the value of the function $f(x)$ at the location x_j, and the term $E(x)$ in Eq. (7.15) is the error in the approximation. If one solves Eq. (7.15) for the term $E(x)$, one sees that it is simply the difference between the exact value of the function $f(x)$ and its Lagrange polynomial approximation $\hat{f}(x)$.

The error term $E(x)$ can be shown to have the form

$$E(x) = \frac{1}{n+1} \frac{d^{(n+1)}}{dx^{(n+1)}} f(\xi) \cdot \prod_{i=0}^{n} (x - x_i), \quad \xi \in [x_0, x_n], \tag{7.17}$$

where, as earlier, n is the degree of the Lagrange polynomial and the notation $\Pi_{i=0}^{n}$ implies multiplication of terms $i = 0, \ldots, n$. Thus, for a quadratic Lagrange polynomial ($n = 2$), there are three nodes in the interval; they are at x_0, x_1, and x_2. Notice that the function is evaluated at $x = \xi$. However, the precise location of ξ is not known, and never is. We only know that it lies somewhere in the interval $\xi \in [x_0, x_n]$. Moreover, we normally do not know the function $f(x)$ either, since it is the solution of the equation. Thus the error expression is useful only in a theoretical context, as will be seen shortly.

7.3   FINITE-DIFFERENCE METHODS

Finite-difference methods represent continuous derivatives such as found in Eq. (7.1) with expressions involving the evaluation of the unknown function at discrete points. The natural consequence of this action is the transformation of a problem involving classical derivatives to one involving algebraic equations. Let us see how this is done using the concepts from Section 7.2.

The first step is to represent $f(x)$ using Eq. (7.16) and to differentiate this expression to approximate the derivative. Differentiation of both sides of Eq. (7.16) yields

$$\frac{d}{dx} \hat{f}(x) = \sum_{j=0}^{n} \left[\frac{d}{dx} \ell_j^n(x) \right] f(x_j). \tag{7.18}$$

Note that the differentiation of the right-hand side of Eq. (7.16) involves only the Lagrange polynomial $\ell_j^n(x)$, where the superscript n denotes the degree of the Lagrange polynomial and not a power of ℓ_j. Why do we not see a derivative of $f(x_j)$? It is because the term $f(x_j)$ is not a function of x; it is a specific value of $f(x)$ at the point x_j.

The next step is to select a family of Lagrange polynomials. We will select the quadratic family, that is the one associated with $n = 2$. The required family is found in Eqs. (7.10)–(7.12). Thus we have for Eq. (7.18)

$$\frac{d}{dx} \hat{f}(x) = \sum_{j=0}^{2} \left[\frac{d}{dx} \ell_j^2(x) \right] f(x_j). \tag{7.19}$$

Consider now the term in Eq. (7.19) associated with $j = 0$. The appropriate Lagrange polynomial is given by Eq. (7.10). Substitution of Eq. (7.10) into Eq. (7.19) yields

$$\left[\frac{d}{dx} \ell_0^2(x) \right] f(x_0) = \frac{d}{dx} \left(\frac{(x - x_1)(x - x_2)}{(x_0 - x_1)(x_0 - x_2)} \right) f(x_0). \tag{7.20}$$

Now completing the differentiation required on the right-hand side of Eq. (7.20), we obtain

$$\left[\frac{d}{dx}\ell_0^2(x)\right] f(x_0) = \frac{(x-x_1)+(x-x_2)}{(x_0-x_1)(x_0-x_2)} f(x_0) \qquad (7.21)$$

or, upon simplification and assuming the increment $(x_1 - x_0) = (x_2 - x_1) = \Delta x$,

$$\left[\frac{d}{dx}\ell_0^2(x)\right] f(x_0) = -\frac{x_1 - 2x + x_2}{2\Delta x^2} f(x_0). \qquad (7.22)$$

Using a similar strategy for $j = 1$ and $j = 2$, we have

$$\sum_{j=0}^{2}\left[\frac{d}{dx}\ell_j^2(x)\right] f(x_j) = -\frac{x_1 - 2x + x_2}{2\Delta x^2} f(x_0) - \frac{x_0 - 2x + x_2}{-\Delta x^2} f(x_1)$$

$$- \frac{x_0 - 2x + x_1}{2\Delta x^2} f(x_2). \qquad (7.23)$$

The next step is to determine where we wish to have the derivative evaluated. Assume that we would like to evaluate $df/dx|_{x_0}$. Substitution of $x = x_0$ into Eq. (7.23) gives

$$\frac{d}{dx}\hat{f}(x_0) = \sum_{j=0}^{2}\left[\frac{d}{dx}\ell_j^2(x)\right]_{x_0} f(x_j) = -\frac{x_1 - 2x_0 + x_2}{2\Delta x^2} f(x_0)$$

$$- \frac{x_0 - 2x_0 + x_2}{-\Delta x^2} f(x_1) - \frac{x_0 - 2x_0 + x_1}{2\Delta x^2} f(x_2) \qquad (7.24)$$

or, upon simplification and introduction of Eq. (7.15),

$$\frac{df(x_0)}{dx} = \frac{1}{2\Delta x}(-3f(x_0) + 4f(x_1) - f(x_2)) + \frac{dE(x_0)}{dx}. \qquad (7.25)$$

Equation (7.25) states that we can approximate the derivative df/dx at the location $x = x_0$ by three discrete values of $f(x)$, namely, those located at x_0, x_1, and x_2 with an error of $dE(x_0)/dx$. In concept, we can determine $dE(x_0)/dx$ through differentiation of $E(x)$ as given in Eq. (7.17). The general result for any location $x = x_k$ is

$$\frac{dE(x_k)}{dx} = \frac{1}{(n+1)!}\frac{d^{(n+1)}}{dx^{(n+1)}} f(\xi) \frac{d}{dx}\left[\prod_{i=0}^{n}(x-x_i)\right]_{x_k}, \qquad (7.26)$$

which simplifies for our case of $n = 2$ to

$$\frac{dE(x_0)}{dx} = \frac{1}{3!}\frac{d^3}{dx^3} f(\xi) \frac{d}{dx}\left[\prod_{i=0}^{2}(x-x_i)\right]_{x_0}. \qquad (7.27)$$

Equation (7.27) shows that the error in our approximation is proportional to the third derivative of the sought-after function $f(x)$ evaluated at some location $\xi \in [x_0, x_2]$ and to $d \left[\Pi_{i=0}^{2} (x - x_i) \right]_{x_0} /dx$, and is inversely proportional to $(n + 1)!$. We now expand the term $d \left[\Pi_{i=0}^{2} (x - x_i) \right]_{x_0} /dx$ to give

$$
\frac{d}{dx} \left[(x - x_0)(x - x_1)(x - x_2) \right]_{x_0}
$$
$$
= \left[(x - x_0)(x - x_1) + (x - x_1)(x - x_2) + (x - x_0)(x - x_2) \right]_{x_0}
$$
$$
= (x_0 - x_1)(x_0 - x_2) = 2\Delta x^2, \tag{7.28}
$$

which states that the error is proportional to the square of Δx. The overall error term is then written as $O \left(\Delta x^2 \right)$, which is read as "order of Δx squared." From this we can see that if Δx is sufficiently small, then the higher the power of Δx, the more accurate the solution. Similarly, due to the effect of the term $d^3 f (\xi) /dx^3$, the smoother the function $f(x)$, the smaller the $d^3 f (\xi) /dx^3$ term will be and the better will be the approximation. In fact, if the function $f(x)$ were a polynomial of degree equal to or less than two, there would be no error in the approximation. Given we are using a quadratic function to interpolate $f(x)$, this result is intuitively reasonable.

The procedure outlined above can be repeated evaluating Eq. (7.23) at $x = x_1$ and $x = x_2$. The evaluation at $x = x_2$ does not reveal any surprises, but evaluation at $x = x_1$ does. Substitution of $x = x_1$ into Eq. (7.23) gives

$$
\frac{d}{dx} \hat{f} (x_1) = \sum_{j=0}^{2} \left[\frac{d}{dx} \ell_j^2 (x) \right]_{x_1} f (x_j) \tag{7.29}
$$
$$
= -\frac{x_1 - 2x_1 + x_2}{2\Delta x^2} f(x_0) - \frac{x_0 - 2x_1 + x_2}{-\Delta x^2} f(x_1) - \frac{x_0 - 2x_1 + x_1}{2\Delta x^2} f(x_2) \tag{7.30}
$$

or, upon simplification,

$$
\frac{d}{dx} \hat{f} (x_1) = -\frac{f(x_0) - f(x_2)}{2\Delta x}.
$$

The unusual aspect of this approximation is that information at the node at which the approximation is being generated, namely, $f(x_1)$, does not appear in the formula. The consequences of this omission are evident in the resulting solutions, as we will see later.

Observation of Eq. (7.1) reveals that, while we have a first-order derivative in time, we have a second-order derivative in space. To find an approximation for the second derivative, we simply differentiate Eq. (7.15) twice, that is,

$$
\frac{d^2}{dx^2} \hat{f} (x) = \sum_{j=0}^{2} \left[\frac{d^2}{dx^2} \ell_j^2 (x) \right] f (x_j). \tag{7.31}
$$

The result is equivalent to differentiating Eq. (7.23) once, which yields

$$
\frac{d^2}{dx^2} \hat{f} (x) = \sum_{j=0}^{2} \left[\frac{d^2}{dx^2} \ell_j^2 (x) \right] f (x_j) \tag{7.32}
$$

$$\frac{d^2}{dx^2} \hat{f}(x) = \frac{2}{2\Delta x^2} f(x_0) - \frac{4}{2\Delta x^2} f(x_1) + \frac{2}{2\Delta x^2} f(x_2)$$

$$= \frac{1}{\Delta x^2} (f(x_0) - 2f(x_1) + f(x_2)) \qquad (7.33)$$

Since the variable x does not appear in this expression, the same formula would apply irrespective of the nodal location where the second derivative is being approximated.

The error of this approximation is obtained by taking the second derivative of $E(x)$ and evaluating it at the node where the approximation is located. The result is

$$\frac{d^2 E(x_0)}{dx^2} = \frac{1}{3!} \frac{d^3}{dx^3} f(\xi) \frac{d^2}{dx^2} \left[\prod_{i=0}^{2} (x - x_i) \right]_{x_0} + \frac{2}{3!} \frac{d^4}{dx^4} f(\xi) \frac{d}{dx} \left[\prod_{i=0}^{2} (x - x_i) \right]_{x_0}$$

$$+ \left[\frac{1}{3!} \frac{d^5}{dx^5} f(\xi) \prod_{i=0}^{2} (x - x_i) \right] \qquad (7.34)$$

or

$$\frac{d^2 E(x_0)}{dx^2} = \frac{1}{3!} \frac{d^3}{dx^3} f(\xi) \frac{d}{dx} [(x - x_0)(x - x_1) + (x - x_1)(x - x_2)$$

$$+ (x - x_0)(x - x_2)]_{x_0} + O(\Delta x^2) \qquad (7.35)$$

$$= \frac{1}{3!} \frac{d^3}{dx^3} f(\xi) [(x - x_0) + (x - x_1) + (x - x_1) + (x - x_2)$$

$$+ (x - x_0) + (x - x_2)]_{x_0} + O(\Delta x^2) \qquad (7.36)$$

$$= \frac{1}{3!} \frac{d^3}{dx^3} f(\xi) [2(x_0 - x_1) + 2(x_0 - x_2)] = -\frac{1}{3!} \frac{d^3}{dx^3} f(\xi) 6\Delta x = O(\Delta x)$$

$$(7.37)$$

Thus, given one uses a quadratic Lagrange polynomial as the interpolating function, the error of the approximation for the second derivative at location x_0 is one order larger than that of the first. Note that a central difference approximation (evaluation at $x = x_1$) is second order accurate.

If one begins with a first-degree Lagrange polynomial, that is,

$$\ell_0 = \frac{x - x_1}{x_0 - x_1} \quad \text{and} \quad \ell_1 = \frac{x - x_0}{x_1 - x_0}, \qquad (7.38)$$

one obtains the first-derivative approximation

$$\frac{d}{dx} f(x) = \frac{d}{dx} \hat{f}(x) + \frac{dE}{dx} \qquad (7.39)$$

$$= f(x_0) \frac{d\ell_0}{dx} + f(x_1) \frac{d\ell_1}{dx} + \frac{dE}{dx}$$

$$= f(x_0) \frac{1}{x_0 - x_1} + f(x_1) \frac{1}{x_1 - x_0}$$

$$+ \frac{1}{(1+1)!} \frac{d^{(1+1)}}{dx^{(1+1)}} f(\xi) \frac{d}{dx} \left[\prod_{i=0}^{1} (x - x_i) \right]_{x_k}, \qquad (7.40)$$

which, when evaluated at the location $x = x_0$, reduces to

$$\frac{d}{dx} f(x_0) = \frac{f(x_1) - f(x_0)}{\Delta x} - \frac{1}{2!} \frac{d^2}{dx^2} f(\xi) \cdot \Delta x. \qquad (7.41)$$

Equation (7.41) is a two-point approximation of $df(x_0)/dx$ as contrasted to the three-point approximation provided in Eq. (7.25). However, the price that is paid for using fewer nodes is a larger truncation error, that is, $O(\Delta x)$ rather than $O(\Delta x^2)$.

In Table 7.1 we provide a sample of notation that facilitates the expression of finite-difference equations. Commonly used formulas, derived from those given in the table, are

$$\mu \delta f(x_i) = \frac{\delta f(x_{i+1/2}) + \delta f(x_{i-1/2})}{2} = \frac{f(x_{i+1}) - f(x_{i-1})}{2}, \qquad (7.42)$$

$$\delta^2 f(x_i) = \delta(\delta f(x_i)) = \delta\left(f(x_{i+1/2}) - f(x_{i-1/2})\right)$$
$$= f(x_{i+1}) - 2f(x_i) + f(x_{i-1}).$$

The finite-difference formulas presented above are easily derived in an alternative way using Taylor series expansions. We have elected to use the polynomial expansion approach because it leads naturally into a discussion of the finite-element and finite-volume methods.

7.3.1 Finite-Difference Representation of the Groundwater Flow Equation

In Eq. (7.1) we presented an expression describing the flow of groundwater in a saturated porous medium. We rewrite it here as

$$\frac{\partial}{\partial z} K\left[\frac{\partial}{\partial z}(h(z, t))\right] - S_s \frac{\partial}{\partial t}(h(z, t)) = 0, \quad z \in [0, L], \ t \in [0, \infty). \qquad (7.43)$$

Using the notation found in Table 7.1, we can approximate this equation as

$$\frac{1}{\Delta z^2} \delta_z(K[\delta_z h(z_i, t_k)]) - \frac{1}{\Delta t} S_s \Delta_t h(z_i, t_k) = 0, \quad i = 0, \ldots, I, \ k = 0, \ldots, K,$$
$$\qquad (7.44)$$

where $z_i = i \Delta z$, $L = I \Delta z$, and $t_k = k \Delta t$.

TABLE 7.1. Operator Notation for Finite-Difference Formulas

Operator	Symbol	Difference Representation
Forward difference	Δ	$\Delta f(x_i) = f(x_{i+1}) - f(x_i)$
Backward difference	∇	$\nabla f(x_i) = f(x_i) - f(x_{i-1})$
Central difference	δ	$\delta f(x_i) = f(x_{i+1/2}) - f(x_{i-1/2})$
Shift	E	$Ef(x_i) = f(x_{i+1})$
Average	μ	$\mu f(x_i) = \dfrac{f(x_{i+1/2}) + f(x_{i-1/2})}{2}$

Upon expansion, and defining $h_{i,k} \equiv h(x_i, t_k)$, we obtain

$$\underbrace{\frac{1}{\Delta z}\left(\left[K_{i+1/2}\frac{h_{i+1,k}-h_{i,k}}{\Delta z}\right]-\left[K_{i+1/2}\frac{h_{i,k}-h_{i-1,k}}{\Delta z}\right]\right)}_{A}$$

$$-\underbrace{S_{si,k}\frac{h_{i,k+1}-h_{i,k}}{\Delta t}}_{B}=0, \quad i=0,\ldots,I, \ k=0,\ldots,K. \qquad (7.45)$$

Associated with this formula is the computational molecule found in Figure 7.3. To understand this molecule, imagine you are located on the node marked i, k. To the left and to the right are nodes at the same time level, k, but at different locations along the z-axis. Above you is a node at the same location along the z-axis, but at the new time level $k+1$.

Now compare this with Eq. (7.43). The term A is the space term and it is identified with time level k. Thus it is written using the three black dots in Figure 7.3. The term B is the time derivative. It is a forward difference and therefore it is located at location i, k and is projecting forward to time level $k+1$.

In making calculations, the time level $k=0$ corresponds to the initial conditions and is therefore known. Thus the three black dots are known values. The grey dot is unknown since it is at the new time level. There is thus one unknown value that appears in this computational molecule, namely, $h_{i,k+1}$. Even when there are many nodes along the x-axis, still there will be but one unknown for each finite-difference equation. After all of the values at the $k+1$ level are calculated, then the molecule shifts upward one Δt value and the whole process is repeated. The result is a very efficient computational scheme. However, algebraically speaking, there is no "free lunch." This very simplistic scheme comes with a restriction on the size of the time step that can be used. If that criterion is exceeded, the algorithm becomes unstable and the computed values are nonsense.

Consider now the following variant on Eq. (7.45):

$$\underbrace{\frac{1}{\Delta z}\left(\left[K_{i+1/2}\frac{h_{i+1,k+1}-h_{i,k+1}}{\Delta z}\right]-\left[K_{i+1/2}\frac{h_{i,k+1}-h_{i-1,k+1}}{\Delta z}\right]\right)}_{A}$$

$$-\underbrace{S_{si,k}\frac{h_{i,k+1}-h_{i,k}}{\Delta t}}_{B}=0, \quad i=0,\ldots,I, \ k=0,\ldots,K. \qquad (7.46)$$

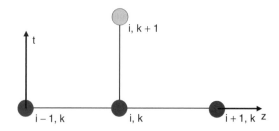

FIGURE 7.3. Computational molecule for the explicit finite-difference scheme for groundwater flow.

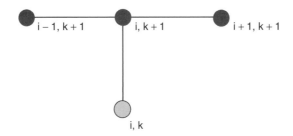

FIGURE 7.4. Computational molecule for a backward difference approximation in time.

The key change appears in term A, where we observe that the space derivative is now identified at the $k + 1$ time level. The corresponding computational molecule is found in Figure 7.4. Note that, in this instance, there are three unknown values at the $k + 1$ level and only one equation. Thus to solve the equations describing this problem, it will be necessary to have a set of equations that must be solved simultaneously.

If we assume we have $I + 1$ nodes lined up along the z-axis, then we have $I + 1$ equations in $I + 3$ unknowns. There are two extra unknowns, one associated with the equation at $i = 0$ (i.e., the unknown at node $i = -1$) and one with the equation at node $i = I$ (i.e., the unknown at node $i = I + 1$). To solve the problem requires the same number of equations as unknowns, so we need two additional equations.

The two additional equations come from boundary conditions. Boundary conditions, introduced earlier, define the interface between the model and the outside world. In groundwater flow there are three types of boundary conditions as described in Section 4.3.3. The simplest of these expressions states that we know, as a function of time, the behavior of the hydraulic head at the boundary of the domain, that is,

$$h(z, t) = f_1(t), \quad z \in \{0, L\} \quad \textit{Dirichlet.} \tag{7.47}$$

In terms of our finite-difference formulas, this translates into the condition

$$h(z_i, t) = h_i(z_i, t), \quad i = 0 \text{ and } I. \tag{7.48}$$

However, we have nodes outside the region $z \in [0, L]$, that is, at node $i = -1$ and node $i = i + 1$. To resolve this dilemma we remove the equations written at nodes and replace them with the two boundary specifications found in Eq. (7.48). Thus we have eliminated the two pesky unknowns outside our region of interest and those at $i = 0$ and $i = I$ and we now have the correct number of equations, namely, $I + 1$ of them, and the correct number of unknowns.

Slightly more complex is the flux condition, which states that

$$\frac{\partial h(z, t)}{\partial z} = f_2(t), \quad z \in \{0, L\} \quad \textit{Neumann.} \tag{7.49}$$

Keeping in mind that

$$q_z(z, t) = -K \frac{\partial}{\partial z} h(z, t), \quad z = [0, L], \tag{7.50}$$

one can see that the derivative $\partial h(z,t)/\partial z$ is given by the specified flux $q(z)$ and the hydraulic conductivity K at the ends of the domain of interest, that is,

$$\frac{\partial}{\partial x}h(z,t) = -\frac{q(z,t)}{K}, \quad z = \{0, L\}. \tag{7.51}$$

The representation of the derivative at the ends of the interval can employ either a centered, backward, or forward difference strategy, depending on the circumstances involved. Consider the location $z = 0$. The backward difference approximation at this point would be

$$\frac{\partial}{\partial z}h(0,t) = \nabla_z h(0,t) = -\frac{q(0,t)}{K}, \tag{7.52}$$

which provides an additional equation to augment the $I+1$ currently available. Assuming a forward difference representation of $\partial h(L,t)/\partial z$ is provided, using a similar strategy, at the other end of our domain, we have now the required $I+3$ equations in the $I+3$ unknowns.

The third type boundary condition (also called a Robbins or leakage condition) is given by Eq. (4.25), that is,

$$\alpha h(z,t) + \gamma \frac{\partial h(z,t)}{\partial z} = C_0(z,t), \quad z \in \{0, L\} \quad Robbins, \tag{7.53}$$

which simplifies to

$$K\frac{\partial}{\partial z}h(t) = \kappa(h_0(t) - h(x,t)) \tag{7.54}$$

for problems normally encountered in groundwater flow. Equation (7.54) describes leakage into an aquifer when κ is considered as the *leakage coefficient*, $h_0(z)$ the head value in the external world (e.g., a lake elevation), and $h(z,t)$ the head in the aquifer. One observes in this case that there are two terms involving the unknown value $h(z,t)$ in the defining equation. To accommodate this boundary condition, we approximate the derivative as in the case of the Neumann boundary condition, which results in one additional unknown and one additional equation. This is the same situation we faced with the Neumann boundary condition so we proceed from this point along the same lines as we did in that case.

The various boundary conditions can be mixed. One does not need to have the same boundary type at each end of the domain. For example, a constant head might be defined at $z = 0$ and a Neumann condition at $z = L$.

Now that we have learned how to create the finite-difference equations, we must address the question of their solution. In the case of the forward difference Eq. (7.45), the problem is easy. For each node at a given time level k you have but one unknown, namely, that at $k+1$, which you can determine by solving one equation in one unknown.

The implicit, backward difference Eq. (7.46) provides a greater challenge. Since we have one equation written at time level $k+1$ and three unknowns at this level (the value at level k is known from earlier calculations or initial conditions), we must consider the equations at all nodes simultaneously. The result is a matrix equation of the form

$$[K]\{h\}^{k+1} + [S]\left(\{h\}^{k+1} - \{h\}^k\right) - \{f\} = 0, \tag{7.55}$$

where typical nonzero elements of the matrices $[K]$ and $[S]$ are

$$k_{ij} = \begin{cases} 0, & |i-j| \geq 2 \\ \dfrac{K_{i+1/2}}{\Delta z^2}, & j = i+1 \\ \dfrac{K_{i+1/2}}{\Delta z^2}, & j = i-1 \\ -\dfrac{K_{i+1/2} + K_{i-1/2}}{\Delta z^2}, & i = j \end{cases} \tag{7.56}$$

and

$$s_{ii} = \begin{cases} 0, & i \neq j \\ -\dfrac{S_{s_i}}{\Delta t}, & i = j \end{cases}. \tag{7.57}$$

The vector of known values $\{f\}$ is composed of information derived from the introduction of boundary conditions into the governing equation. The introduction of boundary conditions also modifies the first and last rows of the $[K]$ matrix.

Equation (7.55) can be rearranged as

$$([K] + [S])\{h\}^{k+1} = [S]\{h\}^k + \{f\} \tag{7.58}$$

or, defining $[C] \equiv ([K] + [S])$, we obtain

$$[C]\{h\}^{k+1} = [S]\{h\}^k + \{f\}. \tag{7.59}$$

Equation (7.59) can be solved formally by taking the inverse of the matrix $[C]$, to yield $[C]^{-1}$. Then via matrix multiplication, one obtains the vector of head values at the nodes at the new time step $k + 1$:

$$\{h\}^{k+1} = [C]^{-1}\left([S]\{h\}^k + \{f\}\right). \tag{7.60}$$

In practice, one does not actually calculate the inverse matrix. Rather, the sparse structure of the matrix is exploited to design more efficient matrix solution methods.

A transient simulation thus consists of solving Eq. (7.60) sequentially for each of the time levels $k = 1, \ldots, K$. It is convenient to now plot the resulting values of head to show the behavior of the system over the period of analysis. We will see such a plot when we consider the two-dimensional problem in Section 7.3.2. If a steady-state solution is needed, the matrix $[S]$ is set to zero, and Eq. (7.60) is solved only once for the steady-state head values.

7.3.2 Finite-Difference Method in Two Space Dimensions

We now extend our analysis to problems with two space dimensions. As our "type" equation we rewrite the vertically averaged groundwater flow equation originally presented as Eq. (4.49) as

$$\nabla_{xy} \cdot \mathbf{T} \cdot \nabla_{xy} h = S\frac{\partial h}{\partial t} + Q, \tag{7.61}$$

where we define the source term $Q = q_T + q_B + q$ and, as earlier, drop the overbars that are used to denote a vertical average. It is helpful at this point to expand the operator notation in Eq. (7.61) to give

$$\frac{\partial}{\partial x}\left(T_{xx}\frac{\partial h}{\partial x}\right) + \frac{\partial}{\partial y}\left(T_{yy}\frac{\partial h}{\partial y}\right) + 2\frac{\partial}{\partial y}\left(T_{yx}\frac{\partial h}{\partial x}\right) = S\frac{\partial h}{\partial t} + Q. \qquad (7.62)$$

We now use our finite-difference operator notation to obtain an approximation for Eq. (7.62), that is,

$$\frac{1}{\Delta x^2}\delta_x\left(T_{xx}\delta_x h_{ijk}\right) + \frac{1}{\Delta y^2}\delta_y\left(T_{yy}\delta_y h_{ijk}\right)$$

$$+ \frac{2}{\Delta x\,\Delta y}\mu\delta_y\left(T_{yx}\mu\delta_x h_{ijk}\right) = \frac{S}{\Delta t}\Delta_t h_{ijk} + Q_{ijk} \qquad (7.63)$$

where $x = i\,\Delta x$, $y = j\,\Delta y$, and $t = k\,\Delta t$. The computational molecule for this finite-difference scheme is given in Figure 7.5.

In general, we often assume that the coordinate axes are aligned with the largest and smallest components of the transmissivity tensor **T**. When this is done, the cross-derivative terms in Eqs. (7.62) and (7.63) vanish and the simplified finite-difference equation is

$$\frac{1}{\Delta x^2}\delta_x\left(T_{xx}\delta_x h_{ijk}\right) + \frac{1}{\Delta y^2}\delta_y\left(T_{yy}\delta_y h_{ijk}\right) = \frac{S}{\Delta t}\Delta_t h_{ijk} + Q_{ijk}. \qquad (7.64)$$

While this may seem like a rather incidental difference, there is a significant impact on the computational effort required to solve the equation. The computational molecule for Eq. (7.64) is given in Figure 7.6. The reason that the computational effort identified with Figure 7.5 is more than that associated with Figure 7.6 lies in the number of unknown head values associated with the node ijk. In the case of the nine-point template, the bandwidth of the resulting matrix is a little greater than is the case for the five-point template. Since, in classical methods, the effort involved in the solution of the matrix

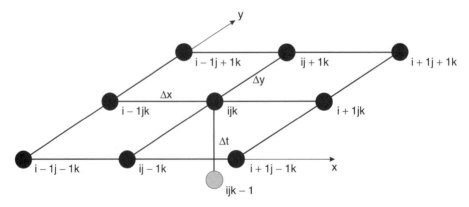

FIGURE 7.5. Finite-difference molecule for backward difference in time approximation.

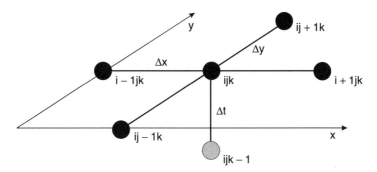

FIGURE 7.6. Finite-difference computational molecule for the case of no cross derivative.

equations is proportional to the square of the bandwidth, the nine-point formulation is slightly more computationally burdensome than the five-point formulation.

Boundary conditions for the two-dimensional (and three-dimensional) model are treated as they are in one dimension because the grid is generally aligned with the coordinate axes. In other words, the derivative approximations appearing in the two-dimensional model appear as they do in the one-dimensional case.

The question naturally arises as to how to treat aquifers that have irregular geometry. In standard finite-difference methods, this is generally (but not exclusively) achieved by approximating the irregular boundary as illustrated in Figure 7.7. In that figure the aquifer boundary is concave toward the interior of the model. To approximate it, the

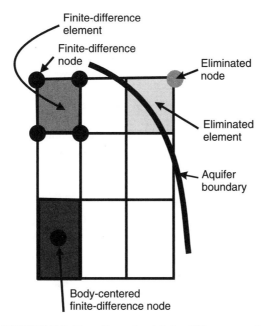

FIGURE 7.7. Two-dimensional finite-difference mesh.

indicated element in the first row and last column is removed along with the associated corner node. The result is a step-like representation of the boundary.

Also illustrated in this figure is the use of rectangular, as distinct from square, elements. In using rectangular elements, some care should be exercised not to make their geometry too extreme. The use of long narrow elements can result in a poorly conditioned coefficient matrix $[K]$, leading to problems in solving the set of approximating algebraic equations.

Figure 7.7 also shows two ways of visualizing a finite-difference mesh. In the upper right-hand corner, a *node-centered element* is shown. The nodes are on the corners. In the lower left-hand corner of the mesh, one observes a *body-centered element*. In this case the node is located in the center of the element. While the two approaches are very similar, they lead to slightly different interpretations of how boundary conditions are implemented using finite-difference approximations.

7.4   FINITE-ELEMENT METHODS

Finite-element methods are a popular alternative numerical strategy to finite-difference methods for the simulation of groundwater flow. While the theory behind finite-element methods is more abstract than finite-difference methods, finite-element methods have significant advantages over finite-difference methods in some practical applications. In the following we use the polynomial approximation theory presented in Section 7.2 as our point of departure in developing these methods. Combination of Eqs. (7.15) and (7.16) yields

$$f(x) = \hat{f}(x) + E(x) = \sum_{j=0}^{n} \ell_j^n(x) f(x_j) + E(x), \tag{7.65}$$

where $f(x)$ is the sought-after function, for example, the groundwater head $h(x)$, and $\hat{f}(x)$ is the polynomial approximation to $f(x)$, which in our case is $\hat{h}(x)$. The term $E(x)$ is the error of the approximation.

The classical approach to development of the finite-element method utilizes the concept of a weighted residual. To make this rather abstract concept more concrete, it is perhaps helpful to select an equation to which the finite-element method is to be applied. In this spirit let us define an operator $\mathcal{L}(x)$ as the steady-state one-dimensional groundwater flow equation with a source $Q(x)$, that is,

$$\mathcal{L}(h(x)) \equiv K \frac{d^2 h(x)}{dx^2} - Q(x) = 0. \tag{7.66}$$

We now define the *residual* $R(x)$ as the difference between the value of the operator acting on the exact solution $h(x)$ and the value obtained when we substitute the approximate solution $\hat{h}(x)$. Thus we obtain

$$R(x) \equiv \mathcal{L}(h(x)) - \mathcal{L}(\hat{h}(x)). \tag{7.67}$$

Since $\mathcal{L}(h)$ is zero by Eq. (7.66), Eq. (7.67) becomes

$$R(x) = -\mathcal{L}(\hat{h}(x)). \tag{7.68}$$

Now consider a weighting function $w(x)$ that is also a function of x. The method of weighted residuals can simply be stated as

$$\int_X \mathcal{L}\left(\hat{h}(x)\right) w(x)\, dx = 0, \quad x \in X. \tag{7.69}$$

Equation (7.69) states that we are seeking to find a value of $\hat{h}(x)$ such that the integral over the domain X of the residual weighted by the function $w(x)$ vanishes. The choice of the function $w(x)$ will dictate the nature of the weighted residual method. We will consider two forms of $w(x)$ in this chapter.

7.4.1 Galerkin's Method

The most commonly used weighting function is the same function that is used for approximating the unknown function, $\hat{h}(x)$. This formulation is called Galerkin's method [1]. From Eq. (7.65) we see this to be the Lagrange polynomial, $\ell_j^n(x)$. Equation (7.69) now becomes

$$\int_X \mathcal{L}\left(\hat{h}(x)\right) \ell_i^n(x)\, dx = 0, \quad x \in X, \ i = 0, 1, \ldots, I. \tag{7.70}$$

The next step is to substitute our approximation for $\hat{h}(x)$ as given by Eq. (7.65) less the error term. The result is

$$\int_X \mathcal{L}\left(\sum_{j=0}^n \ell_j^n(x) h\left(x_j\right)\right) \ell_i^n(x)\, dx = 0, \quad x \in X, \ i = 0, 1, \ldots, I. \tag{7.71}$$

Keep in mind that the value $h\left(x_j\right)$ is a number representing the function $\hat{h}(x)$ at $x = x_j = j\,\Delta x$. If we assume that $I = n = 2$ (a quadratic Lagrange polynomial), we can use the information provided in Figure 7.2 (reproduced here for convenience as Figure 7.8).

Given these assumptions, for this one-element, three-node problem, Eq. (7.71) becomes

$$\int_{x_0}^{x_2} \mathcal{L}\left(\sum_{j=0}^2 \ell_j^2(x) h\left(x_j\right)\right) \ell_i^2(x)\, dx = 0, \quad i = 0, 1, 2, \tag{7.72}$$

which becomes, upon substitution of our definition of

$$\mathcal{L}(x) = K\frac{d^2 h(x)}{dx^2} - Q(x), \tag{7.73}$$

$$\int_{x_0}^{x_2} K\frac{d^2}{dx^2}\left(\sum_{j=0}^2 \ell_j^2(x) h\left(x_j\right)\right) \ell_i^2(x) - Q(x)\ell_i^2(x)\, dx = 0, \quad i = 0, 1, 2 \tag{7.74}$$

However, we know from Eq. (7.32) that

$$\frac{d^2}{dx^2}\left(\sum_{j=0}^2 \ell_j^2(x) h\left(x_j\right)\right) = \frac{1}{\Delta x^2}\left(h(x_0) - 2h(x_1) + h(x_2)\right). \tag{7.75}$$

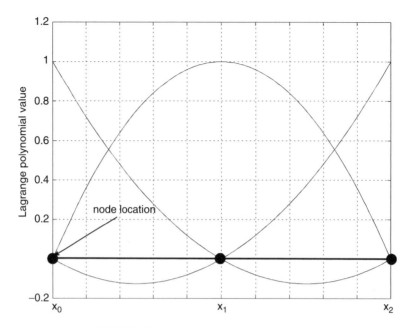

FIGURE 7.8. Quadratic Lagrange polynomial.

Substitution of Eq. (7.75) into Eq. (7.74) yields

$$\int_{x_0}^{x_2} \left[K \frac{1}{\Delta x^2} (h(x_0) - 2h(x_1) + h(x_2)) \right] \ell_i^2(x) - Q(x)\ell_i^2(x) \, dx = 0, \quad i = 0, 1, 2.$$
(7.76)

We observe at this point that we have three equations ($i = 0, 1, 2$) in three unknowns ($h(x_0)$, $h(x_1)$, and $h(x_2)$).

With malice of forethought, let us write the equation for $i = 1$. We obtain

$$\int_{x_0}^{x_2} \left[K \frac{1}{\Delta x^2} (h(x_0) - 2h(x_1) + h(x_2)) \right] \ell_1^2(x) - Q(x)\ell_1^2(x) \, dx = 0$$
(7.77)

or

$$\int_{x_0}^{x_2} \left[K \frac{1}{\Delta x^2} (h(x_0) - 2h(x_1) + h(x_2)) \right] \frac{(x - x_0)(x - x_2)}{(x_1 - x_0)(x_1 - x_2)}$$
$$- Q(x) \frac{(x - x_0)(x - x_2)}{(x_1 - x_0)(x_1 - x_2)} \, dx = 0.$$
(7.78)

Since the term in square brackets in Eq. (7.78) is independent of x, and assuming for convenience that Q is constant in space, we have

$$\left(\left[K \frac{1}{\Delta x^2} (h(x_0) - 2h(x_1) + h(x_2)) \right] - Q \right) \int_{x_0}^{x_2} \frac{(x - x_0)(x - x_2)}{(x_1 - x_0)(x_1 - x_2)} \, dx = 0.$$
(7.79)

But one can divide both sides by the integral, leaving

$$\left[K \frac{1}{\Delta x^2} \left(h(x_0) - 2h(x_1) + h(x_2) \right) \right] - Q = 0, \tag{7.80}$$

which is just the finite-difference approximation for the equation. It seems like we made little progress.

However, if we take the same three nodes, but employ linear Lagrange polynomials using a slightly different strategy, a new formulation emerges that might give a new approximation. Let us substitute our definition of $\mathcal{L}\left(\hat{h}(x) \right)$ into Eq. (7.70) to give

$$\int_X \left(K \frac{d^2 \hat{h}(x)}{dx^2} - Q(x) \right) \ell_i^n(x) \, dx = 0, \quad x \in X, \ i = 0, 1, \ldots, I. \tag{7.81}$$

Now define $\hat{h}(x)$ using linear Lagrange polynomials as shown in Figure 7.9.

The approximation to $h(x)$ is now written

$$\hat{h}(x) = \ell_0(x) h(x_0) + \ell_1(x) h(x_1) + \ell_2(x) h(x_2), \tag{7.82}$$

where

$$\ell_1 |x| = \begin{cases} \dfrac{x - x_0}{x_1 - x_0}, & x_0 \leq x \leq x_1 \\ \dfrac{x_2 - x}{x_2 - x_1}, & x_1 \leq x \leq x_2 \\ 0, & \text{all other } x \end{cases} \tag{7.83}$$

Substitution of Eq. (7.82) into Eq. (7.81) yields, for the equation generated by the weighting function ℓ_1,

$$\int_X \left(K \frac{d^2}{dx^2} \left(\ell_0(x) h(x_0) + \ell_1(x) h(x_1) \right. \right.$$

$$\left. \left. + \ell_2(x) h(x_2) \right) - Q(x) \right) \ell_1(x) \, dx = 0, \quad x \in X. \tag{7.84}$$

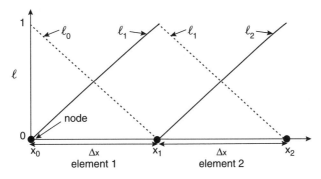

FIGURE 7.9. Two elements, each spanned by linear Lagrange polynomials.

A problem immediately arises. The second derivative of a linear function is zero almost everywhere so the approximation $d^2\hat{h}(x)/dx^2$ vanishes except at the node points, where it becomes unbounded.

To circumvent this problem, we apply integration by parts to the second-derivative terms. The result for one weighting function is

$$-\int_X K\frac{d}{dx}\left(\ell_0(x)h(x_0)+\ell_1(x)h(x_1)+\ell_2(x)h(x_2)\right)\frac{d}{dx}\ell_1\,dx$$
$$+\left(K\frac{d}{dx}\left(\ell_0(x)h(x_0)+\ell_1(x)h(x_1)+\ell_2(x)h(x_2)\right)\right)\ell_1\Bigg|_{x=x_0}^{x=x_2}$$
$$-\int_{X_1}Q(x)\ell_1(x)\,dx=0,\quad x\in[x_0,x_1].\tag{7.85}$$

Notice that because $\ell_1(x)$ is zero at $x=x_0$ and also at $x=x_2$, the second term in Eq. (7.85) vanishes.

From Eq. (7.41) we have the approximation

$$\frac{d}{dx}\left(\ell_0(x)h(x_0)+\ell_1(x)h(x_1)\right)=\frac{d\hat{h}(x)}{dx}=\frac{h(x_1)-h(x_0)}{\Delta x},\quad x\in[x_0,x_1]\tag{7.86}$$

and from Eq. (7.83) we have

$$\frac{d}{dx}\ell_1=\frac{1}{\Delta x},\quad x\in[x_0,x_1].\tag{7.87}$$

Similarly,

$$\frac{d}{dx}\left(\ell_1(x)h(x_1)+\ell_2(x)h(x_2)\right)=\frac{d\hat{h}(x)}{dx}=\frac{h(x_2)-h(x_1)}{\Delta x},\quad x\in[x_1,x_2]\tag{7.88}$$

and

$$\frac{d}{dx}\ell_1=-\frac{1}{\Delta x},\quad x\in[x_1,x_2].\tag{7.89}$$

Combination of Eqs. (7.85) through (7.89) yields

$$-\int_{x_0}^{x_1}K\frac{h(x_1)-h(x_0)}{\Delta x}\frac{1}{\Delta x}\,dx+\int_{x_1}^{x_2}K\frac{h(x_2)-h(x_1)}{\Delta x}\frac{1}{\Delta x}\,dx$$
$$-\int_{x_0}^{x_2}Q(x)\ell_1(x)\,dx=0\tag{7.90}$$

or

$$-K\frac{h(x_1)-h(x_0)}{\Delta x}+K\frac{h(x_2)-h(x_1)}{\Delta x}-Q\,\Delta x=0,\tag{7.91}$$

where we have assumed, as earlier, that Q is a constant. Rewriting Eq. (7.91) we have

$$K\frac{h(x_0) - 2h(x_1) + h(x_2)}{\Delta x^2} - Q = 0. \tag{7.92}$$

It appears that, whether we use finite-difference methods or finite-element methods, we always end up with the same numerical approximation to the steady-state groundwater flow equation in the presence of a constant source Q. While this is indeed the case in this example, it is not true for more general forms of the groundwater flow equation, or even for this simple form considered in higher space dimensions. The formulation would also have been different from than the finite-difference counterpart had we assumed $Q = Q(x)$.

In our use of the quadratic formulation in the above example considered earlier in this subsection, we considered a special case, namely, the equation that is generated when the weighting function at the center node of the element is used. If alternatively we use either of the end-node weighting functions, that is, at $x = x_2$ or $x = x_0$, we will find once again that it is necessary to increase the number of equations. The reason is that the end nodes are shared with the adjacent elements. Thus the resulting equations at the end nodes have five unknowns associated with them (three in each element with the middle node being common to both elements). The resulting formulation is a five-point approximation quite different from any we have heretofore encountered.

7.4.2 Two-Dimensional Finite-Element Formulation

The extension of the preceding finite-element theory to two space dimensions requires relatively few new concepts. We begin by considering the two-dimensional time-independent groundwater flow equation (we will consider the transient case shortly):

$$\frac{\partial}{\partial x}\left(T_{xx}\frac{\partial h(x, y)}{\partial x}\right) + \frac{\partial}{\partial y}\left(T_{yy}\frac{\partial h(x, y)}{\partial y}\right) + 2\frac{\partial}{\partial y}\left(T_{yx}\frac{\partial h(x, y)}{\partial x}\right) = Q. \tag{7.93}$$

Next, we extend our earlier representation of the hydraulic head to a two-dimensional form $\hat{h}(x, y)$, where the finite elements are assumed to be rectangles as shown in Figure 7.10.

Given this mesh geometry, the following holds:

$$\hat{h}(x, y) = \sum_{i,j} h(x_i, y_j)\ell_{ij}(x, y), \quad i = 0, 1, \ldots, I, \; j = 0, 1, \ldots, J, \tag{7.94}$$

where $\ell_{ij}(x, y) \equiv \ell_i(x)\ell_j(y)$ and we assume for clarity of presentation that the Lagrange polynomials are linear. The function $\ell_{ij}(x, y)$ is shown in Figure 7.11 for the element D and the node identified by the letter a. Note that it is linear in both the x and y directions. However, along a line drawn from node a to node c, the function is not linear.

The weighted residual equation now reads

$$\int_\Omega \mathcal{L}\left(\hat{h}(x, y)\right)\ell_{ij}(x, y)\, dx\, dy = 0, \quad x, y \in \Omega, \; i = 0, 1\ldots, I, \; j = 0, 1, \ldots, J, \tag{7.95}$$

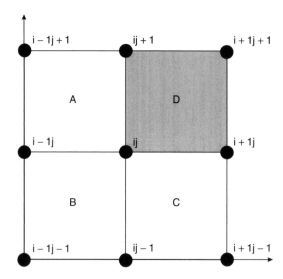

FIGURE 7.10. Four finite elements (A–D) associated with the node i, j.

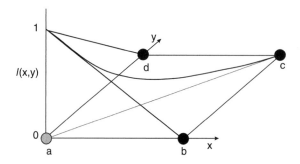

FIGURE 7.11. Two-dimensional Lagrange polynomial that is defined for node a and is linear in x and in y.

where Ω is the two-dimensional domain that represents the area to be modeled (e.g., the area represented in Figure 7.7) and the operator $\mathcal{L}\left(\hat{h}\left(x, y\right)\right)$ is now defined as

$$\mathcal{L}\left(\hat{h}\left(x, y\right)\right) \equiv \frac{\partial}{\partial x}\left(T_{xx}\frac{\partial \hat{h}\left(x, y\right)}{\partial x}\right) + \frac{\partial}{\partial y}\left(T_{yy}\frac{\partial \hat{h}\left(x, y\right)}{\partial y}\right) + 2\frac{\partial}{\partial y}\left(T_{yx}\frac{\partial \hat{h}\left(x, y\right)}{\partial x}\right) - Q.$$

$$(7.96)$$

It is convenient at this point to introduce the notation

$$\nabla_{xy}\hat{h}\left(x, y\right) \equiv \frac{\partial \hat{h}\left(x, y\right)}{\partial x}\mathbf{i} + \frac{\partial \hat{h}\left(x, y\right)}{\partial y}\mathbf{j}, \qquad (7.97)$$

where $\mathbf{i}$ and $\mathbf{j}$ are the *unit vectors* in the x and y coordinate directions, respectively. We can now write, Eq. (7.96) as

$$\mathcal{L}\left(\hat{h}\left(x, y\right)\right) \equiv \nabla_{xy} \cdot \mathbf{T} \cdot \nabla_{xy}\hat{h}\left(x, y\right) - Q, \qquad (7.98)$$

where $\mathbf{T}$ is the *transmissivity tensor*.

Combination of the weighted residual equation (Eq. (7.95)) and the definition provided in Eq. (7.98) yields

$$\int_{\Omega} \left(\mathbf{\nabla}_{xy} \cdot \mathbf{T} \cdot \mathbf{\nabla}_{xy} \hat{h}\,(x,\,y) - Q \right) \ell_{kl}\,(x,\,y)\,dx\,dy = 0,$$

$$x,\,y \in \Omega, \quad k = 0, 1, \ldots, K, \quad l = 0, 1, \ldots, L. \tag{7.99}$$

We now apply *Green's second theorem*, which is the multidimensional extension of *integration by parts*, to give

$$\int_{\Omega} - \left(\mathbf{T} \cdot \mathbf{\nabla}_{xy} \hat{h}\,(x,\,y) \right) \cdot \mathbf{\nabla}_{xy} \ell_{kl}\,(x,\,y) - Q \ell_{kl}\,(x,\,y)\,dx\,dy$$

$$+ \int_{\partial\Omega} \left(\mathbf{T} \cdot \mathbf{\nabla}_{xy} \hat{h}\,(x,\,y) \right) \cdot \mathbf{n}\,dl = 0, \quad x,\,y \in \Omega, \quad k = 0, 1, \ldots, K, \quad l = 0, 1, \ldots, L.$$

$$\tag{7.100}$$

Although this equation is analogous to Eq. (7.85), it is different in important ways. To see this we expand the first term of the surface integral in Eq. (7.100) for a weighting function located at point (x_k, y_l) to give

$$\int_{\Omega} \left(\mathbf{T} \cdot \mathbf{\nabla}_{xy} \hat{h}\,(x,\,y) \right) \mathbf{\nabla}_{xy} \ell_{kl}\,(x,\,y)\,dx\,dy$$

$$= \int_{\Omega} \left(T_{xx} \frac{\partial \hat{h}\,(x,\,y)}{\partial x} + T_{xy} \frac{\partial \hat{h}\,(x,\,y)}{\partial y} \right) \frac{\partial \ell_{kl}\,(x,\,y)}{\partial x}$$

$$+ \left(T_{xy} \frac{\partial \hat{h}\,(x,\,y)}{\partial y} + T_{yx} \frac{\partial \hat{h}\,(x,\,y)}{\partial x} \right) \frac{\partial \ell_{kl}\,(x,\,y)}{\partial y}\,dx\,dy \tag{7.101}$$

or, substituting for $\hat{h}\,((x,\,y))$, we have

$$\int_{\Omega} \left(\mathbf{T} \cdot \mathbf{\nabla}_{xy} \hat{h}\,(x,\,y) \right) \cdot \mathbf{\nabla}_{xy} \ell_{kl}\,(x,\,y)\,dx\,dy$$

$$= \int_{\Omega} \left(T_{xx} \sum_{i,j} h\,(x_i,\,y_j) \frac{\partial}{\partial x} \ell_i\,(x)\,\ell_j\,(y) \right.$$

$$+ T_{xy} \sum_{i,j} h\,(x_i,\,y_j) \frac{\partial}{\partial y} \ell_j\,(y)\,\ell_i\,(x) \Bigg) \ell_l\,(y) \frac{\partial \ell_k\,(x)}{\partial x}$$

$$+ \left(T_{xy} \sum_{i,j} h\,(x_i,\,y_j) \frac{\partial}{\partial y} \ell_j\,(y)\,\ell_i\,(x) \right.$$

$$+ T_{yx} \sum_{i,j} h\,(x_i,\,y_j) \frac{\partial}{\partial x} \ell_i\,(x)\,\ell_j\,(y) \Bigg) \ell_k\,(x) \frac{\partial \ell_l\,(y)}{\partial y}\,dx\,dy. \tag{7.102}$$

Equation (7.102) cannot be easily simplified because of the products of the various Lagrange polynomials and their derivatives. Thus in two dimensions the finite-element formulation does not reduce or simplify to the corresponding finite-difference form, even when $T_{xy} = T_{yx} = 0$.

The question remains as to how to handle the line integral around the four-element region. The answer lies in the fact that the weighting function for node ij vanishes at the boundary of this region. Thus the equation provided by Eq. (7.102) stands alone. However, on closer inspection one realizes that one has one equation in nine unknowns. Thus the equations generated by the weighting functions defined for each of the unknowns must be solved simultaneously. The weighting functions for nodes on the boundary of the entire region do not vanish. To accommodate these equations one must specify the required boundary conditions.

As mentioned on page 273 there are three types of boundary conditions that can potentially come into play. The first type boundary condition, or *Dirichlet* condition, that defines a constant head is accommodated by simply replacing the unknown head value at the boundary node by the known value. In practice this means replacing the unknown head value in the vector of unknown head values with the specified value and doing the appropriate matrix coefficient multiplications needed to transfer the resulting known information to the right-hand side of the matrix equation. The result is a reduction in the number of rows in the matrix, the reason being that the equation associated with the known head value is now of no use and therefore can be eliminated. The column of coefficients associated with this known value can now be removed also since these coefficients are no longer needed. The final matrix is again square with the same number of equations as unknowns. The total number of equations (and unknowns) is now the number of nodes less the number of Dirichlet boundary conditions.

The second type, or *Neumann*, boundary condition, is handled quite differently. For nodes located at the boundary of the model region the flux into the element may be specified, assuming that a Dirichlet condition is not defined along this boundary segment. In this case one replaces the term $\mathbf{T} \cdot \nabla_{xy} \hat{h}(x, y)$ with the known flux value and then performs the required line integrations. Thus one creates a term that represents the mass flux through the boundary along the outside perimeter. The resulting number is then placed on the right-hand side of the matrix equation in the row associated with the node identified with boundary node location kl, (also the location identified with the weighting function for this equation, $\ell_{kl}(x, y)$).

The most popular and in some ways most useful two-dimensional finite elements are *triangular* rather than rectangular in shape. The advantage of triangles and their three-dimensional counterparts, *tetrahedrons*, is that they provide increased flexibility in locating nodes. No longer must one focus on a regular, evenly distributed array of nodes, but rather the nodes can be located where they are needed to increase solution accuracy, such as around wells, or to better define the model perimeter.

Triangular elements are not conveniently identified using the ij notation normally used in rectangular meshes. Rather each node is numbered sequentially such as shown in Figure 7.12. Equation (7.102), and the theory that led up to its presentation, holds for triangular elements. The difference lies only in the definition of the basis and weighting functions $\phi_i(x, y)$, which now are identified via a single subscript consistent with the nodal identification in Figure 7.12. We use $\phi_i(x, y)$ rather than $\ell_i(x, y)$ in recognition of the fact that the triangular shape function is not a Lagrange polynomial in the classical sense. However, it is in the sense that it satisfies $\phi(\mathbf{x}_i) = \delta_{ij}$ and thus one can consider $\phi_i(x, y)$ as a piecewise planar Lagrange polynomial.

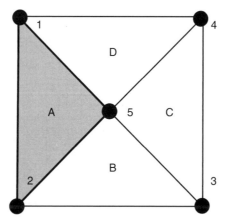

FIGURE 7.12. Node and element numbering strategy for a triangular finite-element mesh.

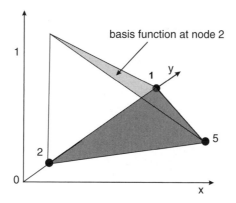

FIGURE 7.13. Triangular basis function for node 2.

Consider the element A in Figure 7.12. The basis function identified with node 2 would appear as shown in Figure 7.13. A similar function would appear at each of the nodes. The important characteristics of this function are that it has a value of one at the node for which it is defined and zero at all other nodes. Note that along each element side the triangular basis function looks like a linear basis function (see page 265). Because the integrations found in Eq. (7.102) are now over a triangle, a coordinate transformation is applied to facilitate the calculations. While the resulting formulas are rather cumbersome, the integrations are easily done exactly on the computer.

An example of the relative advantages of the flexibility of triangular finite elements is shown in Figures 7.14 and 7.15. Figure 7.14 shows a rectangular mesh, in this case generated for a finite-difference application, and Figure 7.15 shows a finite-element mesh for the same problem. A comparison of the two meshes illustrates two features. First, in the neighborhood of the well located in the lower left quadrant of the model, both meshes require a dense array of nodes to assure adequate solution accuracy. However, in the case of the rectangular mesh, geometrical considerations require that a line of nodes

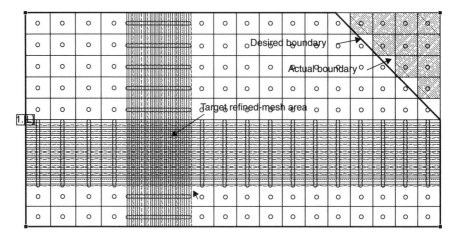

FIGURE 7.14. *MODFLOW* grid generated to illustrate boundary approximations and refinement (from [2]).

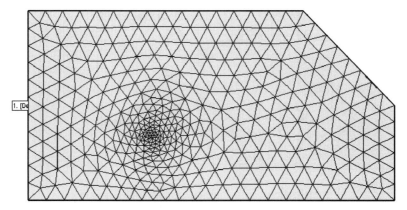

FIGURE 7.15. Finite-element mesh automatically generated for cut-corner example problem (from [2]).

once started must continue to the boundary of the model. The result is an array of very elongated elements. In contrast, in the case of the finite-element mesh, small triangles reflect the large number of nodes near the well, but there is a gradual decrease in the density of nodes as one moves away from the well and a concomitant gradual increase in the size of the finite elements.

The second interesting feature is found in the upper right-hand quadrant of the model. The top right-hand corner is beveled. The bevel is accommodated using the rectangular mesh by using a series of steps. The more steps used, the more accurate the representation and the more nodes required (remember that each new node creates a new line of nodes). The triangular mesh addresses the bevel by locating element sides along the bevel, thereby exactly representing this geometrical feature.

The number of nodes found in the rectangular mesh is about 1600 and in the triangular mesh about 400. However, the factor of 4 in nodes does not translate directly into a corresponding increase in computational effort. The reason for this is that the regular arrangement of the nodes in the rectangular mesh gives rise to the implementation of very efficient equation-solving algorithms. In the end, in practical applications, the amount of computational effort required using rectangles and a specially designed efficient solver, and using triangular elements and a standard solver, is approximately the same for the same level of accuracy.

To generate flow lines, one can utilize the fact that, for a triangle, the gradient is a constant. Thus one can begin to create a flow line from an arbitrary point in the following way. From the point of selection calculate the gradient of the head in the triangle containing the initial point. Follow this direction until the element side is encountered. Move across the boundary and enter the adjacent element at an angle representing the head gradient in that element. By moving from one element to another in this way, a piecewise linear flow line is generated. Such flow lines are especially helpful in contaminant transport problems.

Note that there exists a discontinuity in the head gradient at the boundaries of elements. Thus the fluid mass flux moving from one element to another generally cannot be locally conserved. However, the finite-element formulation does conserve mass globally.

The lack of fluid mass conservation is particularly important when solving the equations for mass transport. In the mass transport equations the fluid velocity is an essential element in the calculation of the convective mass flux. Since the local fluid mass is not conserved, neither is the local contaminant mass. On the other hand, just as global fluid mass is conserved, so also is global contaminant mass. This leads to a number of interesting numerical issues regarding contaminated transport simulations.

7.5   FINITE-VOLUME FORMULATION

An increasingly popular approach to solving problems in computational fluid dynamics, the finite-volume method, can be developed using the polynomial approximation theory and the method of weighted residuals presented earlier. To illustrate how this is done, consider the equation

$$\mathcal{L}(x) = K\frac{d^2 h(x)}{dx^2} + \kappa h(x) = \kappa h^*(x), \tag{7.103}$$

which describes steady one-dimensional groundwater flow with leakage, where κ is the leakage coefficient and $h^*(x)$ the head in the surface-water body.

We now introduce the method of weighted residuals, in combination with the linear Lagrange polynomial based method of approximation of the unknown head value $h(x)$, to obtain (see the development on page 279)

$$\int_{x_0}^{x_2} \left[\mathcal{L}\left(\sum_{j=0}^{2} \ell_j^1(x) h(x_j) \right) - \kappa h^*(x) \right] w_i(x)\, dx = 0, \quad i = 0, 1, 2, \tag{7.104}$$

where $w_i(x)$ is the weighting function. Let us assume that the weighting function $w_i(x)$ is given for node x_0 as w_0 as shown in Figure 7.16. The *Heaviside unit step function* is

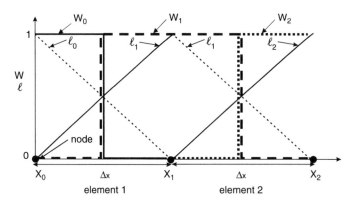

FIGURE 7.16. Approximating functions for the finite-volume formulation. The linear Lagrange polynomials are designated as ℓ_j^1 and the Heaviside unit step function is represented by $w_i(x)$.

defined for element 1 as

$$
w_0(x) = \begin{cases} 0, & x < \Delta x/2 \\ \frac{1}{2}, & x = \Delta x/2 \\ 1, & x > \Delta x/2 \end{cases} .
\tag{7.105}
$$

Note that $w_0(x)$ has a unit value over the first half of element 1 and a value of zero over the second half. The function $w_1(0)$, on the other hand, is zero over the first half of element 1, and unity over the second half. The pattern is repeated for successive elements.

Substitution of Eq. (7.103) into Eq. (7.104) gives

$$
\int_{x_0}^{x_2} \left[K \frac{d^2}{dx^2} \left(\sum_{j=0}^{2} \ell_j^1(x) h(x_j) \right) \right.
$$
$$
\left. + \kappa \left(\sum_{j=0}^{2} \ell_j^1(x) h(x_j) \right) - \kappa h^*(x) \right] w_i(x) \, dx, \quad i = 0, 1, 2.
\tag{7.106}
$$

We observe that, as was the case with the finite-element method, we again have in the first term of Eq. (7.106) a second derivative of a linear function, namely, $\ell_j^1(x)$. We again apply integration by parts to obtain

$$
\int_{x_0}^{x_2} -K \frac{d}{dx} \left(\sum_{j=0}^{2} \ell_j^1(x) h(x_j) \right) \frac{d}{dx} w_i(x) \, dx
$$
$$
+ \int_{x_0}^{x_2} \left(\kappa \left(\sum_{j=0}^{2} \ell_j^1(x) h(x_j) \right) - \kappa h^*(x) \right) w_i(x) \, dx
$$
$$
+ K \left(\frac{d}{dx} \hat{h}(x) \right) w_i(x) \Bigg|_{x=x_0}^{x=x_2}, \quad i = 0, 1, 2.
\tag{7.107}
$$

The appearance of $dw_i(x)/dx$, $i = 0, 1, 2$ in Eq. (7.107) is indeed curious. These are step function derivatives, which at $x = \Delta x/2$ are apparently not defined. However, there is a definition for such a derivative, although it is generally considered nonstandard. The derivative of $w_0(\Delta x/2)$ is given as the *Dirac delta function* $-\delta_0(x - \Delta x/2)$, where the negative sign reflects the fact that the sense of the slope of the step is negative. Similarly, the derivative of $w_1(\Delta x/2)$ is $\delta_1(x - \Delta x/2)$. Thus we have for the case of $w_0(x)$

$$\int_{x_0}^{x_2} K \frac{d}{dx} \left(\sum_{j=0}^{1} \ell_j^1(x) h(x_j) \right) \delta_0(x - \Delta x/2) \, dx$$

$$+ \int_{x_0}^{x_2} \left(\kappa \left(\sum_{j=0}^{1} \ell_j^1(x) h(x_j) \right) - \kappa h^*(x) \right) w_0(x) \, dx$$

$$+ K \left(\frac{d}{dx} \hat{h}(x) \right) w_0(x) \Big|_{x=x_0}^{x=x_2} = 0. \tag{7.108}$$

Notice that the integral is complete for this expression because the $w_0(x)$ function is found only in the first element.

By definition of the Dirac delta function, the first term in this integral becomes the integrand evaluated at $\Delta x/2$. Equation (7.108) can then be written

$$K \left(\sum_{j=0}^{1} \frac{d}{dx} \ell_j^1(x) h(x_j) \right) \Bigg|_{x=\Delta x/2}$$

$$+ \int_{x_0}^{x_1} \left(\kappa \left(\sum_{j=0}^{1} \ell_j^1(x) h(x_j) \right) - \kappa h^*(x) \right) w_0(x) \, dx$$

$$+ K \left(\frac{d}{dx} \hat{h}(x) \right) w_0(x) \Big|_{x=x_0}^{x=x_2} = 0. \tag{7.109}$$

From the definition of $\ell_0(x)$ found in Eq. (7.8), we obtain

$$\frac{d}{dx} \ell_0(x) = \frac{-1}{x_1 - x_0}$$

and similarly

$$\frac{d}{dx} \ell_1(x) = \frac{1}{x_1 - x_0}.$$

Given these definitions we can rewrite Eq. (7.109) as

$$K \frac{h_1 - h_0}{x_1 - x_0} + \int_{x_0}^{x=\Delta x/2} \left(\kappa \left(\frac{x_1 - x}{x_1 - x_0} h_0 + \frac{x - x_0}{x_1 - x_0} h_1 \right) - \kappa h^*(x) \right) dx$$

$$- K \left(\frac{d}{dx} \hat{h}(x) \right) \Bigg|_{x=x_0} = 0,$$

where we have used the fact that $w_0(x_2) = 0$ to simplify the boundary term. Performing the integrations, we obtain

$$K\frac{h_1 - h_0}{x_1 - x_0} + (x_1 - x_0)\kappa\left[\left(\frac{3h_0 + h_1}{8}\right) - \frac{h^*}{2}\right] - K\left(\frac{d}{dx}\hat{h}(x)\right)\bigg|_{x=x_0} = 0.$$

This provides us with our first equation (which is associated with the first node).

The next step is to make a second similar calculation using the weighting function $w_1(x)$. Because the weighting function $w_1(x)$ is found in both element 1 and element 2, the integration is performed over the two elements to give

$$\underbrace{-K\frac{h_1 - h_0}{x_1 - x_0} + (x_1 - x_0)\kappa\left[\left(\frac{3h_1 + h_0}{8}\right) - \frac{h^*}{2}\right]}_{\text{element 1 contribution}} + \underbrace{K\frac{h_2 - h_1}{x_2 - x_1} + (x_2 - x_1)\kappa\left[\left(\frac{3h_1 + h_2}{8}\right) - \frac{h^*}{2}\right]}_{\text{element 2 contribution}} = 0,$$

which can be written

$$K\frac{h_0 - 2h_1 + h_2}{\Delta x} + \Delta x\left[\kappa\frac{h_0 + 6h_1 + h_2}{8} - h^*\right] = 0. \tag{7.110}$$

The third and final equation is obtained by using $w_2(x)$, which is defined only in element 2. The result is

$$-K\frac{h_2 - h_1}{x_2 - x_1} + (x_2 - x_1)\kappa\left[\left(\frac{3h_2 + h_1}{8}\right) - \frac{h^*}{2}\right] + K\left(\frac{d}{dx}\hat{h}(x)\right)\bigg|_{x=x_2} = 0.$$

To make the result consistent with the approach normally used in finite-volume methods, the terms in the expression $\kappa(h_0 + 6h_1 + h_2)/8$ are lumped together to form the approximate counterpart κh_1.

If the boundary conditions are of first type, that is, the head values are specified at each end of the interval, for example, $h(x_0)$ and $h(x_2)$ are provided, then the equations defined for $x = x_0$ and $x = x_2$ can be discarded and replaced by the specified values for the heads determined from the boundary conditions. The remaining expression derived from Eq. (7.110) becomes

$$K\frac{h(x_0) - 2h_1 + h(x_2)}{\Delta x} + \Delta x\left[\kappa\frac{h(x_0) + 6h_1 + h(x_2)}{8} - \kappa h^*\right] = 0,$$

which constitutes one equation in the unknown value h_1.

In the event a second type, or flux, boundary condition is specified at either end of the interval, the equation defined for the appropriate end of the interval is employed.

7.5.1 Two-Dimensional Finite-Volume Formulation

Consider two-dimensional groundwater flow with vertical leakage as described by the following equation:

$$K\nabla^2 h(x, y) + \kappa h(x, y) - \kappa h^*(x, y) = 0.$$

First approximate the unknown function $h(x, y)$ as was done in the finite-element method; that is,

$$h(x, y) \approx \hat{h}(x, y) = \sum_{j=1}^{n} h_j \phi_j(x, y), \qquad (7.111)$$

where the basis functions $\phi_j(x, y)$ can be defined on rectangular or triangular subregions. In our example we will consider triangular elements such as shown in Figure 7.17 and in Figure 7.13.

The lines forming the sides of the *finite-volume polygon* (the shaded area in Figure 7.17) are drawn such that they connect the circumcenters of each triangle. The resulting mesh is called a *Voronoi mesh*. Other mesh-generation schemes are also possible [3]. Thus the lines forming the finite-volume polygon are orthogonal to the sides of the triangles and are continuous across them.

Now write the weighted residual formulation as

$$\int_{\Omega} \left(K \nabla^2 \hat{h}(x, y) + \kappa \hat{h}(x, y) - \kappa h^*(x, y) \right) w_i(x, y) \, dx \, dy = 0, \quad i \in \{1, \ldots, N\},$$

where Ω is the domain of the model, in this case the area A in Figure 7.18. The weighting function is now equal to 1 everywhere as shown in Figure 7.18 for the function identified with the center node, a. The function is unity within the polygon and zero from the polygon boundary C to the polygon boundary formed by the triangles that collectively

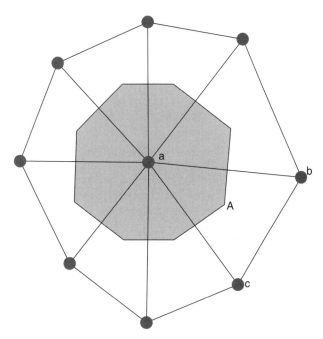

FIGURE 7.17. Illustration of relationship between a finite volume and triangular finite elements. The shaded area is the finite volume.

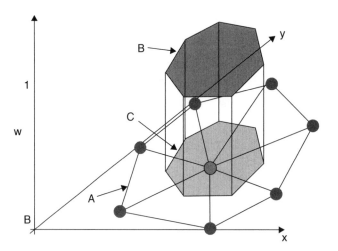

FIGURE 7.18. Step-function weighting function for development of two-dimensional finite-volume method.

define perimeter A. Application of Green's theorem to the second-order term yields

$$\int_{\Omega} K \, \nabla \hat{h}\,(x, y)\, \delta_i\,(x - x_i, y - y_i)\, dx\, dy + \int_{\Omega} \left(\kappa \hat{h}\,(x, y) - \kappa h^*\,(x, y)\right) w_i\,(x, y)\, dx\, dy$$

$$+ \int_{\partial \Omega} K \, \nabla h\,(x, y)\, w_i\,(x, y)\, ds = 0, \quad i = 1, 2, \ldots, N. \tag{7.112}$$

It is important to emphasize that the contour integral, the last term in Eq. (7.112), is taken over the outer perimeter of the elements, that is, the perimeter of Ω, shown as A in Figure 7.18.

The first term in Eq. (7.112) reduces to a line integral with the line defined by the perimeter of the weighting function, line C in Figure 7.18. Now define the normal flux to boundary C as

$$\nabla \hat{h} \cdot \mathbf{n} \equiv \frac{\partial \hat{h}}{\partial n},$$

where $\mathbf{n}$ is the outward directed normal to C. Equation (7.112) can now be written

$$\int_{\partial \Omega_C} K \frac{\partial \hat{h}}{\partial n}\, ds + \int_{\Omega} \left(\kappa \hat{h}\,(x, y) - \kappa h^*\,(x, y)\right) w_i\,(x, y)\, dx\, dy \tag{7.113}$$

$$+ \int_{\partial \Omega} K \nabla \hat{h}\,(x, y)\, w_i\,(x, y)\, ds = 0, \quad i \in 1, \ldots, N, \tag{7.114}$$

where $\partial \Omega_C$ denotes integration along line C.

Let us now examine the equation for node a and element A in Figure 7.17. We first differentiate Eq. (7.111) to obtain

$$\frac{\partial \hat{h}\,(x, y)}{\partial n} = \sum_{j=a}^{c} h_j \frac{\partial \phi_j\,(x, y)}{\partial n}. \tag{7.115}$$

Now combine Eqs. (7.113) through (7.115) to give

$$
\underbrace{\int_{\partial \Omega_{C_{abc}}} K \sum_{j=a}^{c} h_j \frac{\partial \phi_j(x,y)}{\partial n} \, ds}_{P} + \underbrace{\int_A \kappa \sum_{j=a}^{c} h_j \phi_j(x,y) \, w_a(x,y) \, dx \, dy}_{Q}
$$

$$
= \underbrace{\int_A \kappa h^* w_a(x,y) \, dx \, dy}_{R} - \underbrace{\int_{\partial \Omega} K \nabla h(x,y) \, w_a(x,y) \, ds}_{S} = 0, \qquad (7.116)
$$

where $\partial \Omega$ represents the perimeter of the region Ω and $\partial \Omega_{C_{abc}}$ is the segment of the line C within the triangle abc (the edge of the polygon contained in this triangle).

Consider term P. The normal derivatives of the basis functions appearing in this term are calculated along lines ab and ca. A derivative of the form

$$
\frac{\partial \phi_a(x,y)}{\partial n_{ab}} = \frac{\phi_b - \phi_a}{b - a}
$$

can be used directly in Eq. (7.116). A similar situation holds for the remaining normal derivatives. The second term, Q, involves the integral of the sum $\sum_{j=a}^{c} h_j \kappa \int_A \phi_j(x,y) \times w_a(x,y) \, dx \, dy$. This integral can easily be computed, but in the standard application of finite-volume methods, this is not done. Rather, this term is lumped. By lumping we mean that one assumes that the only relevant value of h_j is that at the location $j = i$. In other words, the summation is given by (see also page 292)

$$
h_i \kappa \sum_{j=a}^{c} \int_A \phi_j(x,y) \, w_a(x,y) \, dx \, dy. \qquad (7.117)
$$

Since the basis functions sum to unity everywhere, and w_a is unity in the finite-volume element and zero outside, the expression in Eq. (7.117) becomes $h_i \kappa$ times the area indicated by the shaded portion of element A. The R term is similar to the Q term in how it is treated. However, in this case, the R term is a known quantity. This leaves the S term, which is zero in this equation since w_a vanishes along $\partial \Omega$.

The function $\hat{h}(x,y)$ is continuous across the boundary of the polygon C. Thus any quantity passing this boundary from one finite volume to another is conserved. The flux entering one finite volume must be exactly balanced by that leaving the adjacent one. Consequently, there is exact local conservation. The final form of the equation then becomes

$$
K \sum_{j=1}^{n} h_j \frac{\partial \phi_j(x,y)}{\partial n} \bigg|_{\partial FV} + h_i \kappa A_{FV} - h^* \kappa A_{FV} = 0,
$$

where ∂FV is the perimeter of the finite volume, for example, C in Figure 7.18 and A_{FV} is the area of this same finite volume.

Adjacent finite volumes share nodes. For example, the approximation of the normal derivative $\partial \phi_j(x,y)/\partial n$ requires nodal values from adjacent elements, as is also the case if the form of the function representation shown as Q in Eq. (7.116) is used. Therefore the equations representing adjacent elements are coupled and must be solved simultaneously. It is also worth noting that if a transient problem were considered, the discretized time derivative (finite-difference form) would have the form of term Q, leaving the opportunity to solve the problem using either the distributive form of this term or employing the lumped version shown in Eq. (7.117).

7.6   FINITE-ELEMENT METHOD AND THE TRANSIENT FLOW EQUATION

The point of departure in the discussion of the finite-element solution of the *transient flow equation* is Eq. (7.1) reproduced here as Eq. (7.118):

$$-\frac{\partial}{\partial x} K \left[\frac{\partial}{\partial x} (h(x,t)) \right] + S_s \frac{\partial}{\partial t} (h(x,t)) = 0, \quad x \in [0, L], \quad t \in [0, \infty), \quad (7.118)$$

which we rewrite using a finite-difference representation of the time derivative as

$$\mathcal{L}h(x,t) = -\frac{\partial}{\partial x} K \left[\frac{\partial}{\partial x} (h(x,t_k)) \right] + S_s \frac{h(x,t_k) - h(x,t_{k-1})}{\Delta t} = 0,$$

$$x \in [0, L], \quad t_k = k \Delta t, \quad k = \{1, \ldots, T\}, \quad (7.119)$$

where T is the total number of time steps. Equation (7.119) now looks rather like Eq. (7.81) with the source term replaced by the time-derivative approximation. In fact, we can treat Eq. (7.119) in exactly the same way we did Eq. (7.81), keeping in mind that we must consider head values at two time levels at every nodal location. The finite-element approximation thus can be written as an extension of Eq. (7.85), that is,

$$I_1 = -\int_X \left(K \frac{d}{dx} (\ell_0(x) h(x_0, t_k) + \ell_1(x) h(x_1, t_k)) \frac{d}{dx} \ell_1 \right) dx$$

$$- \int_X (\ell_0(x) h(x_0, t_k) + \ell_1(x) h(x_1, t_k)) \ell_0(x) \, dx$$

$$- \int_X (\ell_0(x) h(x_0, t_{k-1}) + \ell_1(x) h(x_1, t_{k-1})) \ell_0(x) \, dx, \quad x \in [x_0, x_1], \quad (7.120)$$

where the element is assumed to be away from the boundaries of the domain, and therefore the boundary evaluations from integration by parts are zero.

Given the initial state of the system, that is, $h(x,0)$, it is possible to obtain values of $h(x_0, t_0)$ and $h(x_1, t_0)$. Substitution of these values into Eq. (7.120) yields one expression in the two unknowns $h(x_0, t_1)$ and $h(x_1, t_1)$. A similar equation is written for each element in the mesh. One then adds equations from adjacent elements. After application of boundary conditions, one has a set of simultaneous equations with the same number of equations as unknown values of $h(x_i, t_k)$. After solving for these head values at t_k, one substitutes these values for $h(x_i, t_{k-1})$. The process is then repeated, marching forward in time by steps of Δt, until the desired period of time is simulated.

While we do not present it here, a similar formulation can be derived for the finite-volume equations.

7.7 SIMULATION UNDER PARAMETER UNCERTAINTY

The parameters that appear in the groundwater flow and transport equations are difficult to measure and their values are inherently uncertain. Of special concern is the hydraulic conductivity. While there is only one value of hydraulic conductivity at a specified location in three dimensional space, for a given volume-averaging length scale (see Section 4.2 for a discussion of volume averaging), that value cannot be measured directly. Rather, it is necessary to obtain this parameter using an indirect method such as considered in Section 6.2, which is devoted to pumping tests. As a result of the fact that the procedures involved require matching data to theoretical curves derived from idealized models, there is unavoidable uncertainty in the values of hydraulic conductivity obtained at measurement points.

In practical applications the parameters of interest are known only at a limited number of locations. However, in modeling groundwater flow and transport it is necessary to interpolate the values of the parameters such that they are available anywhere within the model domain. The interpolation of parameter values from a limited number of locations to the entire domain of interest leads to further uncertainty in the conductivity values.

In summary, the parameters needed to model groundwater flow and transport, while formally unique, are uncertain and in some applications this uncertainty needs to be explicitly accommodated in the modeling protocol. In this section we review some basic statistical concepts, show how these concepts can be used to describe uncertainty in hydraulic conductivity, and finally indicate how a model with uncertain hydraulic conductivity values can be solved.

7.7.1 Introduction to Probability

The theory of probability is the mathematical apparatus that describes *random phenomena*. Random events are distinguished from *deterministic events* in the following context. If, given a specific set of conditions, an anticipated event always occurs, it is called a deterministic event. If, on the other hand, the event may or may not occur every time the specific set of conditions is realized, it is called a random event. If a measurement, such as that for hydraulic conductivity, may, for the same location, come up different depending on the conduct of the experiment or its interpretation, it is random in the sense of imprecise. Imprecision is the nature of the uncertainty with which we deal in this section.

The classic conceptual experiment that is used to introduce the notion of probability is the coin toss. If a coin is tossed the outcome of the toss, namely, whether it is a head or a tail, is one replication of the experiment. The result is an observation. If the experiment is conducted repeatedly, the frequency with which a tail is realized is found to converge to one-half. The probability that a tail will occur in a replication of the experiment is thereby defined as 0.5. If we define the occurrence of a head as event A, then we write the probability of a head in any given toss as

$$P(A) = 0.5.$$

The probability of the occurrence of an event is described using the distribution function. If we consider, for example, the hydraulic conductivity relative to its mean value as a random variable and call it X, the distribution function $F_X(x)$ is defined by

$$F_X(x) = P\,(X \le x)\,. \tag{7.121}$$

Equation (7.121) says that the probability that the value of the random variable X will take on a value less than or equal to x is $F_X(x)$. The function $F_X(x)$ vanishes as $x \to -\infty$ since there is no chance that X can be smaller than this bound. Similarly, $F_X(x)$ approaches unity as $x \to \infty$ because X is certainly less than this upper bound. Symbolically, we write

$$\lim_{x \to -\infty} F_X\,(x) = 0,$$

$$\lim_{x \to \infty} F_X\,(x) = 1.$$

Using this notation, it is possible to express the probability that a given value of the hydraulic conductivity lies between two bounds a and b. The relationship is

$$P\,(a \le X < b) = F_X\,(b) - F_X\,(a)\,.$$

The most commonly utilized form of distribution function is the cumulative normal distribution. It has the form

$$F_X\,(x) = \frac{1}{\sqrt{2\pi\sigma_X^2}} \int_{-\infty}^{x} \exp\left(-\frac{x - \mu_X}{2\sigma_X^2}\right)\,dx, \tag{7.122}$$

where μ_X is the true *population mean* value of X and σ_X^2 is the true *population variance*. The square root of the variance, σ_X, is the standard deviation. An example of a cumulative normal distribution as described by Eq. (7.122) is shown graphically in Figure 7.19. The value of the mean μ_X is zero and the variance σ_X^2 is unity.

Another important function is the *probability density function* $f_X(x)$, which can be determined by taking the derivative of Eq. (7.122) with respect to x. One obtains

$$\frac{dF_X}{dx} = f_X\,(x)\,, \tag{7.123}$$

where

$$f\,(x) = \frac{1}{\sqrt{2\pi\sigma_X^2}} \exp\left(-\frac{(x - \mu_X)^2}{2\sigma_X^2}\right)\,. \tag{7.124}$$

From Eq. (7.123) it is clear that the following holds:

$$F_X\,(x) = \int_{-\infty}^{x} f\,(\lambda)\,d\lambda.$$

A graphical representation of $f_X(x)$ as given by Eq. (7.124) is provided in Figure 7.20. The normal probability density function has several interesting properties. The grey area indicated in Figure 7.20 is determined by selecting those values of x bounded by one

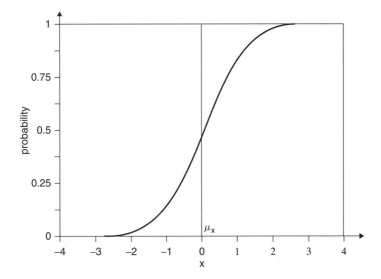

FIGURE 7.19. Cumulative normal distribution.

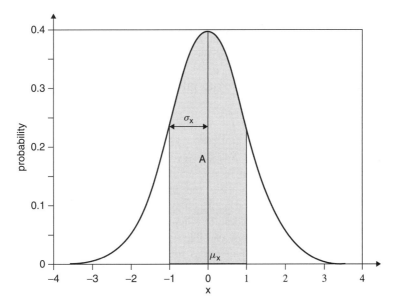

FIGURE 7.20. Probability density function for the normal distribution.

standard deviation, in this case $x \in [-1, 1]$. The shaded area represents 68% of the area underlying $f_X(x)$ and therefore 68% of the observations of X lie within one standard deviation of the mean. In other words, in this example, there is a 68% probability that a value of x drawn at random from X will lie between -1 and 1. Similarly, 95% of the observations of X lie within two standard deviations and 99.7% within three standard deviations.

We now look into several statistical parameters that help to describe a random variable. The expected value, also called the *ensemble mean* and *ensemble average,* of the random variable X is denoted as $E(X)$ and given by

$$\mu(X) = \int_{-\infty}^{\infty} x f_X(x)\, dx = E(X).$$

The function $\mu(X)$ is also called the first moment. Perhaps a more intuitive statement is given by [4]

$$E(X) = \lim_{N \to \infty} \frac{1}{N} \sum_{i=1}^{N} x_i,$$

where N is the number of observations of the random variable X.

The second moment is the *variance* $\sigma_X^2(x)$ and is given by

$$\sigma_X^2(X) = \int_{-\infty}^{\infty} (x - \mu_X)^2 f_X(x)\, dx$$
$$= E\left[(X - \mu_X)^2\right]. \tag{7.125}$$

The variance is a measure of the central tendency of the observations around the mean. It defines the spread of the distribution. As can be seen in Figure 7.20, the smaller the variance, and therefore the smaller the standard deviation, the more pointed the distribution becomes.

The third moment is the skewness $s_X(X)$ defined as

$$s_X(X) = \int_{-\infty}^{\infty} (x - \mu_X)^3 f_X(x)\, dx$$
$$= E\left[(X - \mu_X)^3\right].$$

The skewness describes the asymmetry in the distribution. In the case of a normal distribution, the skewness is zero—that is, it is a symmetric distribution.

To address the uncertainty in hydraulic conductivity, we require theory that accommodates spatially distributed random variables. In the case of hydraulic conductivity, not only are the values at different locations uncertain, but also values at neighboring locations are correlated.

The required theory to describe spatial correlation is found in the mathematics of jointly distributed variables. The first step is to define the concept of an *array of random variables*, which we will denote as **X**. Following Herrera [5], we write

$$\mathbf{X} = \begin{pmatrix} X_1 \\ \vdots \\ X_n \end{pmatrix}. \tag{7.126}$$

The joint distribution of **X** is

$$F_X(x) = P(X_i \le x_i, \ldots, X_n \le x_n), \tag{7.127}$$

where x_i is a measurement taken from X_i (not necessarily the first measurement) and $P(X_1 \leq x_1, \ldots, X_n \leq x_n)$, in the context of Eq. (7.127), can be interpreted to mean the probability of $X_1 \leq x_1$ and $X_2 \leq x_2, \ldots$, and $X_n \leq x_n$ simultaneously. The logical consequence of this is the definition of a joint density function, which can be written

$$F_X(x) = \int_{-\infty}^{x_1}, \ldots, \int_{-\infty}^{x_n} f_X(\lambda_1, \ldots, \lambda_n) \, d\lambda_1 \cdots d\lambda_n.$$

Hydraulic conductivity is generally assumed to be lognormally rather than normally distributed. A lognormal distribution has the important attribute of having only positive values. In this case the logarithm of the hydraulic conductivity is normally distributed. The above expressions can be used by substituting $X \leftarrow \ln X$. Thus, in the above, one can identify the random variable X_i with the value of the logarithm of the hydraulic conductivity at a location $\mathbf{s}_i$. The observations (or measurements) of the hydraulic conductivity at the location $\mathbf{s}_i$ (namely, X_i) are given by x_i. Thus the vector $\mathbf{X}$ can be thought of as the array of hydraulic conductivity values measured at all locations where measurements are available. In other words, Eq. (7.127) describes the interdependence of the values of hydraulic conductivity at the various measurement points.

Since we are working with geological materials wherein regional as well as local processes were at work during their formation, we would expect to find a significant interdependence between values in different spatial locations. For example, one would expect that material deposited along the course of an ancient stream would be of similar grain size, whereas the material outside the bank of the stream would be significantly different, yet correlated with nearby out-of-bank deposits.

The covariance of two random variables, X_k and X_l, is defined as

$$\text{cov}(X_k, X_l) \equiv \int_{-\infty}^{\infty} \int_{-\infty}^{\infty} (x_k - \mu_k)(x_l - \mu_l) f_X(x_k, x_l) \, dx_k \, dx_l$$
$$= E\left[(X_k - \mu_k)(X_l - \mu_l)\right]. \tag{7.128}$$

The main diagonal of the covariance matrix, that is, $\text{cov}(X_k, X_k)$ is the variance $\sigma^2(X_k)$, as can be seen by comparing Eqs. (7.125) and (7.128). In the event the two random variables X_k and X_l are defined as different physical quantities, $\text{cov}(X_k, X_l)$ is called the cross covariance. For example, consider the relationship between hydraulic conductivity and hydraulic head.

The normalized form of the covariance is called the correlation coefficient and is given as

$$\rho(X_k, X_l) \equiv \frac{\text{cov}(X_k, X_l)}{\sigma(X_k)\sigma(X_l)}.$$

In the event that two random variables are not related one to another, such as might be the case in our earlier example of a hydraulic conductivity value inside a buried stream bed versus one obtained from measurements made outside the stream bed, the two variables are called independent. The mathematical statement of this condition is

$$E\left[f_{X_k}(x_k) f_{X_l}(x_l)\right] = E\left[f_{X_k}(x_k)\right] E\left[f_{X_l}(x_l)\right].$$

An uncorrelated pair of jointly distributed random variables is characterized by zero values in the off-main-diagonal elements of the correlation matrix, that is,

$$\text{cov}(X_k, X_l) = 0, \quad k \neq l.$$

The foregoing development that focuses on pairs of individual random variables can be extended to consider the random vector $\mathbf{X}$ introduced earlier in Eq. (7.126). The expected value of a vector is equal to the expected value of its components, that is,

$$E(\mathbf{X}) = \begin{pmatrix} E(X_1) \\ \vdots \\ E(X_n) \end{pmatrix}.$$

The covariance of a random vector is given by the covariance matrix $\mathbf{P_X}$ given by

$$\mathbf{P_X} = E\left[(\mathbf{X} - E(\mathbf{X}))(\mathbf{X} - E(\mathbf{X}))^T\right],$$

where the superscript T denotes the transpose of a matrix.

7.7.2 Random Fields

We used in an earlier example the values of hydraulic conductivity at various locations as an illustration of a set of random variables. In this subsection we formalize this concept of spatial dependence of the random variable. We denote the family of random variables that depends on the location $\mathbf{s}$ as a spatial random field. Spatial random fields can be either continuous or discontinuous depending on whether the locations associated with the random variable are continuous or discontinuous.

The mean value $\mu(\mathbf{s})$ is now a function of $\mathbf{s}$. Thus we have

$$\mu_X(\mathbf{s}) = E[X(\mathbf{s})],$$

where, in the case of a scalar random variable, such as isotropic hydraulic conductivity, the result is a number. If the random variable is a tensor, such as encountered in anisotropic hydraulic conductivity, then the resulting mean is also a tensor.

The covariance of a scalar random field, given two locations $\mathbf{s}$ and $\mathbf{s} + \mathbf{r}$, where $\mathbf{r}$ is a distance vector, is given by

$$\text{cov}_X(\mathbf{s}, \mathbf{s} + \mathbf{r}) = E[(X(\mathbf{s}) - \mu_X(\mathbf{s}))(X(\mathbf{s} + \mathbf{r}) - \mu_X(\mathbf{s} + \mathbf{r}))] \tag{7.129}$$

and the corresponding correlation function is

$$\rho(\mathbf{s}, \mathbf{s} + \mathbf{r}) = \frac{\text{cov}_X(\mathbf{s}, \mathbf{s} + \mathbf{r})}{\sigma_X(\mathbf{s})\,\sigma_X(\mathbf{s} + \mathbf{r})}.$$

Since the statistical parameters describing, in our example, hydraulic conductivity are now a function of space, two new concepts need to be considered. The first is *stationarity*. The random field $X(\mathbf{s})$ is considered to be stationary in the wide sense if it has (1) finite second moments, (2) a constant mean (independent of position), and (3) a covariance that depends only on the distance between the locations of the random variables. Thus we have

$$\mu_X (\mathbf{s}) = \mu \tag{7.130}$$

and

$$\text{cov}_{\mathbf{X}} (\mathbf{s}, \mathbf{s} + \mathbf{r}) = \text{cov}_{\mathbf{X}} (\mathbf{r}). \tag{7.131}$$

The second property is that of *isotropy*. The random field is isotropic in the wide sense if, in addition to the attributes presented above, the covariance depends only on the magnitude of the difference between the locations of the random field values, not on direction.

A function that is related to but different from the covariance is the *variogram*. It is defined as the variance of the difference between two random field values a distance r apart. Note that it does not depend on the point of reference from which the distance r is measured. The variogram relationship is normally written

$$2\gamma (r) = \sigma^2 (X (\mathbf{s} + r) - X (\mathbf{s})) = E \left[(X (\mathbf{s} + r) - X (\mathbf{s}))^2 \right], \tag{7.132}$$

where the function $\gamma (r)$ is defined as the *semivariance*. It is related to the covariance by

$$\gamma (r) = \text{cov}_X (0) - \text{cov}_X (r). \tag{7.133}$$

Logic tells us that as the distance between two points increases, the statistical properties measured at the two points should become less correlated. Two measures of this tendency are the *correlation scale* and the *correlation range*. The correlation scale λ_X is defined as

$$\lambda_X = \frac{1}{\sigma_X^2} \int_0^\infty \text{cov}_X (r) \, dr.$$

At a distance $r = \lambda_X$ the covariance is approximately half the variance [5]. The *range* ε is the distance r at which the correlation is 5% of the variance. Let us now see how these statistical concepts can be used in practice.

7.7.3 The Semivariogram

While we now have some basic nomenclature associated with a random field, such as hydraulic conductivity, we do not yet know how to generate a continuous random field from a set of discrete measurements. In other words, how does one determine the statistical properties of hydraulic conductivity at a location where measurements do not exist. The most popular method for achieving this goal is an interpolation algorithm called *kriging*, which we will now discuss.

The semivariance provided by Eq. (7.132) is clearly a function of distance. The closer the two values are in a regional sense, the smaller will be $\gamma (\mathbf{r})$. A plot of the semivariance as a function of distance r, is called a *semivariogram*. Two semivariograms are shown in Figure 7.21. The semivariogram can be computed directly from data using the relationship

$$\gamma^* (r) = \frac{\sum_{i=1}^N (X_i (\mathbf{s}) - X_i (\mathbf{s} + r))^2}{2N}, \tag{7.134}$$

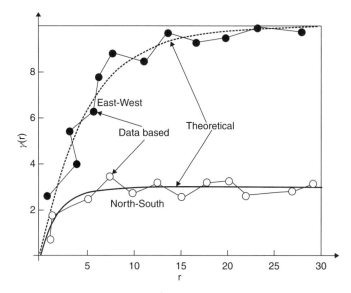

FIGURE 7.21. Idealized semivariograms showing the effect of anisotropy in the data. The range and the sill for the east–west direction is 5 and 10 and in the north–south direction 2 and 3 (adapted from [6]).

where the superscript $*$ indicates the calculation is made using direct observations and N is the number of pairs. The semivariogram can be interpreted as one-half the sum of the squares of the differences in the N observed pairs, each with separation distance r from one another, divided by N.

Of course, there is a problem with this concept. If the data are irregularly spaced, the chances of finding two points exactly a distance r apart is virtually zero. Thus there is a *magnitudinal* and *directional tolerance* built into the value of r that is included in Eq. (7.134). A generally accepted approach is to define a tolerance that is half the value of r. Thus, for a value of r of 100 ft, there would be a tolerance of ± 50 ft. In other words, one simply takes all pairs of points separated by 100 ± 50 ft for making the calculation required in Eq. (7.134). The value of r used in the semivariogram plot will be the average of all the distances between the N pairs falling within the specified tolerance. In our example, that would be 100 ft. The larger the tolerance, the more pairs that will be defined and the smoother the semivariogram will be. The smaller the tolerance the fewer the points, and therefore the more irregular the semivariogram.

The piecewise linear curves on Figure 7.21 are determined from data. Two sets of data have been used (open and closed circles), and the difference between these two curves is the direction in which the distances were measured. Therefore the data set exhibits anisotropy in its statistical structure.

While, in concept, the direct data-derived variograms could be used in kriging, this is not what is normally done in practice. Rather, theoretical kernel variograms that exhibit the desirable geometric properties are employed. The most commonly used form is the exponential model, which is given by

$$\gamma(r) = c_0 + c\left(1 - \exp\left(-\frac{r}{a}\right)\right). \tag{7.135}$$

Terminology associated with this variogram mode, as well as others, is as follows:

- The value of c_0 is identified with the concept of a *nugget*. This term accommodates the unresolved measurement error. It is identified with the error at a measuring point. It is seen on the variogram as the intercept; in the example presented in Figure 7.21 this value is zero.
- The *range*, mentioned earlier, is the factor that describes the degree of correlation between data points, usually denoted as a distance. It is the distance at which the variogram reaches the *sill* or a value that is 95% of the sill. The variable a in Eq. (7.135) represents this value.
- The *sill* is the value of the variogram as the value of $\mathbf{r} \to \infty$. It is equal to the total variance of the data set and is given by c in Eq. (7.135).

The two smooth curves in Figure 7.21 are exponential variograms. For the east–west curve the range is 15 and the sill is 10. For the north–south curve, the range is 5 and the sill is 3. Other functions can play the role of Eq. (7.135) and each exhibits different properties. The analyst is faced with the challenge of selecting the function that best represents the data.

7.7.4 Kriging

Kriging is an estimation procedure developed by G. Matheron [7] (as the "theory of regionalized variables") and D. G. Krige and used as an optimal interpolation method in the mining industry. It was called kriging by Matheron in recognition of Krige. Kriging comes in many flavors, and we will consider only the simplest in this section.

In *ordinary kriging*, a basic assumption is that the sought-after random variable, in our case the hydraulic conductivity distribution, is second-order stationary. The requirements of stationarity are presented in Eqs. (7.130) and (7.131). However, in practice, the mean value of the hydraulic conductivity is seldom constant. Rather, it is likely to have a regional trend. One approach to circumventing this problem is to first create a best-fitting interpolated surface for the data using a simple polynomial function and all available data points. The resulting function is called a *drift*. The residuals, computed as the difference between the actual data values and the drift, should in concept be stationary. Kriging is therefore performed on the residuals. The interpolated (kriged) residuals are then added to the drift to obtain the sought-after function (e.g., the hydraulic conductivity field) at any desired location. The above procedure for accommodating a "nonstationary" field is often called *universal kriging*.

The first step in developing the kriging interpolation approach is to represent the stationary statistics as defined above using the relationship

$$x^* (\mathbf{s}_0) = \sum_{i=1}^{N} w_i (\mathbf{s}_0) x_i, \tag{7.136}$$

where $w_i (\mathbf{s})$ are weighting functions and x_i, $i = 1, \ldots, N$, are the observed data values. Thus the approximate value of the interpolated variable is the weighted sum of existing data. The notation $\mathbf{s}_0$ implies a specific but arbitrary location and the use of the superscript asterisk implies an estimate. Note that the weighting functions are identified with the location $\mathbf{s}_0$.

The challenge is to define a methodology that will provide the weights $w_i(\mathbf{s}_0)$. If we consider the data points x_i as one realization of a random variable, we can perhaps use the properties of random variables to obtain the weights. Two conditions are imposed on the selection of the weights, $w_i(\mathbf{s})$. The first is that the estimator be unbiased, that is, that the expected value of the random variable $X^*(\mathbf{s})$ equal μ. The resulting constraint is

$$\sum_{i=1}^{N} w_i(\mathbf{s}_0) = 1. \tag{7.137}$$

Second, it is required that the variance of the estimation of $X(\mathbf{s}_0)$, the kriging variance, be minimal. To determine the kriging variance we first define the quantity

$$e_i(\mathbf{s}_0) = \left(X^*(\mathbf{s}_0) - X(\mathbf{s}_0) \right),$$

which is the difference between the interpolated estimate of the random variable X, namely, X^*, and the true value X at the location $\mathbf{s}_0$. The kriging variance is now defined as

$$\sigma_e^2(\mathbf{s}_0) = E(e_i(\mathbf{s}_0)) = E\left[\left(X^*(\mathbf{s}_0) - X(\mathbf{s}_0) \right)^2 \right]. \tag{7.138}$$

To simplify notation, we now redefine terms of the form $X(\mathbf{s}_0)$ as X_0.

We now follow the development of DeMarsily [8] in the formulation of the kriging equations. Having substituted Eq. (7.136) into Eq. (7.138), we obtain

$$
\begin{aligned}
E\left[(X_0^* - X_0)^2 \right] &= E\left[\left(\sum_{i=1}^{N} w_{i0} X_i - X_0 \right)^2 \right] \\
&= E\left[\left(\sum_{i=1}^{N} w_{i0} X_i - \sum_{i=1}^{N} w_{i0} X_0 \right)^2 \right] \\
&= E\left[\left(\sum_{i=1}^{N} w_{i0} (X_i - X_0) \right)^2 \right] \\
&= E\left[\left(\sum_{i=1}^{N} w_{i0} (X_i - X_0) \right) \left(\sum_{j=1}^{N} w_{j0} (X_j - X_0) \right) \right] \\
&= \sum_{i=1}^{N} \sum_{j=1}^{N} w_{i0} w_{j0} E\left[(X_i - X_0)(X_j - X_0) \right].
\end{aligned}
\tag{7.139}
$$

From Eq. (7.132) we have

$$
\begin{aligned}
2\gamma(\mathbf{s}_i - \mathbf{s}_j) &= E\left[(X(\mathbf{s}_i) - X(\mathbf{s}_j))^2 \right] \equiv E\left[(X_i - X_j)^2 \right] \\
&= E\left[((X_i - X_0) - (X_j - X_0))^2 \right]
\end{aligned}
$$

$$= E\left[(X_i - X_0)^2 + (X_j - X_0)^2\right] - 2E\left[(X_i - X_0)(X_j - X_0)\right]$$

$$= 2\gamma\,(\mathbf{s}_i - \mathbf{s}_0) + 2\gamma\,(\mathbf{s}_j - \mathbf{s}_0) - 2E\left[(X_i - X_0)(X_j - X_0)\right]. \qquad (7.140)$$

Notice that the expected value term appearing in Eq. (7.140) also appears in Eq. (7.139). Thus substituting the former equation into the latter we obtain

$$E\left[(X_0^* - X_0)^2\right] = -\sum_{i=1}^{N}\sum_{j=1}^{N} w_{i0}w_{j0}\gamma\,(\mathbf{s}_i - \mathbf{s}_j) + \sum_{i=1}^{N}\sum_{j=1}^{N} w_{i0}w_{j0}\gamma\,(\mathbf{s}_i - \mathbf{s}_0)$$

$$+ \sum_{i=1}^{N}\sum_{j=1}^{N} w_{i0}w_{j0}\gamma\,(\mathbf{s}_j - \mathbf{s}_0). \qquad (7.141)$$

One can now factor out $\sum_{j=1}^{N} w_{j0}$ in the second term on the right-hand side of Eq. (7.141) and $\sum_{i=1}^{N} w_{i0}$ in the third term. Since both of these summations equal one by Eq. (7.137), the last two terms become

$$\sum_{i=1}^{N} w_{i0}\gamma\,(\mathbf{s}_i - \mathbf{s}_0) + \sum_{j=1}^{N} w_{j0}\gamma\,(\mathbf{s}_j - \mathbf{s}_0),$$

which are identical. Thus Eq. (7.141) reduces to

$$E\left[(X_0^* - X_0)^2\right] = -\sum_{i=1}^{N}\sum_{j=1}^{N} w_{i0}w_{j0}\gamma\,(\mathbf{s}_i - \mathbf{s}_j) + 2\sum_{i=1}^{N} w_{i0}\gamma\,(\mathbf{s}_i - \mathbf{s}_0). \qquad (7.142)$$

Recall that our goal was to minimize the kriging variance (Eq. (7.138)) subject to the constraints imposed by Eq. (7.137). Our objective function (see Section 10.7 on optimization) becomes

$$f\,(w_1, w_2, \ldots, w_N, \eta) = E\left[(X_0^* - X_0)^2\right] - \eta\left[\sum_{i=1}^{N} w_{i0} - 1\right],$$

where η is called a *Lagrange multiplier*. The Lagrange multiplier is introduced because there is one more equation than there are unknowns. The Lagrange multiplier adds an unknown without upsetting the equality in Eq. (7.142). The minimum value for $f\,(w_1, w_2, \ldots, w_N, \eta)$ is achieved by differentiation of $f\,(w_1, w_2, \ldots, w_N, \eta)$ with respect to the $N + 1$ unknown values $w_1, w_2, \ldots, w_N, \eta$ and setting each of the resulting $N + 1$ equations to zero. The result is

$$\sum_{j=1}^{N} w_{j0}\gamma\,(\mathbf{s}_i - \mathbf{s}_j) + \eta = \gamma\,(\mathbf{s}_i - \mathbf{s}_0), \quad i = 1, \ldots, N$$

$$\sum_{i=1}^{N} w_{i0} = 1. \qquad (7.143)$$

Equation (7.143) can now be solved for the weights and the coefficient η. Substitution of the weights into Eq. (7.136) allows one to obtain a value of the random variable X at any point $(\mathbf{s}_0)$.

It is also possible to estimate the value of the variance of the error. Recall that the variance of the error given by Eq. (7.138) is

$$\sigma_{e0}^2 = E\left[\left(X_0^* - X_0\right)^2\right].$$

Therefore, by substituting the solution to Eq. (7.143) into Eq. (7.142), the variance of the error can be computed at the point $\mathbf{s}_0$.

7.7.5 Solving Problems Involving Random Hydraulic Conductivity Fields

To this point we have described how to obtain a random field of hydraulic conductivity values from field data. The open question is how we use this information in practical applications. The application of interest to us in this text is forecasting state-variable behavior using a mathematical model, which, for generality, we assume to be numerical in nature.

There are two main approaches to the solution of the groundwater flow and transport equations given the parameters of interest are defined using random fields. One involves the expansion of the uncertain parameters and the state variables in terms of a series [9]. This mathematical technique is often called first-order analysis. The challenge with this approach is that there is a limitation on the size of the parameter variances relative to their means that can be accommodated, and this limitation is smaller than the variance values often encountered in field situations.

The second approach is based on *Monte Carlo* methods. While this strategy does not suffer from the same limitations as the perturbation series approach, it has the disadvantage that solving the resulting equations requires a large amount of computational effort. Nevertheless, due to its generality, we focus here on the Monte Carlo approach. A good discussion of this topic is found in Ripley [10].

We assume at this point that the statistical characteristics of the hydraulic conductivity field are known. For convenience we will assume that the field is homogeneous, stationary, and isotropic. As noted earlier, by this assumption we require that the mean and standard deviation are constant over space and that the covariance at two locations depends only on the distance between them. Because we are going to employ numerical methods, which are by their nature discrete, we will be considering a discrete random field.

In the Monte Carlo approach, a sample of hydraulic conductivity values, one for each node in the numerical mesh, is drawn from a population of values consistent with the statistics of the field. Such a set of values is called a realization. A simulation is then performed using the numerical model and the realization. The resulting computed state-variable values constitute one realization of the random state-variable field. The procedure is repeated for additional realizations generated from the known field parameter statistics and the resultant state-variable simulations recorded. The statistics of the state variable can then be determined. Given sufficient realizations, the resultant state-variable statistics should be consistent with the statistical description of the parameter field.

The question that arises is how one generates a series of field-parameter realizations, in our case hydraulic conductivity values, that are consistent with the known parameter statistics. Of the numerous methods that have been developed over the years, those that have been used in practice for groundwater problems are (1) *sequential Gaussian simulation*, (2) the *turning bands* method, (3) *Latin hypercube sampling*, and (4) *LU decomposition*. Since we do not have the space available to address all of these methods, we focus on the relatively intuitive Latin hypercube sampling approach.

The Latin hypercube sampling method will be illustrated via an example problem. The classical "five-point" groundwater flow and transport problem is shown in Figure 7.22. A constant head of 500 m is located in the center of the square. On each corner a constant head of 0 m is defined. The edges of the square are all no-flow and no-diffusive-flux boundaries. The mesh contains 27 blocks along each side and each block is 100 m × 100 m. In the center of the square, a constant concentration source of 10 is provided. The flow field is such that water with a concentration of 10 enters through the injection well at the center of the square and exits the corners. The hydraulic conductivity is lognormally distributed with a mean ln K of 5 and a variance of 1. The relationship for the covariance is

$$\text{cov}_K\,(\mathbf{x}, \mathbf{x} + \mathbf{r}) = \text{cov}_K\,(\mathbf{x}, \mathbf{x})\exp\left[-\sqrt{\left(\frac{r_x}{\lambda_{K_x}}\right)^2 + \left(\frac{r_y}{\lambda_{K_y}}\right)^2}\right], \qquad (7.144)$$

where $\text{cov}_K\,(\mathbf{x}, \mathbf{x} + \mathbf{r})$ is the covariance between two points separated by a distance r, r_x and r_y are the distances between the two points in the x and y directions, respectively (100 m between nearest neighbors in this example), and λ_{K_x} and λ_{K_y} are the correlation lengths in the x and y directions, respectively (224 m for each direction).

The first step in solving this problem is to create a realization of the hydraulic conductivity random field described above for this problem. To do this, we first examine the normal distribution of ln K values as shown in Figure 7.23. The source of such distributions at each nodal location could have been obtained using the kriging algorithm described in Section 7.7.4. In the Latin hypercube approach the probability density

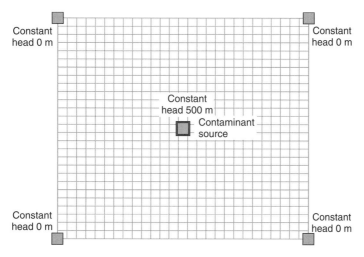

FIGURE 7.22. Definition sketch for the illustration of simulation using a hydraulic conductivity random field (from [11]).

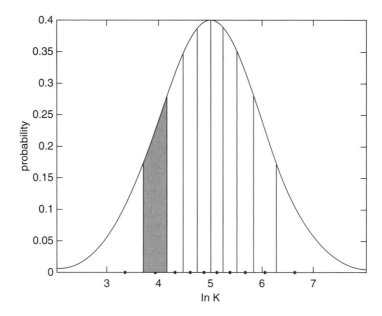

FIGURE 7.23. Distribution of the values of ln K.

function at each nodal location is subdivided into equal areas. Thus the probability of selecting a value of ln K is the same in each interval.

To illustrate how one proceeds from this point forward in the realization selection process, we will consider a much simpler problem than the one just introduced so that the various steps are more transparent. We will return to the original problem shortly.

Consider a random field with four variables; see Table 7.2. The four variables would each represent the ln K value at a particular nodal location. Each of these four variables has a lognormal distribution with a mean of 5 and variance of 1, such as illustrated in Figure 7.23. The target correlation matrix of this random field is determined from the known statistics regarding the field and is given in Table 7.3. We call this matrix R^*.

Let us assume that we are going to do 10 Monte Carlo simulations, such that we will need 10 realizations of the random field, each containing four values of hydraulic conductivity (one for each of the four elements).

TABLE 7.2. Random Field

K1	K2
K3	K4

TABLE 7.3. Target Correlation Matrix R^*

R^*	1	2	3	4
1	1.0	0.702	0.702	0.606
2	0.702	1.0	0.606	0.702
3	0.702	0.606	1.0	0.702
4	0.606	0.702	0.702	1.0

TABLE 7.4. K Matrix

K1	K2	K3	K4
4.615	5.126	4.874	6.036
3.355	5.385	4.326	4.874
5.126	3.964	6.036	4.326
6.645	6.036	6.645	4.615
5.674	4.874	5.385	5.674
4.874	6.645	3.355	5.126
6.036	4.615	4.615	3.964
5.385	4.326	3.964	5.385
3.964	3.355	5.674	6.645
4.326	5.674	5.126	3.355

TABLE 7.5. Correlation Matrix R

R	1	2	3	4
1	1.0	0.135	0.327	−0.227
2	0.135	1.0	−0.293	−0.361
3	0.327	−0.293	1.0	−0.048
4	−0.227	−0.361	−0.048	1.0

The first step is to generate 10 realizations using the standard Latin hypercube sampling technique for each variable, that is, to draw one value from each of the 10 equal areas represented in Figure 7.23, then put them in random order, as shown in Table 7.4. Next, calculate the correlation matrix shown in Table 7.5 using a variant on Eq. (7.128) and the realizations in Table 7.4. Call this matrix R.

Next, perform Cholesky decomposition for both R and R^*; that is, calculate factors Q and Q' such that $R = QQ'$, and P and P' such that $R^* = PP'$. Now calculate a new matrix $K' = K(PQ^{-1})'$, which is shown in Table 7.6 and is, in some sense, an estimate of the values needed for the realizations. Its correlation matrix is calculated in Table 7.7. A comparison of this correlation matrix and the target correlation matrix R^* proves that K' has the same correlation matrix as the target one.

While this set of values has the correct correlation structure, it does not have the correct values as originally drawn from the four probability distributions of ln K.

TABLE 7.6. K' Matrix

K1	K2	K3	K4
4.615	6.475	7.829	12.527
3.355	5.899	7.023	11.047
5.126	5.949	8.441	11.228
6.645	8.357	10.400	14.063
5.674	6.935	8.533	12.854
4.874	7.723	7.389	12.557
6.036	6.967	7.954	11.344
5.385	6.366	7.065	11.478
3.964	4.808	7.444	11.588
4.326	6.694	8.152	11.076

TABLE 7.7. Correlation Matrix of K'

	1	2	3	4
1	1.0	0.702	0.702	0.606
2	0.702	1.0	0.606	0.702
3	0.702	0.606	1.0	0.702
4	0.606	0.702	0.702	1.0

TABLE 7.8. Ranks for K' Matrix

	Rank		Rank		Rank		Rank
4.615	4	6.475	5	7.829	5	12.527	7
3.355	1	5.899	2	7.023	1	11.047	1
5.126	6	5.949	3	8.441	8	11.228	3
6.645	10	8.357	10	10.400	10	14.063	10
5.674	8	6.935	7	8.533	9	12.854	9
4.874	5	7.723	9	7.389	3	12.557	8
6.036	9	6.967	8	7.954	6	11.344	4
5.385	7	6.366	4	7.065	2	11.478	5
3.964	2	4.808	1	7.444	4	11.588	6
4.326	3	6.694	6	8.152	7	11.076	2

TABLE 7.9. Ranks for K Matrix

K1	Rank	K2	Rank	K3	Rank	K4	Rank
4.615	4	5.126	6	4.874	5	6.036	9
3.355	1	5.385	7	4.326	3	4.874	5
5.126	6	3.964	2	6.036	9	4.326	3
6.645	10	6.036	9	6.645	10	4.615	4
5.674	8	4.874	5	5.385	7	5.674	8
4.874	5	6.645	10	3.355	1	5.126	6
6.036	9	4.615	4	4.615	4	3.964	2
5.385	7	4.326	3	3.964	2	5.385	7
3.964	2	3.355	1	5.674	8	6.645	10
4.326	3	5.674	8	5.126	6	3.355	1

To circumvent this problem we rearrange the original correct values (K matrix) and the fabricated ones (K' matrix) in the form found in Tables 7.8 and 7.9. The values identified under the column rank in these tables indicate the relative magnitude in the numbers in the column to the left. For example, the smallest value in column one is 3.355 and this is given the rank value of 1. The largest value in the first column has a value of 6.935 and has a rank value of 10.

Finally, we rearrange the realizations in the K matrix using the ranks in the K' matrix. In other words, the values in K are arranged such that their rank order is the same as in K'. The new matrix D is formed as shown in Table 7.10.

Each row of the D matrix forms one set of Latin hypercube sampling (Lhs) realizations. The correlation matrix of D derived from the rearranged values is shown in Table 7.11. It is very close to the target correlation matrix. Now we have a set of samples that has the correct values from the distributions and nearly the correct correlation.

To conduct our Monte Carlo simulation, we would solve the four-element groundwater flow problem ten times, using one row of the D matrix for each simulation. We would

TABLE 7.10. D Matrix

	Rank		Rank		Rank		Rank
4.615	4	4.874	5	4.874	5	5.385	7
3.355	1	3.964	2	3.355	1	3.355	1
5.126	6	4.326	3	5.674	8	4.326	3
6.645	10	6.645	10	6.645	10	6.645	10
5.674	8	5.385	7	6.036	9	6.036	9
4.874	5	6.036	9	4.326	3	5.674	8
6.036	9	5.674	8	5.126	6	4.615	4
5.385	7	4.615	4	3.964	2	4.874	5
3.964	2	3.355	1	4.615	4	5.126	6
4.326	3	5.126	6	5.385	7	3.964	2

TABLE 7.11. Correlation Matrix Associated with the D Matrix

	1	2	3	4
1	1.0	0.759	0.705	0.672
2	0.759	1.0	0.550	0.610
3	0.705	0.550	1.0	0.587
4	0.672	0.610	0.587	1.0

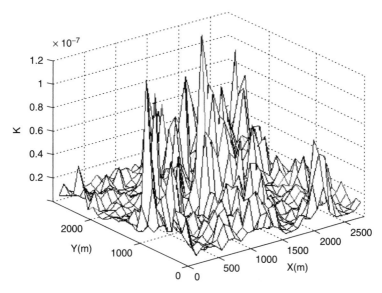

FIGURE 7.24. One realization of hydraulic conductivity K generated using the Latin hypercube sampling technique.

record the values associated with each of these runs and from these compute the state-variable statistics.

Returning now to the larger-scale problem defined by Figure 7.22, we present in Figure 7.24 one realization of the random field for this problem generated using the Latin hypercube sampling strategy.

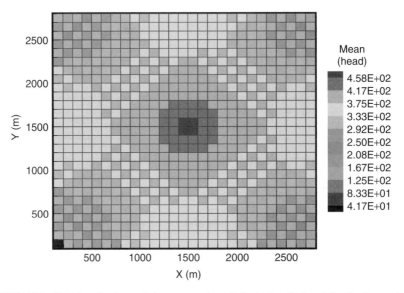

FIGURE 7.25. Calculated values of the mean value of the hydraulic head for the five-point problem (from [12]).

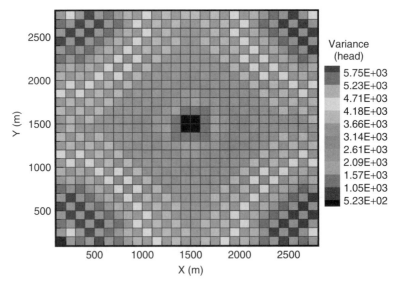

FIGURE 7.26. Calculated variance for the head values for the five-point problem (from [12]).

The solution for the mean values of the hydraulic head is given in Figure 7.25 and the variance of the head values is found in Figure 7.26.

For each realization of the ln K field, a head value field is calculated. From Darcy's law, a velocity field is then calculated. Finally, with the velocity field in hand, a concentration field is calculated. The mean concentration field for the five-point problem is shown in Figure 7.27 and the concentration variance field is shown in Figure 7.28.

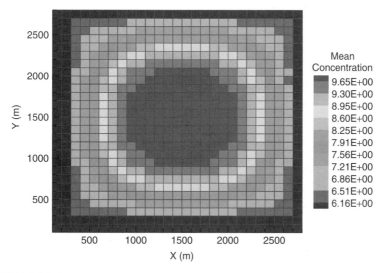

FIGURE 7.27. Calculated mean concentration field for the five-point problem (from [12]).

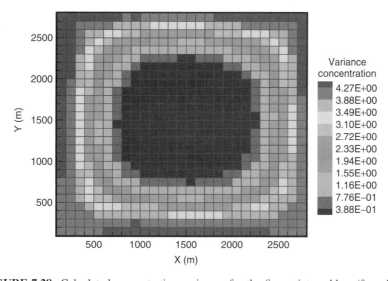

FIGURE 7.28. Calculated concentration variances for the five-point problem (from [12]).

How well does Monte Carlo work with the various schemes used for generating realizations? In other words, if we use the four main techniques for creating realizations, what is the relative rate of convergence of the Monte Carlo solution to the correct solution as the number of realizations increases?

Because of the manner in which the Latin hypercube sampling is done, there is no error in the calculation of the mean value of the realizations. However, the covariance calculation is more difficult to compute as seen by the above example, so we consider a comparison of the covariance computed with the Latin hypercube sampling (Lhs) method

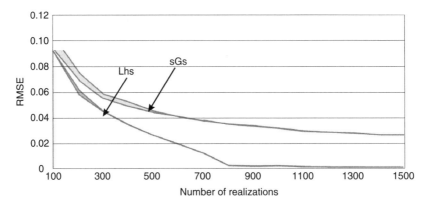

FIGURE 7.29. A comparison of the rate of convergence of the calculated covariance to the exact covariance as the number of realizations increases. RMSE is the root mean square error of the deviation of the computed value relative to the exact (from [11]).

and the popular sequential Gaussian simulation (sGs) method. In Figure 7.29 is shown a measure of the error between the computed and exact correlation values for the two methods calculated using the root mean square error of all calculated values as the measure. One observes that both the sGs and Lhs methods appear to converge to the correct solution as the number of realizations increases. However, at least in this example, the Lhs technique converges more rapidly, reaching a very accurate representation in about 800 realizations. The shaded area represents the variation observed with different initial random seed values. Ideally one would like to see no dependence on the choice of seed value.

Let us now examine how quickly the calculated concentration covariance values obtained by solving the flow and transport equations approach their correct values. In this case there is no known solution to the problem so an inexact reference solution must be used. We have chosen the solution obtained after 9000 realizations as the reference and the errors shown in Figure 7.30 are the differences between the values calculated at

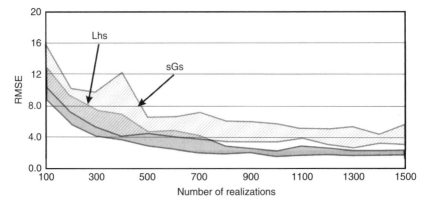

FIGURE 7.30. A comparison of the rate of convergence of the calculated concentration values to the reference concentration field with increasing realizations (from [11]).

the indicated number of realizations and the reference. Again we use the sGs–Lhs comparison. Both methods appear to converge with an increase in the number of realizations. The Lhs approach is again more accurate, but the difference between the two methods is less dramatic than in the case of the covariance of the hydraulic conductivity.

7.8 SUMMARY

The chapter was dedicated to a discussion of the use of numerical simulation for the representation of groundwater flow. The finite-difference, finite-element, and finite-volume methods were described using polynomial approximation and the method of weighted residuals theories as the mathematical underpinnings. The advantages of the various methods for practical application were considered. Finally, the methodology was extended from consideration of deterministic hydraulic conductivity to consideration of stochastic hydraulic conductivity as described by random fields.

7.9 PROBLEMS

7.1. Write a computer code to solve the explicit finite-difference approximation given by Eq. (7.45) and Figure 7.3. Assume the parameters K and S_s are constant, and consider simple boundary and initial conditions given by $h(z, 0) = 0$, $h(0, t) = 1$, and $h(L, t) = 0$. Calculate solutions for a range of parameter values, and for a range of time and space step sizes (Δt and Δz). Show that the solution becomes unstable whenever the dimensionless group $K \Delta t / S_s (\Delta z)^2$ is greater than 0.5.

7.2. Derive the quadratic finite-element approximation to the one-dimensional groundwater flow equation given in Eq. (7.66). However, as opposed to the formulation presented in the text, you should assume that the source term Q varies in space, so that $Q = Q(x)$. How does this change the finite-element approximation (as compared to Eq. (7.80))? How does the finite-element approximation now compare to the finite-difference approximation?

7.3. Repeat Problem 7.2 for the piecewise linear finite-element approximation, where now the appropriate equation for comparison is Eq. (7.90).

7.4. Both the *Heaviside step function* and the *Dirac delta function* were introduced in Section 7.5. Let the Heaviside step function be defined by

$$H(x - x_i) = \begin{cases} 0, & x < x_i \\ 1, & Yx > x_i \end{cases}$$

with the Dirac delta function then defined by

$$\delta(x - x_i) = \frac{d}{dx} H(x - x_i).$$

Note that integration of the Dirac delta function recovers the Heaviside step function. Use the concepts of the Heaviside step function and Dirac delta function to show that the standard piecewise linear representation used in finite-element analysis can be written as follows, where E is the number of elements and $N = E + 1$, where N is the number of nodes.

$$\widehat{h}(x) = \sum_{e=1}^{E} \left[h_e \left(\frac{x_{e+1} - x}{x_{e+1} - x_e} \right) + h_{e+1} \left(\frac{x - x_e}{x_{e+1} - x_e} \right) \right] \left[H(x - x_e) - H(x - x_{e+1}) \right]$$

$$= \sum_{e=1}^{E} \widehat{h}_e(x) \left[H(x - x_e) - H(x - x_{e+1}) \right],$$

$$\frac{d\widehat{h}}{dx} = \sum_{e=1}^{E} \frac{d}{dx} \left\{ \widehat{h}_e(x) \left[H(x - x_e) - H(x - x_{e+1}) \right] \right\}$$

$$= \sum_{e=1}^{E} \frac{d}{dx} \left(\widehat{h}_e(x) \right) \left[H(x - x_e) - H(x - x_{e+1}) \right]$$

$$= \sum_{e=1}^{E} \left(\frac{h_{e+1} - h_e}{x_{e+1} - x_e} \right) \left[H(x - x_e) - H(x - x_{e+1}) \right],$$

$$\frac{d^2\widehat{h}}{dx^2} = \sum_{e=1}^{E} \left(\frac{h_{e+1} - h_e}{x_{e+1} - x_e} \right) \frac{d}{dx} \left[H(x - x_e) - H(x - x_{e+1}) \right]$$

$$= \sum_{e=1}^{E} \left(\frac{h_{e+1} - h_e}{x_{e+1} - x_e} \right) \left[\delta(x - x_e) - \delta(x - x_{e+1}) \right]$$

$$= \frac{d\widehat{h}}{dx} \bigg|_{x_1}^{x_N} + \sum_{j=2}^{N-1} \left[\left[\frac{d\widehat{h}}{dx} \right] \right]_{x_j} \delta(x - x_j),$$

where the double-bracket notation denotes a *jump operator*, defined by

$$[[f]]_{x_j} \equiv \lim_{\epsilon \to 0} \left[f(x_j + \epsilon) - f(x_j - \epsilon) \right].$$

7.5. Use the expressions given in Problem 7.4 to derive the finite-element approximations for the one-dimensional equation given by Eq. (7.66). You should perform the finite-element derivation without using integration by parts, instead using the expression for the second derivative derived in Problem 7.4. Compare the result to the equation derived using integration by parts, Eq. (7.90). (The especially curious student/reader may wish to consider the origin and meaning of the boundary flux terms and how they compare in the two different approaches.)

7.6. Estimate the correlation length for each of the two semivariograms given in Figure 7.21. Explain how you made your estimation, and explain (in words) the meaning of the term correlation length.

7.7. Section 7.7.5 deals with random hydraulic conductivity fields. Assume you are given such a field. Write the associated numerical approximations for the two-dimensional steady-state groundwater flow equation using (a) finite-difference methods, (b) finite-element methods, and (3) finite-volume methods. Be sure to explain clearly how you treat the spatially variable coefficient, which can be either hydraulic conductivity or transmissivity (because the equation is assumed to be two-dimensional, you may wish to use the transmissivity as the spatially varying parameter).

BIBLIOGRAPHY

[1] B. G. Galerkin, Rods and plates. Series in some problems of elastic equilibrium of rods and plates, *Vestn. Inzh. Tekh. (USSR)*, **19**: 897, 1915.

[2] G. F. Pinder, *Groundwater Modeling*, John Wiley & Sons, Hoboken, NJ, 2002.

[3] H. Jianguo and X. Shitong, On the finite volume element method for general self-adjoint elliptic problems, *SIAM J. Numer. Anal.* **35**(5): 1762, 1998.

[4] G. Dagan, *Flow and Transport in Porous Formations*, Springer Verlag, Berlin, 1989.

[5] G. Herrera de Olivares, *Cost Effective Groundwater Quality Sampling Network Design*, Thesis, Department of Mathematics and Statistics, University of Vermont, 1998.

[6] D. Glover, "Lecture 5: Section 3, Objective Mapping and Kriging," http://w3eos.whoi.edu/12.747/notes/lect05/lect05s03.html.

[7] G. Matheron, Principles of geostatistics, *Economic Geol.* **58**: 1246, 1963.

[8] G. DeMarsily, *Quantitative Hydrogeology*, Academic Press, San Diego, CA, 1986.

[9] D. H. Tang and G. Pinder, Analysis of mass transport with uncertain physical parameters, *Water Resour. Res.* **15**(5): 1147, 1979.

[10] B. D. Ripley, *Stochastic Simulation*, John Wiley & Sons, Hoboken, NJ, 1987.

[11] Y. Zhang and G. F. Pinder, Latin-hypercube sample-selection strategies for correlated random hydraulic-conductivity fields, *Water Resour. Res.* **39**(8): 11-1, 2003.

[12] Y. Zhang, personal communication, 2004.

CHAPTER 8

CONTAMINATION OF SUBSURFACE WATER

8.1 TYPES OF CONTAMINANTS

In Section 1.8 we introduced the concept of groundwater contamination. In that section we considered both naturally occurring and anthropogenic (human-generated) contaminants. Contaminants were further subdivided into dissolved and separate phase. An example of a dissolved naturally occurring contaminant would be salt in a coastal aquifer. Oil floating on the water table is an example of a separate phase contaminant. Separate phase contaminants were found to occur as floating products such as oil or as a separate phase that was heavier than water and thus located below the water table. *Nonaqueous phase liquid* contaminants were denoted as NAPLs, those NAPLs lighter than water were called *lighter-than water nonaqueous phase liquids* (LNAPLs) and those heavier than water *denser-than-water nonaqueous phase liquids* (DNAPLs). In Section 1.8 we also discussed the geographical location and frequency of occurrence of contaminants.

In Section 2.6 we talked briefly about the equations that describe the movement of contaminants. We found that the balance equations that describe the conservation of energy, mass, and momentum must be supplemented with constitutive or "experimental" relationships before they can be solved. In addition, to completely define a problem, initial and boundary conditions are needed.

Section 2.6 included a discussion of advection, diffusion, and dispersion. The phenomenon of retardation, which tends to influence the rate of migration of dissolved contaminants, was also considered.

In this chapter we look further into the matter of groundwater contamination, not only addressing some of those topics introduced earlier in greater detail, but also exploring more abstract concepts such as interphase mass transfer and chemical and biological reactions.

Subsurface Hydrology   By George F. Pinder and Michael A. Celia
Copyright © 2006 John Wiley & Sons, Inc.

8.2 MASS CONSERVATION

The groundwater system can be very complex. At a minimum it consists of two phases, a solid matrix and a fluid. One can approach *mass conservation* in such a system in either of two ways—from the most general or from the least complex. We begin with the simplest system and expand the sophistication of our development as the need arises.

The strategy for formulating the mass-conservation equation for dissolved contaminants begins with the species balance equation. One obtains this equation using a strategy similar to that outlined in Section 4.2. The approach in that section was to derive the equations for homogeneous fluids and then to use averaging concepts to arrive at an equivalent expression for a porous medium containing one or more homogeneous fluids (remember that all fluids are not liquid, air is also a fluid). The resulting equations are then applicable to aquifers containing multiple phases such as water, air, and NAPL (non-aqueous phase liquid). In this section we extend the concept of multiple homogeneous fluids to consider dissolved species.

To proceed, one must decide on a definition for the term concentration. In this chapter we use as our definition of concentration the mass concentration, which is defined as mass of species i per unit volume of solution, and write it as ρ_i. We also use the concept of the *mass fraction,* ω_i, which is defined as the mass concentration of species i divided by the mass density of the solution, that is,

$$\omega_i \equiv \frac{\rho_i}{\sum_{i=1}^{I} \rho_i} = \frac{\rho_i}{\rho},\tag{8.1}$$

where I is the total number of species. The mass fraction is a convenient measure because it is, in essence, a normalized concentration such that $0 \leq \omega_i \leq 1$, and it is dimensionless.

We begin with the species balance equation for a single phase. Such an equation describes the movement of dissolved species in a single fluid phase. Because it is derived in a straightforward way using the same concepts introduced in Section 4.1, we present only the final result, that is,

$$\frac{\partial}{\partial t}(\rho \omega_i) + \nabla \cdot (\rho \mathbf{v} \omega_i) - \nabla \cdot \mathbf{i}_i - \rho Q_i = 0,\tag{8.2}$$

where $\mathbf{i}_i$ is the change in mass due to the *nonconvective flux* of species i, in our case diffusion, and Q_i is a source of species i defined in terms of mass fraction per unit time. The convective velocity term is the *mass-average velocity* defined as

$$\mathbf{v} = \frac{\sum_{i=1}^{I} \rho_i \mathbf{v}_i}{\sum_{i=1}^{I} \rho_i}.\tag{8.3}$$

As stated here, the nonconvective flux, $\mathbf{i}_i$, describes the velocity of species i with respect to the mass-average velocity $\mathbf{v}$.

8.3 MASS CONSERVATION IN A POROUS MEDIUM

The species balance equation for a porous medium can be formulated using the averaging strategies presented in Section 4.2. Application of these principles to the point-balance equation (8.2) yields

$$
\frac{\partial}{\partial t}\underbrace{\left(\langle\rho\rangle_W\,\overline{\omega}_i^W\right)}_{A} + \underbrace{\nabla\cdot\left(\overline{\mathbf{v}}^W\overline{\omega}_i^W\,\langle\rho\rangle_W\right)}_{B}
$$

$$
-\underbrace{\nabla\cdot\mathbf{i}_i^W}_{C} - \underbrace{\langle\rho\rangle_W\,e_i^W\,(\rho\omega_i)}_{D} - \underbrace{\langle\rho\rangle_W\,\hat{I}_i^W}_{E} - \underbrace{\langle\rho\rangle_W\,\overline{Q}_i^W}_{F} = 0, \tag{8.4}
$$

where the superscript W implies that the species i is in the water (W) fluid phase. The overbar and angular brackets both imply volume averages as described in Section 4.2.

The terms in Eq. (8.4) have been identified with the letters A through F. The term identified as A is the mass accumulation term. It describes the total rate of change of the mass of species i at a point $\mathbf{x}$ at time t. It is helpful to keep in mind that since $\overline{\omega}$ is a mass average over the representative elementary volume (REV), the product $\langle\rho\rangle_W\,\overline{\omega}_i^W$ is equivalent to $\langle\rho\omega_i\rangle_w$. Also, the average density is written more as an intrinsic phase average, $\langle\rho\rangle_w^w$, which is related to $\langle\rho\rangle_w$ by the volume fraction n_w. Therefore this term can be written $n_W\,\langle\rho\rangle_W^W\,\overline{\omega}_i^W$, where $\langle\rho\rangle_W^W$ is the average fluid density per volume of fluid (the commonly used measure) and $n_W = V_W/V$. In the case of one fluid phase n_W is equivalent to the porosity. Thus one observes that, in saturated porous media, the mass accumulation term for the species i includes the time rate of change of the porosity as well as the concentration of species i.

The term B is the convective mass term. It describes the movement of the species i by virtue of the average fluid movement. In fact, the velocity $\overline{\mathbf{v}}^W$ is a mass-average velocity analogous to that described in the point equation, Eq. (8.3). The formal relationship is given as

$$
\overline{\mathbf{v}}^W(\mathbf{x},t) \equiv \frac{\langle\rho\mathbf{v}\rangle_W(\mathbf{x},t)}{\langle\rho\rangle_W(\mathbf{x},t)},
$$

which is analogous to the mass-average point velocity defined in Eq. (8.3).

The term denoted as C is the nonconvective mass flux term. It is perhaps the most interesting of the terms found in Eq. (8.4) and we will now demonstrate why this is the case.

We described in the case of the point equation a term $\mathbf{i}_i$, which describes the velocity of a species relative to the average fluid velocity. A similar concept exists for a porous medium inasmuch as one can define a particle velocity in terms of the mass-average velocity of the fluid, for example (see Figure 4.5 for a description of the relevant coordinate systems),

$$
\mathbf{v}^W(\mathbf{x}+\boldsymbol{\eta},t) \equiv \overline{\mathbf{v}}^W(\mathbf{x},t) + \widetilde{\mathbf{v}}^W(\mathbf{x}+\boldsymbol{\eta},t). \tag{8.5}
$$

The term $\mathbf{v}^W(\mathbf{x}+\boldsymbol{\eta},t)$ describes the fluid (water) velocity at a point at the microscopic level. We know this because it is defined in the local coordinate $\boldsymbol{\eta}$ as defined in Section 4.2. It is different from the mass-average velocity $\overline{\mathbf{v}}^W(\mathbf{x},t)$ by a quantity $\widetilde{\mathbf{v}}^W(\mathbf{x}+\boldsymbol{\eta},t)$, which is also a microscopic quantity. In other words, the difference between the average velocity $\overline{\mathbf{v}}^W(\mathbf{x},t)$ in a REV and a point velocity $\mathbf{v}^W(\mathbf{x}+\boldsymbol{\eta},t)$ is the deviation $\widetilde{\mathbf{v}}^W(\mathbf{x}+\boldsymbol{\eta},t)$.

Note, however, that the term appearing in Eq. (8.4) is a product, that is, $\overline{\mathbf{v}}^W \overline{\omega}_i^W$. It arises from the average of the product of v^w and ω_i^w, which we can be expand as

$$\mathbf{v}^W(\mathbf{x} + \boldsymbol{\eta}, t)\omega_i^W(\mathbf{x} + \boldsymbol{\eta}, t)$$
$$= \left(\overline{\mathbf{v}}^W(\mathbf{x}, t) + \widetilde{\mathbf{v}}^W(\mathbf{x} + \boldsymbol{\eta}, t)\right) \cdot \left(\overline{\omega_i}^W(\mathbf{x}, t) + \widetilde{\omega}_i^W(\mathbf{x} + \boldsymbol{\eta}, t)\right). \qquad (8.6)$$

If we average both sides of Eq. (8.6) the products involving deviation quantities and average quantities vanish and we are left with

$$\left\langle \mathbf{v}^W(\mathbf{x} + \boldsymbol{\eta}, t)\omega_i^W(\mathbf{x} + \boldsymbol{\eta}, t)\right\rangle_W$$
$$= \left\langle \overline{\mathbf{v}}^W(\mathbf{x}, t)\,\overline{\omega_i}^W(\mathbf{x}, t)\right\rangle_W + \left\langle \widetilde{\mathbf{v}}^W(\mathbf{x} + \boldsymbol{\eta}, t)\widetilde{\omega}_i^W(\mathbf{x} + \boldsymbol{\eta}, t)\right\rangle_W$$

or

$$\left\langle \mathbf{v}^W(\mathbf{x} + \boldsymbol{\eta}, t)\omega_i^W(\mathbf{x} + \boldsymbol{\eta}, t)\right\rangle_W$$
$$= \overline{\mathbf{v}}^W(\mathbf{x}, t)\,\overline{\omega_i}^W(\mathbf{x}, t)_W + \left\langle \widetilde{\mathbf{v}}^W(\mathbf{x} + \boldsymbol{\eta}, t)\widetilde{\omega}_i^W(\mathbf{x} + \boldsymbol{\eta}, t)\right\rangle_W$$

since the average of an average is the original average.

Observe that the term B in Eq. (8.4) does not include the perturbation terms $\left\langle \widetilde{\mathbf{v}}^W(\mathbf{x} + \boldsymbol{\eta}, t)\widetilde{\omega}_i^W(\mathbf{x} + \boldsymbol{\eta}, t)\right\rangle_W$. The reason for this is that these terms are imbedded in the nonconvective flux term, term C. The transport due to the variability of the microscopic flow field within the REV is captured by the term $\nabla \cdot \mathbf{i}_i^W$. The steps required to make this transformation are quite complex and beyond the scope of this book. Suffice it to say that the *nonconvective flux is due, at least in part, to the tortuous motion of the water flow at the sub-REV scale*. The phenomenon is called *hydrodynamic dispersion,* and it is essential to conserving the sub-REV scale physical behavior during the averaging process.

The two terms denoted as D and E represent the *movement of mass across the interfaces* between the different phases. The term $\langle \rho \rangle_W e_i^W (\rho \omega_i)$ denotes movement of solute by virtue of the movement of the phase interface. For example, if one has a glass of ice made from water containing a red dye, the dye particles will move across the interface as the ice melts, although the dye molecules could be considered, in some sense, as stationary. The term $\langle \rho \rangle_W \hat{I}_i^W$ describes the movement of species i across the phase interface. An example of this phenomenon would be the movement of salt ions from oil to water given the oil and water phases coexisted in the porous medium.

The final term is a source term. It represents the creation or destruction of species i not accounted for by the other terms in the equation. The creation of radioactive daughter products could be accommodated using this source term.

In the simplified example we are considering, where there is but one fluid phase and one solid phase, the system can be considered using Eq. (8.4) with terms D and E set to zero. However, if there is adsorption (movement of species i from the liquid to the solid phase) or dissolution of the solid phase to provide a source for species i, this assumption would not be justified. For this case, let us simplify notation by removing the averaging decoration while recognizing that the volume-averaging process is still in play:

$$\underbrace{\frac{\partial}{\partial t}(\varepsilon \rho \omega_i)}_{A} + \underbrace{\nabla \cdot (\mathbf{v} \omega_i \varepsilon \rho)}_{B} - \underbrace{\nabla \cdot \mathbf{i}_i}_{C} - \underbrace{\rho e\,(\varepsilon \rho \omega_i)}_{D} - \underbrace{\varepsilon \rho \hat{I}_i}_{E} - \underbrace{\varepsilon \rho \overline{Q}_i}_{F} = 0. \qquad (8.7)$$

For the case of a nonadsorbing solute and an inert solid matrix, one could further simplify Eq. (8.7) to

$$\underbrace{\frac{\partial}{\partial t}(\varepsilon \rho \omega_i)}_{A} + \underbrace{\nabla \cdot (\mathbf{v} \omega_i \varepsilon \rho)}_{B} - \underbrace{\nabla \cdot \mathbf{i}_i}_{C} - \underbrace{\varepsilon \rho \overline{Q}_i}_{F} = 0. \tag{8.8}$$

It is interesting at this point to illustrate how the fluid flow equation can be generated from the species transport equation. Since, from Eq. (8.1), we know that $\rho = \sum_{i=1}^{I} \rho_i$—that is, the fluid is made up of the species components that constitute it—it would seem reasonable to assume that by summing the species transport equations for all species in solution, one would obtain the fluid flow equation.

Let us test this hypothesis. Summing Eq. (8.8) for the I species in solution we obtain

$$\sum_{i=1}^{I} \frac{\partial}{\partial t}(\varepsilon \rho \omega_i) + \nabla \cdot (\mathbf{v} \omega_i \varepsilon \rho) - \nabla \cdot \mathbf{i}_i - \varepsilon \rho \overline{Q}_i = 0, \tag{8.9}$$

which we can rewrite as

$$\frac{\partial}{\partial t}\left(\varepsilon \rho \sum_{i=1}^{I} \omega_i\right) + \nabla \cdot \left(\mathbf{v} \sum_{i=1}^{I} \omega_i \varepsilon \rho\right) - \nabla \cdot \sum_{i=1}^{I} \mathbf{i}_i - \varepsilon \rho \sum_{i=1}^{I} \overline{Q}_i = 0.$$

We now introduce the definition of Eq. (8.1), that is, $\omega_i = \rho_i / \rho$, to obtain

$$\frac{\partial}{\partial t}\left(\varepsilon \sum_{i=1}^{I} \rho \frac{\rho_i}{\rho}\right) + \nabla \cdot \left(\mathbf{v} \varepsilon \sum_{i=1}^{I} \frac{\rho_i}{\rho} \rho\right) - \nabla \cdot \sum_{i=1}^{I} \mathbf{i}_i - \varepsilon \rho \sum_{i=1}^{I} \overline{Q}_i = 0. \tag{8.10}$$

To complete our development it is necessary to examine the last two terms in Eq. (8.10). The nonconvective mass flux is the deviation of the flux of species i relative to the mass-average flux. The sum of the divergence of the nonconvective flux of all the species is written

$$\sum_{i=1}^{I} \nabla \cdot \mathbf{i}_i \equiv -\sum_{i=1}^{I} \nabla \cdot \left(\varepsilon \rho \left\langle \widetilde{\mathbf{v}}^W(\mathbf{x} + \boldsymbol{\eta}, t) \widetilde{\omega}_i{}^W(\mathbf{x} + \boldsymbol{\eta}, t)\right\rangle_W\right). \tag{8.11}$$

Equation (8.11) can be written

$$-\sum_{i=1}^{I} \nabla \cdot \left(\varepsilon \rho \left\langle \widetilde{\mathbf{v}}^W(\mathbf{x} + \boldsymbol{\eta}, t) \widetilde{\omega}_i{}^W(\mathbf{x} + \boldsymbol{\eta}, t)\right\rangle_W\right)$$

$$= -\nabla \cdot \left(\varepsilon \left\langle \widetilde{\mathbf{v}}^W(\mathbf{x} + \boldsymbol{\eta}, t) \rho \sum_{i=1}^{I} \widetilde{\omega}_i{}^W(\mathbf{x} + \boldsymbol{\eta}, t)\right\rangle_W\right)$$

$$= -\nabla \cdot \left(\varepsilon \left\langle \widetilde{\mathbf{v}}^W(\mathbf{x} + \boldsymbol{\eta}, t) \widetilde{\rho}(\mathbf{x} + \boldsymbol{\eta}, t)\right\rangle_W\right) = 0, \tag{8.12}$$

where it is assumed that neither diffusion nor tortuosity will systematically cause a perturbation in the fluid density.

The last term in Eq. (8.10) is the source term and since it can be expressed as a function of ω_i, we have

$$\varepsilon\rho \sum_{i=1}^{I} \overline{Q}_i = \varepsilon\rho f\left(\sum_{i=1}^{I}\left(\frac{\rho_i}{\rho}\right)\right) = \varepsilon f\left(\rho\right) \equiv \varepsilon Q, \tag{8.13}$$

where $f\left(\cdot\right)$ is a function descriptive of the source and Q is the fluid source. Combination of the information in Eq. (8.10) through (8.13) yields

$$\frac{\partial}{\partial t}\left(\varepsilon\rho\right) + \nabla \cdot \left(\mathbf{v}\varepsilon\rho\right) - \varepsilon Q = 0, \tag{8.14}$$

which is the flow equation for a single phase fluid in an inert porous medium. We are now at the end of this short diversion, and it is now time to return to the species transport equation.

8.3.1 The Dispersion Relationship

We noted earlier that the term C in Eq. (8.8) describes the nonconvective (species) mass flux: that is, the movement of the species not attributable to the mass-average flow of the water. In this section we examine how to represent this term using a *constitutive relationship*.

The concept of nonconvective mass flux has received considerable attention in the porous-medium literature. In essence, it describes molecular diffusion superimposed on a mixing process. The mixing process is due to the structure of the porous medium and its resultant impact on the microscopic (sub-REV) flow field. Greenkorn [1] identified several mechanisms that contribute to the mixing process:

1. *Molecular Diffusion*: This process can play a role in low-permeability formations over long time frames.
2. *Mixing Due to Obstructions*: As seen in Figure 1.24, the paths taken by particles in a porous medium can be very tortuous. A particle starting out relatively close to another may, after both have moved the same distance, be located a significant distance away from its original neighbor. The result is a spreading both longitudinally and transversely.
3. *Presence of Autocorrelation in the Flow Paths*: Preferential pathways through a porous medium can result in flow paths that bypass a subset of other paths in the porous medium. The resulting diversion around such areas may result in additional mixing.
4. *Recirculation Caused by Local Regions of Reduced Pressure*: Under unusual fluid dynamical circumstances, fluids can recirculate and thereby modify the concentration distribution in a manner that enhances mixing.
5. *Macroscopic or Megascopic Dispersion*: Variability in packing and grain size causes streamlines to vary significantly from those one might anticipate when using the average hydraulic conductivity and pore structure. The resulting variability in flow pattern causes mixing. The phenomena can occur at different scales.

6. *Hydrodynamic Dispersion*: Variability in flow pattern at the pore level can result in mixing. Basic fluid mechanical principles dictate that flow at the grain boundary is very different from that at the pore center. The variability of the flow field results in a macroscopic mixing.

7. *Eddies*: While flow in porous media is perceived as laminar, within pores turbulent mixing may, at least in concept, occur. The resulting mixing phenomenon would lead to enhanced dispersion.

8. *Dead-End Pores*: When a fluid containing solute passes a pore that is isolated (a dead end), diffusion into the pore fluid may occur. The contaminants in the dead-end pore will then proceed to diffuse out as the solute concentration in the fluid convected past the pore decreases. The resulting concentration in the convected fluid may appear to have increased mixing.

9. *Adsorption*: Adsorption causes dissolved species to adhere, in a statistical sense, to grains while in transit. The adsorbed molecules continually adhere and release from the grains. The resulting deposition and removal of contaminants as the concentrations in the convected fluid change may enhance mixing.

The dispersion coefficient is normally identified with "fluid mechanical" phenomena, leaving processes 8 and 9 on this list for incorporation elsewhere in the transport equation.

Returning to the nonconvective flux term $\nabla \cdot \mathbf{i}_i$, we see that the nonconvective flux is a vector quantity. It is generally agreed that this quantity in a saturated medium is of the general form

$$\mathbf{i}_i = -\varepsilon \rho \mathbf{D} \cdot \nabla \omega_i, \tag{8.15}$$

where, as written, $\mathbf{D}$ is a tensorial quantity of rank two. A similar equation can also be written assuming D is a scalar quantity. The parameter $\mathbf{D}$ (or D) is called the *dispersion coefficient* and its functional form has been the source of much research. The most generally accepted representation of scalar dispersivity D is $D = \alpha v$, where α is the dispersivity and v is the longitudinal fluid velocity. The tensorial form of the dispersion tensor is

$$D_{\alpha\beta} = \alpha_T |v| \delta_{\alpha\beta} + (\alpha_L - \alpha_T) v_\alpha v_\beta / |v| + D_m \delta_{\alpha\beta}, \tag{8.16}$$

where α_L and α_T are the longitudinal (direction of flow) and transverse (orthogonal to flow direction) dispersivities, respectively. The subscripts α and β represent the x, y, and z coordinate directions. Substitution of x, y, and z for α and β yields nine values for the dispersion tensor. Thus, as was the case for the tensorial form of the hydraulic conductivity, the nonadvective flux is given by

$$\left\{ \begin{array}{c} i_x \\ i_y \\ i_z \end{array} \right\} = - \left[\begin{array}{ccc} D_{xx} & D_{xy} & D_{xz} \\ D_{yx} & D_{yy} & D_{yz} \\ D_{zx} & D_{zy} & D_{zz} \end{array} \right] \left\{ \begin{array}{c} \dfrac{\partial \omega_i}{\partial x} \\ \dfrac{\partial \omega_i}{\partial y} \\ \dfrac{\partial \omega_i}{\partial z} \end{array} \right\}. \tag{8.17}$$

In Eq. (8.17) the dependence of the nonconvective flux in the x-coordinate direction depends not only on the derivative of ω_i in the x direction, but also on the ω_i derivatives in the y and z directions. For example,

$$i_x = -D_{xx} \frac{\partial \omega_i}{\partial x} - D_{xy} \frac{\partial \omega_i}{\partial y} - D_{xz} \frac{\partial \omega_i}{\partial z}.$$

Note that, in this formulation, there is but one transverse dispersivity—not two that are orthogonal to one another as one might expect.

8.3.2 One-Dimensional Transport

The solution of the one-dimensional transport equation provides helpful insight into the behavior of contaminants in the subsurface. To illustrate the significance of the various transport phenomena, we will solve the transport equation for some simple initial boundary-value problems of practical interest. Since D is, for all practical purposes, a constant in one-dimensional transport, we will assume the appropriate transport equation is of the form

$$\underbrace{\frac{\partial}{\partial t} \varepsilon \rho \omega_i}_{a} + \underbrace{\frac{\partial}{\partial x} (\varepsilon v_x \rho \omega_i)}_{b} - \underbrace{\frac{\partial}{\partial x} \left(\varepsilon \rho D \frac{\partial \rho \omega_i}{\partial x} \right)}_{c} = 0$$

or, in terms of the concentration ρ_i,

$$\underbrace{\frac{\partial}{\partial t} \varepsilon \rho_i}_{a} + \underbrace{\frac{\partial}{\partial x} (\varepsilon v_x \rho_i)}_{b} - \underbrace{\frac{\partial}{\partial x} \left(\varepsilon D \frac{\partial \rho_i}{\partial x} \right)}_{c} = 0. \tag{8.18}$$

In Eq. (8.18), term a is the rate of change of the species i, the term b is the change in the value of i due to convection, and the term c is the change of species i due to nonconvective flux.

In one space dimension, several initial boundary-value problems are of interest. In the first problem we define a constant concentration boundary on the left-hand side of a semi-infinite domain. The initial and boundary conditions are

$$\rho_i (x, t = 0) = \rho_i, \qquad x > 0,$$

$$\rho_i (x = 0, t) = \rho_u, \qquad t \geq 0,$$

$$\rho_i (x \to \infty, t) = \rho_l = 0, \quad t \geq 0.$$

The solution to this equation was first presented by Ogata and Banks [2]:

$$\frac{\rho}{\rho_u} = \frac{1}{2} \left[\text{erfc} \left(\frac{x - vt}{\sqrt{4Dt}} \right) + \exp \left(\frac{xv}{D} \right) \text{erfc} \left(\frac{x + vt}{\sqrt{4Dt}} \right) \right], \tag{8.19}$$

where erfc, the *complimentary error function*, is a tabulated function. The solution of this equation for four values of the dispersion coefficient D is given in Figure 8.1, assuming that $v_x = 1$. The solutions plotted are ρ_i / ρ_u as a function of time, observed at the fixed location $x = 0.5$. Such curves are often referred to as breakthrough curves.

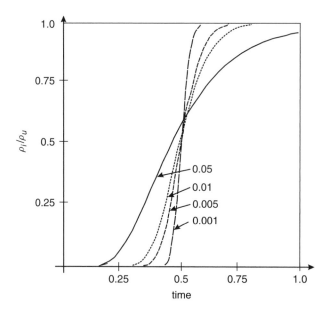

FIGURE 8.1. Solutions for one-dimensional contaminant transport generated by Eq. (8.19). The different curves are generated by using different values of the dispersion coefficient D, that is, $D = 0.001, 0.005, 0.01$, and 0.05. The curves represent the change in concentration at the fixed point $x = 0.5$ when the velocity is 1.0.

Several features of the solutions presented in Figure 8.1 are worthy of note. Observe that the value of the concentration is 0.5 at time $t = 0.5$, for $D = 0.001, 0.005$, and 0.01. To explain this, recall that the distance the front would travel in the absence of dispersion (a sharp front) would be $x = vt$. Thus in a time of $t = 0.5$, the distance traveled given $v = 1.0$ would be $vt = 1.0 \times 0.5 = 0.5$. We observe that the concentration at this observation location, that is, $x = 0.5$, is 0.5. We conclude that the location of the front, if defined as the 0.5 value, is at the same location one would expect to find a sharp front (the case of zero dispersion). The 0.5 concentration location is therefore a good indicator of the average groundwater flow velocity.

Another interesting point is that the curves generated by the different diffusion coefficients are symmetric around the 0.5 location. The increase in mass downstream of this location is equal to the decrease in mass upstream of this location. One can conclude from this observation that the phenomenon of dispersion is symmetric, occurring equally efficiently in the upstream and downstream directions. Equation (8.18) suggests this might be the case since the dispersive term c in that equation involves a second-order derivative that is associated with symmetric solutions. An additional observation regarding this symmetry is that it demonstrates that mass is being conserved. The upstream decrease and downstream increase exactly balance.

We observe also that the solution for $D = 0.05$ behaves oddly. It does not follow the general pattern illustrated by the other curves. The reason for this is that the increased dispersion using this larger value of D is generating a solution that is impacted by the upstream boundary condition.

An alternative way to look at this solution is to view the front as it moves over time. The curves shown in Figure 8.2 represent the concentration front at four different points

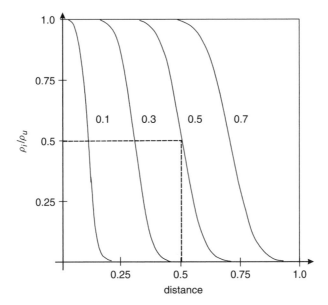

FIGURE 8.2. Concentration profiles as a function of time. The curves represent times of $t = 0.1$, 0.3, 0.5, and 0.7. The dispersion coefficient is 0.005 and velocity $v = 1.0$.

in time, $t = 0.1$, 0.3, 0.5, and 0.7 for the case of $D = 0.005$. Note that with $v = 1.0$, the concentration at $t = 0.5$ at the spatial location $x = 0.5$ is 0.5, indicating as before that the 0.5 concentration value identifies the location of the front. Also evident from this figure is the increasing dispersion that is occurring as time elapses. The evidence is found in the fact that the front at $t = 0.7$ is more spread out or more dispersed (less steep) than that at $t = 0.1$.

A quite different perspective is provided when a pulse is introduced as input and one follows the behavior of the peak over time. The formulation of this problem requires that the domain be infinite. The auxiliary conditions are

$$\rho_i(x, t) \rightarrow 0 \quad \text{as } x \rightarrow \pm\infty,$$

$$\rho_i(x, 0) = \frac{M}{\varepsilon A}\delta(x - x_i),$$

where $\delta(x - x_i)$ is the Dirac delta function, x_i is the location of the peak, M is the amount of mass present, and A is the cross-sectional area of the one-dimensional region. The solution to this problem (including the effects of retardation, where R is the retardation coefficient) is

$$\rho_i(x, t) = \frac{M/A}{2\varepsilon R\sqrt{\pi(D/R)t}}\exp\left[-\frac{\left(x - \frac{v}{R}t\right)^2}{4(D/R)t)}\right]$$

and is shown in Figure 8.3.

Note that the area under each curve is the same, indicating that the amount of mass in the system remains constant from the point in time of injection. Also of interest is the

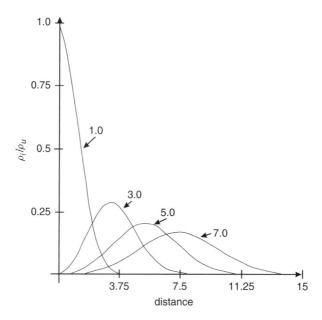

FIGURE 8.3. Concentration profiles calculated using a pulse input of 1.0, $v = 1.0$, and $D = 0.5$. The elapsed time is indicated for each profile.

decrease in the magnitude of the peak and the increase in the breadth of the base. This is due to dispersion. The 0.5 concentration has no significance in terms of the distance traveled by the pulse. However, the pulse peak is indicative of the velocity of the curve and thus the velocity of the convecting fluid.

8.4 RETARDATION

As we saw demonstrated in Section 1.6.1, dissolved species can adsorb to solid particles, especially those containing organic carbon. In this section we will establish how this phenomenon is accommodated in the species transport equation and how it influences contaminant behavior. Consider rewriting Eq. (8.4), but now for the solid phase:

$$
\underbrace{\frac{\partial}{\partial t}\left(\langle\rho\rangle_S\,\overline{\omega}_i^S\right)}_{A} + \underbrace{\nabla\cdot\left(\overline{\mathbf{v}}^S\overline{\omega}_i^S\,\langle\rho\rangle_S\right)}_{B}
$$

$$
- \underbrace{\nabla\cdot\mathbf{i}_i^S}_{C} - \underbrace{\langle\rho\rangle_S\,e_i^S\,(\rho\omega_i)}_{D} - \underbrace{\langle\rho\rangle_S\,\hat{I}_i^S}_{E} - \underbrace{\langle\rho\rangle_S\,\overline{Q}_i^S}_{F} = 0. \tag{8.20}
$$

Since we are considering the solid phase, terms B, C, and F can be neglected. We are left with

$$
\underbrace{\frac{\partial}{\partial t}\left(\langle\rho\rangle_S\,\overline{\omega}_i^S\right)}_{a} - \underbrace{\langle\rho\rangle_S\,e_i^S\,(\rho\omega_i)}_{b} - \underbrace{\langle\rho\rangle_S\,\hat{I}_i^S}_{c} = 0, \tag{8.21}
$$

where the subscript and superscript S refer to the solid phase. The first term in this equation describes the change in species i concentration on the solid phase. The second describes the movement of the species i from the fluid to the solid phase by virtue of the change in the volume of the solid phase. Term c describes the movement of the species from the fluid to the solid phase by virtue of diffusion.

Experiments have shown that it is possible to relate the concentration of the species i on the solid phase to the concentration of the species i in solution. A common expression for this is the linear adsorption isotherm, which states that at equilibrium

$$\overline{\omega_i}^S = \frac{\langle \rho \rangle_W \, k_d \overline{\omega_i}^W}{\varepsilon}, \tag{8.22}$$

where k_d is the partition coefficient and as noted earlier

$$\langle \omega_i \rangle_\alpha (\mathbf{x}, t) \equiv \frac{1}{dV} \int_{dV} \omega_i (\mathbf{x} + \boldsymbol{\eta}, t) \, \gamma_\alpha (\mathbf{x} + \boldsymbol{\eta}, t) \, dv_\xi,$$

$$\overline{(\omega_i)}^\alpha (\mathbf{x}, t) \equiv \frac{\langle \rho \omega_i \rangle_\alpha (\mathbf{x}, t)}{\langle \rho \rangle_\alpha (\mathbf{x}, t)}, \tag{8.23}$$

and

$$\varepsilon \langle \omega_i \rangle_\alpha^\alpha = \langle \omega_i \rangle_\alpha .$$

In our case α takes the values W (water) and S (soil grains). The term $\langle \rho \rangle_W$ is the fluid density averaged over the entire REV (representative elementary volume). The mass-averaged mass fraction for the grains describes the mass of species i per unit mass of solid grains. The product $\langle \rho \rangle_W \overline{\omega_i}^W$ is a measure of the concentration of i per unit volume of solution. Multiplication of both sides of Eq. (8.23) by $\langle \rho \rangle_\alpha (\mathbf{x}, t)$ illustrates this to be the case. The division of the right-handside by ε is to change the average from a volume average to one defined over only the fluid phase (intrinsic phase average $\langle \omega_i \rangle_W^W$).

Substituting Eq. (8.22) into Eq. (8.21) we have

$$\underbrace{\frac{\partial}{\partial t} \left(\langle \rho \rangle_S \frac{\langle \rho \rangle_W \, k_d \overline{\omega_i}^W}{\varepsilon} \right)}_{a} - \underbrace{\langle \rho \rangle_S \, e_i^S \, (\rho \omega_i)}_{b} - \underbrace{\langle \rho \rangle_S \, \hat{I}_i^S}_{c} = 0. \tag{8.24}$$

If there are but two phases present, the terms D and E in Eq. (8.20) must be equal and opposite to terms b and c in Eq. (8.24). Therefore, if we sum Eqs. (8.20) and (8.24), we have

$$\underbrace{\frac{\partial}{\partial t} \left(\langle \rho \rangle_W \overline{\omega_i}^W \right)}_{A} + \underbrace{\frac{\partial}{\partial t} \left(\langle \rho \rangle_S \frac{\langle \rho \rangle_W \, k_d \overline{\omega_i}^W}{\varepsilon} \right)}_{a}$$

$$+ \underbrace{\nabla \cdot \left(\overline{\mathbf{v}}^W \overline{\omega_i}^W \langle \rho \rangle_W \right)}_{B} - \underbrace{\nabla \cdot \mathbf{i}_i^W}_{C} - \underbrace{\langle \rho \rangle_W \overline{Q}_i^W}_{F} = 0. \tag{8.25}$$

Combination of terms A and a yields

$$\frac{\partial}{\partial t} \left(\langle \rho \rangle_W \overline{\omega_i}^W + \langle \rho \rangle_S \frac{\langle \rho \rangle_W \, k_d \overline{\omega_i}^W}{\varepsilon} \right)$$

$$+ \nabla \cdot \left(\overline{\mathbf{v}}^W \overline{\omega_i}^W \langle \rho \rangle_W \right) - \nabla \cdot \mathbf{i}_i^W - \langle \rho \rangle_W \overline{Q}_i^W = 0. \tag{8.26}$$

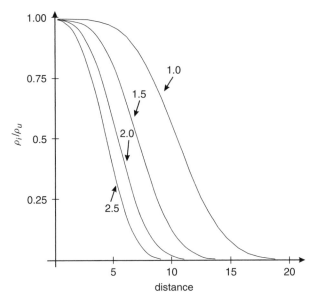

FIGURE 8.4. Concentration profile at time of 10.0, $v = 1.0$, $D = 0.5$, and retardation values of $R = 1.0$, 1.5, 2.0, and 2.5.

The retardation coefficient, R, is defined as

$$R \equiv 1 + \langle \rho \rangle_S \frac{k_d}{\varepsilon},$$

such that Eq. (8.26) becomes

$$\frac{\partial}{\partial t} \left(R \langle \rho \rangle_W \overline{\omega}_i^W \right) + \nabla \cdot \left(\overline{\mathbf{v}}^W \overline{\omega}_i^W \langle \rho \rangle_W \right) - \nabla \cdot \mathbf{i}_i^W - \langle \rho \rangle_W \overline{Q}_i^W = 0. \qquad (8.27)$$

We get to the final form of the equation by introducing the *constitutive relationship* found in Eq. (8.15) and removing the averaging notation:

$$\frac{\partial}{\partial t} \left(\varepsilon R \rho \omega_i \right) + \nabla \cdot \left(\varepsilon \mathbf{v} \omega_i \rho \right) - \nabla \cdot \varepsilon \rho \mathbf{D} \, \nabla \omega_i - \varepsilon \rho Q_i = 0. \qquad (8.28)$$

Written in term of concentration we have

$$\frac{\partial}{\partial t} \left(\varepsilon R \rho_i \right) + \nabla \cdot \left(\varepsilon \mathbf{v} \rho_i \right) - \nabla \cdot \varepsilon \mathbf{D} \, \nabla \rho_i - \varepsilon Q_i = 0, \qquad (8.29)$$

where Q_i is now expressed in terms of mass per unit volume per unit time.

One way to see the impact of retardation is to examine its impact on a concentration profile calculated at a specific time. Such a curve is presented in Figure 8.4. Note that the increasing retardation results in an apparent decrease in solute velocity as judged from the location of the 0.5 concentration location. Also apparent is the decrease in effective dispersion with increase in retardation. If one considers dividing through Eq. (8.29) by the retardation, R, one observes that, given R must be greater than unity, the apparent velocity and the apparent dispersion are reduced.

An alternative approach for seeing the impact of retardation is to consider the problem illustrated in Figure 8.3, this time with retardation added. Initially no mass is adsorbed

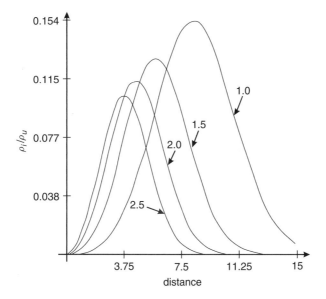

FIGURE 8.5. Concentration profiles calculated using a pulse input of 1.0, $v = 1.0$, and $D = 0.5$, and a series of retardation coefficients as indicated for each curve.

to the solid; then as the pulse begins to propagate, adsorption is initiated. The results are shown in Figure 8.5. The curve with $R = 1$ (no retardation) has its peak at approximately 8.0 units as seen in the earlier example. With an increase in the value of R, the curve is lagged. The distance the peak has traveled in the case of $R = 2$ is half that of the peak associated with $R = 1$. The increased retardation has also caused the peak to be smaller. However, in this case the area under the curves is not the same. The reason for this is that part of the dissolved mass is being absorbed on the solid phase.

To show this we present Figure 8.6. In this case the problem is modified such that a third-type (Robbins) boundary condition is used at the inlet side of a semi-infinite domain (see Eq. (8.30)). For a period t_f the combined convective and dispersive flux equals vC_f, where C_f is a prescribed constant. The interpretation of this kind of boundary condition is that there exists a standing body of water adjacent to the aquifer which has a concentration C_f and infiltrates the aquifer for a period t_f, after which it stops. Beyond t_f there is no mass flux entering the inlet boundary. The auxiliary conditions for this problem are written

$$\rho_i\,(x, t = 0) = 0,$$

$$\rho_i\,(x \to \infty, t) = 0,$$

$$\left[v\rho_i - D\frac{\partial \rho_i}{\partial x} \right]_{x=0} = vC_f, \quad 0 < t \le t_f,$$

$$\left[v\rho_i - D\frac{\partial \rho_i}{\partial x} \right]_{x=0} = 0, \quad t > t_f. \tag{8.30}$$

The solution to this problem can be found in Torride et al. [3].

The concentration profiles in terms of the soil grain concentration (panel B) and the solvent concentration (panel A) are essentially the same, except for the scale on the

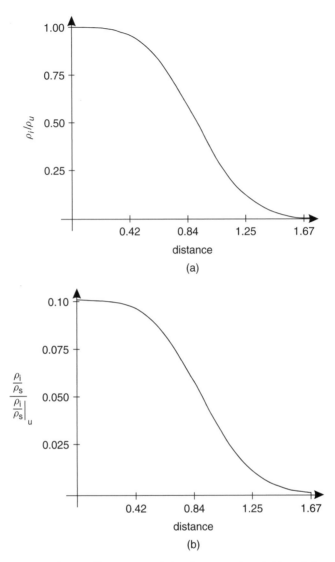

FIGURE 8.6. Concentration profiles computed for the water (panel A) and the soil (panel B). Note that the concentration scales of the soil and water profiles are different.

concentration axis. Since the equilibrium constant between the fluid and solid is 0.1, it is not surprising that the solid concentration is one-tenth that of the liquid phase. The reason for this is that the relationship between $\overline{\omega_i^s}$ and $\overline{\omega_i^w}$ (Eq. (8.22)) assumes instantaneous equilibrium between the solid and liquid species concentrations.

8.5 CHEMICAL REACTIONS

Chemical reactions occur both in the dissolved and adsorbed species, so for readers who are unfamiliar with, or have forgotten, elementary chemistry, an introduction to basic chemical concepts is found in Section 10.4. Because chemical reactions are found in the subsurface, it is necessary to modify Eq. (8.21) to include the chemical reaction source and sink term, that is,

$$\frac{\partial}{\partial t} (\varepsilon R \rho \omega_i) + \nabla \cdot (\varepsilon \mathbf{v} \omega_i \rho) - \nabla \cdot \varepsilon \rho \mathbf{D} \, \nabla \omega_i - \varepsilon \rho Q_i^W - (1 - \varepsilon) \rho_S Q_i^S = 0, \qquad (8.31)$$

where ρ is the fluid density and ρ_s is the solid density. The source terms are identified with the liquid phase, Q_i^W, and the solid phase, Q_i^S.

Until now, we have not defined what is meant by the term species. As used here, a species is synonymous with a chemical species, which by definition is an ensemble of chemically identical molecular entities such as atoms, molecules, ions, ion pairs, radicals, radical ions, and complexes, identifiable as a separately distinguishable entity. A species can be either the solvent or the solute. The *solvent is that species that is predominant.* In general, we are interested primarily in molecular and ionic species.

Chemical reaction kinetics is the study of the speed with which a chemical reaction occurs and the factors that affect this speed. The concentration measures used in reaction kinetic relationships are normally either mass per unit volume, that is, ρ_i in our notation, or *moles.* In either case, the protocol used to determine the rate expression or rate constant is developed through a plot of the change in concentration versus time. When the logarithm of the concentration is plotted against linear time, a straight-line relationship denotes a first-order rate expression, for example, $r = -\lambda \rho_i$. In general, a relationship of the form $r = -\lambda_n \rho_i^n$ indicates an nth-order rate equation. In groundwater problems where the primary mechanism for chemical change is bioremediation, the first-order rate expression is usually used with a zero-order rate expression used with high concentration of substrates. The various rate expressions used in bioremediation, their functional forms, and their limitations are presented in Steven [4].

Consider now Eq. (8.31). The reaction terms can be considered collectively or individually. If, as is commonly the case, the experiments used to determine the rate expression and its associated rate constant are conducted on soil, then the reactions are occurring on both the soil grains and in solution and it is not possible to extract the expression for the soil and the water separately. With this in mind, we combine the two source terms appearing in Eq. (8.31):

$$\varepsilon \rho Q_i^W + (1 - \varepsilon) \rho_s Q_i^S = r_i^W + r_i^S = \lambda_w \rho_i + \lambda_s \rho_i^s = \lambda_w \rho_i + \lambda_S \, \langle \rho \rangle_S \, \frac{k_d}{\varepsilon} \rho_i$$

$$= (\lambda_W + \lambda_S \rho_s k_d) \, \rho_i. \qquad (8.32)$$

If we assume that only one rate constant can be observed for the experiment, that is, the experiment is conducted on the saturated soil, then

$$(k_W + k_S \rho_s k_d) \, \rho_i = R \lambda \rho_i,$$

and we have

$$\frac{\partial}{\partial t} (\varepsilon R \rho \omega_i) + \nabla \cdot (\varepsilon \mathbf{v} \omega_i \rho) - \nabla \cdot \varepsilon \rho \mathbf{D} \, \nabla \omega_i - R \lambda \rho_i = 0. \qquad (8.33)$$

In Figure 8.7 we provide calculated concentration profiles for a fixed time and a set of first-order reaction rate constants. The auxiliary conditions are as found in Eq. (8.30). For the case of no ongoing reactions, the profile is that of transport with no retardation and no chemical reactions. As λ increases, the concentration is markedly reduced due to the

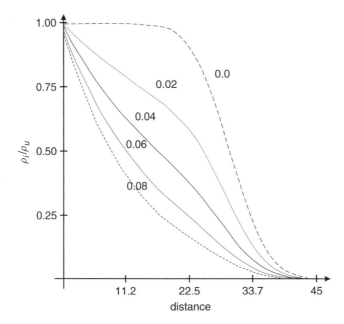

FIGURE 8.7. Concentration profiles computed using $R = 1$, $D = 0.5$, $t = 30$, and $\lambda = 0.0$, 0.02, 0.04, 0.06, 0.08.

reaction term. The overall area under the curves with increased chemical degradation is reduced due to the loss of species through chemical or biochemical reactions. The curve is maintained at unity at the source due to the first-type boundary condition employed.

We can now examine the combined affects of retardation and chemical reaction. Using the same auxiliary conditions found in Eq. (8.30), we obtain the profiles shown in Figure 8.8. The curve $R = 1$, $\lambda = 0$ is the reference. Neither retardation nor chemical reaction (decay) is represented. The addition of retardation ($R = 1.5$, $\lambda = 0$) causes the front to slow down and the dispersion to decrease (front gets steeper). This behavior is similar to what we observed in Section 8.4.

The effect of decay ($\lambda = 0.05$) is to decrease the concentration value and to distort the front, in terms of both sharpness and location. As would be expected, the greatest impact is observed when both retardation and decay are present ($R = 1.5$, $\lambda = 0.05$). Note that the concentration at the inlet is not 1.0 when decay is active. The reason for this is the fact that a third-type, not first-type, condition was specified and the decay is sufficient to influence the inlet concentration.

8.6   NUMERICAL SOLUTION TO THE GROUNDWATER TRANSPORT EQUATIONS

In Chapter 7 we considered the numerical solution of the groundwater flow equation. In this section we extend these concepts to consider the numerical solution of the species transport equation.

The basic techniques we introduced in Chapter 7 can be extended to the simulation of contaminant transport, with a few provisos. To understand these limitations, consider Eq. (8.29), which we rewrite now for convenience as Eq. (8.34) with the addition of a

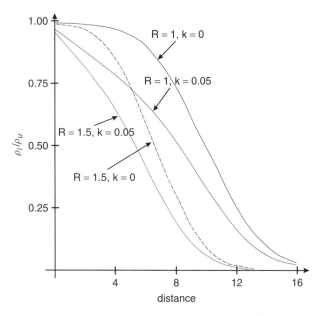

FIGURE 8.8. Concentration profiles calculated for $t = 10$, $v = 1$, $D = 0.5$, $t_f = 1000$, $R = 1$ and 1.5, and $\lambda = 0$ and 0.5.

source term Q_i:

$$\underbrace{\frac{\partial}{\partial t} (\varepsilon R \rho_i)}_{1} + \underbrace{\nabla \cdot (\varepsilon \mathbf{v} \rho_i)}_{2} - \underbrace{\nabla \cdot \varepsilon \mathbf{D} \, \nabla \rho_i}_{3} - \underbrace{\varepsilon Q_i}_{4} = 0. \tag{8.34}$$

Comparison of this expression with the groundwater flow Eq. (7.61) reveals that the terms 1, 3, and 4 in Eq. (8.34) have a counterpart in Eq. (7.61). In other words, if the unknown state variable ρ_i were replaced by the hydraulic head h, the equations would have a corresponding time derivative (term 1), second space derivative (term 3), and source term (term 4). Only term 2 is unique to Eq. (8.34), and it is this term that causes the difficulty when solving this equation. This first-order space derivative changes the character of the equation from a diffusion equation to a convective-diffusion equation. As the convective term (1) becomes large relative to the dispersive term (2), the equation begins to behave like a hyperbolic equation as is used to represent the movement of sharp fronts.

The challenges of representing sharp fronts using classical finite-difference and finite-element methods are well known. As the dispersion term (3) becomes small relative to the convective term (2), oscillations begin to appear at the location of the sharp front (see Figure 8.9). The solution is stable, in that the oscillations do not grow indefinitely, but it is not physically realistic inasmuch as values of the state variable (in this case concentration) greater than the boundary values as well as negative state-variable values are generated via the oscillations.

The oscillations can be suppressed by adding artificial dispersion through the use of a low-order accurate approximation to the first-order time or space derivative appearing in Eq. (8.34). The truncation error in this approximation appears as a dispersion-like term. The smoothing effect generated by this error term is called numerical dispersion (note that the term numerical dispersion has a quite different meaning in wave mechanics).

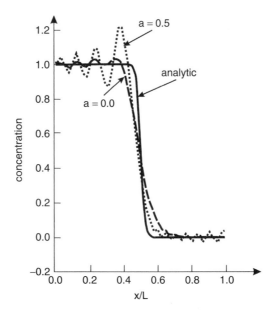

FIGURE 8.9. Concentration solutions obtained using a finite-difference approximation and two values of the coefficient a in the time derivative (adapted from [5]).

Thus one can have a sharp front with oscillations, or an artificially diffused front that is oscillation free.

Recall that in Chapter 7 (on page 271) a number of different finite-difference formulas to represent the first derivative were introduced. Consider the general form

$$\left. \frac{\partial \rho_i}{\partial x} \right|_{x_j} = \alpha \frac{\rho_i \left(x_j + \Delta x \right) - \rho_i \left(x_j \right)}{\Delta x} + (1 - \alpha) \frac{\rho_i \left(x_j \right) - \rho_i \left(x_j - \Delta x \right)}{\Delta x}, \quad \alpha \in [0, 1],$$

(8.35)

which is a weighted average of a forward and backward difference approximation to $(\partial \rho_i / \partial x)|_{x_j}$. Note that when $\alpha = 1$, we have a forward difference and when $\alpha = 0$, we have a backward difference at the point x_j. A value of $\alpha = 0.5$ yields the central difference. It can be shown that, but for the case of $\alpha = 0.5$, the truncation error in Eq. (8.35) is of the diffusive type. However, when $\alpha > 0.5$, the diffusion term has a positive sign, which may lead to instabilities. Therefore the first derivative should always have an approximation that is weighted "upstream" instead of "downstream." To assure convergence, a value of $\alpha \leq 0.5$ is used.

In the finite-element method, the same philosophy holds. If an asymmetrical approximation to the first derivative can be achieved, then a truncation error of the dispersive form is generated. The question is how to create such an approximation. As a point of departure, consider Eq. (8.36) rewritten for the state variable ρ_i:

$$\int_X \left[\mathcal{L} \left(\sum_{j=0}^n \ell_j^n (x) \rho_i \left(x_j \right) \right) - \varepsilon Q_i \right] \ell_i^n (x) \, dx = 0, \quad x \in X, i = 0, 1, \ldots, I, \quad (8.36)$$

where

$$\mathcal{L} (\cdot) = \frac{\partial}{\partial t} (\varepsilon R \cdot) + \frac{\partial}{\partial x} (\varepsilon v \cdot) - \frac{\partial}{\partial x} \cdot \varepsilon D \frac{\partial}{\partial x} (\cdot).$$

The term of interest to us, that does not appear in the groundwater flow equation, is

$$\int_X \left[\frac{\partial}{\partial x} \left[\varepsilon v \left(\sum_{j=0}^n \ell_j^n (x) \, \rho_i \left(x_j \right) \right) \right] \right] g \, (x)_i^m (x) \, dx,$$

where we have generalized the approach to use a *weighting function* $g \, (x)_i^m$ that is different from the *basis function* $\ell_j^n (x)$. Let us consider a linear basis function as we did earlier and a quadratic weighting function. The linear basis function is curve b in Figure 8.10B and the quadratic weighting function is curve a in Figure 8.10B. Panel (A) of this figure shows the weighting functions for both nodes defined for one element. The arrow in panel (A) shows the directional sense of the velocity term. The functional forms of the asymmetric weighting function is

$$g_1 \, (\xi, \alpha) = 1 - \xi + 3\alpha \left(\xi^2 - \xi \right),$$

$$g_2 \, (\xi, \alpha) = \xi - 3\alpha \left(\xi^2 - \xi \right),$$

where

$$\xi = (x - x_1) / \Delta x.$$

Notice that for $\alpha = 0$, one has the standard linear functions, that is, curve b in Figure 8.10B. As α increases, the quadratic nature of $g_j \, (\alpha, x)$ increases. Therefore, from a practical perspective, to reduce oscillatory behavior as the velocity increases, α should also increase to enhance the upstream weighting effect.

Aside from this change, the application of the finite-element method proceeds as presented earlier.

An alternative strategy for solving the transport equation involves eliminating, in some sense, the troublesome convective term in the equation. The approach utilizes the concept of a substantial derivative, which can be defined as

$$\frac{D \, (\cdot)}{Dt} \equiv \frac{\partial \, (\cdot)}{\partial t} + \mathbf{v} \cdot \nabla \, (\cdot) \tag{8.37}$$

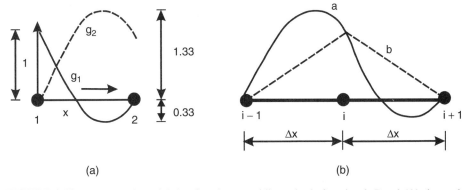

FIGURE 8.10. Asymmetric weighting function a and linear basis function b. Panel (A) shows the two weighting functions defined for one element and panel (B) the weighting function and basis function for the node i defined over two elements (adapted from [6]).

where $D(\cdot)/Dt$ is the substantial derivative and $\mathbf{v}$ is the fluid velocity. The substantial derivative describes the rate of change in a Lagrangian (moving) coordinate system. With this in mind, let us rewrite Eq. (8.34) neglecting for this illustration the effects of retardation,

$$\frac{\partial}{\partial t}(\varepsilon\rho\omega_i) + \nabla\cdot(\varepsilon\mathbf{v}\rho\omega_i) - \nabla\cdot\varepsilon\mathbf{D}\,\nabla\rho\omega_i - \varepsilon Q_i = 0. \tag{8.38}$$

and then expand this as

$$\varepsilon\omega_i\frac{\partial\rho}{\partial t} + \rho\frac{\partial\varepsilon\omega_i}{\partial t} + \varepsilon\omega_i\nabla\cdot(\rho\mathbf{v}) + \rho\mathbf{v}\cdot\nabla(\varepsilon\omega_i) - \nabla\cdot\varepsilon\mathbf{D}\,\nabla\rho\omega_i - \varepsilon Q_i = 0.$$

If we regroup the terms in this expression we have

$$\varepsilon\omega_i\left(\frac{\partial\rho}{\partial t} + \nabla\cdot(\rho\mathbf{v})\right) + \rho\frac{\partial\varepsilon\omega_i}{\partial t} + \rho\mathbf{v}\cdot\nabla(\varepsilon\omega_i) - \nabla\cdot\varepsilon\mathbf{D}\,\nabla\rho\omega_i - \varepsilon Q_i = 0,$$

which, assuming the mass balance equation

$$\frac{\partial\rho}{\partial t} + \nabla\cdot(\rho\mathbf{v}) - Q = 0,$$

gives

$$\varepsilon\omega_i Q + \rho\frac{\partial\varepsilon\omega_i}{\partial t} + \rho\mathbf{v}\cdot\nabla(\varepsilon\omega_i) - \nabla\cdot\varepsilon\mathbf{D}\,\nabla\rho\omega_i - \varepsilon Q_i = 0.$$

Using Eq. (8.37) we obtain

$$\rho\frac{D}{Dt}(\varepsilon\omega_i) - \nabla\cdot\varepsilon\mathbf{D}\,\nabla\rho_i + \varepsilon(\omega_i Q - Q_i) = 0. \tag{8.39}$$

In this form, it is clear that if you are moving with the fluid at a velocity $\mathbf{v}$, the equation appears as a diffusion equation, which, like the groundwater flow equation, is very easy to solve numerically.

However, to solve Eq. (8.39) one must move with the fluid. Two distinctly different approaches are used to represent the fluid motion. One method is called the *method of characteristics*. In this approach mathematical particles are introduced into the modeled system. The particles move with the fluid. Each particle is assigned a characteristic concentration, which it convects according to the groundwater velocity. The particles are integrated at the end of each time step and the information they contain is transferred to a fixed finite-difference grid. The dispersion equation is then solved on the fixed grid and the moving particles are adjusted for the dispersive effect. The whole procedure is repeated for each time step.

A quite different approach is used in finite-element techniques, where a regular grid (rectangular) may not be desirable. In this case, values at a new time level are calculated at the nodes by projecting backward to the preceding time level along the fluid path. Information from that time level is then brought forward to the new time level to complete the time-derivative approximation. The diffusion equation is now solved for the new time step using a standard finite-element formulation. The process is repeated for each time step until the end of the period of analysis.

8.7 SUMMARY

The focus of this chapter is the nature and behavior of contaminants in the subsurface. The chapter begins with a discussion of natural and anthropogenic contaminants. The equations that describe the evolution of these contaminants in the subsurface are subsequently developed and their characteristic behavior is illustrated through their solutions to selected initial boundary-value problems. Finally, the numerical solution to these equations is considered, and the difficulties associated with their solution using finite-difference and finite-element methods and the ways of dealing with these computational challenges are described.

8.8 PROBLEMS

8.1. Show that the application of volume averaging to Eq. (8.2) produces Eq. (8.4). Explain why Eq. (8.4) has six terms while Eq. (8.2) has only four terms.

8.2. Consider a system with one-dimensional flow, such that the velocity vector $\mathbf{v} = (v_x, 0, 0)$. For this case, with the representation of the dispersion tensor as given in Eq. (8.16), show that $i_x = -(\alpha_L v_x + D_m)\, \partial \omega_i / \partial x$, $i_y = -(\alpha_T v_x + D_m)\, \partial \omega_i / \partial y$, and $i_z = -(\alpha_T v_x + D_m)\, \partial \omega_i / \partial z$.

8.3. In the expansion and subsequent averaging of Eq. (8.6), averages of products involving deviation quantities and average quantities were assumed to vanish. Explain why this is the case, and then speculate on any conditions under which this may not be true.

8.4. Consider the one-dimensional transport equation, whose solution is given by Eq. (8.19). For the case of $\rho_i(x, 0) = 0$, $\rho_i(0, t) = 1$, and $\rho_i(x \to \infty, t) = 0$, with parameters $v = 1$ and $D = \alpha v = 0.1$, calculate and graph the breakthrough curve at location $x = 2$, that is, $\rho_i(2, t)$.

8.5. Explain why the terms D and C in Eq. (8.20) are equal and opposite to the terms b and c in Eq. (8.24) when only two phases are present. Speculate on what would happen if three phases (say, water, air, and solid) were present in the system.

8.6. Write a computer code to compute the numerical approximation to the equation

$$\frac{\partial \rho_i}{\partial t} + V \frac{\partial \rho_i}{\partial x} - D \frac{\partial^2 \rho_i}{\partial x^2} = 0, \quad 0 < x < L, \ t > 0,$$

$$\rho_i(x, 0) = 0, \quad 0 < x < L,$$

$$\rho_i(0, t) = 1, \quad t > 0,$$

$$\frac{\partial \rho_i}{\partial x}(L, t) = 0, \quad t > 0.$$

Use central finite differences for both of the spatial derivatives, so that

$$\left. \frac{\partial \rho_i}{\partial x} \right|_{x_j}^{t^n} \simeq \frac{(\rho_i)|_{x_{j+1}}^{t^n} - (\rho_i)|_{x_{j-1}}^{t^n}}{2\Delta x},$$

$$\left. \frac{\partial^2 \rho_i}{\partial x^2} \right|_{x_j}^{t^n} \simeq \frac{(\rho_i)|_{x_{j+1}}^{t^n} - 2(\rho_i)|_{x_j}^{t^n} + (\rho_i)|_{x_{j-1}}^{t^n}}{(\Delta x)^2}.$$

For the time derivative, use the simple Euler approximation given by

$$\frac{\partial \rho_i}{\partial t}\bigg|_{x_j}^{t^n} \simeq \frac{(\rho_i)|_{x_j}^{t^{n+1}} - (\rho_i)|_{x_j}^{t^n}}{\Delta t}.$$

Notice that this approximation is *explicit in time* and centered in space. The spatial step and the time step are assumed constant ($\Delta x = x_{j+1} - x_j = x_j - x_{j-1}$ and $\Delta t = t^{n+1} - t^n = t^n - t^{n-1}$). (a) Show that the dimensionless groupings that arise from these approximations are the Courant number ($Cu \equiv (V\,\Delta t)/\Delta x$), the Peclet number ($Pe \equiv (V\,\Delta x)/D$), and the dimensionless diffusion number ($Df \equiv (D\,\Delta t)/(\Delta x)^2 = Cu/Pe$). Write the approximation so that these groupings appear explicitly in the algebraic equations. (b) To observe the numerical behavior of this equation, choose several different values of the Courant and Peclet numbers, for example, in the range of $0 < Cu < 1$ and $0 < Pe < 10$. Examine the solution at time $t^* = L/2V$, which corresponds to the time when the value of $\rho_i = 0.5$ will reach the spatial location $x = L/2$. Be sure to explain how you have implemented the boundary condition at $x = L$.

8.7. Repeat the calculations of Problem 8.6, except now replace the approximation for the first spatial derivative with an *upstream weighted approximation* given by

$$\frac{\partial \rho_i}{\partial x}\bigg|_{x_j}^{t^n} \simeq \frac{(\rho_i)|_{x_j}^{t^n} - (\rho_i)|_{x_{j-1}}^{t^n}}{\Delta x}.$$

Note in particular the behavior of the solution when the Courant number is equal to 1.

8.8. Finally, repeat the calculation of Problem 8.6, except now use an *implicit approximation* in time, such that the spatial derivatives are now evaluated at the later time t^{n+1}. You will notice that in each finite-difference equation, there are now three unknowns instead of one. Therefore, to obtain a solution for this implicit approximation, a matrix equation must be solved at each time step in the solution.

BIBLIOGRAPHY

[1] R. A. Greenkorn, *Flow Phenomena in Porous Media*, Marcel Dekker, New York, 1983.

[2] A. Ogata and R. B. Banks, *A Solution of the Differential Equation of Longitudinal Dispersion in Porous Media*, U.S. Geological Survey, Professional Paper No. 411-A, 1961.

[3] N. Torride, F. Leij, and M. Th. van Genuchten, A comprehensive set of analytical solutions for nonequilibrium solute transport with first-order decay and zero-order production, *Water Resour. Res.* **29**: 2167, 1993.

[4] D. K. Steven, Mathematical modeling in bioremediation of hazardous wastes, in *Bioremediation of Contaminated Soils*, Agronomy Monograph No. 37, 1999, p. 631.

[5] G. F. Pinder and W. G. Gray, *Finite Element Simulation in Surface and Subsurface Hydrology*, Academic Press, San Diego, CA, 1977.

[6] P. S. Huyakorn and G. F. Pinder, *Computational Methods in Subsurface Flow*, Academic Press, San Diego, CA, 1983.

CHAPTER 9

GROUNDWATER–SURFACE-WATER INTERACTION

9.1 INTRODUCTION

At the beginning of this book, especially in Section 1.2, we learned that groundwater interacts with surface-water bodies such as lakes, streams, seas, and oceans. In this chapter we investigate the nature of this interaction and the role it plays in the quantity and quality of groundwater and surface-water supply. We begin with a discussion of the how groundwater and surface-water bodies interact.

In humid climates the surface-water body most often acts as a sink or discharge point for groundwater. Figure 9.1 illustrates an example of this kind of groundwater–surface-water interaction wherein the movement of groundwater through the subsurface discharges into a gaining stream. The view presented in panel (A) shows groundwater entering a stream along the soil–water interface where the stream is in contact with the groundwater. In essence, this is a close-up view of the regional groundwater flow behavior in the neighborhood of surface-water bodies that we introduced in Figure 5.16. Note that the flow moves radially into the stream.

In Figure 9.1B we see a contour map of a water table, drawn in the areal plane, that is consistent with the gaining stream model. To reveal an interesting characteristic of a gaining stream, first draw a line that connects the two points where the 40 contour intersects the vertical sides of the panel, that is, points a and b Figure 9.1B. In the absence of the stream we would expect that the groundwater flow line would be approximately collinear with the line we have drawn. However, if one follows the line from point a toward the stream, one encounters a deviation of the observed contour from the line we have drawn. The observed contour deviates such that it is pointing upstream as it crosses the stream. At the point at which our line crosses the stream, the water table is at a lower elevation than the water table observed at points a and b. Indeed there is a depression in the water table as we approach the stream following our line. In this case there exists

GAINING STREAM

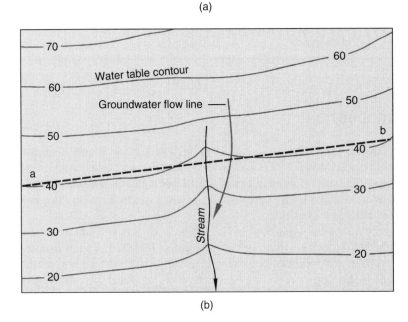

(a)

(b)

FIGURE 9.1. Diagrammatic representation of a gaining stream (from [1]).

a depression of perhaps 4 feet relative to other points along the line $a-b$. The existence of this depression might have been anticipated since we observed that groundwater flow is toward the river in panel (A) and groundwater must flow downgradient. *It is a general rule that groundwater contours will point upstream in a gaining stream.*

Consider now the case of a losing stream as presented in Figure 9.2. In this instance water is moving from the surface-water body to the groundwater. As we will see a little later, this situation can be encountered during a period of surface-water rise, such as experienced subsequent to a heavy precipitation event, or circumstances when the groundwater has been lowered as a result of exploitation for water supply.

In the case of a losing stream, the elevation of the stream is higher than the hydraulic head observed in the aquifer, thereby inducing water movement from the stream to the aquifer. Water flows radially out of the stream in contrast to the earlier example wherein there was radial flow toward the stream.

LOSING STREAM

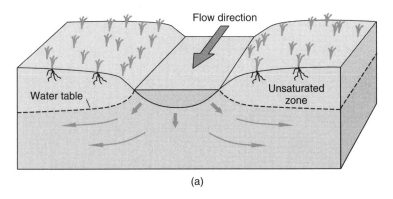

(a)

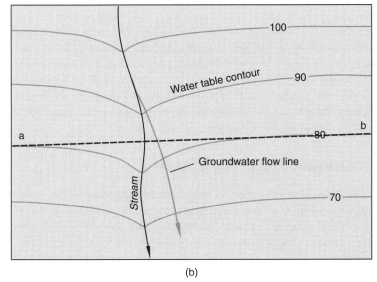

(b)

FIGURE 9.2. Diagrammatic representation of a losing stream (from [1]).

If we now draw a line that connects the two points *a* and *b*, locations where the 80 contour cuts the vertical sides of panel (B), we can demonstrate a different water-table topology. In this case, as we move from point *b* toward the stream, the water table rises. At the stream it forms a ridge with a relief of approximately 5 feet relative to other points on the line *a–b*. The water-table rise is consistent with the fact that water is flowing from the stream and must move down a potential gradient. *The contours now point downstream, a phenomenon generally associated with losing streams.*

In addition to gaining and losing steams (or other surface-water bodies), there is a special case of losing streams that warrants attention. In Figure 9.2 the aquifer below the stream is completely saturated. While this may generally be true, one can conceive of situations where the water table below the river is at an elevation below the river bed (see Figure 9.3). This may happen when the water table is especially depressed due to excessive pumping. In this case, the water moving from the stream to the water table encounters an unsaturated flow condition. Unsaturated flow is influenced by the fact that

DISCONNECTED STREAM

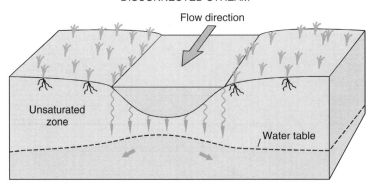

FIGURE 9.3. Disconnected stream separated from the groundwater system by an unsaturated zone (from [1]).

the hydraulic conductivity is reduced relative to the saturated case due to the presence of the air phase. In addition, gravitational effects play a different role in unsaturated flow such that fluid flow in the unsaturated zone is largely vertical. In essence, the regional fluid potential in the aquifer below the stream in the unsaturated case generally plays a much less important role in determining the amount of water that leaves the stream than in the case of saturated aquifer conditions.

9.2 FLUID DYNAMICS NEAR THE STREAM

In the preceding we were introduced to gaining and losing streams. Let us now consider more carefully the nature of groundwater flow in the neighborhood of the stream as shown in Figure 9.4. The numbered contours describe the fluid potential. In this idealized case, the top of the saturated zone is represented by a sharp transition shown as a line separating the unsaturated and saturated zones. Thus the elevation of the water table corresponds to the value of the fluid potential contour at the point where the contour intersects the water table.

The flow system has been subdivided, somewhat arbitrarily, into three regions denoted by the letters D, E, and F. In the F region the contours decrease in magnitude with depth, indicating a significant downward flow component. In region E the contours are nearly vertical with no significant change with depth, indicative of horizontal flow. Finally, in region D the contours increase in magnitude with depth as expected in the case of upward flow.

Also shown on Figure 9.4 are a series of nested wells. At locations A and B there are two wells of different depths, and at location C there are three wells of different depths. Also noted at each location is the water-level rise expected in each well, assuming that the wells are open (have access to groundwater) only at their deepest points. Because the contours represent a constant fluid potential along their length, by tracing the contour closest to the bottom of the well to the surface one can determine the elevation to which the water in the well will rise. For example, at well A the deep well is located at a contour with a value of 20. By tracing this contour to the surface of the ground one can determine the elevation to which the water will rise (this is indicated for each well in this figure by the horizontal dashed lines).

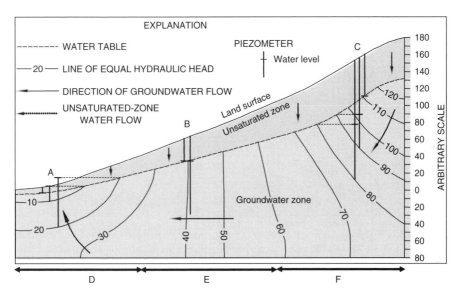

FIGURE 9.4. Idealized representation of groundwater flow in the near-stream environment (from [1]).

The point that we wish to illustrate by this example involves the water-level elevations in the observation wells. Wells in region F (those in nest C) that are drilled deeper have lower water levels than those that are drilled to shallower depths, those drilled in zone E (those are wells in nest B) change little with little change in water levels with changes in completion depth; and those in zone D (those are wells in nest A) have higher water levels with increasing completion depth. In fact, the deep well in nest A has a water level higher than the ground surface. What is the meaning of this? It implies that the water will flow spontaneously from the well without pumping. In fact, this is exactly what will happen is such a case. Deep wells in topographic depressions, such as associated with river valleys, can flow spontaneously. They are called flowing or *artesian wells*. When the water discharges naturally to the surface, the discharge point is called a *spring*.

9.2.1 Bank Storage Effects

Bank storage is a near-stream hydrodynamic behavior that can have significant practical ramifications. The importance of this phenomenon was recognized and reported at least as early as 1944 (see [2]). Bank storage occurs when the water-level elevations in a surface-water body, such as a stream, increase beyond the groundwater elevation in the adjacent banks. At this point there is an outward potential gradient from the stream to the groundwater and flow out of the stream is increased. Figure 9.5 illustrates the initial response of the system to the rise in stream-water elevation. Water will move into the groundwater system as long as the stream-water elevation exceeds that of the groundwater in contact with the stream bed (or adjacent flood plain). When the stream-water elevation begins to decrease in its return to normal flow, the whole process is reversed. The water level in the stream bank is now higher than that of the stream. The response is movement of groundwater from the bank into the stream. The water levels in the bank adjacent to the stream begin to decline as drainage to the stream occurs. Further away from the stream, however, a complex flow pattern is observed. Water near the stream moves toward the

BANK STORAGE

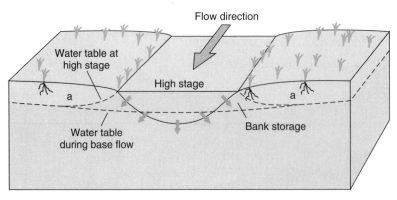

FIGURE 9.5. Diagrammatic representation of bank storage. When the elevation of the surface of a stream rises above the elevation of groundwater adjacent to the stream, water moves into the stream bank (from [1]).

stream, but water at the leading edge of the bank storage wave, at locations marked *a* in Figure 9.5, continues to move away from the stream. The result is a groundwater "hill" that propagates inland. Eventually the effect of the stream elevation is dissipated and the normal flow to the stream from the groundwater is restored.

Consider, as an example of the bank storage phenomenon, an experiment conducted at Musquodoboit Harbor, Nova Scotia. Four wells with automatic recorders were located along a line orthogonal to the Musquodoboit River as shown in Figure 9.6. A flood wave generated in response to a brief but intense storm provided the rise in water level in the river that generated the water-level response observed in wells 3 and 4, which are located on Figure 9.6. The change in head values presented in Figure 9.7 are relative to the projected recession curve for the river as determined from river stage measurements prior to the storm. Figure 9.7 reveals that the water levels in wells 3 and 4, located approximately 200 and 400 feet from the river, respectively, rise in response to the river stage within a short period of time after initiation of the flood-wave event. The water levels peak at approximately 1.8 feet above the projected recession curve near time *a* and then begin a long decline back to the pre-flood-wave state of the system. Notice that while the differences between the two curves are relatively small, they are consistent. For any point in time prior to approximately time *b*, the water level in well 3 is higher than that in well 4, indicating water movement from the river inland. Alternatively, one can observe that the peak of the bank storage "hill" arrives earlier at well 3 than at well 4. Both of these observations are consistent with what would be expected during the period of bank infiltration. One would expect that after time *b* well 4 would have a sightly higher water level than well 3 as drainage toward the river takes place.

The above discussion provides but a transient one-space dimensional picture of the bank storage phenomenon. To obtain insight into the areal aspects of this phenomenon and the impact it has on stream flow, we employ mathematical modeling. The idealized cross section is presented in Figure 9.8. The stream channel is assumed to have rectangular cross section and the flow in the stream is considered one-dimensional along the length of the channel. The aquifer is modeled using a vertically integrated areal finite-difference based model.

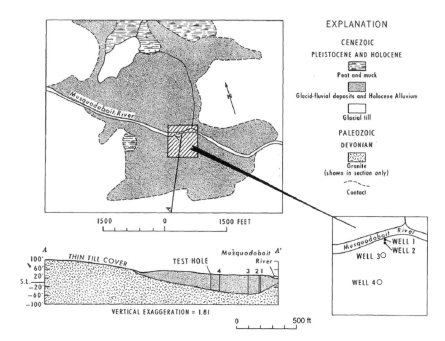

FIGURE 9.6. Geological map and cross section of the alluvial aquifer at Musquodoboit Harbor, Nova Scotia. In the inset are located wells 3 and 4 for which water-level data are provided in Figure 9.7 (from [3]).

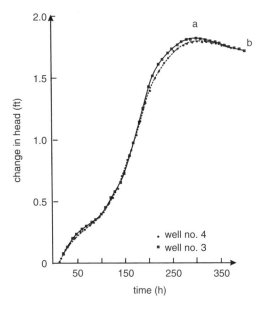

FIGURE 9.7. Water-level observations at wells located on flood plain and responding to flood wave in Musquodoboit River, Nova Scotia.

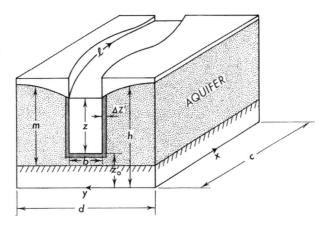

FIGURE 9.8. Idealized cross section of a river-aquifer system used to model the impact of bank storage on the groundwater system and the surface water body (from [4]).

The equation that describes the flow of water in the stream is given by

$$z\frac{\partial v}{\partial \ell} + v\frac{\partial z}{\partial \ell} + \frac{\partial z}{\partial t} = \frac{q_\ell + q_v}{b}, \tag{9.1}$$

$$v\frac{\partial v}{\partial \ell} + g\frac{\partial z}{\partial \ell} + v\frac{q_\ell + q_v}{bz} + \frac{\partial v}{\partial t} = g\left(S_0 - S_f\right), \tag{9.2}$$

where (referring to Figure 9.8) b is the channel width, g is gravitational acceleration, l is the space coordinate, q_v is the flow into the channel per unit length through its wetted perimeter, q_ℓ is lateral inflow per unit length over the channel banks and from tributaries, S_f is the friction slope, S_0 is the slope of the channel bottom, t is time, v is velocity of flow, and z is the depth. These equations are sometimes referred to as the shallow water equations.

The groundwater flow equation used in this analysis is given by

$$\nabla \cdot (\mathbf{T}\nabla h) = S\frac{\partial h}{\partial t} + \frac{q_v}{b + 2z} \tag{9.3}$$

below the channel, and elsewhere the governing equation is

$$\nabla \cdot \mathbf{K}m\nabla h = S_y\frac{\partial h}{\partial t}, \tag{9.4}$$

where h is the hydraulic head, m is the saturated thickness of the aquifer (a linear function of h), $\mathbf{K}$ is the conductivity, S is the storage coefficient, S_y is the specific yield, and $\mathbf{T}$ is transmissivity.

The coupling between Eqs. (9.1), (9.2), (9.3), and (9.4) is a form of Darcy's law, that is,

$$\frac{q_v}{b + 2z} = -K_p\frac{z + z_0 - h}{\Delta z'},$$

where K_p is the hydraulic conductivity of the bottom sediments of the channel, $\Delta z'$ is the thickness of the bottom sediments along the wetted perimeter of the channel, and z_0 is the elevation of the stream bottom measured from the same datum as h.

The open-channel flow boundary conditions applicable to Eqs. (9.1) and (9.2) are specification of either the stage or discharge at the upstream end of the channel and a definition of the stage–discharge relationship at the downstream boundary. The initial conditions on stream flow are the depth and velocity of flow in the channel (see Franz and Melching [5] for a detailed discussion).

The groundwater flow equations are solved for a square domain with no-flow conditions specified around the perimeter (see Figure 9.9). The initial condition for the groundwater equations is the equilibrium solution for the stream aquifer system with steady flow in the stream.

The solution to the above set of equations for a flood wave introduced into the upstream boundary of the stream (left boundary) is presented in Figure 9.9 after an elapsed time of 21.25 hours. Among the interesting features of this simulation result are the following:

1. The maximum elevation of the hydraulic head of the groundwater is located away from the stream. Thus the flood wave has passed through the channel at this point in time and the groundwater "hill" is moving both away from the stream at its furthest extent and toward the stream at its nearest point.

2. The hill has propagated further inland on the left (west) side of the model in comparison with the location of the "hill" on the right-hand side and is of higher elevation.

Let us now examine how bank storage has impacted the flood wave. Consider the behavior over time of the flood wave at a location 50,000 feet from the left boundary of

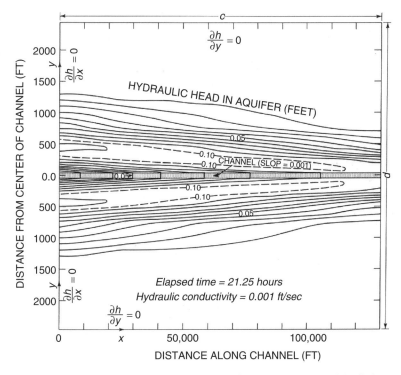

FIGURE 9.9. Water-level contours generated by a finite-difference model of the combined surface-water–groundwater system (from [4]) .

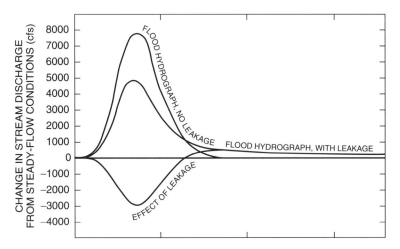

FIGURE 9.10. Attenuation of flood wave at a point 50,000 ft from upstream boundary when no bank storage is occurring (from [4]).

the model as shown in Figure 9.10. Three curves describe the change in stream discharge at this location as a function of time. The curve labeled "flood hydrograph, no leakage" shows the behavior of the routed flood wave assuming no bank storage; the one identified as "flood hydrograph with leakage" illustrates the evolution of the flood wave given bank storage exists; and the curve specified as "effect of leakage" shows the net effect of bank storage. Thus, at this location, water begins to enter the bank as soon as the elevation of the stream is increased, reaches a peak at the same time as the peak of the flood wave arrives, and then after the flood wave passes, drains back into the stream. Notice that the drainage is very slow and continues for the duration of the simulation.

The behavior of the flood wave at a location 140,000 feet downstream from the model boundary is shown in Figure 9.11. At this point the impact of bank storage is greater than at 50,000 feet, indicating that the effect of bank storage is cumulative. Thus bank storage may be of considerable importance in regulating flood discharge in the lower reaches of the stream. Bank storage has attenuated the flood wave, decreasing the peak discharge and extending the hydrograph base time.

9.2.2 Influence of Pumping Wells

When water is supplied by wells located near surface-water bodies, several hydrodynamic scenarios are possible. Consider the case of flow to a stream as depicted in Figure 9.12. In panel A discharge to the stream is not influenced by the presence of a pumping well. When a well is active near a stream, two scenarios are possible. Panel B shows the case where the well draws water that would normally reach the stream, thereby reducing aquifer discharge to the stream. However, no water from the stream actually finds its way into the well. Finally, in panel C the well discharge is sufficiently large that it not only reduces discharge from the aquifer to the steam but actually draws water directly from the stream.

Under some circumstances wells located near streams or rivers may draw water from the side of the stream or river opposite from that of the well. Such occurrences are generally associated with wells pumping large amounts and rivers that are shallow relative to the depth of the well. A classic case of this phenomenon is found in Figure 5.10. In

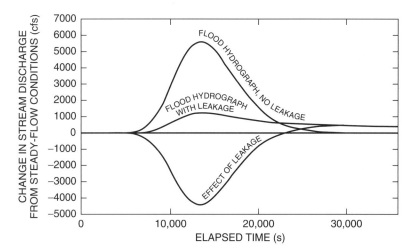

FIGURE 9.11. Attenuation of flood wave at a point 140,000 ft downstream from model boundary when no bank storage is occurring (from [4]).

this instance the public supply wells for the town of Woburn, Massachusetts, located on the east side of the Aberjona River, drew water from the west side of the river.

With the concepts presented in Section 9.2.1 combined with those found in this section, it is possible to introduce the concept of conjunctive use. The term conjunctive use describes the strategy whereby surface water and groundwater are used in combination to satisfy water-use demands, especially irrigation. Typically, wells are located adjacent to rivers and water is either drawn directly from the stream or from the installed wells, depending on the availability of surface-water supplies.

The issue of conjunctive use is particularly important in the arid areas of the United States where water is allocated based on the doctrine of *prior appropriation*, which can be loosely translated into *first in time is first in right*. The entity that first makes claim on a water resource is the owner. However, there is an important corollary to this legal strategy. To have a right to water it must be "beneficially" or "reasonably" used on the land identified by the applicant. Nevertheless, the water is considered as a resource unrelated directly to land. Some call this strategy the "water law of the West."

Prior appropriation is different from riparian water rights, which is the water law applicable in the Eastern United States. Under the riparian principle, all landowners whose property is proximate to a body of water have the right to make reasonable use of it. In the event there is not enough water to satisfy all users, allotments are typically fixed in proportion to frontage on the water source. Unlike rights associated with prior appropriation, riparian rights cannot be sold or transferred other than in conjunction with the adjoining land. In addition, water cannot be transferred out of the watershed.

Because, under the doctrine of prior appropriation, the owner of the water rights cannot exceed his/her allotment, in times of low surface-water availability the farmer risks being unable to adequately irrigate his available acreage. To minimize the uncertainty inherent in using a surface-water supply that may vary markedly from season to season, or even within a season, water users may seek to develop groundwater supplies because their availability is less volatile. While water levels in wells vary from year to year, they tend to change over longer periods of time than is the case for surface-water discharge.

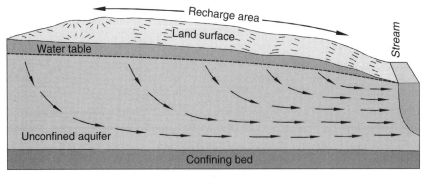

(A)

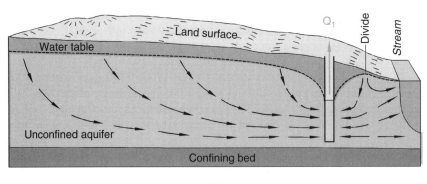

(B)

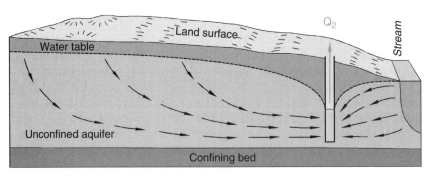

(C)

FIGURE 9.12. Effect of pumping well on aquifer dynamics near a stream. In panel (A) the discharge pattern to the stream is unaffected by a well. In panel (B) a well captures a portion of the water that would normally reach the stream, thereby reducing aquifer discharge to the stream. Finally, in panel (C) the well is pumping at a rate such that it actually draws water directly from the stream (from [1]).

The difficulty with this strategy lies in the fact that it is less expensive to draw water directly from surface-water bodies that to utilize groundwater. The costs associated with well construction, maintenance, and operation are the source of the additional expense. Thus the prudent investor would like to utilize surface water to the largest degree possible and minimize dependence on groundwater.

As we learned earlier, water extracted by wells in the neighborhood of surface-water bodies results in less discharge to the surface-water body and therefore depletion of its flow. Even if the well does not draw water directly from, for example, a stream, it still reduces stream flow because less water reaches the stream than is normally the case. In fact, in 1966 the State Engineer of Colorado ordered 39 pumping wells shut down in the Arkansas Valley because their pumping was reducing the flow of the river and thereby damaging the surface-water rights of those downstream who had legal claim to the water [6].

Bredehoeft and Young [6] analyzed the financial trade-offs associated with conjunctive use for the South Platte River system in Colorado. They employed optimization techniques similar to, but also different from, those we discuss briefly in Section 10.7. They found, among other things, that for the system they studied the following holds:

> Our results suggest that under the current economic condition existing in the South Platte Valley in Colorado, the most reasonable groundwater capacity is a total capacity capable of irrigating all the available acreage with groundwater. Our results suggest that the additional costs of pumping ground water is not significant. The best strategy is to totally discount surface water in considering what pumping capacity to install, assuming, of course, that one has the capability for augmenting streamflow to the downstream users in periods of low flow.

9.2.3 Hydrograph Separation

The flow observed in a flowing stream or river is derived from a number of sources. Hydrograph separation is the science of determining the relative contribution of each of these sources. A particularly detailed analysis is captured in Figure 9.13. The ability to deconvolve a total runoff hydrograph is limited by the techniques that must be employed. In essence, only the total runoff is directly measurable; all other components are determined indirectly.

One approach to determining the groundwater discharge component is through examination of the stream-water chemistry in conjunction with a knowledge of the groundwater chemistry. One such attempt is shown in Figure 9.14. The basic idea is to recognize the relationship

$$C_{tr}Q_{tr} = Q_{dr}C_{dr} + Q_{gw}C_{gw}. \tag{9.5}$$

Solving for groundwater discharge we obtain

$$Q_{gw} = (C_{tr}Q_{tr} - C_{dr}Q_{dr})/C_{gw}$$

or

$$Q_{gw} = \left[(C_{tr} - C_{dr})/(C_{gw} - C_{dr})\right]Q_{tr}, \tag{9.6}$$

where

$$Q_{tr} = Q_{gw} + Q_{dr},$$

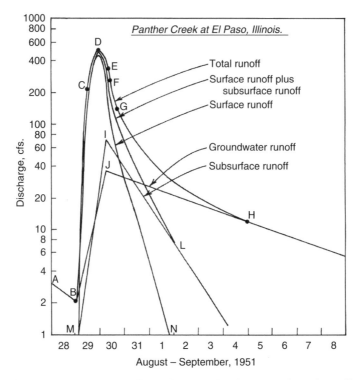

FIGURE 9.13. Semilogarithmic plot of a hydrograph showing separation of runoff components (adapted from [7]).

and C_{tr} is the total dissolved solids concentration in total runoff, C_{dr} is the total dissolved solids concentration in direct runoff, C_{gw} is the total dissolved solids concentration in groundwater runoff, Q_{tr} is the total runoff, Q_{dr} is the direct runoff, and Q_{gw} is the groundwater runoff.

To solve Eq. (9.6) a number of quantities need to be determined. The quantity of total runoff, Q_{tr}, is obtained from the stream discharge hydrograph at the same time as a sample is taken to obtain the chemistry of total runoff (C_{tr}).

The determination of the direct runoff concentration, C_{dr}, is a little more subtle. It is known from Section 9.2.1 that during flood events the water moves from the stream into the aquifer to form bank storage. During this period, the water in the stream is essentially direct runoff. Thus samples of the stream flow taken from this period should reflect the chemistry of direct runoff. However, the Fraser Brook basin does not have an extensive flood plain, so this strategy does not appear justified.

The Fraser Brook basin is one of three in Nova Scotia that were carefully studied in an effort to determine the groundwater component of stream discharge. The basin has considerable relief and poorly developed flood plains. A plot of water quality versus discharge in the upper segment of the Fraser Brook basin showed that the water chemistry was not a strong function of discharge. The conclusion to be drawn from this analysis is that the groundwater and direct runoff components do not vary significantly in the upper reach of the basin (this was also observed for two other basins that were studied in the same area). Thus samples selected during peak discharge in the upper reach of the brook were used as an estimate of C_{dr}.

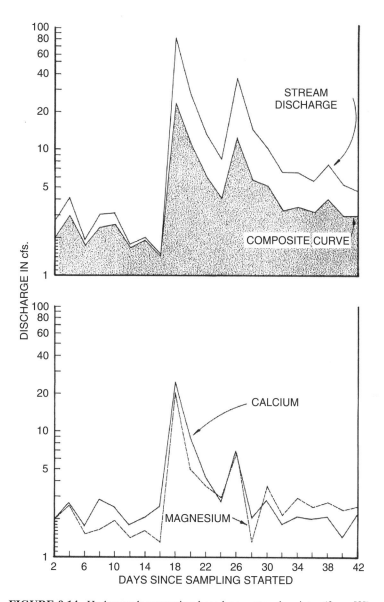

FIGURE 9.14. Hydrograph separation based on water chemistry (from [8]).

The chemistry of the groundwater $\left(C_{gw}\right)$ was taken as the value of C_{tr} at the lowest recorded flow over the period of analysis (assumed a constant during the flow wave event). The rationale for this assumption is that the water during lowest stream flow is baseflow, which is attributed to groundwater. At this point we have measurements or estimates of C_{tr}, C_{dr}, C_{gw}, and Q_{tr}, which allows us to solve for Q_{tr}. The calculated values of the groundwater component of the hydrograph for the Fraser Brook basin in Nova Scotia as determined using calcium and magnesium concentrations are shown in the bottom panel of Figure 9.14. The composite hydrograph shown in the upper panel is determined using bicarbonate, calcium magnesium, and sodium.

In viewing Figures 9.14 and 9.13, it is important to recognize that a logarithmic representation of the discharge is presented. The groundwater component accounted for only 42% of total flow at the peak of the stage. The analysis of Fraser Brook basin demonstrates that, with an increase in total runoff, there is a significant increase in groundwater discharge. This is contrary to what we stated to be the case earlier in our discussion of bank storage effects. However, in the case of Fraser Brook, there is a poorly developed flood plain. The conclusion to be drawn is that, in basins with little potential for bank storage, groundwater infiltration to streams increases significantly during flood events.

9.3 CHEMICAL INTERACTION

Surface-water quality can be influenced by the quality of groundwater discharge and the quality of surface water can impact that of adjacent groundwater. An example is the classic problem of irrigation return flow. In this situation water extracted from wells or introduced through ditches is used for irrigation (see Figure 9.15). A portion of the water applied to the soil surface evaporates, a portion is used in plant transpiration, and the remainder enters the groundwater. Through the evaporation and transpiration processes, which *remove only water and not solutes*, the infiltrate becomes more concentrated with soluble salts. The result is groundwater with varying degrees of salinity as shown in Figure 9.15. The saline groundwater makes its way to the river, often discharging water to the river that is more concentrated in salt than the resident surface water. The result is an increase in salinity in the stream water as it passes through irrigated areas.

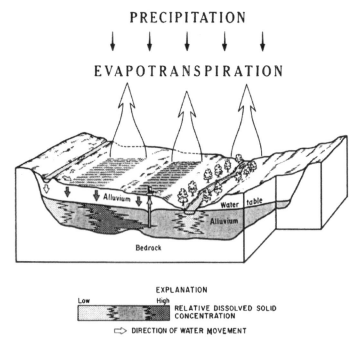

FIGURE 9.15. Diagrammatic representation of the evolution of water used in irrigation in an arid climate (from [9]).

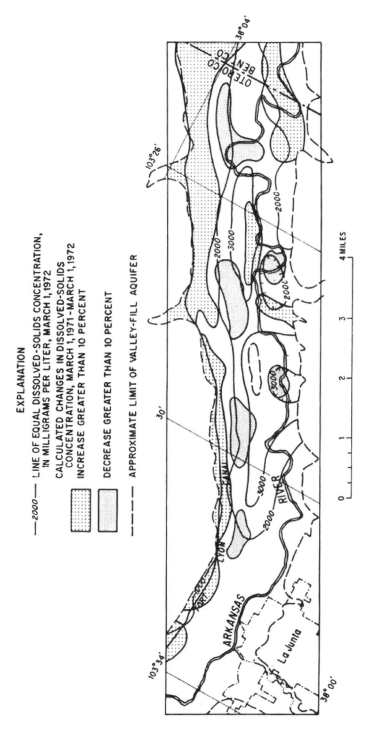

FIGURE 9.16. Total dissolved solids concentration in the Arkansas River aquifer March 1, 1971 (from [9]). Shaded areas indicate change in TDS over the period from March 1, 1971 to March 1, 1972.

As an example of the water quality changes that can occur in a system such as illustrated in Figure 9.15, consider the 11 mile stretch of the Arkansas River Valley in southeastern Colorado between La Junta and the Bent–Otero county line as described in Konikow and Bredehoeft [9]. The following discussion draws heavily from their report.

The area of interest and the concentration of dissolved solids found at the beginning of their investigation is shown in Figure 9.16. Features of particular interest in this figure are (1) the concentration contours of dissolved solids in mg/L; (2) the Arkansas River that flows west to east; (3) the Fort Lyon Canal located to the north of the Arkansas River and flowing west to east; (4) the approximate limits of the valley-fill aquifer denoted by the dashed lines; and (5) the municipality of La Junta and the Bent–Otero county line, which approximately delineate the western and eastern boundaries, respectively, of the study area.

As a point of reference regarding the reported concentrations, it is worthy of note that values greater than 500 mg/L are generally not recommended for drinking water. In addition, from an irrigation perspective, Peterson [10] reported that "TDS levels below 700 mg/L are considered safe; TDS between 700 mg/L and 1750 mg/L are considered possibly safe, while levels above these levels are considered hazardous to any crop."

Also illustrated in Figure 9.16 is the change in TDS concentration over the one year period from March 1, 1971 to March 1, 1972.

Specific conductance and approximate TDS values in the Arkansas River near the upstream inlet (Fort Lyon Canal) and downstream outlet (Bent–Otero county line) are shown in Figure 9.17. During the period recorded, the TDS concentrations were consistently higher at the inlet of the reach of interest than at the outlet and were near the upper limit of acceptability for irrigation. A trend toward an increasing difference as one moves through time is also suggested by this figure. Let us now investigate the reason for the high concentrations in TDS, its distribution, and its change over time.

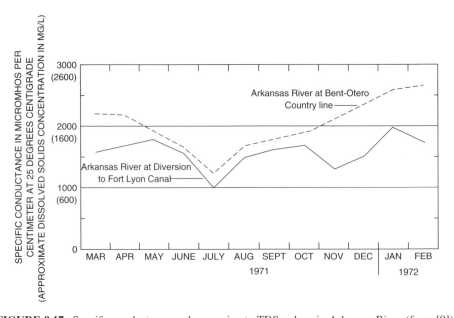

FIGURE 9.17. Specific conductance and approximate TDS values in Arkansas River (from [9]).

The agricultural area in Figure 9.16, which is most of the valley-fill aquifer outside the municipality of La Junta, receives natural precipitation, surface-water irrigation, and groundwater irrigation. The quality of the water from these three sources is different. Precipitation is assumed to contain no dissolved solids. Irrigation water is obtained from the Lyon Canal for application to more than 50% of the study area and its concentration of TDS is thought to be approximately that of the river at the point of diversion from the river into the canal that is located upstream of the reach of interest.

The irrigation water is normally applied to a field through ditches and furrows. In general, the surface-water irrigation was lowest in the western part of the study area. Water applied from the irrigation wells is assumed to be applied with concentrations consistent with what is found in the aquifer at the time of pumping. It is reasonable to assume that the water applied reflects the average concentration of the three sources: precipitation, Lyon Canal water, and well water.

The question now arises as to why the groundwater (and surface water) contains such high concentrations of TDS. First, consider the fact that the total applied water is either (1) evaporated, (2) consumed by crops, (3) stored in the soil, (4) recharged to the groundwater, or (5) transferred as runoff to the river. To obtain an estimate of the amount recharged to groundwater, equations are available that have the following properties [9];

1. The ratio of an increment of recharge to an increment of applied water equals 1 when the total applied water exceeds the potential evapotranspiration.
2. This same ratio is less than 1 when the total applied water is less than the potential evapotranspiration.
3. When the total applied water is less than the potential evapotranspiration, recharge increases as total application approaches potential evapotranspiration.
4. It is assumed that the recharge rate does not exceed the infiltration capacity of the soil.

The resulting calculations determined that the recharge within the study reach was 32% of the total applied during the one-year study period. The monthly computed values are presented in Figure 9.18.

A key observation in the pursuit of our goal of understanding the water chemistry is that the dissolved solids concentration of the recharge water can be calculated by assuming that the *total mass of dissolved solids in the recharge water is the same as that in the total applied water. Thus there is an increase in concentration in the recharge water that is proportional to the decrease in volume of the applied water due to evapotranspiration. Because the recharged water is higher in concentration than the resident groundwater, the TDS concentration in the groundwater increases.*

Since the groundwater discharges, in large part, to the river, the river concentration will increase as the groundwater concentration increases. However, in the area of study there are two other sources of TDS contributing to the TDS in the river: (1) the Anderson Arroyo and the Crooked Arroyo and (2) effluent from the municipal treatment plant at La Junta (TDS of 5200 mg/L). It is estimated that about one-half of the increase in TDS observed along the study reach is attributable to these other sources.

In summary, the observed increases in groundwater concentrations of TDS, and concomitant river concentrations, are due to upstream sources and to a concentration of TDS due to evapotranspiration. It is worth noting that, in some areas of the world where irrigation has been practiced for very long times, the accumulation of salts in the soil

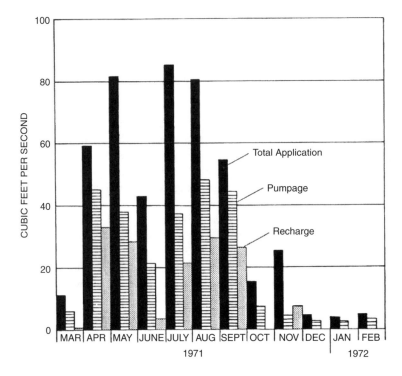

FIGURE 9.18. Monthly total application, groundwater pumpage, and calculated recharge to the aquifer within the study reach (from [9]).

due to increases in soil water TDS threaten the agricultural potential of the land. Julian Martinez Beltran, with the Food and Agriculture Organization (FAO) in Italy, said "soil salinity limits crop production and consequently has negative effects on food security worldwide. In 2002, FAO estimated that about 20–30 million hectares of irrigated land were seriously damaged by the build-up of salts and 0.25–0.50 million hectares were to be lost from production every year as a result of salt build-up" [11].

While agricultural activities may impact the quality of surface water, hazardous waste can also play a role. An example of such an occurrence is provided in Figure 9.19. Groundwater, contaminated by materials from a waste site, forms a plume that extends to and discharges into a river. Remediation implemented at this site has largely ameliorated the problem. In this instance, as is often the case, the volume of water in the river is so great that groundwater contaminants entering the river are diluted such that the concentrations are found to be at or below target levels in the river.

Looking at the contamination problem from a different perspective, an example of the impact of surface-water quality on municipal well-water quality is presented via a study conducted by Duncan et al. [12] and presented in Winter et al. [1]. The following description presented in Winter et al. [1] summarizes the situation.

In a second study, atrazine was detected in ground water in the alluvial aquifer along the Platte River near Lincoln, Nebraska. Atrazine is not applied in the vicinity of the well field, so it was suspected that ground-water withdrawals at the well field caused contaminated river water to move into the aquifer. To define the source of the atrazine, water samples

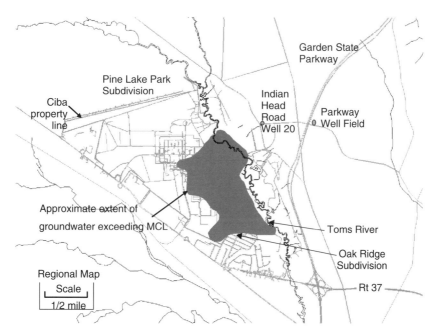

FIGURE 9.19. Contamination moves from groundwater to surface water at this site.

were collected from monitoring wells located at different distances from the river near the well field. The pattern of concentrations of atrazine in the ground water indicated that peak concentrations of the herbicide showed up sooner in wells close to the river compared to wells farther away [Figure 9.20]. Peak concentrations of atrazine in ground water were much higher and more distinct during periods of large ground-water withdrawals (July and August) than during periods of much smaller withdrawals (May to early June).

In this unusually well-documented example, it is demonstrated via the sampling that the source of the contaminant was the river. Thus one must conclude that well fields pumping large quantities of water, derived in significant part from infiltration from surface-water bodies, may be susceptible to surface-water contamination.

9.4 SUMMARY

In this chapter we have addressed several aspects of the interaction of groundwater and surface water. We initially focused on the fluid dynamics near the streams, especially streams where there are significant flood plains and therefore bank storage effects are important. The interesting question of the amount of groundwater entering a stream during a flood event was also considered with a particular orientation toward the use of water-quality data for hydrograph separation. Next, the phenomenon of surface-water–groundwater interaction was considered with particular emphasis on the role of agricultural practices in increasing total dissolved solids concentrations in both groundwater and surface water. Finally, the impact of surface-water contamination on groundwater quality was considered.

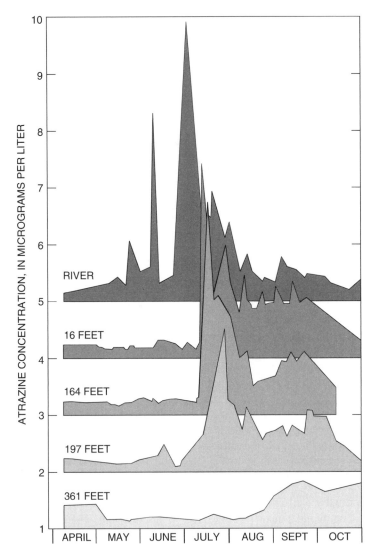

FIGURE 9.20. Atrazine detected in groundwater in the alluvial aquifer along the Platte River near Lincoln, Nebraska (from [1]).

9.5 PROBLEMS

9.1. A gaining stream is characterized by water-table contours that point upstream (see Figure 9.1). Explain the hydrodynamic reasons for this contour geometry.

9.2. When a stream is separated from the underlying water table—that is, there is a partially saturated zone beneath the stream—what is the influence on the stream leakage of lowering the water table?

9.3. Observe that in Figure 9.9 the maximum head elevation contour in the aquifer is located upstream and significantly distant from the river. Compare this to the behavior downstream and explain the observed topology.

9.4. Consider the hydrograph separation shown in Figure 9.13. Explain under what circumstances the groundwater component to the hydrograph would be negative; that is, water is actually lost from the stream and gained by the aquifer.

9.5. Explain what physical–chemical process is taking place that results in increased salt concentration in irrigated soils.

9.6. In Figure 9.19 the contaminant plume has not moved beyond the flood plain to the west. Describe the circumstances that would be required to have the contaminant plume reach and enter the production wells at Indian Head Road (well 20).

9.7. Assume that an aquifer is located adjacent to and in contact with a surface-water body that is influenced by tidal action. Assume further that a groundwater contaminant plume has reached and is discharging into this surface-water body. Describe the effect the tidal action will have on the concentration of the water in the aquifer just before entering the surface-water body.

BIBLIOGRAPHY

[1] T. C. Winter, J. W. Harvey, O. L. Franke, and W. M. Alley, *Ground Water and Surface Water: A Single Resource*, U.S. Geological Survey Circular 1139, p. 79, 1998.

[2] C. R. Hursh, Appendix B. Report of sub-committee on subsurface flow, *Trans. Am. Geophys. Union*, **25**: 743, 1944.

[3] G. F. Pinder, J. D. Bredehoeft, and H. H. Cooper Jr., Determination of aquifer diffusivity from aquifer response to fluctuations in river stage, *Water Resour. Res.* **5**(4):850, 1969.

[4] G. F. Pinder and S. P. Sauer, Numerical simulation of flood wave modification due to bank storage effects, *Water Resour. Res.* **7**(1):63, 1971.

[5] D. D. Franz and C. S. Melching, *Full Equations (FEQ) Model for the Solution of the Full, Dynamic Equations of Motion for One-Dimensional Unsteady Flow in Open Channels and Through Control Structures,* U.S. Geological Survey Water-Resources Investigations Report 96-4240, http://il.water.usgs.gov/proj/feq/feqdoc/contents_1.html.

[6] J. D. Bredehoeft and R. A. Young, Conjunctive use of groundwater and surface water for irrigated agriculture: risk aversion, *Water Resour. Res.* **19**(5):1111, 1983.

[7] V. T. Chow (ed.), *Handbook of Applied Hydrology*, McGraw-Hill, New York, 1964, p. 14-1.

[8] G. F. Pinder and J. F. Jones, Determination of the ground-water component of peak discharge from the chemistry of total runoff, *Water Resour. Res.* **5**(2):438, 1969.

[9] L. F. Konikow and J. D. Bredehoeft, Modeling flow and chemical quality changes in an irrigated stream-aquifer system, *Water Resour, Res.* **10**(3):546, 1974.

[10] H. G. Peterson, *Irrigation and Salinity,* WateResearch Corp. and Agriculture and Agri-Food Canada–Prairie Farm Rehabilitation Administration, 1999.

[11] Yahoo Finance, "Increased Salt in Soil Poses Worldwide Threat to Society," April 26, 2005. http://www.waterresources.ucr.edu/news_events/intlsf_meeting/Increased_Salt_in_Soil.html.

[12] D. Duncan, D. T. Pederson, T. R. Shepherd, and J. D. Carr, Atrazine used as a tracer of induced recharge, *Ground Water Monitoring Rev.* **11**(4):144, 1991.

CHAPTER 10

REMEDIATION

In this chapter we consider methodologies for the remediation of contaminated soil and groundwater. Remediation, as used in this discussion, refers to the reduction in contaminant levels to below target values established by regulatory agencies, generally through human intervention.

Remediation strategies can be subdivided into two categories. In one the contaminants are removed from the subsurface and treated in facilities designed for this purpose. In the other the contaminants are treated where they occur, either in the soil or in the groundwater. The latter form of treatment is referred to as in situ treatment.

10.1 PUMP-AND-TREAT

10.1.1 Inward-Gradient Approach

In the classical approach to groundwater contaminant remediation, contaminated groundwater is pumped from the subsurface and then transported to treatment facilities, where the contaminants are removed using standard water-quality engineering methodologies. While each pump-and-treat design is unique and dependent on the physical and chemical attributes of the contaminated site, all involve pumping contaminated water and disposal of the contaminant-free effluent.

Pump-and-treat designs seek to either contain or reclaim a contaminated aquifer. In containment, the goal is to preclude expansion of the contaminant plume. A sufficient condition to realize this goal is to have the groundwater move inward toward the plume interior everywhere along the plume perimeter. The resulting strategy is called the inward-gradient approach.

To achieve an effective inward-gradient design, wells are placed in the interior of the plume and pumped. If properly designed, the pumping rates will create a cone of

Subsurface Hydrology   By George F. Pinder and Michael A. Celia
Copyright © 2006 John Wiley & Sons, Inc.

depression around each well and collectively for all the wells; that is, there will be a large cone of depression within which the plume resides and therefore will be contained.

The cone will contain the locations of the pumping wells. As long as the wells are active, the groundwater head gradient will be toward the plume interior along its perimeter. Since groundwater flows, for all intents and purposes, in the direction of the head gradient, clean groundwater is flowing into the plume through the perimeter of the cone of depression. Thus the plume cannot expand but for the effects of molecular diffusion, which are of minor importance in this physical situation.

A variant on the above described strategy is to use *injection wells* (or *recharge basins*) as well as *withdrawal wells*. Clean water is injected into wells, usually external to the plume, to hydrodynamically push water external to the plume inward toward and across the original plume perimeter. One advantage to this strategy is that the inward gradient at the plume boundary is increased due to the superposition of the cone of depression of the withdrawal wells and the cone of impression of the injection wells (see Section 6.2.8). If carefully and cleverly designed, the resulting system could reduce overall pumping costs.

The second advantage to the use of injection wells is that it provides a vehicle for disposing of the pumped water after treatment. Water from the treatment plant can, in concept, be reinjected into the subsurface, thereby minimizing the amount of water being wasted through nonbeneficial use discharge strategies.

As an example of the use of the inward-gradient approach, we consider a site in Toms River, New Jersey. An air photograph of the site is provided in Figure 10.1. The area of interest is the open region to the west of the Toms River bounded by the heavy line.

The simulated groundwater flow pattern prior to implementation of inward-gradient control pumping is shown in Figure 10.2 for the shallow aquifer. The shallow aquifer is the one of primary concern. The natural flow pattern is from locations of high topographic

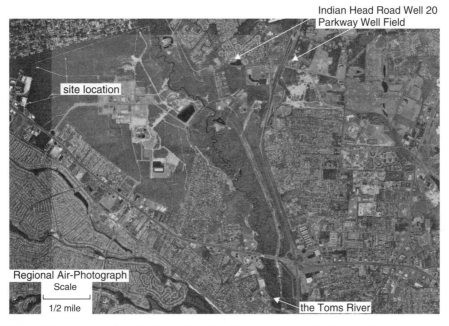

FIGURE 10.1. Air photo of the Toms River site. The open area with the bounding line is the site location.

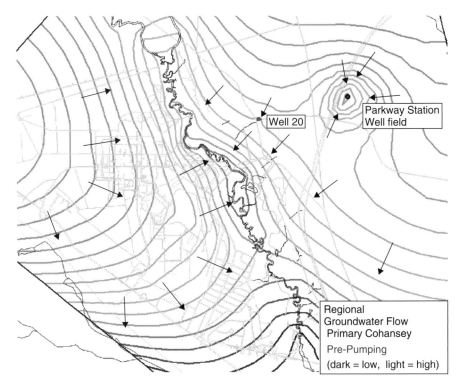

FIGURE 10.2. Groundwater head contours and approximate direction of flow at Toms River site.

elevation toward and into the Toms River. On the east side of the Toms River, the Parkway Station well and, to a lesser extent, well 20 generate cones of depression.

The flow pattern is markedly changed with the implementation of the inward-gradient remediation strategy (Figure 10.3). Groundwater that discharged to the Toms River earlier now is captured, along with any dissolved constituents.

The shaded area in Figure 10.4 is the area of concern since herein the groundwater exhibits contaminant concentrations above the target levels (*maximum contaminant levels*, MCLs). The zone of capture generated by the remedial pumping strategy is shown by the black line and associated arrows. One observes that the majority of the area of concern is captured by the proposed strategy.

To the north of the site, the northeast recharge basin recharges treated groundwater to the aquifer. The impact of the recharge area is an enhanced inward gradient to the west and south of this location. In addition to facilitating the capture of the plume, the recharge area is a convenient location to dispose of the treated water. As is inevitable, the pumping strategy also captures clean water entering the system from the northwest.

10.1.2 Risk-Based Approach

While the inward-gradient strategy is effective in preventing expansion of the contaminant plume, it does not directly address groundwater reclamation. Discharging wells used in an inward-gradient design remove contaminants, but the wells are not located, or their discharge rates calculated, to specifically address contaminant removal. An alternative approach based on achieving a priori specified concentration goals at selected locations

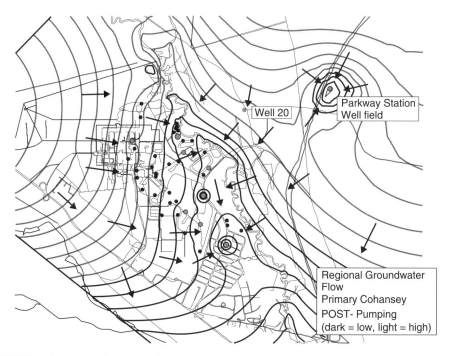

FIGURE 10.3. Groundwater heads and flow pattern after inward-gradient control pumping is initiated.

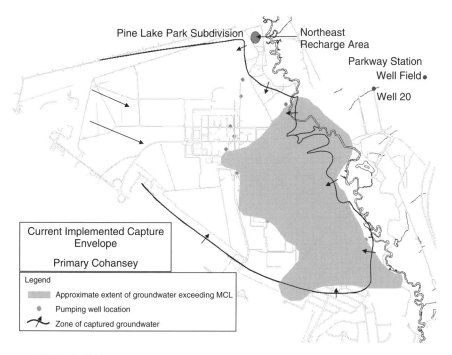

FIGURE 10.4. Zone of capture generated by inward-gradient control pumping.

is more appropriate when water-quality targets are to be addressed. Designs based on water-quality objectives are called *risk-based designs*.

The risk-based design seeks to employ pump-and-treat technology to reach, at the end of a design period, prespecified concentration targets at prespecified locations. The concentration targets are normally the government mandated water-quality standards. The locations where these targets are to be met may coincide with existing pumping or monitoring wells or may be identified with especially sensitive areas of concern where no accessible wells exist.

The goal is to design a pump-and-treat strategy that will achieve prespecified concentration-based goals at minimum cost. A combination of pumping and recharge wells can be employed to achieve this goal. To assure that the contaminant is removed, rather than allowed to move off-site, intermediate concentration objectives should be incorporated into the pump-and-treat design. While the final goal may be to achieve the target concentration values at selected locations at a future time, the satisfaction of concentration objectives throughout the duration of the project will assure that the plume does not escape the monitoring well network prior to end of the designated design period.

Consider, as an example, the problem formulation presented in Figure 10.5. The objective is to achieve, at least cost, concentration values no greater than 0.05 ppb at the locations denoted by the squares and identified as the "o" wells. The locations that can be used to install pumping wells to achieve this goal at the end of 20 years are shown as circles and designated as the "w" wells. Based on projections made using a flow and transport model of the aquifer, the concentration goals can be met using the pumping strategy presented in Table 10.1.

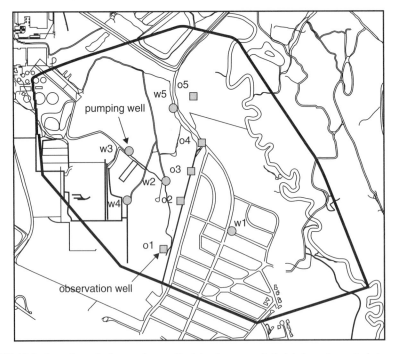

FIGURE 10.5. Location of observation wells (o) where $c = 0.5$ risk-based constraints are to be met, and potential pumping wells (w) at the Toms River site.

**TABLE 10.1. Optimal Pumping Rates
for Risk-Based Remedial Design,
Toms River, New Jersey**

Well	Pumping Rate (L/min)
w1	610
w2	284
w3	284
w4	118
w5	0

10.2 SOIL VAPOR EXTRACTION (SVE)

A variant on pump-and-treat is *soil vapor extraction* (SVE), also called *soil venting* and *vacuum extraction*. The basic idea is to circulate air through the *vadose zone* containing volatile contaminants, especially petroleum-based compounds. Air is induced to enter the subsurface via vacuum pumps connected to either vertical or horizontal wells. The most effective strategy is to locate the extraction wells near the water table. In the event the water table is elevated due to the pressure reduction caused by the vacuum pumps, water-level depression pumps may be required to retain the necessary vadose zone depth.

As the air moves past the contaminants resident in the vadose zone, either in dissolved or separate phase form, the contaminants are transferred from the liquid to the gas. The gas containing the contaminants is transported to the surface where it is treated. Treatment usually involves the use of activated carbon filtration (we were introduced to activated carbon in Section 1.6.2). The treated air is typically released to the atmosphere.

The concentrations of many volatile contaminants in the soil gas are in equilibrium, or near equilibrium, with the contaminants in the liquids. Clean air entering the subsurface generates a concentration gradient across the fluid–air interface, which in turn induces volatile contaminant flow from the liquid to air phase. The transfer of contaminant i from one fluid phase to another is described, assuming local chemical equilibrium, by

$$c_{ia} = K_h c_{iw} \tag{10.1}$$

and

$$c_{io} = K_o c_{iw}, \tag{10.2}$$

where c_{ia}, c_{iw}, and c_{io} are the concentrations of species i in the air, water, and oil phases, respectively, defined in units of g/cm^3. The partition coefficient K_h is often defined as H, the Henry's law constant, and is unique to each constituent. The coefficient K_o is the oil–water partition coefficient and is a function of the composition of the oil phase. The oil–water partition coefficient is defined as

$$K_o = \frac{c_{io}}{S_i X_i},$$

where S_i is the effective solubility of component i in the water phase. The coefficient X_i is the mole fraction of compound i in the oil phase. Equation (10.2) is derived from the behavior of ideal liquids as described by *Raoult's law* [1].

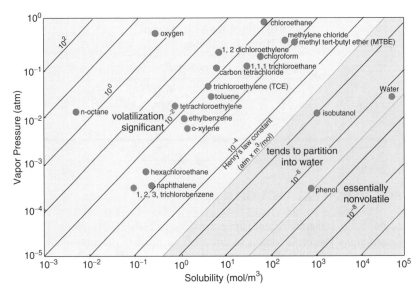

FIGURE 10.6. Graphical representation of the relationship between Henry's law, solubility, and vapor pressure for selected compounds (from [2] and [3]).

In Figure 10.6 are provided the *water solubility*, *vapor pressure*, and *Henry's law* constants (diagonal lines) for compounds of practical interest in the groundwater remediation of volatile compounds. The chart is divided into three areas signifying different degrees of volatilization for the compounds selected. Compounds to the lower right tend to be nonvolatile while those in the upper left are volatile.

Because the SVE approach depends on the ability of air to move through the contaminated soil, factors that reduce the permeability to air are going to be detrimental to the effectiveness of this approach. Since the presence of liquid water reduces the relative permeability to air, and therefore impedes its flow, high water saturations result in less efficient SVE projects. Soil characteristics, such as variations in intrinsic permeability, also play a role. Areas of low permeability and heterogeneous media are found to inhibit air flow and therefore reduce contaminant removal.

A side benefit of SVE is the enhancement of the biodegradation of offending compounds (see Section 10.6). The circulation of the air enhances biological activity and therefore bioremediation.

The SVE process is shown diagrammatically in Figure 10.7. Air moves laterally toward the vapor extraction well, whereupon it is removed and sent to the surface. Pipes carry the contaminated vapor to a treatment facility from which the treated gases are exhausted to the atmosphere.

When the air reaches the surface, it contains liquid water drawn into the vapor stream from the soil. A separator is used to remove this water, which, if excessive, may require treatment.

The air is often treated through oxidation of the volatile organic compounds. Oxidation causes the target compounds to break up into innocuous compounds. The oxidation process uses heat to break apart the target compounds and the new innocuous compound forms spontaneously. The process can also employ a catalytic material that facilitates breakdown of the target compound at a lower temperature.

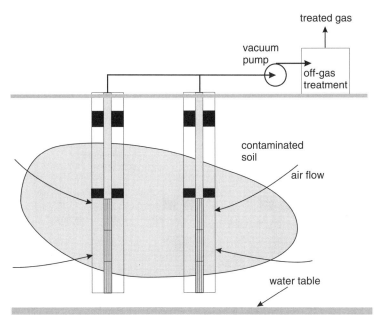

FIGURE 10.7. Diagrammatic representation of the soil vapor extraction process. The shaded area is the contaminated soil.

An alternative treatment strategy is the use of activated carbon similar to that used in the carbon filtration of contaminated groundwater (recall we used this material in an experiment in Section 1.6.2).

To assure that there is adequate lateral flow through the vadose zone, the land surface in the neighborhood of the contaminated soil may be covered to impede the vertical flow of air. When dictated by the hydrogeology of the site, air injection wells can be used to increase the air flow through the vadose zone and thereby improve the removal rate of the contaminant vapor.

10.3 AIR SPARING

As described previously, the soil vapor extraction process cannot be used in the saturated zone. However, a variant on this process called air sparing can be used in certain hydrogeological environments. In this process air is bubbled into the saturated zone as shown in Figure 10.8. The air enters the subsurface through an air-injection well. It moves as bubbles through the saturated soil as a separate air phase. At the air–water interface defining the bubble, there is a transfer of volatile contaminants from the water to the air as defined by Henry's law (see Eq. (10.1) and Figure 10.6). The air bubbles have a velocity component vertically upward due to the air buoyancy. As a result the majority of the air eventually reaches and crosses the water table. Once in the vadose zone, the contaminant-containing air is drawn to the air-extraction wells, whereupon it is transferred to the surface for treatment.

As in the case of SVE, air sparing requires that significant volumes of air move through the soil. Soil with higher permeability, such as coarse sand and gravel, are therefore more effectively treated using the air sparing approach.

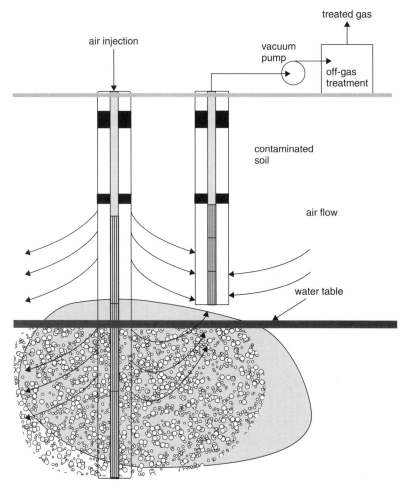

FIGURE 10.8. Diagrammatic representation of the air sparging strategy. Air is bubbled into the saturated zone, whereupon it encounters the contaminated saturated soil. Volatile components move from the water to the gas, which, upon reaching the water table, is drawn into the extraction wells.

10.4 SOME BASIC CHEMISTRY

As a preamble to further discussion of in situ remediation techniques, we will review some elementary concepts in chemistry. We begin with the concept of oxidation–reduction reactions.

10.4.1 Oxidation–Reduction Reactions

Oxidation–reduction (redox) reactions are those involving changes of oxidation states of reactants and they play a central role in many of the reactions occurring in natural waters and water and wastewater treatment processes. The behavior of compounds containing carbon, nitrogen, sulfur, iron, and manganese in natural water (and in treatment processes) is largely influenced by redox reactions.

All atoms are electrically neutral, although they are themselves made up of positive and negative subatomic particles. The oxidation state of an atom or ion is the sum of the negative and positive charges. Thus, since the atom is neutral, its oxidation state is zero. The loss of electrons from an atom produces a positive (increased) oxidation state and concomitantly the gain of electrons by an atom reduces the oxidation state to a negative value.

The representative elements of the periodic table can be divided into two classes, *metals* and *nonmetals*. Metals are found on the left side of the periodic table and are separated from the nonmetals on the right-hand side by a series of elements called *metalloids* (boron, silicon, germanium, arsenic, antimony, tellurium, and astatine). Metal atoms are characterized by their tendency to be oxidized through the loss of one or two electrons to the more electronegative, nonmetallic elements on the far right side of the periodic table. The resulting positively charged ion is called the *cation*, which now has an increased oxidation state. The number of electrons that are lost depends on the group number of the metal in the periodic table. A typical *oxidation reaction* is given by

$$\underbrace{Ca}_{\text{calcium atom}} \rightarrow \underbrace{Ca^{2+}}_{\text{calcium cation}} + \underbrace{2e^-}_{\text{number of electrons lost}} \; ,$$

wherein the oxidation state of calcium has increased from zero to two.

The nonmetal ion is now negatively charged and called an *anion*. The process of acquiring electrons is called *reduction*. The number of electrons an atom acquires is again dependent on its location in the periodic table. A typical reduction reaction is given by

$$\underbrace{O}_{\text{oxygen atom}} + \underbrace{2e^-}_{\text{electrons gained}} \rightarrow \underbrace{O^{2-}}_{\text{oxygen anion}} \; .$$

The exchange of electrons to form anions and cations always occurs together. Thus an oxidation reaction requires also a reduction reaction—therefore the nomenclature *oxidation–reduction reactions*. A simple example of an oxidation–reduction reaction is the formation of sodium chloride, the white table salt. When an electron from the soft metal solid sodium is transferred to the green chlorine gas, a violent reaction occurs that creates an entirely new white crystalline substance called sodium chloride. The powerful electrostatic bond that develops between these ions causes them to rearrange themselves into a *lattice*, which is called an ionic compound. Since sodium is from group 1 it will lose one electron and, since chlorine is from group 17, it will gain an electron. The equation for this reaction would be

$$\underbrace{Na^{1+}}_{\text{sodium ion}} + \underbrace{Cl^{1-}}_{\text{chloride ion}} \rightarrow \underbrace{NaCl}_{\text{sodium chloride}} .$$

Between the metals on the left of the periodic table and the nonmetals on the right are the *transition metals*. Unlike the representative elements, the transition metals can have multiple oxidation states; that is, they can form cations with different positive charges. A typical example, and one that will be of considerable interest to us shortly, is iron. Iron can form a cation with either a $+2$ or $+3$ charge by losing two or three electrons, respectively. The choice of oxidation state will depend on the conditions under which the reaction takes place, such as the amount of nonmetal present and thermodynamic conditions.

TABLE 10.2. Electronegativity Values for Selected Elements

Metallic Elements			Nonmetallic Elements			
Li (1.0)	Be (1.5)		C (2.5)	N (3.0)	O (3.5)	F (4.0)
Na (1.0)	Mg (1.2)	Al (1.5)		P (2.1)	S (2.5)	Cl (3.0)
K (0.9)	Ca (1.0)	Sc (1.3)			Se (2.4)	Br (2.8)

Source: Asato et al. [4].

Determination of the oxidation state of a transition metal requires knowledge of the compound being formed. For example, $FeCl_3$ contains Fe^{3+} iron. How do we know? Because chlorine always forms an anion with oxidation state -1 and there are three chlorine atoms in this compound; then for electrical neutrality, the iron must have a valence state of $+3$.

Not all reactions are of the form presented above. In some instances, the electrons do not actually transfer between atoms. Rather, each kind of atom is attracted to the other in an effort to form a chemical bond. The result is a *redox reaction*. The electron attracting ability of each atom is reported on an electronegativity scale. Fluorine is given the largest electronegativity value and all other atoms are scaled to it. Electronegativity values of selected elements are provided in Table 10.2.

The elements with the larger electronegativity values will appear to be pulling the electrons away from the elements with the smaller electronegativity values. The degree to which this takes place depends on the difference between the two electronegativity values. Note that the electronegativity values of the nonmetallic elements are larger than those of the metallic elements. Thus one could argue that the nonmetallic elements in some sense "gain" the electrons, as a general guideline.

Reactions between metals and nonmetals will usually result in the formation of ionic compounds [4].

Perhaps the simplest form of redox reaction is the *combination reaction* in which two elements combine to form a chemical compound. In this process, one element is always oxidized while the other is necessarily reduced. A simple example is the following in which two free elements combine to form a compound. In this reaction the hydrogen is oxidized and the oxygen is reduced [4]. The oxidation state is shown beneath each reactant.

$$2H_2 + \underset{0}{O_2} \rightarrow 2\underset{+2}{H_2}\underset{-2}{O} .$$

A combination reaction is reversible, that is, the resulting compound can be decomposed into the components from which it was formed. The result is a *decomposition reaction* such as presented below [4]:

$$2\underset{+1}{K}\underset{+5}{Cl}\underset{-6}{O_3} \rightarrow 2\underset{+1}{K}\underset{-1}{Cl} + 3\underset{0}{O_2} .$$

Notice that chlorine ion does not have its characteristic charge. The reason for this is that in compounds containing oxygen the standard rule regarding chlorine does not apply.

In a *single displacement redox reaction*, one element replaces another from a compound. In this process the replacing element is always oxidized and the replaced element is reduced. Consider the following example wherein hydrogen is displaced by metallic iron [4]:

$$2 \underbrace{Fe}_{0} + 6 \underbrace{H}_{+1} \underbrace{Cl}_{-1} \rightarrow 2 \underbrace{Fe}_{+3} \underbrace{Cl_3}_{-3} + 3 \underbrace{H_2}_{0}.$$

10.4.2 Organic Chemistry

Organic chemistry plays a major role in groundwater contamination. Organic compounds are those that contain carbon. Many of the problems currently being addressed in groundwater remediation involve organic compounds. Although organic compounds were studied intensively in the mid-nineteenth century, they were known to exist for centuries. For example, references to the physiological effects of ethyl alcohol, a well known and much appreciated organic compound, can be found in the Old Testament. However, because organic compounds tend to be found in nature as complex mixtures, and methods of separation have been realized only in the last few centuries, the science of organic chemistry is relatively recent.

The term organic chemistry reflects the observation that the early organic compounds were derived from natural sources and living organisms. The term now embraces that subarea of chemistry devoted to the study of carbon-containing compounds.

Two aspects of carbon make it unique. First, carbon can unite with itself indefinitely to form compounds. Second, it virtually always has a valence state of 4. In addition, unlike in the case of inorganic chemical compounds, organic compounds do not utilize either ionic or electrovalent bonding. Rather, in the case of organic compounds, *electrons are shared* between atoms rather than transferred. The resulting bonds are called covalent and are typical of most carbon compounds. In contrast with the ionic bond, where the atoms are held together by electrostatic forces between ions, the covalent bond utilizes electrostatic attractions between the nuclei of both atoms and the bonding electron. Covalent bonds can be modified to exhibit some ionic bond character.

Compounds that utilize covalent bonding have properties that are different from ionic compounds. For example, unlike inorganic salts that have high melting points, organic compounds exhibit low melting points, that is, below $400\,^{\circ}$C. Of considerable importance from a water-quality perspective is the observation that while inorganic salts are readily soluble in water, organic compounds are not water soluble unless they have polar groups (such as hydroxyl [—OH] or carboxyl [—COOH]) attached to them. However, organic compounds can be quite soluble in nonaqueous solvents in contrast to inorganic compounds, which are sparingly soluble in nonaqueous phase liquids.

With a few exceptions (e.g., carbon tetrachloride (CCl_4)) organic compounds are combustible, while inorganic compounds are not. This property can be used to determine whether a compound is organic or inorganic. If the unknown compound burns readily and leaves a residue, it is most likely organic. If the compound burns partially but leaves a residue, it may be the salt of an organic compound. If the compound is completely unaffected by heat, it is probably inorganic.

Because of the complexity and abundance of organic compounds, a complex nomenclature has developed. The nomenclature is still evolving, but certain conventions for the description of the structure of organic compounds have been more or less standardized. Thus a shared electron bond involving a single pair of electrons is shown diagrammatically as a single line and is called a *single bond*, a bond that shares two pairs of electrons is denoted by a double line and is called a *double bond*, and one that shares three pairs of electrons is denoted by three lines and is called a *triple bond*.

An organic compound that exhibits only single bonds between carbon atoms is called *saturated*. A molecule made up of carbon atoms that contains one or more multiple bonds is called *unsaturated* [5]. A *homologous series of compounds,* in the context used here, is distinguished by the fact that each member of the series differs from the next lower member by the increment CH_2. The nomenclature CH_4 denotes methane whose molecular structure is

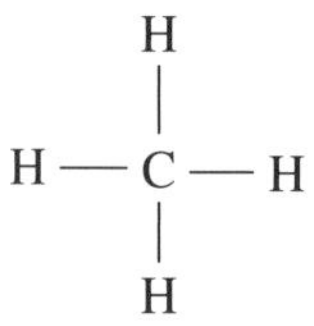

$$
\begin{array}{ccc}
 & H & \\
 & | & \\
H - & C & - H \\
 & | & \\
 & H & \\
\end{array}
$$

Methane

The homologous series beginning with methane can then be written as *methane* CH_4, *ethane* CH_3CH_3 or C_2H_6, and *propane* $CH_3CH_2CH_3$ or C_3H_8.

The structure of carbon atoms is variable. Long chains, both straight and branched, and rings are encountered. In addition, atoms other than carbon (*heteroatoms*) can be covalently bonded into carbon rings. Based on their structure and the existence of heteroatoms, organic molecules can be classified into three main groups:

Compound	Structure
Acyclic	No ring-structural arrangements of atoms
Carbocyclic	One or more rings made up only of carbon atoms
Heterocyclic	One or more rings containing atoms other than carbon

Examples of the three classes are [5]:

$$CH_3 - CH_2 - CH_2 - CH_2 - CH_3$$

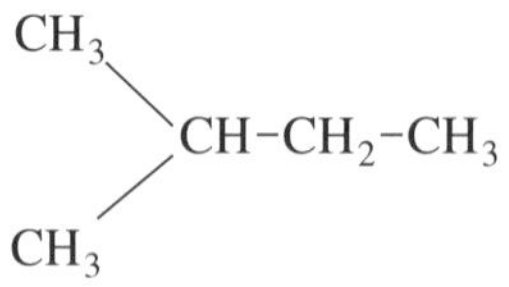

$$
\begin{array}{l}
CH_3 \\
\quad \diagdown \\
\qquad CH - CH_2 - CH_3 \\
\quad \diagup \\
CH_3 \\
\end{array}
$$

Acyclic

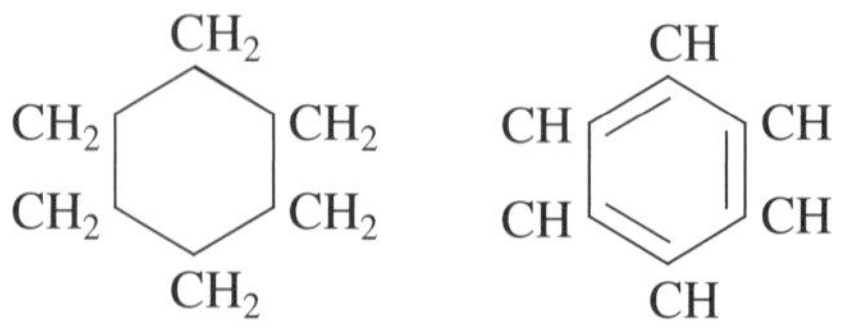

Carbocyclic

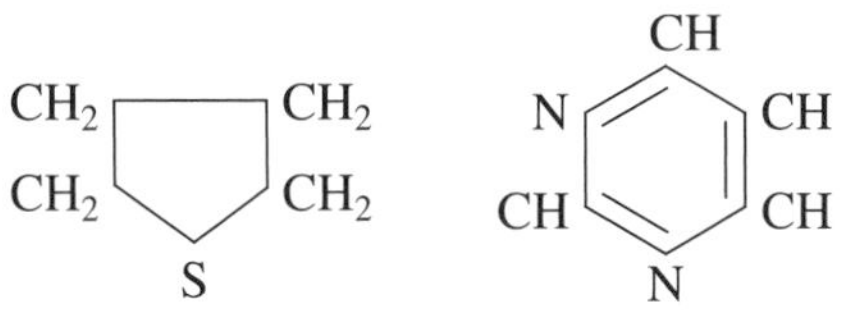

Heterocyclic

A compound made up only of carbon and hydrogen is called a *hydrocarbon*. In concept, any of the hydrogen atoms in the hydrocarbon can be replaced with any atom that can enter into a covalent bond. The following example shows the concept:

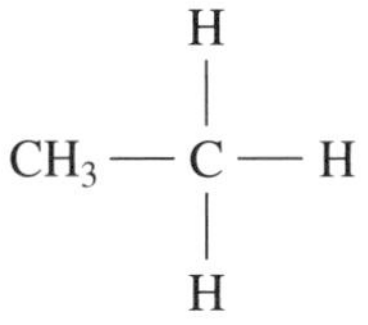

parent compound

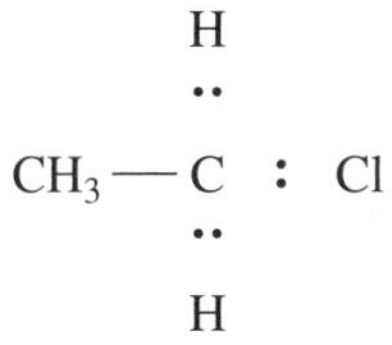

substitution of H by chlorine

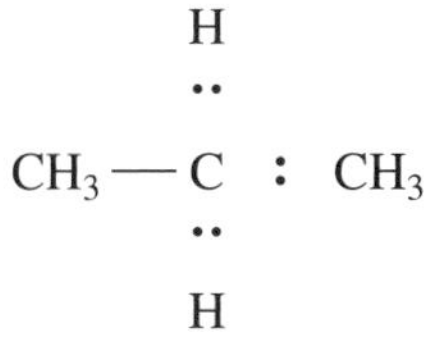

substitution of H by —CH_3

In the above figure the *substituent*[1] *group* of atoms —CH_3 was formed from the molecule methane, CH_4, through the loss of one atom of hydrogen. The double dot notation indicates a pair of electrons not involved in bond formation (lone pair). When a substituent group of atoms is formed through the loss of one or more atoms of hydrogen from a member of any of the three main classes of organic compounds, the resulting group is called a *radical*. Thus —CH_3, having been formed from methane, is the *methyl radical*.

When the substituent group is a single atom or group of atoms other than a radical, it is called a *functional group*. Functional groups change the character of compounds. Since double or triple carbon-to-carbon bonds also influence the properties of compounds in which they are found, they are also considered as functional groups. The common functional groups are found in Table 10.3 [5].

---

[1]A substituent is an atom or group of atoms that replaces another atom or group in a compound.

TABLE 10.3. Common Functional Groups

Name	Structure
Halo (chloro, bromo, etc.)	$-Cl, -Br$
Hydroxyl	$-OH$
Aldehyde	$\begin{matrix} H \\ \vert \\ -C=O \end{matrix}$
Carboxyl	$\begin{matrix} O \\ \parallel \\ -C-OH \end{matrix}$
Ketone	$\diagdown C=O$
Ether	$-O-$
Amino	$-NH_2$
Cyano	$-C\equiv N$
Thiol or mercapto	$-SH$
Sulfonic acid	$-SO_3H$

Source: Elderfield [5].

Acyclic Compounds

We noted earlier that there were three organic compound classifications—*acyclic, carbocyclic*, and *heterocyclic*. In addition, the *aliphatic* compounds consist of the acyclic or cyclic saturated or unsaturated compounds. Rules exist for naming compounds within each of these classifications. While we cannot detail here the strategy associated with the naming conventions, we can give some insight into the origin of the compound names we are likely to encounter in contaminant hydrogeology.

The *acyclic* compounds that contain no multiple bonds are called *alkanes* (or *paraffins*), those containing multiple bonds are called *alkenes* (or *olefins*), and those with triple bonds are denoted as *alkynes* (or *acetylenes*). The most simplistic of the alkanes are methane (one carbon), ethane (two carbons), propane (three carbons), and *butane* (four carbons). The *halogen compounds* with prefixes *chloro, fluoro, bromo*, and *iodo* are used with the parent alkane name to yield, for example, chlorobutane.

The *alcohols* are generated via the presence of the *hydroxyl functional group*, that is, $-OH$. Thus one has, for example, *methyl alcohol*, CH_3OH. *Ethers* are named via the *alkyl* radicals present (named by replacing the suffix *ane* of the molecule by suffix *yl*) plus the ether. Thus one has the compound ethyl methyl ether ($CH_3CH_2-O-CH_3$).

Carbocyclic Compounds

Carbocyclic compounds are subdivided into *aromatic* and *alicyclic*. Aromatic compounds are characterized by an alternating sequence of single and double bonds within a ring structure. Because of the simplicity of the aromatic compounds, their structural representation is abbreviated. Rather than show the carbon and hydrogen atoms at each location, the *carbon positions are numbered* and it is assumed that there are an adequate number of hydrogen atoms present to complete the molecular structure. One of the important environmental compounds is *benzene*, the structural molecule for which follows:

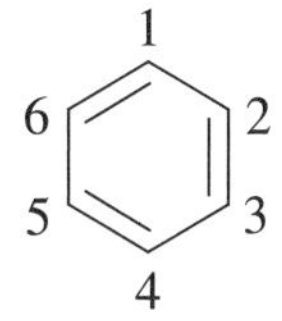

Benzene

The loss of one or more hydrogen atoms from a carbocyclic compound creates a *phenyl radical*. Functional group substituents are denoted by an appropriate prefix. For example, one could replace a hydrogen with a chlorine atom to get *chlorobenzene*:

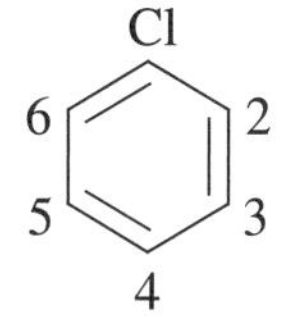

Chlorobenzene

When more than one substitution occurs, that is, more than one carbon atom is replaced, the location of the substitution is denoted as *ortho* when substitution is on adjacent atoms, *meta* when substitution is on carbon atoms once removed, and *para* when the substituted atoms are opposite one another on the ring. Alternatively, the number of the atom can be used, for example, one can refer to *o-dichlorobenzene* or *1,2-dichlorobenzene*:

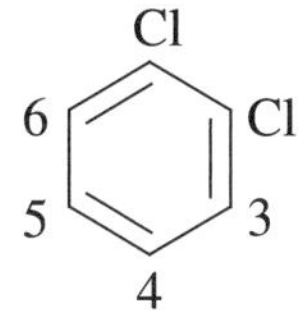

o-Dichlorobenzene or
1,2-Dichlorobenzene

Several members of the benzene series are encountered in an environmental context. Members of the benzene series of compounds are considered as carcinogens, that is, they are implicated in cancer. Three examples are *toluene, ethylbenzene,* and *xylene*. Together these compounds in combination with benzene are called BTEX. In addition to *ortho-xylene, meta-xylene* and *para-xylene* also exist.

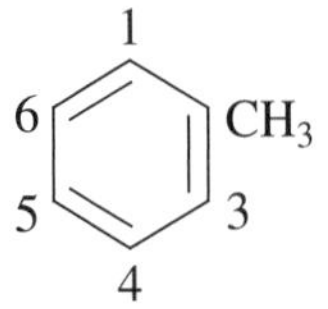

Toluene 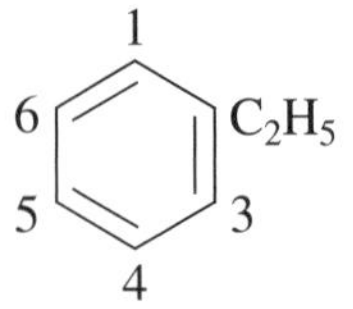   Ethylbenzene 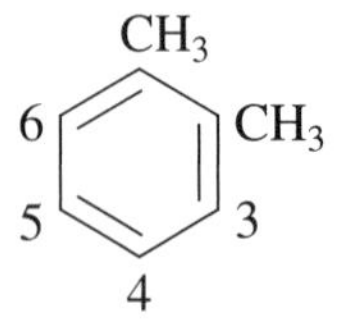   *o*-Xylene

Polyring or *polycyclic aromatic hydrocarbons* (PAHs) are also important from an environmental perspective. A very important example is *benzo[a]pyrene*, which is a potent carcinogen:

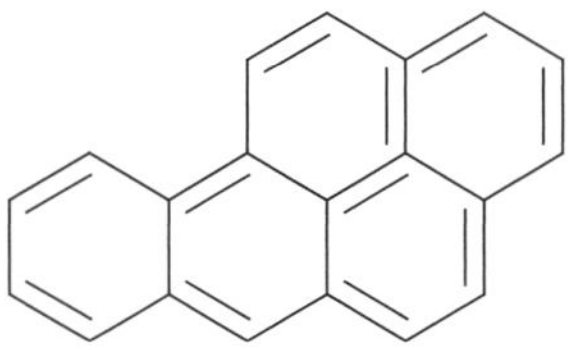

Benzo[*a*]pyrene

Polychlorinated biphenyls (PCBs) are double-ring compounds with the following general structure [6]:

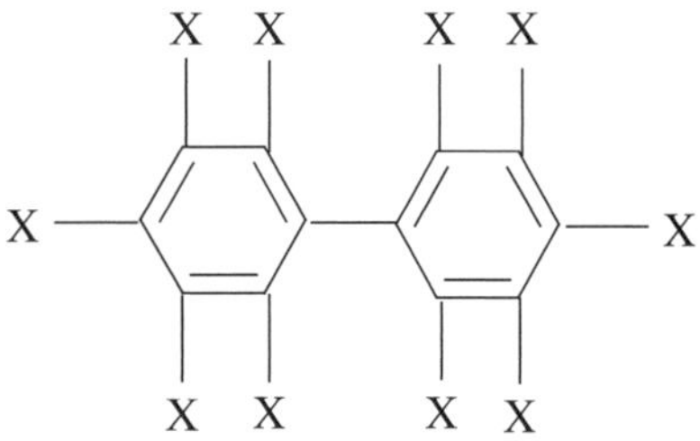

where the X atoms can be either chlorine or hydrogen. The 209 possible PCB compounds (called *cogeners*) are grouped together into *Aroclors*, which each contain PCBs with similar numbers of chlorine atoms. Historically, they were widely used as coolants in transformers and capacitors. Production ceased in the United States in 1977. PCBs are of concern because they have been reported to have been associated with adverse reproductive and developmental effects in humans.

The replacement of a carbon group in benzene by a *hydroxyl group*, that is, —OH, yields *phenol*:

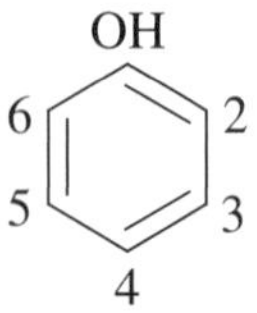

Phenol

Phenol is a derivative of coal tar, which is found in many areas of the United States. It is difficult to treat, but recent advances have shown that it can be destroyed aerobically by bacteria.

Alicyclic carbocyclic compounds are compounds that exhibit properties that are more similar to cyclic compounds than aromatic compounds. Their structure can extend from the simple to the relatively complex. Both single and double bonds can exist, provided they do not occur as an aromatic arrangement. An example of an alicyclic compound is *cyclohexane.*

Heterocyclic Compounds The heterocyclic compounds have one or more carbon atoms replaced by heteroatoms (noncarbon), sometimes in combination with hydrogen atoms. Some are aliphatic in character and others are aromatic. Of the various heterocyclic compounds, among the most important is *purine* and *pyrimidine.* Derivatives of these compounds play several key roles in microbiology and are the bases that form the major components of nucleic acids, which contain the genetic blueprints for all life. The structures of these two molecules are as follows:

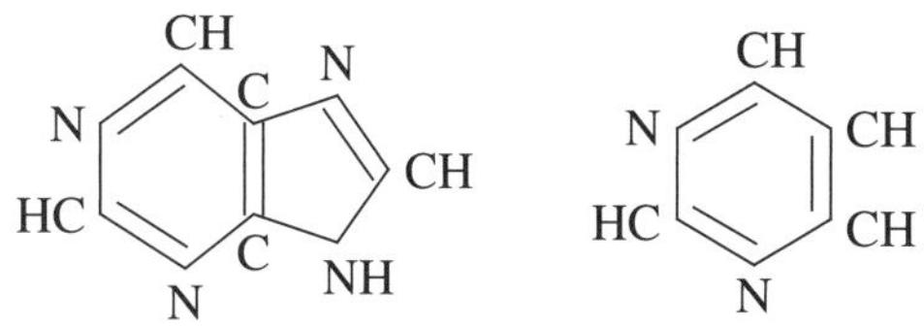

Purine   Pyrimidine

In the following figures, we present the structure of compounds selected primarily because of their notoriety. DDT was a widely used insecticide until banned in the United States in 1972. However, it is still popular outside the United States for the control of malaria. It is a chlorinated compound with alternative structures, the most common of which is

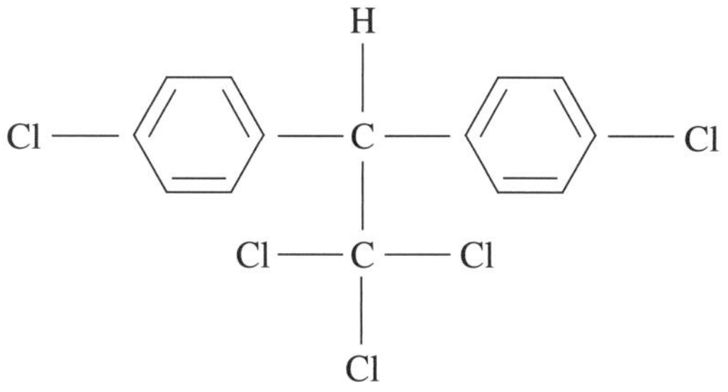

DDT

Dioxin is a general term applied to a group of organic compounds that are very persistent in the environment. It is formed as a by-product in chlorinated waste-product incineration and in the manufacture of chemicals and pesticides. According to the EPA, dioxin is very toxic and there is no safe level of exposure to it. The most toxic form of dioxin is *2,3,7,8-tetrachlorodibenzo-p-dioxin* (TCCD). The form of this molecule is

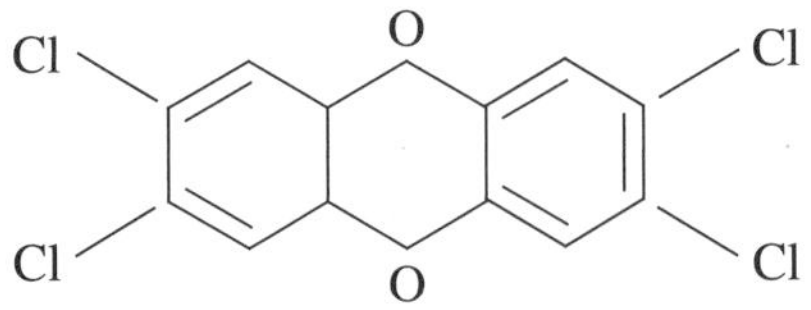

TCDD

Transformation Reactions From an environmental perspective, there are five organic chemical transformations of interest: *photochemical, hydrolysis, oxidation, reduction*, and *biotransformation*. An especially useful presentation of chemical transformations is found in Sawyer et al. [6] and we follow their development in the following discussion.

In photochemical reactions light energy (photons) is absorbed by either the compound of interest, an intermediate compound, or a compound separate and distinct from the compound of interest. When the photon is absorbed by the compound of interest and that compound is then transformed with a concomitant release of energy, the process is called *direct photolysis*. In the event that light is absorbed by an intermediate compound such as Fe^{3+} with the resultant production of oxidants, such as O_3 (ozone), or free radicals such as —OH (hydroxyl), the process is called *oxidation* or *free-radical oxidation*. The result oxidants (oxidizing agents are discussed in Section 10.4.1) are then available to oxidize organic compounds. Such reactions can be rapid. If the photon is absorbed by a nontarget compound such as a humic substance and that resultant excited molecule transmits its energy to the target organic causing it to be transformed, we say that *indirect photolysis* has taken place.

Nucleophilic substitution reactions are those in which nucleophiles (normally electron-rich ions such as OH^- with water as an exception) replace atoms on a molecule without changing the oxidation state of the organic compound. An example of importance is *hydrolysis*, wherein water is the added compound. An example is the hydrolysis of chlorobenzene, that is,

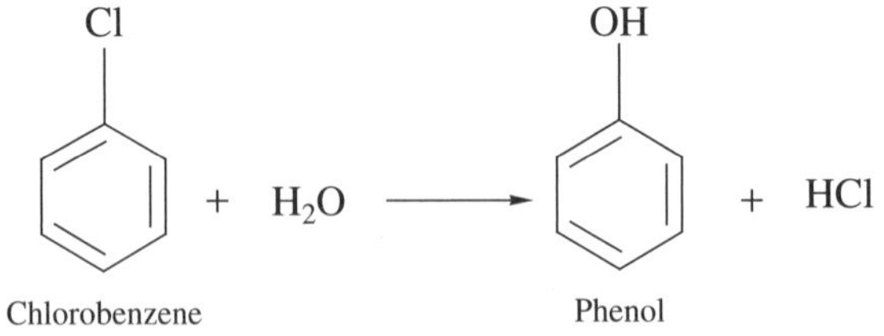

Elimination reactions result in the removal of atoms from adjacent carbon atoms such that a double bond remains between them. When the elimination target is HX (where X is a halogen)[2] the process is called *dehydrohalogenation*. An example is the transformation of trichloroethane to dichloroethene, that is,

---

[2]Halogens are the very reactive, nonmetallic atoms found in group 17 of the periodic table and include F, Br, Cl, and I.

1,1-Trichloroethane   1,1-Dichloroethene

As noted in Section 10.4.1 oxidation is the process whereby the compound loses one or more electrons to an oxidizing agent. The resulting abiotic reactions are relatively slow when compared to those involved in photochemical reactions. The important naturally occurring oxidants are oxygen, iron, and manganese. The role of zero valent iron will be considered shortly. Biologically enhanced oxidation reactions will be considered in Section 10.6 devoted to bioremediation.

Reduction reactions result in molecules gaining one or more electrons from a reducing agent. Highly oxidized organics such as some of the chlorinated organics can be reduced, especially in the absence of oxygen (reducing conditions). Both abiotic and microbially enhanced reduction are common. An example of a reduction reaction is the transformation of tetrachloroethene to trichloroethene. This is an example of reductive dehalogenation and shows the potential of modifying chlorinated contaminants to less noxious compounds. In this instance, the sequential application of reduction would result in ethene as the final product. The transformation of tetrachloroethene to trichloroethene with the resultant release of a chlorine ion is as follows:

Tetrachloroethene   Trichloroethene

10.5 PERMEABLE REACTIVE BARRIERS

Permeable reactive barriers are an in situ methodology for remediation of contaminated groundwater. The barriers are actually not physical barriers to flow but rather they are permeable subsurface walls composed of materials capable of reacting with groundwater contaminants. To enhance the efficiency of the reactive walls, impermeable subsurface walls may be constructed to funnel groundwater to the reactive media. The resulting arrangement is called a *funnel-and-gate design* and is illustrated in Figure 10.9.

The gate contains compounds that will react with the groundwater contaminants. The contaminants will be either reduced to innocuous compounds or immobilized. The target compounds can be either organic or inorganic. To illustrate the concept, we consider as contaminants hexavalent chromium (inorganic compound) and trichloroethylene (organic compound).

Blowes and Ptacek [7] suggested, based on laboratory experiments, that under suitable pH conditions, certain iron-bearing solids could be used in a porous reactive wall to reduce and remove hexavalent chromium Cr^{6+}. The concept involved the use of *zero-valent iron* $Fe^0(ZVI)$ to reduce hexavalent chromium to trivalent chromium.

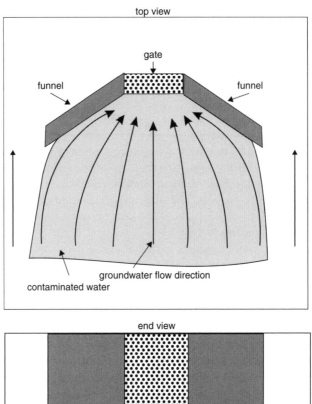

FIGURE 10.9. Diagrammatic representation of a funnel-and-gate design.

Certain compounds involving Cr^{6+} appear to be carcinogenic while insufficient evidence is available to establish the carcinogenicity of Cr^{3+}. It is estimated that Cr^{6+} is on the order of 1000 times more toxic than Cr^{3+}.

In the reduction of Cr(VI) to Cr(III), Fe^0 is oxidized to Fe^{3+}. The reaction is the following:

$$Cr^{6+} + Fe^0 \iff Cr^{3+} + Fe^{3+}$$
$$(1-x)Fe^{3+} + xCr^{3+} + 3H_2O \iff (Cr_xFe_{(1-x)})(OH)_3 + 3H^+$$
$$(1-x)Fe^{3+} + xCr^{3+} + 2H_2O \iff Fe_{(1-x)}Cr_xOOH + 3H^+.$$

The compound *goethite* (FeOOH) and the *chromium-substituted goethite* ($Fe_{(1-x)}$ Cr_xOOH) have been identified as the solid precipitates in the oxidizing reaction. These compounds have a solubility in water with a pH of between 7 and 10 that generates aqueous solutions with concentrations of Cr^{3+} that are within government recommended maximum concentration levels. Laboratory experiments indicate that the above reactions take place rapidly.

Consider now the organic compound trichloroethylene (TCE). The compound reacts with the iron as it comes in contact with the metal surface. In the following figure the reductive chlorination of TCE is illustrated. The final products are either ethene or ethane, either of which is easily biodegraded. The electrons appearing in these equations are provided by the oxidation of the zero valence iron via the reaction

$$Fe^0 \rightarrow Fe^{2+} + 2e^-.$$

The dechlorination can proceed along two competing pathways [8]. The equation describing the sequential hydrogenolysis pathway is

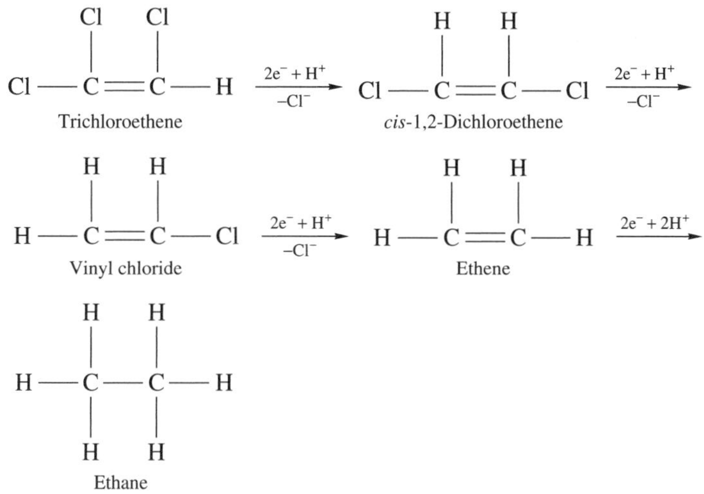

The reductive elimination pathway is given by

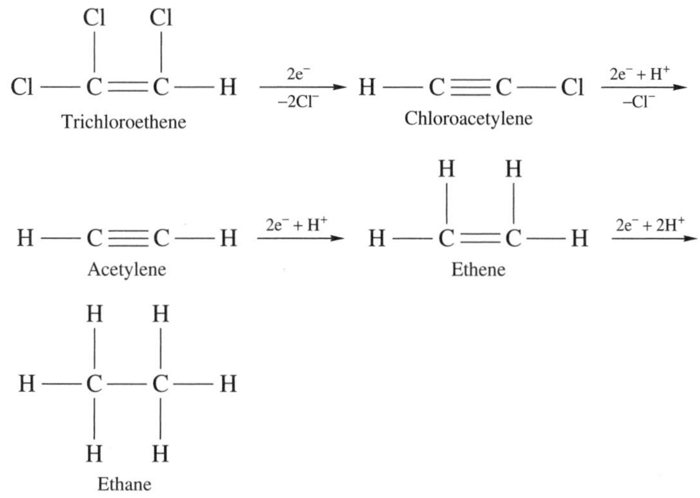

As an application example of the zero-valent iron wall concept, we examine the results of a field scale test facility at Moffet Field in northern California near San Francisco [9]. The contaminant plume at this site is more than 10,000 ft long and 5000 ft wide and includes waste oils, solvents, cleaners, and jet fuels. We will focus on the solvent trichloroethylene.

The sediments in the area consist of a complex mixture of fluvial-alluvial clay, silt, sand, and gravel that extends to depths greater than 200 ft.

The entire barrier extends 50 ft, but the gate portion shown in Figure 10.10 is but 10 ft wide. The iron used for this demonstration was selected through bench-scale testing and that which was found most efficient was employed. The particle size was −8 to +40 mesh (see Section 1.5.1, page 12). The thickness of the iron wall is 6 ft as determined using the bench-scale data. The objective of the design was to reduce the TCE concentration to below MCLs, which is 5 ppb for TCE and 70 ppb for *cis*-1,2-DCE. There is no MCL for 1,1-DCA.

The pea gravel sections shown in Figure 10.10 were added to better distribute the influent and effluent flow in the cell. A geosynthetic liner was used to bound the top and bottom of the cell. Backfill placed above the upper liner was used to make up the grade.

Based on data obtained in January 1997, after approximately seven months of operation, a determination of the effectiveness of the iron wall was made. The approach was to average the samples collected over each of five sections of the cell. The sections were each 2 feet across oriented orthogonal to the direction of flow. The first and fifth

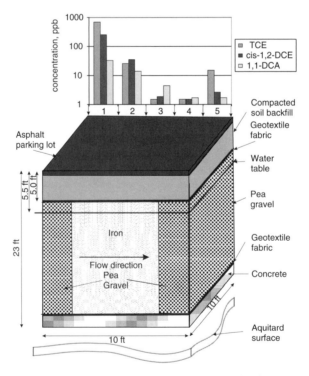

FIGURE 10.10. Iron wall design at Moffet Field in northern California near San Francisco (from [9]).

sections were the pea gravel sections. The graph in Figure 10.10 shows the change in concentration in each section for three contaminants: TCE, *cis*-1,2-DCE, and 1,1-DCA. Note that the concentration scale is logarithmic to better illustrate the behavior of the system at low concentrations.

The plot shows that there was significant decline in the concentration of all of the contaminants moving from section one to section four. A slight increase in the concentrations in section five is due to the fact that this is the exit to the cell and there has been infiltration of untreated water into this section.

10.6   BIOREMEDIATION

10.6.1   The Process

Bioremediation can be thought of as the utilization of biological (more specifically *fungal*, *bacterial*, or other *microbial processes*) to convert environmental contaminants into innocuous substances. The breakdown of contaminants due to microbial activity is called biodegradation. While such processes have been known to be effective for the treatment of hydrocarbons for decades [10], the concept of using this strategy in the more general context of subsurface remediation is more recent [11]. Since its introduction, its promise of potential financial savings in the remediation of contaminated groundwater has resulted in a flurry of both research and development activity in the search for effective bioremedial design strategies [12],[13].

The fundamental principles needed to understand the concept of bioremediation are summarized in a publication by the National Research Council [13], and the following discussion draws from that work. An especially helpful discussion of biodegradation is found in Maier et al. [14].

In essence, the bioremediation process involves the utilization of bacteria to change the chemical nature of contaminants such that the resulting compounds are harmless. The bacteria must be capable of degrading the contaminant, the contaminant must be accessible to the bacteria, and there must be the necessary compounds present to sustain the bacteria while they are destroying the contaminants.

In general, bacteria transform contaminants in an effort to survive and reproduce. However, the existence of an initial population of bacteria depends on several factors. One important factor is the availability of organic matter, the primary source of carbon for most microorganisms. In general, there is abundant organic matter in the near-surface environment, especially in the shallow soil zone (root zone). Lesser amounts are found in the deeper vadose and saturated zones.

Oxygen availability is generally essential to the support of a diverse microbiological population capable of achieving *aerobic bioremediation*. As would be expected, oxygen availability is greatest near the land surface. In the vadose zone both air and water coexist. Consequently, this is an ideal environment of microbial populations. In the saturated zone there is no gaseous oxygen available and the resident water is generally oxygen starved. Reflecting the dual need of water and oxygen, the optimal degree of saturation in support of soil microorganisms is between 38% and 81% [16].

Bacteria are not limited to the use of oxygen as an electron acceptor. In the absence of oxygen some bacteria can use other compounds, such as nitrate NO_3^-, sulfate SO_4^{2-}, and certain metals, such as iron Fe^{3+} and manganese Mn^{4+}. When bacteria use oxygen as the electron acceptor, they are exhibiting *aerobic respiration*. When using other than oxygen, they are utilizing *anaerobic respiration*.

A variant on the above strategy is to have the bacteria use inorganic chemicals as electron donors. In this case the inorganic compound is oxidized and the compound receiving the electrons, usually oxygen, is reduced. Since carbon is not released in this scenario, those bacteria that utilize this strategy must obtain the carbon needed for growth from an alternative source.

Maier et al. [14] suggest, based on the availability of water, organic matter, and oxygen, the following generalizations:

1. Biodegradation in surface soils is primarily aerobic.

2. Biodegradation in the vadose zone is also primarily aerobic, but significant acclimation times may be necessary for significant biodegrading populations to build up.

3. Biodegradation in the deep groundwater region is also initially slow because of low numbers and can rapidly become anaerobic because of lack of available oxygen. Biodegradation in shallow groundwater regions is initially more rapid because of higher microbial numbers but is similarly slowed by low oxygen availability.

From the point of view of the bacteria, the contaminants provide a source of carbon and a source of energy. Energy is released through chemical reactions catalyzed by the bacteria. Chemical bonds are broken such that electrons are released as described in Section 10.4.1. The contaminant releases electrons (is oxidized) and another molecule, often oxygen, O_2, accepts the electrons (is reduced). The compound releasing the electrons is called the *electron donor* and the one receiving the electrons the *electron acceptor*. When oxygen is the electron donor the chemical by-products are CO_2, H_2O, and increased microbial mass.

Bacteria may generate compounds (*enzymes*), as a by-product of their metabolic activity, that transform contaminants even though the bacteria do not metabolize the contaminant directly. Although such activity is of little or no benefit to the responsible bacteria, the process can lead to destruction of contaminants. Such a process is called *cometabolism*. An example of cometabolism is the destruction of chlorinated hydrocarbon by enzymes (monooxygenase and dioxygenase) produced by *methanotrophic bacteria* via the oxidation of methane, although the bacteria cannot metabolize the chlorinated hydrocarbon directly. Thus the methane is the primary food source for the bacteria (primary electron donor) and the chlorinated hydrocarbon is a secondary substrate because it does not directly support the growth of the bacteria [13].

In addition to destroying toxic chemicals through chemical transformation to less noxious compounds, bacteria can also induce changes in the properties of compounds, such as their solubility. For example, soluble uranium (U^{6+}) can act as an electron acceptor to produce (U^{4+}), which is insoluble. The resultant (U^{4+}) then forms a relatively immobile precipitate and thereby minimizes the transport of the uranium in groundwater [13]. Another important example of immobilization through precipitation is the reduction of the toxic hexavalent chromium ion Cr^{6+} to the less nocuous trivalent form Cr^{3+}, which can precipitate in the form of compounds such as chromium oxide.

Bacteria can also facilitate the destruction of contaminants, especially halogenated organic compounds, through the process of *reductive dehalogenation*, a concept introduced earlier. In this process, the bacteria catalyze a reaction whereby a halogen atom is replaced with a hydrogen atom. On page 385 we illustrate how this process could be used to reduce tetrachloroethene to trichloroethene. In concept, the process could be continued until the remaining product was ethene. Reductive dehalogenation usually occurs in an anaerobic environment.

For reductive dehalogenation to take place, an electron donor substance other than the halogenated compound is needed. Low molecular weight organic compounds, such as acetate and methanol, are possibilities. While in general the organisms do not derive energy from the dehalogenation process, counterexamples have been reported [13].

To survive in the subsurface, microbes require nutrients for cell growth. Because a typical bacterial cell is composed of 50% carbon, 14% nitrogen, 3% phosphorus, 2% potassium, 1% sulfur, 0.2% iron, and 0.5% calcium, magnesium, and chloride [13], cell growth may be restricted by a lack of any one of these compounds. Consequently, a bioremedial design that utilizes bacteria may need to supplement existing levels of these compounds to assure effective growth and contaminant removal. In practice, a ratio of carbon to nitrogen to phosphorus of 100:10:1 is used to augment existing subsurface nutrients. In some instances much larger amounts of carbon are used because much of the carbon that is metabolized is released as carbon dioxide.

10.6.2 Bioremediation as a Remediation Concept

In Section 10.4.2 we introduced the concept of organic molecules and their classifications. Among the compounds to be found in this group are petroleum hydrocarbons and their derivatives. Considerable success has been achieved in bioremediating many of these compounds, especially gasoline, fuel oil, alcohols, ketones, and esters.

BTEX As noted on page 381 an important family of molecules from an environmental perspective are the gasoline components *benzene, toluene, ethylbenzene*, and *xylene*, collectively know as BTEX. There are several reasons why these compounds have been especially easy to remediate [13]:

- They are relatively soluble compared to other common organic contaminants and other gasoline components.
- They can serve as the primary electron donor for many bacteria widely distributed in nature.
- They are rapidly degraded relative to other contaminants shown in Table 10.4.
- The bacteria that degrade BTEX grow readily if oxygen is available.

BTEX can be degraded aerobically or anaerobically. While each compound in BTEX follows different biodegradative pathways, a common endpoint is the development of a catechol. The following catechol molecule is along the aerobic metabolic pathway:

TABLE 10.4. Biodegradability of Hydrocarbons and Derivatives[a]

Compound	Frequency of Occurrence	Status of Bioremediation
Gasoline (BTEX), fuel oil	Very frequent	Established
Polycyclic aromatic hydrocarbons (PAHs)	Common	Emerging
Creosote	Infrequent	Emerging
Alcohols, ketones, esters, ethers	Common	Emerging

[a]Data from [13].

where the molecule R is H, CH_3, CH_2CH_3, and CH_3 in the case of benzene, toluene, ethyl-benzene, and *m*-xylene, respectively. It is interesting to compare this molecule with those presented earlier on page 381 to see the biologically induced changes in the molecular structure.

In the case of benzene, the cleavage of the aromatic ring of the catechol by the dioxygenase enzyme follows one of two pathways, ortho or meta. The ortho pathway is given below. The notation CoA denotes *coenzyme A*.[3] The acronym TCA in this context means *tricarboxylic acid* cycle, which results in complete mineralization to CO_2 and H_2O.

Anaerobic degradation is an important degradation pathway for BTEX because the demand for oxygen by the microbe often exceeds the available supply. As noted earlier, in the case of groundwater flow in the saturated zone, oxygen concentrations can be very low.

The pathway of benzene and xylene anaerobic degradation is not well known. Both toluene and ethylbenzene have been shown to have the common intermediate benzoyl-CoA. The catechols[4] resulting from the biological degradation are subsequently cleaved by dioxygenase as noted in the preceding aerobic case.

---

[3] A coenzyme is a nonproteinaceous organic molecule that plays an accessory but essential role in the catalytic action of an enzyme.

[4] A colorless crystal ($C_6H_6O_2$), soluble in water, alcohol, ether, benzene, and alkalis. An alkaline solution gives a coloration with ferric chloride, which turns brown on standing in air. It can be obtained from catechu, a natural dye, or prepared by fusing orthobenzenedisulfonic acid with caustic soda. Catechol is the principal constituent of the condensed (catechol) tannins (from [15]).

Benzoyl-CoA

The aromatic ring of benzoyl-CoA is subsequently reduced and transformed to acetyl-CoA, an important metabolic intermediate in cells.

The ability of organisms to anaerobically metabolize xylene appears to be limited to a few strains of denitrifying bacteria that use this compound as a growth substrate. The associated biodegradation pathways are poorly understood.

Chlorinated Solvents   Chlorinated aliphatic hydrocarbons (CAHs) (see page 380) are among the most frequently encountered contaminants in soil and groundwater. These organic CAHs are typically manufactured via the conversion of naturally occurring hydrocarbons such as methane, ethane, or ethene through the substitution of one or more hydrogen atoms by a chlorine atom. Alternatively, chlorinated compounds can be dechlorinated to a less chlorinated state [16]. Due to their wide use as solvents, commonly encountered CAHs are tetrachloroethene (PCE), trichloroethene (TCE), carbon tetrachloride (CT), chloroform (CF), and methylene chloride (MC).

Biodegradation of CAH compounds occurs primarily through oxidation and reductive halogenation. While their nonchlorinated counterparts are readily degraded aerobically, the presence of chlorine atoms inhibits aerobic degradation. Direct mechanisms are more likely to occur with compounds that are less chlorinated. More chlorinated compounds, which provide no direct energy contribution to the organism and are dependent on cometabolism for their degradation, are degraded relatively slowly.

To achieve effective cometabolism requires that the responsible bacteria have a readily available supply of substrate. Systems that have reportedly been effective in the cometabolism of chlorinated hydrocarbons include the enzyme methane monooxygenase produced by methanotrophic bacteria using substrates such as methane or formate and toluene dioxygenase produced by bacteria grown on toluene.

The aerobic transformation of trichloroethene via cometabolic oxidation converts TCE to TCE epoxide intermediate by a methane monooxygenase enzyme. The resulting epoxide breaks down spontaneously in water into a TCE diol. [5] The TCE diol breaks down according to two possible reaction pathways. Under acid conditions. the TCE diol forms glyoxylic acid. Alternatively, under basic conditions the TCE diol forms formate with the release of a CO molecule [14].

---

[5]Diol means the molecule has two hydroxy groups.

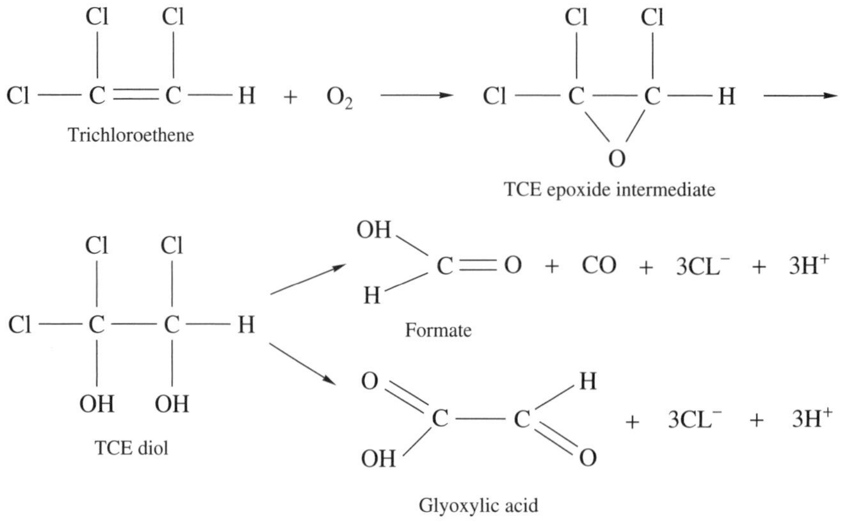

Trichloroethene

TCE epoxide intermediate

TCE diol

Formate

Glyoxylic acid

A second biodegradational mechanism for addressing chlorinated hydrocarbons involves reductive dehalogenation. In contrast to the oxidation of TCE, which requires an aerobic environment, dehalogenation generally takes place under anaerobic conditions. It is particularly effective, relative to aerobic processes, when the number of halogen substitutes is high. The following figure illustrates the reaction path that leads from TCE to ethylene. A dehalogenase enzyme is involved in each of the indicated steps. The TCE–ethylene reductive dehalogenation pathway presented is not unique and others can be found in the literature.

A concern about the use of anaerobic biodegradation is the observation that complete biodegradation of highly halogenated aliphatics under anaerobic conditions may not be realized, and the derivatives of the biodegradation may be more toxic than TCE itself. A process, for example, that terminates at vinyl chloride, would be problematic since the EPA has determined that vinyl chloride is a human carcinogen.

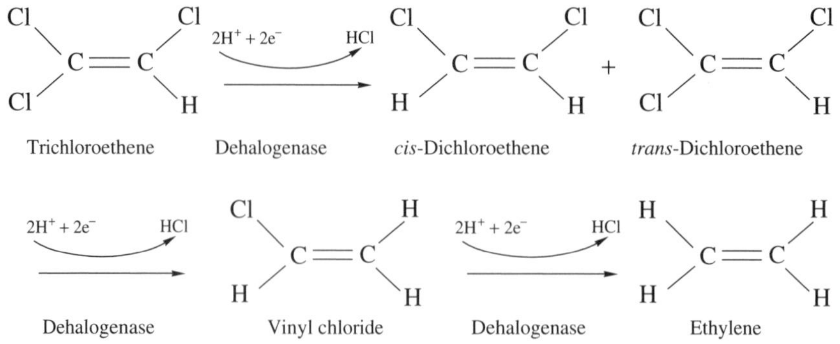

Trichloroethene   Dehalogenase   *cis*-Dichloroethene   *trans*-Dichloroethene

Dehalogenase   Vinyl chloride   Dehalogenase   Ethylene

The determination of whether an aerobic or anaerobic biodegradation strategy will be effective is a function of the history of the targeted site. The microorganisms

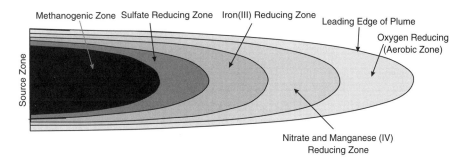

FIGURE 10.11. Redox zones of a typical petroleum plume located in an aerobic aquifer (adapted from [16]).

associated with a contaminant plume will use the electron acceptor that generates the most energy relative to other electron acceptors. Once the optimal acceptor is depleted, another acceptor that generates less energy will be utilized. The result is a hierarchy of groundwater zones that reflect the history of the plume. Figure 10.11 shows diagrammatically the evolution of a petroleum plume located in an initially aerobic aquifer. The leading edge of the plume is characterized by an oxygen-rich environment. However, nearer the source area, the oxygen has been depleted and an anaerobic environment conducive to anaerobic biodegradation exists.

In this section we have examined only the surface of the general topic of bioremediation of chlorinated hydrocarbons. Various other reaction pathways are possible. Some of these are summarized in Tables 10.5 and 10.6. More comprehensive tables, from which these have been abstracted, can be found in [16].

Implementation   The concepts presented above can be implemented in several different engineering design configurations. In selecting a design, several preliminary steps are required. Figure 10.12 illustrates a sequence of events that will assist in assuring that a suitable design is realized.

TABLE 10.5. Aerobic Oxidation Contaminants and Products

Degradation Mechanism	Compound[a]	Conditions	Product
Aerobic oxidation (direct)	DCE, VC	Aerobic	CO_2
	DCE, VC, DCA, CA, MC, CM	Aerobic	CO_2
Aerobic oxidation (cometabolic)	TCE	Aerobic, electron donor (phenol, toluene, benzene)	CO_2
	TCE	Aerobic, electron donor (toluene)	Not reported
	TCE, DCE, VC, TCA, CF, MC	Aerobic, electron donor (methane, aromatics, ammonia)	CO_2

[a] Acronyms: DCE, dichloroethene; VC, Vinyl chloride; DCA, dichloroethane; CA, chloroethane; MC, methylene chloride; CM, chloromethane; TCE, trichloroethene; CF, chloroform; MC, methylene chloride.
Source: U.S.EPA [16].

TABLE 10.6. Anaerobic Reduction Contaminants and Products

Degradation Mechanism	Compound[a]	Conditions	Product
Anaerobic reductive dechlorination (dehalorespiration)	PCE, TCE, DCE,VC, DCA	Anaerobic, electron donor (hydrogen or fermentative hydrogen source)	Ethene, ethane
	TCE	Anaerobic, electron donor (lactate, methanol butyrate, glutamate, 1,2-propanediol toluene)	Ethene
	PCE, TCE, c-DCE,VC	Anaerobic, electron donor (hydrogen, propionate or lactate)	Not reported
	PCE	Anaerobic, electron donor (methanol)	Not reported
Anaerobic reductive dechlorination (cometabolic)	PCE, TCE, DCE, VC, DCA	Anaerobic, electron acceptor (nitrate, sulfate), electron donor (hydrogen)	Ethene, ethane
	PCE, TCE, CT	Anaerobic, electron acceptor (nitrate, sulfate), electron donor (hydrogen)	Ethene, methane
	CT	Anaerobic, electron acceptor Fe(III)	CF, MC

[a]Acronyms: CT, carbon tetrachloride; c-DCE, *cis*-DCE.
Source: U.S. EPA [16].

The proposed point of departure is the evaluation of the site. Site evaluation should take into account the physical, chemical, and biological parameters of the site. The key physical parameters are those that impact subsurface processes such as fluid flow and mass transport, namely, permeability, porosity, degree of saturation, and the organic content of the soil. The design of the nutrient and electron acceptor delivery system therefore depends on these subsurface parameters.

Chemical parameters help to define the site and its potential for bioremediation. Determination of the concentration distribution at the site provides a baseline against which the change in contaminant concentration achieved using an implemented bioremediation strategy can be compared. Examination of the various contaminant species present, especially degradation products of biochemical reactions, provides insight into existing degradation mechanisms. Knowledge of substrate levels are also helpful in assessing the potential success of an existing or introduced microbial colony. The concentration of electron donors, for example, toluene or phenol, will help to assess the success of spontaneous (not enhanced) cometabolic degradation of contaminants of concern. Oxygen content (indirectly indicated by the redox potential)[6] will determine whether anaerobic or aerobic conditions are present at a site. The concentration of electron acceptors other than oxygen, such as nitrate or sulfate, will further indicate whether electron acceptors are

[6]The redox potential is a measure (in volts) of the affinity of a substance for electrons when compared with hydrogen (which is set at 0).

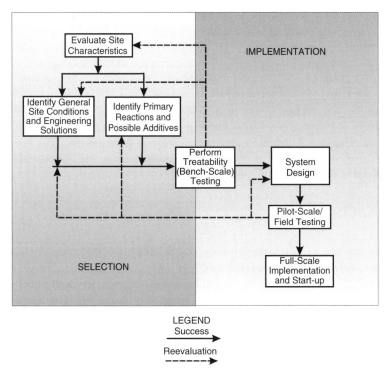

FIGURE 10.12. Typical selection and implementation steps for in situ bioremediation (adapted from [16]).

adequate for aerobic biodegradation. Nutrient levels will indicate whether the microbial environment requires the addition of nutrients or is self-sufficient.

Because the biodegradation of compounds such as chlorinated hydrocarbons is influenced by the abundance of specific and nonspecific microbes, direct microbial analysis at a site can be helpful. Techniques, such as oxygen uptake rate analysis, are also available for determining the rate of activity of the targeted microbes.

10.7 OPTIMAL DESIGN OF REMEDIATION AND MANAGEMENT STRATEGIES

In this section we bring together modeling concepts described earlier, and some new technology borrowed from operations research, to show how optimal (usually in the sense of least cost) remediation designs can be developed. We begin with the case of contaminant plume containment using pump-and-treat technology and then tackle the more challenging problem of risk-based (concentration-based) cleanup designs.

10.7.1 Gradient-Constrained Pump-and-Treat Designs

Consider a pump-and-treat inward-gradient design such as described in Section 10.1.1. The design objective is to develop a hydraulic gradient that is everywhere inward along the perimeter of the contaminant plume (see Figure 10.4). While we assume in this section that the only technology available to the groundwater professional is pump-and-treat employing pumping and recharging wells, other hydrodynamic containment strategies could be considered.

Using this approach, the goal is to locate pumping and recharging wells such that the inward gradient can be realized at least cost. The cost consists of two elements: (1) construction and (2) operation and maintenance. Construction costs are those associated with the construction of the wells, the piping network, and the treatment plant. The operating and maintenance costs encompass those that are reoccurring, such as the replacement of activated carbon in an activated-carbon filtration unit. The equation that describes these costs is called in operations research parlance the objective function. A typical objective function would have the form

$$f\left(\alpha_i, q_i, \alpha_i^0\right) = \sum_{i=1}^{n} \alpha_i^+ q_i^+ + \alpha_i^- q_i^- + \alpha_i^{0+} \delta\left(\mathbf{x}_i - \mathbf{x}_i^{0+}\right) + \alpha_i^{0-} \delta\left(\mathbf{x}_i - \mathbf{x}_i^{0-}\right), \quad (10.3)$$

where n is the number of potential wells; α_i^+ and α_i^- are the costs of operating an injection or withdrawal well, respectively, per unit volume of injection or discharge, per unit time; α_i^{0+} and α_i^{0-} are the installation costs of injection well i and discharge well i, respectively; q_i^+ and q_i^- are the well recharge or discharge, respectively, in units of volume per unit time; and $\delta\left(\mathbf{x}_i - \mathbf{x}_i^{0-}\right)$ is the Dirac delta function that locates well i in the region defined by $\mathbf{x}$. The Dirac delta function is zero everywhere except at the location $\mathbf{x}_i^{0-}$. The goal is to locate the n wells or a subset of the n wells at optimal locations and to pump these wells at an optimal rate so as to minimize the overall cost defined by the objective function (Eq. (10.3)). Mathematically we state this as

$$\min\left(f(\alpha_i, q_i, \alpha_i^0)\right). \quad (10.4)$$

In addition to identifying an objective function, it is necessary to define a set of physical and design constraints. Without such constraints, the obvious solution to Eq. (10.4) would be to set all the decision variables $(\mathbf{q}\,(\mathbf{x}))$ to zero. The zero recharge–discharge alternative does not normally result in a satisfactory design.

A typical physical design constraint would be

$$0 \leq q_i^+ \leq q_i^{+up},$$

$$0 \leq q_i^- \leq q_i^{-up},$$

where q_i^{+up} and q_i^{-up} are the maximum injection and withdrawal values achievable at well location $\mathbf{x}_i$.

Design constraints restrict the choice of the decision variables such that the specified design goals are met. Physical constraints assure that the choice of decision-variable values does not result in physically impossible solutions. For example, without constraints on the amount of water to be pumped from a target well, practically unachievable pumping rates might be assigned to that well in the optimal solution.

In the case of the gradient-constrained pump-and-treat design, the groundwater potential surface is required to have an inward gradient in excess of a prespecified design value γ all along the plume perimeter. The resulting constraint is written

$$-\frac{\partial h\,(\mathbf{x})}{\partial \mathbf{n}} \geq \gamma, \quad \mathbf{x} \in \partial\mathbf{\Omega}, \quad (10.5)$$

where $\mathbf{n}$ is the outward directed normal to the plume boundary defined by $\partial\Omega$. Other constraints on the head could be added, such as a maximum permitted drawdown at a specified location. Design constraints on discharge might be

$$q_i^+ - q_i^- = 0, \tag{10.6}$$

which describes the requirement that all of the water discharged must, upon decontamination, be recharged to the aquifer.

Figure 10.13 provides a geometric representation of the optimization problem for the case of two wells, q_1 and q_2. On the $q_1 \times q_2$ plane, one finds a polygonal surface. The perimeter of this surface is defined by a series of straight lines. The lines represent the intersection of the various constraints and the zero plane. In other words, the lines represent the location where the constraints defined by Eq. (10.5) are exactly satisfied. Along these lines the constraints are indeed equalities, that is,

$$-\frac{\partial h(\mathbf{x})}{\partial \mathbf{n}} = \gamma, \quad \mathbf{x} \in \partial\mathbf{\Omega}.$$

On the interior of this polytope (formally a bounded polyhedron), the gradient is larger than γ. Outside this polytope the gradient is smaller than needed. Thus any point outside the region does not satisfy the constraints, and any point inside exceeds the constraints.

The planar surface above the $q_1 \times q_2$ plane represents the objective function. The optimal solution to this problem is the point on the $q_1 \times q_2$ plane where the objective function is smallest yet is still within the polytope of constraints. The optimal location will always be at a vertex of the polytope. In this case the point is indicated by the black dot, which is represented on both the $q_1 \times q_2$ plane and on the objective function. In essence, one can imagine being located on the objective function and sliding down the function in the direction of maximum gradient until one encounters a point on the polytope. That point will represent the optimal solution to the problem.

The question arises as to how one obtains the information provided in Figure 10.13. Notice that the objective function is defined in terms of the decision variables q_1 and

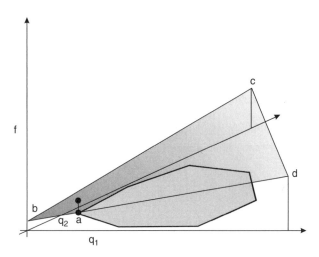

FIGURE 10.13. Objective function surface f and decision variables q_1 and q_2. The polygonal surface defines the feasible region.

q_2, and the constraints are defined in terms of the state variable head, h. How does one determine the values of q_1 and q_2 that correspond to a state wherein the constraints are exactly satisfied? The answer lies in the solution of the groundwater flow equation, Eq. (4.22), reproduced here with the addition of a source term q (in this case a well), that is,

$$S_s \frac{\partial h}{\partial t} - \nabla \cdot (\mathbf{K} \cdot \nabla h) - q = 0. \tag{10.7}$$

In the groundwater flow equation (Eq. (10.7)) we see both the state variable h and the decision variable q. Thus, for any combination of q values, one can determine the head and the gradient in head from this equation. Thus it is the groundwater flow equation that couples the state and decision variables and therefore permits the creation of the objective function and constraints found in Figure 10.13.

In is interesting to note from Section 6.2 that the head change due to pumping is a linear function of the pumping rate. Thus the relationship between the constraints that are defined in terms of head and the pumping rate values q_i is linear. This explains why the polytope has straight sides. Similarly, the objective function is the sum of two linear functions (rate of increase of pumping) and therefore also is linear. Taken together, the system is defined by linear functions, one the objective function and the others the constraints.

Many algorithms exist for solving this optimization problem. However, we will not consider these methods here. A midlevel text that discusses both linear programming, such as presented above, and nonlinear programming, which we will describe shortly, is that by Luenberger [17]. A book dedicated specifically to the use of optimization for the design of plume containment using pump-and-treat technology is that by Ahlfeld and Mulligan [18].

10.7.2 Risk-Constrained Pump-and-Treat Designs

In the preceding section the goal was essentially contaminant containment. While the process of containment also implies mass removal, in the preceding section mass removal was not explicitly part of the remediation design. However, since the duration of cleanup has a marked impact on the overall costs, the question arises as to whether one could use the optimization approach to design a pump-and-treat strategy that is based on concentration constraints, that is, a risk-based design. We use the term risk-based because the design focuses on contaminant concentration, which, in turn, can be related to human health risk.

The answer is yes, although the problem formulation is more complex. A relatively simple form of the risk-based remediation-design problem is

$$\min \sum_{i=1}^{n} \alpha_i^+ q_i^+ + \alpha_i^- q_i^- + \alpha_i^{0+} \delta \left(\mathbf{x}_i - \mathbf{x}_i^{0+} \right) + \alpha_i^{0-} \delta \left(\mathbf{x}_i - \mathbf{x}_i^{0-} \right), \tag{10.8}$$

such that

$$c_j (\mathbf{q}) \le c_j^*, \qquad j \in J, \tag{10.9}$$

$$0 \le q_i^+ \le q_i^{+up}, \tag{10.10}$$

$$0 \le q_i^- \le q_i^{-up}. \tag{10.11}$$

While Eq. (10.8) is structurally the same as Eq. (10.3), the definitions of the coefficients α_i are different. In this instance the α_i coefficients incorporate the cost of the treatment of the extracted fluid as well as the costs associated with the gradient-constrained problem described in Section 10.7.1. The α_i^0 coefficients remain as the fixed costs of well construction.

The major difference between the risk-constrained and gradient-constrained problems lies in the concentration constraint given by Eq. (10.9). This constraint states that the concentration at specified locations $\mathbf{x}_j$, $j \in J$ must be less than a target concentration c_j^* at the end of the remediation period. The challenge now is to once again find a mechanism of expressing the concentration constraint in terms of the well discharge or recharge q_i. The mathematical vehicle for doing this is the transport equation, Eq. (8.29), reproduced as

$$\frac{\partial}{\partial t}(\varepsilon R c) + \nabla \cdot (\varepsilon \mathbf{v} c) - \nabla \cdot \varepsilon \mathbf{D} \, \nabla c - \varepsilon q = 0. \tag{10.12}$$

This equation contains the state variable concentration, c, and the decision variable discharge, q. However, this equation also contains the groundwater velocity, $\mathbf{v}$. To obtain the velocity, we must use Darcy's law in combination with the groundwater flow equation. Because of the dependence of the concentration on the velocity, there is not a linear relationship between the discharge q and the concentration c. In other words, equal increments of change in q do not generate equal incremental changes in concentration. The result is that the constraints, which appear as straight lines in Figure 10.13, are now curved. While this may appear as a rather incidental change, it makes a great deal of difference in the degree of complexity of the optimization problem and the sophistication of the algorithms required to solve the resulting optimization algorithm. Once again Luenberger [17] is a good source of information, in this case on nonlinear programming algorithms.

In this discussion we have only touched the tip of the iceberg regarding the use of optimization methods in the minimization of cost in groundwater remediation. Published extensions involve the inclusion of the surface treatment facility in the design (see [19]) and long-term monitoring.

10.8 SUMMARY

Remediation of groundwater can be considered in terms of containment, decontamination, and natural attenuation. Containment encompasses methodologies that assure that a contaminant plume does not expand beyond the area defined at the time of remediation. Pump-and-treat technologies are especially effective in realizing this objective.

Decontamination technologies seek to remove contaminants from an aquifer system, usually within a specified time period and to a specified target contaminant value. The target value is normally based on government criteria, which are often, although not always, established via risk to public health. Currently, pump-and-treat methodology is the vehicle of choice to achieve these engineering goals. However, new strategies that incorporate several methodologies such as bioremediation and reactive walls, either employed separately or in combination, are candidate design elements.

Natural attenuation relies heavily on the effectiveness of bioremediation and long-term monitoring. The concept usually employs resident microorganisms to degrade dissolved contaminants. Since many contaminant problems must be addressed in the short term, the enhancement of biodegradation strategies via the introduction of nutrients and other compounds is currently being investigated, and in some instances implemented.

10.9 PROBLEMS

10.1. Use the concepts of well hydraulics and image theory to illustrate how a combination of pumping and injection can be used to contain a contamination plume via the inward-gradient approach.

10.2. Which of the following contaminants is most suitable for soil vapor extraction (SVE): phenol, isobutanol, or TCE? Explain your answer. For the contaminant that is least suitable for SVE, propose an alternate remediation strategy.

10.3. Consider the funnel-and-gate system depicted in Figure 10.9. Explain how the tangent law applies to this system.

10.4. Consider the redox zones shown in Figure 10.11. If you were to write a mathematical model of this system, what processes would need to be included? Propose a set of transport equations that could model the development and evolution of the different redox zones shown in the figure.

BIBLIOGRAPHY

[1] S. Feenstra, Evaluation of multi-component DNAPL sources by monitoring of dissolved-phase concentration, in K. U. Weyer (ed.), *Subsurface Contamination by Immiscible Fluids*, Balkema, Rotterdam, Netherlands, 1992, p. 65.

[2] W. J. Lyman, W. F. Reehl, and D. H. Rosenblatt, *Handbook of Chemical Property Estimation Methods*, McGraw-Hill, New York, 1982.

[3] R. J. Charbeneau, *Groundwater Hydraulics and Pollutant Transport*, Prentice Hall, Englewood Cliffs, NJ, 2000.

[4] R. Asato, Y. Sakai, B. Kimura, and C. Shiroma, "Internet Chemistry," http://library.kcc.hawaii.edu/external/chemistry/redox_title.html, 2003.

[5] R. C. Elderfield, Organic chemistry, in *McGraw-Hill Encyclopedia of Science and Technology*, Vol. 9, McGraw-Hill, New York, 1982.

[6] C. N. Sawyer, P. L. McCarty, and G. F. Parkin, *Chemistry for Environmental Engineering Science*, McGraw-Hill, New York, 2003, p. 752.

[7] D. W. Blowes and C. J. Ptacek, *Geochemical Remediation of Groundwater by Permeable Reactive Walls: Removal of Chromate by Reaction with Iron Bearing Solids*, in *Proc. Subsurface Restoration Conference*, Third International Conference on Groundwater Quality Research, Dallas, TX, June 21–24, 1992, pp. 214–216, 1992.

[8] U.S. Environmental Protection Agency, *Permeable Reactive Subsurface Barriers for the Interception and Remediation of Chlorinated Hydrocarbons and Chromium(VI) Plumes in Ground Water*, U.S. EPA Remedial Technology Fact Sheet, EPA/600/F-97/008, 1997.

[9] U.S. Department of Defense, *Permeable Reactive Wall Remediation of Chlorinated Hydrocarbons in Groundwater*, ESTCP Cost and Performance Report, 1999.

[10] W. C. Tausson, Naphthalin als kohlenstoffquelle für bakterien, *Planta* **4**:214, 1927.

[11] R. L. Raymond, V. W. Jamison, and J. O. Hudson, Beneficial stimulation of bacterial activity in groundwater containing petroleum hydrocarbons, *Am. Inst. Chem. Eng. Symp. Ser.* **73**(166):390, 1977.

[12] D. C. Adriano, J. M. Bollag, W. T. Frankenberger, Jr., and R. C. Sims, Bioremediation of contaminated soils, *Agronomy* **37**:820, 1999.

[13] National Research Council, *In Situ Bioremediation*, National Academy Press, Washington, DC, 1993.

[14] R. M. Maier, I. L. Pepper, and C. P. Gerba, *Environmental Microbiology*, Academic Press, San Diego, CA, 2000.

[15] M. T. Roberts and D. Etherington, "Bookbinding and the Conservation of Books, A Dictionary of Descriptive Terminology," http://palimpsest.stanford.edu/don/dt/dt0612.html, 2002.

[16] U.S. EPA, *Engineered Approaches to in Situ Bioremediation of Chlorinated Solvents: Fundamentals and Field Applications,* EPA 542-R-00-008, 2000.

[17] D. G. Luenberger, *Linear and Nonlinear Programming*, Addison-Wesley, Boston, 1984.

[18] D. P. Ahlfeld and A. E. Mulligan, *Optimal Management of Flow in Groundwater Systems*, Academic Press, San Diego, CA, 2000.

[19] A.A. Spiliotopoulos and G. P. Karatzas, A multiperiod approach to the solution of groundwater management problems using an outer approximation method, *J. Eur. Operations Res.* **157**:514, 2004.

CHAPTER 11

MULTIFLUID FLOW AND TRANSPORT

While groundwater is almost always extracted from aquifers, in which all of the pore space is filled with water, there are zones of the subsurface, and situations involving certain kinds of contaminants, for which nonaqueous fluids occupy part of the pore space. These include the unsaturated soil zone, in which air occupies some of the pore space, and contamination problems that involve fluids like oil or chlorinated solvents that are only slightly miscible with water. It is usual in groundwater hydrology to refer to these systems as *multiphase* systems, where *phase* is used to denote a particular substance, independent of its state (solid, liquid, or gas). Sometimes use of the term phase is confusing, because in other areas of science the term phase refers to the state of a particular substance (solid, liquid, and gaseous phases). To avoid confusion, and for consistency of presentation, we will use the more consistent terminology *multifluid* to describe systems in which more than one fluid occupies the pore space. For our purposes, a fluid can be either a liquid or a gas.

Porous-media systems are characterized by a solid phase within which interconnected pore space allows fluids to flow. When the pore space is completely filled with a single fluid, for example, water, then the entire space is filled with a combination of solid particles and one fluid, with surfaces of separation between the solid and fluid corresponding to the boundary of the solid particles. We already know that any transfer of mass between the solid and the fluid, such as sorption processes, occurs along the areal boundary separating the solid and fluid. In general, any transfer of mass occurs along these boundaries, because they are the only locations of contact between the fluid and solid. Multifluid porous-media systems are characterized by additional boundaries, or interfaces, that separate the different fluids from one another, and from the solid particles that comprise the porous matrix. By the same argument as that for single-fluid systems, any exchange of mass between different fluids will take place across the shared interfaces that separate them. Figure 11.1 shows a schematic of a multifluid system, resolved at the pore scale.

Subsurface Hydrology By George F. Pinder and Michael A. Celia
Copyright © 2006 John Wiley & Sons, Inc.

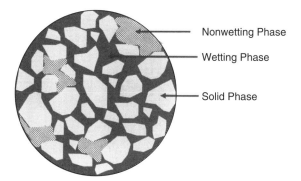

FIGURE 11.1. Schematic of fluid and solid phases of a porous medium, resolved at the pore scale.

As in other cases, our interest will be to write equations for the porous medium system at a length scale larger than the pore scale. However, because the interfaces that exist at the pore scale play an important role in understanding the behavior of these systems, we will need to think about pore-scale processes as we proceed through this chapter.

11.1 PROBLEMS INVOLVING MULTIPLE FLUIDS

11.1.1 Unsaturated Soils

The unsaturated zone has been defined earlier as that zone between the land surface and the top of the capillary fringe. We have already observed that the water pressure in this zone is less than atmospheric, and that the percent of void space occupied by liquid water can vary from 100% (at the top of the capillary fringe) to values that approach zero. The solid materials that comprise the soil can be highly variable, but they are usually unconsolidated with a zone of high organic content at and just below the land surface. A special attribute of the unsaturated zone is the possible existence of plants, with root systems that are part of the unsaturated zone. Plant roots not only provide mechanical stability for the plant but also are used by plants to extract water from the soil. Dissolved nutrients in the extracted water provide essential building blocks for plant growth. The soil is also teeming with microbial and small-animal life. Microorganisms play an important role in nutrient cycling and general life cycles in the soil. Earthworms and other larger organisms also play important roles in the soil ecosystem. Complex chemical and biological reactions form the foundation of fairly complex cycling of important materials like carbon, nitrogen, oxygen, and, of course, water. The unsaturated zone is a conduit for recharge of aquifer systems through infiltration of precipitation and subsequent movement through the unsaturated zone and into the saturated zone.

Our interest will be with the basic hydrological processes that occur in the unsaturated zone. These include infiltration driven by rainfall, movement of water within the unsaturated zone, fluxes of water from the unsaturated zone to the saturated zone, nutrient or contaminant transport associated with the water movement, and water extraction due to plant root uptake. We will view the unsaturated zone as a special case of a two-fluid porous medium, in which the two fluids are water and air. We will take advantage of the fact that, relative to water, air has low density and viscosity and will derive simplified equations to describe fluid movement in this part of the subsurface.

11.1.2 Nonaqueous-Phase Liquids

There are many other examples of multifluid porous-media systems. When both water and (at least) one other liquid exist within the pore space, then we have a system with an aqueous (water-based) liquid and a nonaqueous liquid. Examples of nonaqueous liquids include hydrocarbon products like heating oil, gasoline, and jet fuel, as well as organic solvents like trichloroethylene (TCE). Accidental spills of these ubiquitous nonaqueous liquids, or deliberate disposal at or below the land surface, have led to a number of groundwater contamination problems. These fluids have become known as *nonaqueous-phase liquids*, or NAPLs. Because density often plays an important role in the movement and eventual fate of these liquids, we distinguish subclasses of NAPLs as *dense nonaqueous-phase liquids*, or DNAPLs, and *light nonaqueous-phase liquids*, or LNAPLs. Hydrocarbons tend to be LNAPLs; for example, the density of gasoline is about 730 kg/m^3, compared to water density of about 1000 kg/m^3. Organic solvents tend to be DNAPLs; TCE has a density of about 1600 kg/m^3. While other factors may influence the movement of fluids, generally speaking, these density differences lead to significant buoyant forces that determine the vertical location of NAPL spills. For spills with sufficient volumes, LNAPLs tend to infiltrate through the unsaturated zone and then "float" on top of the water table, while DNAPLs can "sink" through the water-saturated zone. These behaviors imply different distributions in the vertical dimension, and different effects on groundwater resources.

If a NAPL is spilled, or otherwise placed, at the land surface, and it infiltrates into the subsurface, it will move into the unsaturated zone. Because water and air already form a two-fluid system in unsaturated soils, introduction of a second liquid results in a *three-fluid porous-medium* system. Three-fluid systems are quite complicated to analyze, and we will spend only a small amount of time considering them. In two-fluid systems, we have two fluids and one solid, and between these three substances we have three possible interface combinations: interfaces between the two fluids, interfaces between one fluid and the solid, and interfaces between a second fluid and the solid. When three fluids are present, we have six possible interface types (interface pairs), so already we see that the system is becoming quite complex. While practical applications involving three fluids almost always involve some set of simplifying assumptions to make the problem more tractable, the system is inevitably difficult to analyze.

11.1.3 Fluid Components and Groundwater Contamination

An additional complication to the overall multifluid problem is the observation that each fluid is often composed of a number of components. For example, the aqueous fluid (we can call the composite fluid "water") is composed mostly of H_2O but will almost always have other substances dissolved within it, such as salts and perhaps other "contaminants." Water is especially prone to accumulation of solutes because one of its many unusual properties is that it is an almost universal solvent [1]. In addition, a NAPL like gasoline has a number of components in it, some of which are lighter fractions like benzene, toluene, ethylbenzene, and xylene (referred to as BTEX). These light fractions tend to dissolve into the aqueous fluid, and if a gaseous phase is present they can also *volatilize*. Partitioning into the aqueous liquid leads to contamination of the water, because these compounds are suspected carcinogens. Volatilization implies addition of new components into the mix of gases that we refer to as "air."

Sometimes we need to analyze specific components of a fluid in detail (e.g., contaminants dissolved in water), and sometimes we do not analyze anything related to components within a fluid but simply deal with the fluid as a single bulk entity (e.g., in the unsaturated zone we usually do not worry about the fact that air is 80% N_2—we just call the whole gaseous mixture "air"). The level of detail used in any problem description depends on the questions being asked, and the level of sophistication needed to answer the questions.

11.2 GOVERNING EQUATIONS

As usual, we take a mathematical approach to analyze these systems. We will write fairly general equations, then simplify them so that we can derive some practical results that provide insights into the behavior of the problems under consideration, and provide quantitative tools to study system responses. The overall procedure we will follow for multifluid systems is the same one we followed for earlier cases involving only a single fluid. We begin by writing mass balance equations to describe the relevant parts of the system, then augment those balance equations with material-dependent constitutive relationships. We will focus our development on two-fluid flow systems and begin by writing a mass balance equation for each fluid in the system. We then write an extended version of Darcy's law for fluid fluxes. We will complete the equation set by adding equations based on laboratory observations and measurements.

11.2.1 Mass Balance Equations

Consider a general two-fluid porous-media system, which is depicted schematically in Figure 11.1. That figure shows the two fluids and the solid. Let us begin by deciding on a designation for the two fluids. For a system containing a pair of fluids and a solid surface, one of the fluids tends to be more strongly attracted to the solid surface than the other. We will say that the fluid which is more strongly attracted by the solid surface preferentially wets the solid surface. That fluid is referred to as the *wetting fluid*. The other fluid then becomes the *nonwetting fluid*. The usual measure of wettability is the *contact angle*, defined as the angle formed between the solid surface and the interface between the two fluids. The fluid occupying the space within the acute angular space between the *fluid–fluid interface* and the solid surface is the wetting fluid (e.g., see [2] for a more detailed discussion). Smaller angles imply stronger surface attraction, and therefore a higher wettability of the surface with respect to the wetting fluid. If we denote the two fluids by the symbols *w* (for wetting) and *nw* (for nonwetting), then we can write a statement of mass balance for each of the fluids, based on a balance like that performed for the single-fluid case. For the system depicted in Figure 11.1, we may write the following mass balance statement defined for a cubic elementary volume:

$$
\left[(\rho_\alpha \varepsilon S_\alpha)^{t+\Delta t/2} - (\rho_\alpha \varepsilon S_\alpha)^{t-\Delta t/2} \right] \Delta x \, \Delta y \, \Delta z
$$

$$
= \left[(\rho_\alpha q_{x\alpha})_{x-\Delta x/2} - (\rho_\alpha q_{x\alpha})_{x+\Delta x/2} \right] \Delta y \, \Delta z \, \Delta t
$$

$$
+ \left[(\rho_\alpha q_{y\alpha})_{y-\Delta y/2} - (\rho_\alpha q_{y\alpha})_{y+\Delta y/2} \right] \Delta x \, \Delta z \, \Delta t
$$

$$
+ \left[(\rho_\alpha q_{z\alpha})_{z-\Delta z/2} - (\rho_\alpha q_{z\alpha})_{z+\Delta z/2} \right] \Delta x \, \Delta y \, \Delta t
$$

$$
+ \rho_\alpha Q^\alpha \, \Delta x \, \Delta y \, \Delta z \, \Delta t. \tag{11.1}
$$

In this equation, α denotes a particular fluid ($\alpha = w$ or nw), ρ_α is the density of fluid α (mass of fluid α per volume of fluid α, [ML^{-3}]), ε is porosity (volume of voids per total volume, [L^{3}L^{-3}] or [L^0]) (see Section 1.6), S_α is the volumetric saturation with respect to fluid α (volume of fluid α per volume of voids, [L^{3}L^{-3}] or [L^0], see Section 2.4), $\mathbf{q}_\alpha$ is the volumetric flux vector for fluid α (volume of fluid α per total area per time, [L^{3}L^{-2}T^{-1}] or [LT^{-1}]), and Q^α denotes a source or sink of mass to or from fluid α (volume of fluid α per total volume per time, [L^{3}L^{-3}T^{-1}] or [T^{-1}]). The fluid mass balance is written around the point $(\mathbf{x}, t) = (x, y, z, t)$, and all terms in the equation are interpreted as already having been averaged over some length and time scale consistent with the *representative elementary volume* (REV) concept. Then, as with standard arguments for these kinds of box balances, we divide all terms in the equation by $\Delta x \, \Delta y \, \Delta z \, \Delta t$ and then take the limit as each of these increments goes to zero. The result is the following partial differential equation for the balance of mass for fluid α:

$$\frac{\partial}{\partial t} (\rho_\alpha \varepsilon S_\alpha) + \frac{\partial}{\partial x} (\rho_\alpha q_{x\alpha}) + \frac{\partial}{\partial y} (\rho_\alpha q_{y\alpha}) + \frac{\partial}{\partial z} (\rho_\alpha q_{z\alpha}) = \rho_\alpha Q^\alpha \qquad (11.2)$$

or, using the shorthand notation of the divergence operator,

$$\frac{\partial}{\partial t} (\rho_\alpha \varepsilon S_\alpha) + \nabla \cdot (\rho_\alpha \mathbf{q}_\alpha) = \rho_\alpha Q^\alpha. \qquad (11.3)$$

We make two observations before continuing. When the system is fully saturated, so that only one fluid exists in the pore space, the volumetric saturation for that fluid (S_α) is equal to one, and Eq. (11.3) reduces to the basic mass balance equation for single-fluid flow (see Eq. (4.22)). The second observation is that in Eq. (11.3), if we assume we know the porosity of the medium, the densities of the two fluids, and the source/sink term (say, it is zero), then there are four unknowns in the equation, for a given α. Those unknowns are the three components of the volumetric flux vector, $\mathbf{q}_\alpha = (q_{x\alpha}, q_{y\alpha}, q_{z\alpha})$, and the saturation S_α. For the two-fluid system, Eq. (11.3) is written twice, once for $\alpha = w$ and once for $\alpha = nw$. That gives two equations with eight unknowns. Clearly we need more than these two equations in order to be able to solve for the eight unknowns. As in the single-fluid case, we look to Darcy for help.

11.2.2 Darcy's Law for Multifluid Flow

Recall that Darcy's experiments of 1856 provided one of the foundational equations for groundwater flow analysis. Darcy performed experiments using saturated, homogeneous sand columns and found that the volumetric flux is proportional to the hydraulic head gradient. The coefficient of proportionality is what we call the *hydraulic conductivity*. In equation form, Darcy's experiments lead to the following expression:

$$\mathbf{q} = -K \, \nabla h. \qquad (11.4)$$

The hydraulic conductivity is composed of the intrinsic permeability k, the fluid density ρ and viscosity μ, and the gravitational acceleration constant g, with $K = k\rho g/\mu$. We have seen (Chapter 2, pp. 86–89) that the intrinsic permeability can have different values for flows in different directions, thereby requiring a tensorial representation to account for anisotropy. In addition, the more general expression of Darcy's law treats the components of the hydraulic head, namely, pressure and elevation, separately. These considerations lead to the following more general form of Darcy's equation, which is taken directly from Eq. (2.56) in Chapter 2:

$$\mathbf{q} = -\frac{\mathbf{k}}{\mu} \cdot (\nabla p - \rho \mathbf{g}).$$

(11.5)

In Eq. (11.5) vector $\mathbf{g}$ is the gravitational acceleration vector, equal to $-g\nabla z$ when z is the vertical coordinate defined to be positive upward. Equation (11.5) is a generalization of Eq. (11.4), although it is still applicable only to systems with a single fluid present.

Now let us consider the case of two-fluid porous-media systems. Can we still use Darcy's equation? The answer is: not directly. Somehow we need to account for impacts caused by more than one fluid residing in the pore space. To understand the impact of multiple fluids in the pore spaces, we need to remember that the permeability coefficient in Darcy's equation is a measure of the ease with which a fluid can flow through the pore space. With all else being equal, a connected pore space that has larger pores will have a higher permeability coefficient. That is because, for a given flow, there is less frictional loss in the flowing fluid when the pores are larger. If the pores are reduced in size, the permeability decreases. This means that a given potential gradient produces less flow, or conversely, a larger gradient is needed to produce the same flow, as compared to the larger permeability case. If we now consider the flow of one of the fluids, for example, the nonwetting fluid, in a porous medium containing both wetting and nonwetting fluids, then the amount of pore space available to the nonwetting fluid is not the entire pore volume, but only a fraction of it (that fraction is the saturation S_{nw}). As the space available for flow is reduced, we expect the permeability to decrease because fewer flow channels are available. Therefore, again with all else being equal, we expect the permeability of the porous medium to the nonwetting fluid to decrease as the saturation of nonwetting fluid decreases, eventually reaching zero when there are no longer any connected pathways of nonwetting fluid through the pore space of the medium.

If we consider a measure of permeability reduction due to saturation changes, it makes sense to take the permeability at full saturation as the baseline case. Assume we have two systems, one fully saturated with nonwetting fluid, and a second with only half of the pores filled with nonwetting fluid. If the same potential gradient is applied to both systems, and the samples are otherwise identical, the flow of nonwetting fluid in the partially saturated case will be less than the flow in the saturated case. The ratio of these two flow rates is a measure of flow reduction due to variable saturation. We refer to this ratio as the *relative permeability*, k_{rnw}, where subscript r denotes relative, and nw refers to the nonwetting fluid. The relative permeability function can be defined by the ratio of fluxes as a function of saturation,

$$k_{r\alpha}(S_\alpha) = \frac{q_\alpha(S_\alpha)}{q_\alpha^{\text{sat}}},$$

where α again denotes fluid, and q_α^{sat} denotes the flux at fully saturated conditions, that is, $S_\alpha = 1$.

Before writing the multiphase version of Darcy's equation, we need to introduce one additional concept, to which we return shortly with more details. In multifluid systems, each of the fluids has its own pressure, related to, but different from, the pressure(s) in the other coexisting fluid(s). For example, in the unsaturated zone, the air pressure tends to be essentially equal to atmospheric pressure, while the water pressure varies significantly while remaining less than atmospheric pressure. So in a two-fluid system, with fluids denoted by w and nw, we have two distinct pressures, p_w and p_{nw}. These pressures lead to distinct values of fluid potential (pressure plus gravitational) for each fluid, and therefore the driving forces for flow in each fluid will be different.

With these concepts of relative permeability and phase pressures in mind, we can return to the Darcy equation and modify its mathematical form to include fluid-specific pressures and the relative permeability function. We first observe that the flux term in the equation must now be evaluated for each fluid present, and we use α to denote each of the fluids of interest. We make the assumption that each individual fluid α is driven to flow by gradients of potential within that fluid, and that the potential in each fluid is the sum of the pressure and elevation heads associated with that fluid. Therefore, if we return to the Darcy equation written in Eq. (11.5), on the right-hand side of the equation, the pressure must be fluid-specific, and the relative permeability function must be introduced into the equation. All of these considerations lead to the following extended form of Darcy's equation that accounts for multifluid flow:

$$\mathbf{q}_\alpha = -\frac{\mathbf{k}k_{r\alpha}}{\mu_\alpha}\left(\nabla p_\alpha - \rho_\alpha \mathbf{g}\right).\tag{11.6}$$

This simple heuristic argument about the dependence of permeability on saturation is born out by many experimental observations. Figure 11.2 shows typical shapes and behaviors for relative permeability functions for a two-fluid system. From these curves we make the following observations. First, the curves are clearly nonlinear, so that k_r is a nonlinear function of S. Sometimes the nonlinearity is captured by fitting a simple polynomial, often a cubic, to the data. More complicated functional fits are also used at times. No matter what functional fitting is used, the point to be remembered is that the coefficient in the equation is nonlinear, therefore the governing equations for these

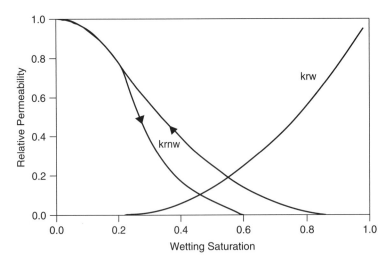

FIGURE 11.2. Typical relative permeability functions, plotted as a function of wetting fluid saturation.

multifluid problems will be nonlinear, which makes the equations more difficult to solve than their linear (single-fluid) counterparts. The second observation is that the relative permeability curves for a given fluid appear to go to zero at saturation values greater than zero. This happens when the fluid of interest still occupies some of the pore space, but the spatial distribution within the pore space lacks significant spatial connections. In that case there are no connected, or continuous, pathways across a length of porous medium that corresponds to the representative elementary volume (REV). This nonflowing saturation is usually referred to as residual saturation. So relative permeability has the characteristic that when saturation is at or below residual, $k_r = 0$. A third observation is that the general shape of the wetting curve is somewhat different from the shape of the nonwetting curve. Upon reduction of saturation from a value of one (full saturation) to a value somewhat less than one (say, 0.9), the wetting relative permeability function, k_{rw}, decreases more rapidly than the analogous decrease in k_{rnw} when the nonwetting fluid saturation is reduced from one to a similar saturation. Finally, we observe that the nonwetting fluid exhibits hysteresis in its relative permeability function; that is, the function measured when S_w is decreasing is different from the function when S_w is increasing. Hysteresis in the relative permeability function is more pronounced for the nonwetting fluid than for the wetting fluid. In Figure 11.2 no hysteresis is shown for k_{rw}.

11.2.3 Capillary Pressure and Governing Equations

If we replace the fluid volumetric flux terms in Eq. (11.3) by substitution of the multifluid version of Darcy's equation (Eq. (11.6)), then we have the following equation for each fluid α:

$$\frac{\partial}{\partial t}(\rho_\alpha \varepsilon S_\alpha) - \nabla \cdot \left(\rho_\alpha \frac{\mathbf{k} k_{r\alpha}}{\mu_\alpha} (\nabla p_\alpha - \rho_\alpha \mathbf{g}) \right) = \rho_\alpha Q^\alpha. \tag{11.7}$$

The primary unknowns in Eq. (11.7) are the two fluid pressures and the two fluid saturations. Assuming we know fluid densities, porosity, and the source/sink term, we now have two equations and four unknowns for a two-fluid system. A third equation arises from a simple volume argument, namely, that by definition all of the void space must be filled with fluid. This leads to the simple volume constraint

$$\sum_{\alpha \text{ fluid}} S_\alpha = 1. \tag{11.8}$$

This provides a third equation. Therefore we need only one more equation to close the system.

The final equation is almost always taken to be a functional relationship between fluid saturation S_α and the difference in pressure between the nonwetting and wetting fluids, which we define as the capillary pressure, P_c. To understand the physical basis of this relationship, we need to examine the multifluid system at the pore scale. Recall that, at the pore scale, distinct fluid–fluid interfaces serve as boundaries between the fluid phases. Interfaces between two fluid phases exist because of small-scale attractive forces within fluid phases, and these forces result in interfaces that have a kind of "mechanical" property, similar to a thin flexible membrane, wherein the interface can support a certain amount of stress without rupturing. Such a stress is generated across a fluid–fluid interface whenever the fluids on either side of the interface have different pressures. To understand how the system behaves, consider a thin capillary tube filled with wetting fluid, with one end submerged in a reservoir of wetting fluid, and the other end exposed to a second fluid that is nonwetting with respect to the solid material from which the tube is constructed. Assume that the two fluids start out at the same pressure. Now, let us increase slightly the pressure in the nonwetting fluid. It turns out that the interface responds to this pressure difference by deforming, such that it attains a nonzero curvature. If the nonwetting pressure is increased again, the curvature of the interface increases. Continued increases in nonwetting pressure eventually require a curvature that is beyond what is possible for the tube and fluid system; at that point the interface becomes unstable and the nonwetting fluid invades the tube, displacing the wetting fluid.

We can perform a simple force balance on this interface system to get some insight into the overall behavior. Consider the system shown in Figure 11.3, where nonwetting fluid pushes against the interface with pressure p_{nw}, wetting fluid pushes back with pressure p_w, and the interface has an attractive force with respect to the solid around its perimeter, where it connects to the solid surface of the tube. That force acts in the direction of the interface, so its directional components will depend on the angle between the interface and the solid. Let that angle be called the contact angle, denoted by $\theta_s^{nw,w}$, where superscripts identify the fluids and the subscript s denotes the solid material, thereby making it clear that the contact angle is a function of both the fluids and the solid. Then we have a system where two pressures act upon the interfacial area, while a force related to the interface acts along the line of contact between the two fluids and the solid, which we call the contact line. If we now perform a force balance in the direction of the tube axis, and assume the tube has radius R, then we have

$$(p_{nw} - p_w)\,\pi R^2 = 2\pi R \sigma^{nw,w} \cos \theta_s^{nw,w}, \qquad (11.9)$$

where we use the symbol $\sigma^{nw,w}$ to denote the force per unit length of contact (around the circumference of the contact line) due to the interface. This term, $\sigma^{nw,w}$, which is a

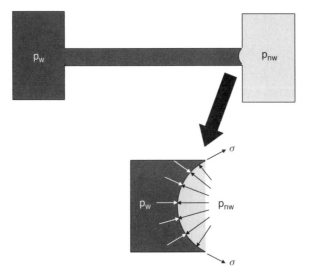

FIGURE 11.3. Behavior of an interface between two fluids in a capillary tube. The pressure difference between the fluids is balanced by the interfacial tension associated with the interface.

measure of "strength" of the interface, is called the *interfacial tension*. Equation (11.9) may be rearranged to give an expression relating capillary pressure to the interface properties:

$$p_{nw} - p_w = P_c = \frac{2\sigma^{nw,w}\cos\theta_s^{nw,w}}{R}. \tag{11.10}$$

For a given pair of fluids, call them α and β, and a given set of environmental variables like temperature and pressure, there exists a characteristic interfacial tension $\sigma^{\alpha\beta}$ for any pair of fluids α and β. For example, in a water–air system, the interfacial tension is about 70 dynes/cm. Contact angle $\theta_s^{\alpha,\beta}$ depends on the fluid pair and on the solid surface material. When contact angles are low, in the range of 0–30 degrees, the system is usually referred to as strongly wetting; larger angles lead to *intermediate* and even *mixed* wettability systems. For systems with more complex shapes, the exact form of Eq. (11.10) changes due to different geometric expressions, but the overall form of the relationship remains the same, with capillary pressure being directly proportional to interfacial tension and contact angle, and inversely proportional to a measure of "effective radius" of the pore opening. More details about interfacial phenomena can be found in Dullien [2], Sahimi [3], or Morrow [4].

We can now return to our earlier experiment and explain the behavior of the two fluids and the interface. When the fluid pressures are equal, there is no net force exerted on the interface by the two fluids, and therefore the interface is flat. As capillary pressure is increased, the interface curves more and more until the maximum curvature is reached. That maximum is the curvature of a spherical section that can fit within the cylindrical tube such that the interface makes an angle with the side of the tube equal to the contact angle $\theta_s^{\alpha,\beta}$. In this regard we interpret the capillary pressure expressed in Eq. (11.10) as the maximum pressure that the interface can support, given an interfacial tension, a contact angle, and a tube radius.

Now let us return to the idea that a relationship exists between capillary pressure and fluid saturation. Consider performance of an experiment using a device called a pressure cell. A schematic of such a device is shown in Figure 11.4. This cell is filled with a porous medium, with the top boundary exposed to a nonwetting fluid whose pressure is controlled, the bottom boundary exposed to wetting fluid whose pressure is also controlled, and all lateral sides constructed as impermeable boundaries.

Let us perform an experiment in which the pore space of the sample is initially filled completely with wetting fluid, and the pressure of the nonwetting fluid at the top boundary is equal to the pressure of the wetting fluid at the top of the sample. Close examination of the top boundary of the sample would reveal a complex distribution of pore spaces and a set of fluid–fluid interfaces that have no curvature (due to the equal phase pressures). Now let us increase the pressure of the nonwetting phase slightly. Then each of the fluid–fluid interfaces along the top boundary begins to deform, but none of them becomes unstable. As we continue to increase the nonwetting fluid pressure, keeping the wetting fluid pressure constant, we eventually reach a critical capillary pressure for which the interface in the largest pore along the boundary surface becomes unstable and drains through that pore. The interface will move through the porous medium, perhaps splitting into multiple interfaces as it encounters the tortuous and branching pathways in the pore space, until it (or they) reach sufficiently small pore(s) to be stable under the applied pressure difference.

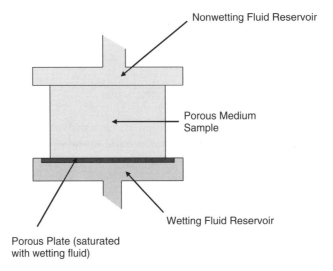

FIGURE 11.4. Schematic of a pressure cell, used to measure the relationship between capillary pressure and saturation for a given porous medium.

Additional increases in nonwetting fluid pressure lead to additional destabilizations of interfaces, subsequent displacement of wetting fluid by nonwetting fluid, and eventual achievement of a new equilibrium state. For each equilibrium state, we can measure the applied pressures as well as the amount of wetting fluid that has flowed out of the system. This provides one data pair in the relationship between fluid saturation and capillary pressure. If we keep track of these data pairs, we can then plot S versus P_c, and from that set of points attempt to derive a functional relationship. We can also perform the experiment in the opposite direction, by either decreasing nonwetting fluid pressure or increasing wetting fluid pressure, so that wetting fluid reenters the sample through displacement of nonwetting fluid. When nonwetting fluid displaces wetting fluid, we refer to the process as *drainage*. When wetting fluid displaces nonwetting fluid, the process is called *imbibition*.

An example of a typical $P_c - S$ relationship is presented in Figure 11.5. In that figure, capillary pressure, taken to be the difference between the imposed pressures in the nonwetting and wetting fluids, and wetting-fluid saturation are plotted, beginning with the fully saturated case ($S_w = 1$, $P_c = 0$).

We can observe three important features of these curves. First, in order to see any significant changes in saturation when beginning with $S_w = 1$, a nonzero capillary pressure must be applied. The capillary pressure at which saturation begins to show significant changes is referred to as the *entry pressure*. The second observation is that the wetting fluid saturation does not go to zero when P_c becomes large, but instead it appears to approach an asymptotic value that is greater than zero. We refer to this saturation value as the *residual wetting-fluid saturation* and denote it by S_w^{res}. Similarly, when the displacement process is reversed, nonwetting fluid saturation does not return to zero when $P_c = 0$, but instead it remains at a residual saturation value that we call *residual nonwetting-fluid saturation* and denote by S_{nw}^{res}. Finally, our third observation is that the path followed during drainage is different from the path followed during imbibition, meaning the relationship exhibits hysteresis. Finally, we note that both curves shown

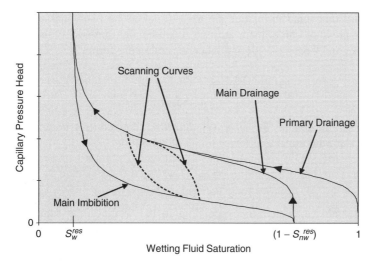

FIGURE 11.5. Typical form of the relationship between capillary pressure and saturation. Shown in this figure are different curves, which illustrate the nonlinear and hysteretic nature of the relationship.

in Figure 11.5 are nonlinear. Therefore the relationship between capillary pressure and saturation is both nonlinear and hysteretic.

Because the relationship between P_c and S is hysteretic, we have multiple curves that define this relationship. Some of these curves are given specific names. When we begin the displacement process at full saturation, the associated curve is referred to as a primary displacement curve. For example, in Figure 11.5, we see a *primary drainage curve*, which by definition begins at the point $P_c = 0$, $S_w = 1$. If we were to begin with full saturation of the nonwetting fluid, the curve is called the *primary imbibition curve*. If a displacement process is carried out until residual saturation of the displaced fluid is reached, the process is reversed and carried out until residual of the other fluid is reached, and the resulting curve is called the main displacement curve. In Figure 11.5, we identify both the *main imbibition curve* and the *main drainage curve*. Notice that these curves form a closed loop. If we move only part way along one of the main displacement curves, then reverse direction, we trace out what is called a *scanning curve*. Scanning curves typically lie within the closed loop formed by the main displacement curves.

There are many interesting aspects of the relationship between P_c and S, and the interested reader might want to consult the more detailed discussions in Dullien [2] or some recent research by, for example, Hassanizadeh and Gray [5], Held and Celia [6], and Hassanizadeh et al. [7] and references therein. For our purposes, we simply accept that some relationship between P_c and S exists, and we note that it is quite complex.

With this final relationship between capillary pressure and saturation, we have a fourth equation and therefore have achieved a system with four equations and four primary unknowns. With appropriate boundary and initial conditions, the equations can be solved.

11.2.4   Functional Forms for Relative Permeability and Capillary Pressure

While the functional relationships between capillary pressure and saturation, and between relative permeability and saturation, are complicated nonlinear functions, there are a few fairly standard functional forms that are used to define the functional relationships. For the capillary pressure curve, there are two functions that are widely used. The first is due to Brooks and Corey [8, 29] and takes the following form:

$$S_{\text{eff}} = 1, \quad P_c < P_{\text{entry}}, \tag{11.11}$$

$$P_c = P_{\text{entry}} (S_{\text{eff}})^{-1/\lambda}.$$

In Eq. (11.11), S_{eff} denotes the *effective saturation*, defined as a normalized saturation measure between the upper and lower limits of saturation:

$$S_{\text{eff}} = \frac{S - S_{\text{min}}}{S_{\text{max}} - S_{\text{min}}}. \tag{11.12}$$

For primary drainage, with S corresponding to the wetting-fluid saturation, we have $S_{\text{max}} = 1$ and $S_{\text{min}} = S_w^{\text{res}}$, while for main drainage or main imbibition, $S_{\text{min}} = S_w^{\text{res}}$ and $S_{\text{max}} = 1 - S_{nw}^{\text{res}}$. The parameter P_{entry} is the *entry pressure*, while the parameter λ is related to the pore-size distribution. These parameters are typically determined by fitting this equation to measured data.

Another popular functional form for the relationship between capillary pressure and saturation is due to van Genuchten [9] and takes the following form:

$$P_c = P^* \left[(S_{\text{eff}})^{-1/m} - 1 \right]^{1-m}, \tag{11.13}$$

where P^* and m are parameters that are fit to measured data. This function is often written with S as a function of P_c, which takes the following form:

$$S_{\text{eff}} = \left[1 + \left(\frac{P_c}{P^*} \right)^n \right]^{-m},$$

where $n = 1/(1 - m)$. Expansion of the expression for S_{eff} and replacement of P^* by $\alpha = 1/P^*$ yields a fairly standard form of the van Genuchten equation,

$$S(P_c) = S_{\text{min}} + \frac{S_{\text{max}} - S_{\text{min}}}{\left[1 + (\alpha P_c)^n \right]^m}.$$

Note that the parameter α has dimension of inverse pressure, $[\text{LT}^2\text{M}^{-1}]$. Sometimes pressure head is used instead of pressure, in which case the dimension for α would be $[\text{L}^{-1}]$. Notice that for both of these functional fits, hysteresis in the P_c–S relationship implies that different parameters will apply to the different curves for drainage and imbibition.

The relative permeability function is more difficult to measure in the laboratory than the P_c–S relationship. Therefore the relative permeability functions are often predicted from the P_c–S relationships. The basis of these derivations comes from the work of Burdine [10] and Mualem [11]. Brooks and Corey [8] used the ideas of Burdine to relate their P_c–S functional form to the relative permeability curves. They arrived at the following functional forms:

$$k_{rw} = (S_{\text{eff}})^{(2+3\lambda)/\lambda},$$

$$k_{rnw} = (1 - S_{\text{eff}})^2 \left[1 - (S_{\text{eff}})^{(2+\lambda)/\lambda} \right], \tag{11.14}$$

where λ is a fitting parameter. van Genuchten [9] used the Mualem [11] approach to relate his P_c-S function to the relative permeability function for the wetting phase (he was actually looking at unsaturated flow and therefore focusing on water), with the following functional result:

$$k_{rw} = (S_{\text{eff}})^{1/2} \left\{ 1 - \left[1 - (S_{\text{eff}})^{1/m} \right]^m \right\}^2 . \tag{11.15}$$

This concept was extended to include the nonwetting fluid by Parker et al. [12]:

$$k_{rnw} = \left[1 - (S_{\text{eff}}) \right]^{1/2} \left[1 - (S_{\text{eff}})^{1/m} \right]^{2m} . \tag{11.16}$$

Finally, for practical purposes, we often use simplified forms of the relative permeability functions, which capture the basic behavior but are functionally quite simple. They take the forms

$$\begin{aligned} k_{rw} &= \left(S_{\text{eff}} \right)^n , \\ k_{rnw} &= \left[1 - (S_{\text{eff}}) \right]^n , \end{aligned} \tag{11.17}$$

where $n = 3$ is often chosen. While these do not relate to the P_c-S relationships, they are often adequate to capture the relative permeability behavior.

11.2.5   Equation for Solid Phase

While our focus is almost always on the fluid phases in a porous medium, it is useful to have at our disposal equations for the solid phase. We will treat the solid as one composite phase, even though the composition of the solid could be quite complex. If we define the density of the solid to have the usual meaning of density, namely, mass of solid per volume of solid, and we note that the volume fraction occupied by the solid is equal to one minus the porosity, then we can use the standard mass balance approach to write the following equation for the solid phase:

$$\frac{\partial}{\partial t} \left(\rho_s \left(1 - \varepsilon \right) \right) + \nabla \cdot \left(\rho_s \mathbf{q}_s \right) = \rho_s Q^s . \tag{11.18}$$

In this equation, the volumetric flux vector for the solid, $\mathbf{q}_s$, is the volume of solid phase passing though a unit area of porous medium per unit time. We usually deal with solids that are essentially stationary, so that $\mathbf{q}_s = \mathbf{0}$, and therefore the balance equation reduces to

$$\frac{\partial}{\partial t} \left(\rho_s \left(1 - \varepsilon \right) \right) = \rho_s Q^s . \tag{11.19}$$

This equation shows that any changes to the mass of the solid occur due to exchanges between the solid and the adjacent fluid phases.

11.2.6   Equations for Components Within a Fluid

Most fluids of environmental concern are mixtures of different components. For example, air is a mixture of gases, including nitrogen, oxygen, and water vapor; crude oil is a complex mixture of components of different molecular weight; and groundwater almost always has solutes dissolved in it and therefore is not pure H_2O. Most solutes in groundwater occur at low concentrations and are considered benign, but some dissolved components are not benign and become identified as contaminants.

When we have more than one fluid in the pore space, each fluid will generally be composed of multiple components, one or more of which may be a contaminant of interest. Therefore we need to understand how components within each fluid behave in the environment. Thus we now turn our attention to mathematical descriptions of components within fluid phases. In much the same way that we wrote general equations for entire fluid phases, we can also write governing equations for individual components within a fluid. We will use measures of concentration of a component within a fluid using both mass-per-mass and mass-per-volume units. Let fluid α be composed of M components, denoted by $i = 1, 2, ..., M$. We define the mass fraction for component i in fluid α as (see discussion in Section 8.2)

$$\omega_i^\alpha \equiv \frac{\text{Mass of component } i \text{ in fluid } \alpha}{\text{Mass of fluid } \alpha} \tag{11.20}$$

and the associated concentration of component i as

$$c_i^\alpha = \frac{\text{Mass of component } i \text{ in fluid } \alpha}{\text{Volume of fluid } \alpha}. \tag{11.21}$$

Finally, we define the mass *flux vector for component i* within fluid α by the symbol $\mathbf{m}_i^\alpha$, where "flux" has the usual meaning of "per total area per time." With these definitions, we can repeat the standard box balance approach to derive a mass balance equation for component i in fluid α. The resulting expression may be written

$$\frac{\partial}{\partial t} \left(\rho_\alpha \varepsilon S_\alpha \omega_i^\alpha \right) + \nabla \cdot \left(\mathbf{m}_i^\alpha \right) = \rho_\alpha \varepsilon S_\alpha Q_i^\alpha. \tag{11.22}$$

In Eq. (11.22), the source/sink term on the right-hand side of the equation now has subscript i to denote that it represents the addition or subtraction of mass of component i to/from fluid α due to exchange with other fluids or with the solid, or due to reactions within fluid α. The variable Q_i^α has dimensions of mass of component i per total mass of fluid α per time, or $[T^{-1}]$. Just as in the single-fluid case (see Chapter 8), the mass flux term is composed of two parts, one due to bulk movement of the fluid, which we call *advection*, and a second due to the relative motion of component i with respect to the bulk fluid movement, which we call the *nonadvective flux*. We therefore expand the mass flux vector as

$$\mathbf{m}_i^\alpha = \rho_\alpha \mathbf{q}_\alpha \omega_i^\alpha + \mathbf{J}_i^\alpha. \tag{11.23}$$

In addition, we will assume that the nonadvective flux may be represented by an analog to *Fick's law*, where the coefficient is understood to represent both diffusion and mechanical dispersion. Therefore we write a more expanded version of Eq. (11.23) as

$$\mathbf{m}_i^\alpha = \rho_\alpha \mathbf{q}_\alpha \omega_i^\alpha - \rho_\alpha \varepsilon S_\alpha \mathbf{D}_i^\alpha \cdot \nabla \omega_i^\alpha. \tag{11.24}$$

In Eq. (11.24), the coefficient $\mathbf{D}_i^\alpha$ is the *effective dispersion coefficient*, generally a 3×3 tensor that includes molecular diffusion and mechanical dispersion, and could be a function of saturation (see Section 8.3.1). We will not explore possible relationships to saturation here; the interested reader is referred to Sahimi [3] and references therein.

Substitution of Eq. (11.24) into Eq. (11.22) gives a fairly general statement of mass balance for component i in fluid α, assuming Fickian behavior of the dispersion term, that is,

$$\frac{\partial}{\partial t} \left(\rho_\alpha \varepsilon S_\alpha \omega_i^\alpha \right) + \nabla \cdot \left(\rho_\alpha \mathbf{q}_\alpha \omega_i^\alpha \right) - \nabla \cdot \left(\rho_\alpha \varepsilon S_\alpha \mathbf{D}_i^\alpha \cdot \nabla \omega_i^\alpha \right) = \rho_\alpha \varepsilon S_\alpha Q_i^\alpha. \qquad (11.25)$$

This equation of course needs to be augmented with additional information, which we discuss shortly. But before moving on to practical simplifications for applications, we note that Eq. (11.25) can be combined with the equation of mass balance for the entire fluid (Eq. (11.3)) to produce a different version of the equation for component i. To do this, we expand the first two derivative terms on the left-hand side of Eq. (11.25), so that the equation is written, in equivalent but expanded form, as

$$\omega_i^\alpha \left[\frac{\partial}{\partial t} \left(\rho_\alpha \varepsilon S_\alpha \right) + \nabla \cdot \left(\rho_\alpha \mathbf{q}_\alpha \right) \right] + \rho_\alpha \varepsilon S_\alpha \frac{\partial \omega_i^\alpha}{\partial t} + \rho_\alpha \mathbf{q}_\alpha \cdot \nabla \omega_i^\alpha$$
$$- \nabla \cdot \left(\rho_\alpha \varepsilon S_\alpha \mathbf{D}_i^\alpha \cdot \nabla \omega_i^\alpha \right) = \rho_\alpha \varepsilon S_\alpha Q_i^\alpha.$$

We now observe that the term in square brackets is equal to $\rho_\alpha Q^\alpha$, from the fluid balance equation. Therefore Eq. (11.25) may be rewritten

$$\rho_\alpha \varepsilon S_\alpha \frac{\partial \omega_i^\alpha}{\partial t} + \rho_\alpha \mathbf{q}_\alpha \cdot \nabla \omega_i^\alpha - \nabla \cdot \left(\rho_\alpha \varepsilon S_\alpha \mathbf{D}_i^\alpha \cdot \nabla \omega_i^\alpha \right) = \rho_\alpha \left(Q_i^\alpha - \omega_i^\alpha \varepsilon S_\alpha Q^\alpha \right). \quad (11.26)$$

Comparison of Eqs. (11.25) and (11.26) reveals that Eq. (11.25) has a form with a time derivative of a mass term being balanced by the divergence of a total flux, plus any source or sink term for the component and fluid of interest. That form is lost when the total fluid equation is combined with the component equation, resulting in a form that does not show a clear balance between accumulation and flux. Because of these differences in form, and the fact that the first form (Eq. (11.25)) corresponds directly to the underlying statement of mass balance, Eq. (11.25) is referred to as the *conservative form* of the component equation, while Eq. (11.26) is referred to as the *nonconservative form* of the equation. Sometimes one form of the equation is more convenient than the other, for example, in numerical simulations and the derivation of numerical approximations, but both are equally correct, and their use is a matter of choice, which depends on the context of the problem and the analysis.

11.2.7   Equations for Components Within the Solid

To complete our presentations of mass balance equations, we include the equation for component i within the solid phase (phase s). If ω_i^s denotes the mass fraction, and if we assume that no diffusion or dispersion takes place within the solid phase, then the appropriate mass balance equation is

$$\frac{\partial}{\partial t} \left(\rho_s \left(1 - \varepsilon \right) \omega_i^s \right) + \nabla \cdot \left(\rho_s \mathbf{q}_s \omega_i^s \right) = \rho_s \left(1 - \varepsilon \right) Q_i^s, \qquad (11.27)$$

where the source/sink term Q_i^s has dimensions of mass of component i per mass of solid s per time, or $[\mathrm{T}^{-1}]$. If we again assume the solid phase to be stationary, so that $\mathbf{q}_s = \mathbf{0}$, then the equation reduces to

$$\frac{\partial}{\partial t} \left(\rho_s \left(1 - \varepsilon \right) \omega_i^s \right) = \rho_s \left(1 - \varepsilon \right) Q_i^s. \tag{11.28}$$

11.2.8   Reactions, Mass Transfer, and Source/Sink Terms

When dealing with components of a fluid, we sometimes need to describe the net transfer of mass from one fluid to another. For example, when a soil dries by evaporation, liquid H_2O moves across the air–water interface and becomes water vapor, which is a component of the air. Similarly, if a hydrocarbon is spilled onto the land surface and seeps into the unsaturated zone, some of the components of the hydrocarbon may volatilize into the air, while some components may dissolve into the water. And some of the components that dissolve into the water may wind up sorbing onto the solid material. In all of these examples, there is a net flux across the fluid–fluid or fluid–solid interfaces that leads to a change in the amount of mass of that component in each of the fluids as well as the solid. These are all examples of mass transfers between fluids, or between fluid and solid, which we will call inter fluid, or inter phase, mass transfer. When viewed from the representative elementary volume (REV) (see Section 8.3) scale, these exchanges of components between fluids or phases appear to be internal source or sink terms, because at the REV scale individual interfaces that separate different phases are not resolved and therefore do not appear as individual boundaries but rather as part of the unresolved internal structure of the system. Therefore these mass exchanges appear in the source/sink term Q_i^α.

In addition to transfer of mass between pairs of fluids or between a fluid and solid, components may be added to, or lost from, a particular fluid due to chemical or biological reactions. An example of this might be the progressive decay of dissolved radioactive components within a fluid (see Section 8.5).

Given the different ways that components can be added to or subtracted from a fluid within the porous medium, it is useful to expand the source/sink term to more explicitly represent these different pathways. Let the net rate of exchange between fluid α and any other fluid be denoted by $Q_i^{\alpha\beta}$, where β denotes the fluid with which fluid α is exchanging component i. In addition, let the net rate of exchange between fluid α and the composite solid phase be denoted by $Q_i^{\alpha s}$. Finally, let any reactions taking place within fluid α, and involving component i, be denoted by $Q_i^{\alpha,rxn}$. Then the source/sink term for component i within fluid α may be expressed as

$$Q_i^\alpha = \sum_{\substack{\beta \text{ Fluid} \\ \beta \neq \alpha}} Q_i^{\alpha\beta} + Q_i^{\alpha s} + Q_i^{\alpha,rxn}. \tag{11.29}$$

Note that there are some constraints on the terms appearing in Eq. (11.29). In particular, the amount of component i that leaves fluid α and enters fluid β must be equal and opposite to the amount of component i that leaves fluid β and appears in fluid α. That is,

$$\begin{aligned}
\rho_\alpha \varepsilon S_\alpha Q_i^{\alpha\beta} &= -\rho_\beta \varepsilon S_\beta Q_i^{\beta\alpha}, \\
\rho_\alpha \varepsilon S_\alpha Q_i^{\alpha s} &= -\rho_s \left(1 - \varepsilon \right) Q_i^{s\alpha}.
\end{aligned} \tag{11.30}$$

This set of constraints is important when we consider equilibrium versus kinetic descriptions of the mass exchange phenomenon.

Recall that when dealing with sorption for the saturated porous-medium ($S_w = 1$) case, we found that it is often reasonable to use an assumption of equilibrium partitioning between the water and the solid. That is consistent with the time scale for solute partitioning between the water and solid being very fast relative to other time scales in the system, such as those associated with advection and dispersion. In that case, we defined sorption isotherms and wrote algebraic relationships between the concentration of solute in solution and the concentration sorbed to the solid. That relationship could be linear, in which case we had linear isotherms, or it could be nonlinear, in which case standard nonlinear algebraic functions like those attributed to Freundlich or Langmuir are often used.

The idea of equilibrium sorption can be extended to more general equilibrium partitioning of a component among different fluids and/or solids present in the system. If we have a simple case of a liquid and a gas, then we often use an equilibrium partitioning equation based on *Henry's law*. More generally, partitioning among multiple fluids and/or solids requires more sophisticated equations, about which we say a few words in the later section dealing with compositional models.

While partitioning equations may be somewhat complicated, there is an important overall point that needs to be understood regarding use of equilibrium partitioning. Recall that equilibrium partitioning is consistent with times scales for partitioning being very fast relative to other time scales for the problem. The assumption of equilibrium partitioning in fact implies that the exchange between or among fluid(s) and solid(s) occurs instantaneously, such that the rate of exchange is infinitely fast.

However, this is a problem from the mathematical point of view, because the exchange term Q_i^α on the right-hand side of our component balance equation is a rate term, and if the exchange is "instantaneous" then the rate becomes infinitely large. This makes the component balance equation written for a specific fluid (in this case fluid α) difficult to deal with. The way we get around this problem is through use of the constraints in Eq. (11.30). For all fluids or solids involved in the equilibrium partitioning, we write the component equation and then sum those equations over all of the fluids and solids involved (see Section 8.3). Because the exchange terms must be equal and opposite, they cancel one another, independent of their magnitudes. Therefore all component equations should be summed over all fluid and solid phases whenever equilibrium partitioning takes place. In the case of equilibrium partitioning, once one of the mass fractions is known (or once the total mass, summed over all phases, of component i is known), then all of the mass fractions can be determined from the equilibrium relationships. Therefore only one transport equation needs to be solved, and from that solution all mass fractions can be determined.

We have already seen this principle applied when we dealt with equilibrium sorption in the fully saturated case ($S_w = 1$). In that case, we summed the component equation for water and for the solid, leading to the combined time-derivative term that in turn leads to the retardation coefficient (see Eq. (8.26)). Once the concentration in the water was determined, the sorbed concentration could be determined easily from the sorption isotherm. If we were to have an air–water system with a volatile and sorbing contaminant dissolved in the water, and we decided equilibrium partitioning between water and solid and between water and air was appropriate, then we would sum the component equations for water, air, and solid, and that would be the equation we write for the component of interest. Equilibrium relationships among the sorbed, dissolved, and volatilized mass fractions for the component would then be used to determine the partitioning of the component among the three phases.

If the time scale for mass exchange is not fast relative to transport processes, then use of equilibrium partitioning is not appropriate, and a kinetic description must be used. Usually this involves writing an expression for mass transfer that is proportional to the difference between the current mass fraction in the fluid and the equilibrium mass fraction. If $\overline{\omega}_i^\alpha$ denotes the *equilibrium value of the mass fraction in fluid* α, then the source/sink term for the rate of mass exchange involving other fluids, and possibly the solid, takes the form

$$Q_i^\alpha = \kappa_i^\alpha \left(\omega_i^\alpha - \overline{\omega}_i^\alpha \right), \tag{11.31}$$

where the coefficient κ_i^α is an effective rate coefficient that has dimension $[T^{-1}]$. When kinetic exchange terms are used, the mass fractions in each of the fluids, as well as the solid, need to be determined through solution of the transport equations. It is no longer possible to solve only one (summed) transport equation and then use equilibrium partitioning relationships to infer all other mass fractions. Of course, the constraints of Eq. (11.30) still apply, so we can choose to sum the equations over fluid and/or solid phases, but this does not eliminate the need to solve a set of transport equations for the individual mass fractions. For more information about these kinds of systems, and example calculations, see Lichtner [13].

Equations of State   In general, equations of state provide equations for fluid properties, like density, as a function of pressure, temperature, and composition. Depending on the problem under consideration, these might be highly simplified expressions, or they might be quite complex. For example, in the simple isothermal groundwater flow problem, without consideration of compositional effects, a simple linear relationship between density and pressure might be used, such as that in Eq. (4.19) in Chapter 4. Combined with matrix compressibility, this equation leads to the standard storativity coefficient. In more complicated problems, which might involve large changes in pressure or temperature, the fluid might change phase, so that water could convert from liquid to vapor, or perhaps vapor under very high pressure (steam) could depressurize and change to a liquid. In these cases, which are applicable to problems involving things like geothermal energy, the equations of state are quite complicated.

We will write the general functional relationships as

$$\rho_\alpha = \rho_\alpha \left(p_\alpha, T, \mathbf{\Omega}^\alpha \right),$$
$$\mu_\alpha = \mu_\alpha \left(p_\alpha, T, \mathbf{\Omega}^\alpha \right) \tag{11.32}$$

where $\mathbf{\Omega}^\alpha$ is used to denote the vector of mass fractions for all components of fluid α, so that $\mathbf{\Omega}^\alpha = \left(\omega_1^\alpha, \omega_2^\alpha, \dots, \omega_M^\alpha \right)$. This represents the dependence of density and viscosity on fluid composition. For many problems, we assume isothermal conditions so that the temperature is assumed constant within the problem domain. Many practical problems also involve very dilute solutions, so that composition dependence is very weak and can be neglected. One exception of hydrological importance is the problem of saltwater intrusion in coastal aquifers, where the density of the salt water causes significant density-dependent fluxes that usually need to be included in the problem analysis.

Depending on the level of complexity of the system being studied, we may need to write equations to model deformation of the porous medium beyond the simple linear elastic response model of Chapter 4. This would provide information about changes in porosity. We may also need to solve equations for energy transport, from which we calculate system temperature as a function of spatial location and time. All of these require additional equations, which are beyond the scope of what we will cover in this book.

11.3 THE UNSATURATED ZONE AND RICHARDS' EQUATION

The unsaturated zone is a two-fluid porous-media system with air and water as the two fluids. We have already noted the importance of the unsaturated zone in the transmission of water from the atmosphere to the saturated zone via infiltration of precipitation, in the support of plant life, and in the active return of water from the subsurface to the atmosphere. In this section we consider the unsaturated zone more closely, noting that the disparity in properties between air and water allows us to make some simplifications to the system of two-fluid equations, and thereby allows for simplified methods for quantitative analysis of this important zone of the environment.

11.3.1 Properties of Air and Water and Richards' Assumption

Air at typical atmospheric conditions (1 atm, 20 °C) is a gas that has density about 1 kg/m^3 and a viscosity of about 1.8×10^{-5} Pa · s. Water at the same conditions has density of about 1000 kg/m^3 and viscosity of about 10^{-3} Pa · s. Therefore air is approximately three orders of magnitude less dense than water, and two orders of magnitude less viscous. With these differences in mind, let us consider the multifluid version of Darcy's equation, Eq. (11.6), where we observe that the flux term is inversely proportional to the fluid viscosity. In order for a certain volumetric flux of fluid to be achieved, if all else were equal, a given horizontal flux of air requires a pressure gradient roughly 100 times less than the pressure gradient needed to achieve the same flux of water. This argument leads us to expect significantly smaller pressure gradients in the air than in the water, in unsaturated soils. Furthermore, because the density of air is very small, the static distribution of pressure in a column of air also changes little over distances applicable to the unsaturated zone. Because the top boundary of the unsaturated zone is the atmosphere, the air pressure at the top of the unsaturated zone is equal to atmospheric pressure. Because density does not cause significant changes in pressure, and because low viscosity allows air to move without significant build up of pressure, we often observe that air pressure in the unsaturated zone does not differ much from atmospheric. Therefore for most problems it is reasonable to assume that air remains essentially at atmospheric pressure. We will call this assumption the *Richards' assumption,* after L. A. Richards, whose paper of 1931 [14] introduced the equations for water flow in soils.

We want to stress that Richards' assumption does not imply a static air phase. It is based on an assumption that the air phase has sufficiently low viscosity that we can assume it to be inviscid. In that case it is infinitely mobile and can move without any appreciable pressure gradients. Logic that argues for a static air phase due to an assumption of constant air pressure is wrong.

11.3.2 Richards' Equation for Water Flow

If we use the Richards' assumption, we eliminate air pressure from the set of primary state variables, leaving water pressure, water saturation, and air saturation as the three unknowns. The volume constraint of Eq. (11.8) allows one of the saturations to be eliminated (say, the air saturation), leaving two equations and two unknowns. The remaining equations are the *water mass balance equation* (already combined with the Darcy equation for the water) and the *capillary pressure–saturation relationship* and the two unknowns are the *water pressure* and *water saturation*. Of course, these equations may be augmented by appropriate equations of state, including compressibility, but the underlying equation forms do not change.

Let us assume we are using a reference pressure equal to atmospheric pressure, so that $p_{atm} = 0$. Richards' assumption implies that capillary pressure is equal to the negative of the water pressure, $P_c = p_a - p_w = 0 - p_w = -p_w$. Therefore the P_c–S relationship implies that water saturation is a function of water pressure. Thus all equations can be written in terms of just these two variables.

If the water mass balance equation is written, with $\alpha = w$, in Eq. (11.7), then we have

$$\frac{\partial}{\partial t}(\rho_w \varepsilon S_w) - \nabla \cdot \left(\rho_w \frac{\mathbf{k} k_{rw}}{\mu_w} (\nabla p_w - \rho_w \mathbf{g}) \right) = \rho_w Q^w. \tag{11.33}$$

Now let us apply some reasonable simplifying assumptions for problems of unsaturated flow. First, assume water density and porosity are constant, and assume there are no source or sink terms. Then Eq. (11.33) reduces to

$$\varepsilon \frac{\partial S_w}{\partial t} - \nabla \cdot \left(\frac{\mathbf{k} k_{rw}}{\mu_w} (\nabla p_w - \rho_w \mathbf{g}) \right) = 0. \tag{11.34}$$

Next, we define the pressure head as the pressure scaled to length units by the specific gravity of water. Denote the pressure head by ψ, so that $\psi_w = p_w / \rho_w g$. Then Eq. (11.34) may be written

$$\varepsilon \frac{\partial S_w}{\partial t} - \nabla \cdot \left[\left(\frac{\mathbf{k} \rho_w g}{\mu_w} \right) k_{rw} (\nabla \psi_w + \nabla z) \right] = 0. \tag{11.35}$$

We recognize the grouping $\mathbf{k} \rho_w g / \mu_w$ as the usual definition of hydraulic conductivity for saturated groundwater flow. Therefore we denote it as $\mathbf{K}_{sat}^w$ and rewrite Eq. (11.35) as

$$\varepsilon \frac{\partial S_w}{\partial t} - \nabla \cdot \left[\mathbf{K}_{sat}^w k_{rw} (\nabla \psi_w + \nabla z) \right] = 0. \tag{11.36}$$

Finally, we note that the more common measure of fluid saturation in the field of soil science is the fluid content, which is the volume of fluid per total volume of porous medium. For water, we have $\theta_w = \varepsilon S_w$, which is often referred to as the moisture content. With θ_w replacing εS_w, Eq. (11.36) becomes

$$\frac{\partial \theta_w}{\partial t} - \nabla \cdot \left[\mathbf{K}_{sat}^w k_{rw} (\nabla \psi_w + \nabla z) \right] = 0. \tag{11.37}$$

We refer to this equation as the three-dimensional Richards' equation, written in *mixed form*. The "mixed form" means that both the saturation (or moisture content) and the pressure appear as primary variables in the equation. We have not used information contained in the $P_c - S$ relationship to replace one of these variables with the other.

Equation (11.37) may be written in several different forms, depending on whether the time or space derivatives are expanded using the $P_c - S$ relationship. For example, the time derivative in Eq. (11.37) can be expanded by noting that θ_w is a function of pressure head ψ_w, and therefore

$$\frac{\partial \theta_w}{\partial t} = \frac{d\theta_w}{d\psi_w} \frac{\partial \psi_w}{\partial t} = C(\psi_w) \frac{\partial \psi_w}{\partial t}. \tag{11.38}$$

In Eq. (11.38), the term denoted by $C(\psi_w)$ is called the *specific moisture capacity*. It is equal to the slope of the θ_w versus ψ_w curve and corresponds to the moisture content change per unit change of pressure head. When this expanded time-derivative term is substituted into Eq. (11.37), we find a governing equation with the primary dependent variable being pressure head,

$$C(\psi_w) \frac{\partial \psi_w}{\partial t} - \nabla \cdot \left[\mathbf{K}_{sat}^w k_{rw} \left(\nabla \psi_w + \nabla z \right) \right] = 0. \tag{11.39}$$

In a similar way, we could expand the spatial derivatives of pressure head and convert them into derivatives of moisture content. Then we have an equation in which both the time and space derivatives apply to the moisture content. The appropriate expansion is

$$\nabla \psi_w = \frac{d\psi_w}{d\theta_w} \nabla \theta_w,$$

which when substituted into Eq. (11.37) leads to

$$\begin{aligned}
\frac{\partial \theta_w}{\partial t} &- \nabla \cdot \left[\mathbf{K}_{sat}^w k_{rw} \left(\frac{d\psi_w}{d\theta_w} \nabla \theta_w + \nabla z \right) \right] \\
&= \frac{\partial \theta_w}{\partial t} - \nabla \cdot \left[\left(\frac{\mathbf{K}_{sat}^w k_{rw}}{d\theta_w / d\psi_w} \right) \nabla \theta_w + \mathbf{K}_{sat}^w k_{rw} \nabla z \right] \\
&= \frac{\partial \theta_w}{\partial t} - \nabla \cdot \left[D(\theta_w) \nabla \theta_w \right] - \nabla \cdot \left(\mathbf{K}_{sat}^w k_{rw} \right) \nabla z = 0. \tag{11.40}
\end{aligned}$$

In Eq. (11.40), the coefficient $D(\theta)$ is called the *soil moisture diffusivity*. We refer to Eq. (11.40) as the "θ-based" or "moisture content-based" form of Richards' equation, while Eq. (11.39) is called the "ψ-based" or "pressure head-based" form of Richards' equation. The θ-based form of the equation is appropriate for unsaturated systems but is not appropriate for systems that include saturated zones. In the saturated zone, the moisture content is equal to the porosity, and gradients of porosity are not suitable as a gradient that drives fluid fluxes. The potential gradient is not recovered when multiplying by D because the coefficient D becomes unbounded ($d\theta_w / d\psi_w \to 0$). Therefore for saturated zones we should use the pressure-based form of the flux term. Thus, if saturated conditions are a possibility, either the mixed form or the pressure head-based form of Richards' equation should be used.

Note also that all forms of Richards' equation that we have written to this point assume that soil and fluid compressibility can be neglected. This is usually reasonable, because changes in moisture content tend to be much larger than any changes in storage volume due to compressibility. However, we can include compressibility effects if we choose to do so. This might be important for cases where saturated zones exist, because in those zones the water saturation is constant at a value of one, so any time derivative terms must include compressibility. If we begin with the general equation for mass balance (Eq. (11.3), with $\alpha = w$), we can use the following expansion of the time derivative:

$$
\frac{\partial}{\partial t} (\rho_w \varepsilon S_w) + \nabla \cdot (\rho_w \mathbf{q}_w) = \rho_w \varepsilon \frac{\partial S_w}{\partial t} + S_w \frac{\partial}{\partial t} (\rho_w \varepsilon) + \nabla \cdot (\rho_w \mathbf{q}_w)
$$

$$
= \rho_w \varepsilon \frac{\partial S_w}{\partial t} + S_w S_s^w \rho_w \frac{\partial \psi_w}{\partial t} + \nabla \cdot (\rho_w \mathbf{q}_w),
$$

where we assume that the storativity coefficient S_s can be applied in the same way we used it for the fully saturated case (page 174), despite the fact that now we have some of the pore space filled with air. If we now assume that density of water does not change much, such that

$$
\mathbf{q}_w \cdot \nabla \rho_w \ll \rho_w \nabla \cdot \mathbf{q}_w,
$$

then the density term can be removed from all of terms in the equation and we have the following *volume balance equation* that includes compressibility:

$$
\varepsilon \frac{\partial S_w}{\partial t} + S_w S_s^w \frac{\partial \psi_w}{\partial t} + \nabla \cdot \mathbf{q}_w = Q^w.
$$

We can then write the three forms of Richards' equation (mixed, ψ-based, and θ_w-based) including compressibility, and assuming no source on sink term:

$$
\frac{\partial \theta_w}{\partial t} + S_w S_s^w \frac{\partial \psi_w}{\partial t} - \nabla \cdot \left[\mathbf{K}_{sat}^w k_{rw} (\nabla \psi_w + \nabla z) \right] = 0, \tag{11.41a}
$$

$$
\left[C(\psi_w) + S_w S_s^w \right] \frac{\partial \psi_w}{\partial t} - \nabla \cdot \left[\mathbf{K}_{sat}^w k_{rw} (\nabla \psi_w + \nabla z) \right] = 0, \tag{11.41b}
$$

$$
S_w S_s^w \frac{\partial \psi_w}{\partial t} + \frac{\partial \theta_w}{\partial t} - \nabla \cdot [D(\theta_w)\nabla \theta_w] - \nabla \cdot \left(\mathbf{K}_{sat}^w k_{rw} \right) \nabla z = 0. \tag{11.41c}
$$

Finally, we note that Richards' equation is often applied in the vertical dimension only, and the saturated hydraulic conductivity and relative permeability are sometimes combined into an overall unsaturated hydraulic conductivity, denoted simply by K^w. Ignoring compressibility, and dropping the subscript and superscript of w (the water is simply implied in the equation), the simple one-dimensional forms of Richards' equation become

$$
\frac{\partial \theta}{\partial t} - \frac{\partial}{\partial z} \left(K \frac{\partial \psi}{\partial z} \right) - \frac{\partial K}{\partial z} = 0, \tag{11.42}
$$

$$
C(\psi) \frac{\partial \psi}{\partial t} - \frac{\partial}{\partial z} \left(K \frac{\partial \psi}{\partial z} \right) - \frac{\partial K}{\partial z} = 0, \tag{11.43}
$$

$$
\frac{\partial \theta}{\partial t} - \frac{\partial}{\partial z} \left(D \frac{\partial \theta}{\partial z} \right) - \frac{\partial K}{\partial z} = 0. \tag{11.44}
$$

11.3.3 Water and Air Dynamics in the Unsaturated Zone

Recall that the basis of Richards' equation is the assumption that air is much more mobile than water, and therefore it can move easily with very little buildup of pressure. This means that air pressure remains close to atmospheric pressure for most cases of flow in unsaturated soils. We can demonstrate this by calculating a solution to the full two-fluid equation set for air and water, and observing both the pressure buildup in the air, and the movement of the air phase. Consider a simple one-dimensional infiltration problem in which a rainfall rate is applied to the top boundary, and the bottom boundary is maintained at fixed fluid pressures so that both air and water are allowed to escape through it. This problem was solved by numerical methods by Celia and Binning [15] and is based on an experimental system used by Touma and Vauclin [16]. The functional parameters, determined experimentally by Touma and Vauclin, were defined as follows:

$$S_w(P_c) = \frac{S_{ws} - S_{wr}}{\left[1 + \left(\alpha \frac{P_c}{\rho_{0w} g}\right)^n\right]^m} + S_{wr},$$

$$k_{rw} = \frac{A_w}{K_{ws}} (\varepsilon S_w)^{B_w},$$

$$k_{ra} = \frac{A_a}{A_a + \left(\dfrac{P_c}{\rho_{0w} g}\right)^{B_a}},$$

where $\alpha = 0.044$ cm^{-1}, $S_{ws} = 0.843$, $S_{wr} = 0.0.072$, $n = 2.2$, $m = 1 - 1/n = 0.545$, $A_w = 18,130$ cm/h, $B_w = 6.07$, $\varepsilon = 0.37$, $K_{ws} = 15.4$ cm/h, $A_a = 3.86 \times 10^{-5}$, $B_a = -2.4$, and ρ_{0w} is a reference density for water. The applied infiltration rate was taken to be 8.3 cm/h. This rate is less than the saturated hydraulic conductivity, so that full saturation is not reached at the top boundary. The initial conditions were taken to be fluid-static, with hydrostatic conditions in the water and air-static conditions for the air pressure.

Results for the pressure in both the water and the air are shown in Figure 11.6. Each curve in the figure represents water and air pressures at a given time, beginning with the solution at $t = 10$ minutes and ending with the solution at $t = 90$ minutes, with the solution plotted at 10 minute intervals. We see that an infiltration front of water forms with water pressure head behind the infiltrating front at about -15 cm (water equivalent height), and pressure ahead of the front corresponding to the initial condition. This produces large pressure gradients of several tens of centimeters of pressure head change over just a few centimeters in the vertical direction. A clear infiltration front for the water can be seen in the results.

Contrary to these relatively large gradients and large pressure changes for the water, the air pressure only increases to a few centimeters, with the maximum air pressure occurring at the location of the infiltrating water front. This traveling maximum location for air pressure causes air to flow both upward and downward, with the flow direction changing at the infiltration front. So air behind the front flows upward and out the top boundary, while air ahead of the front flows downward and out the bottom boundary. This flow occurs with very little buildup of air pressure.

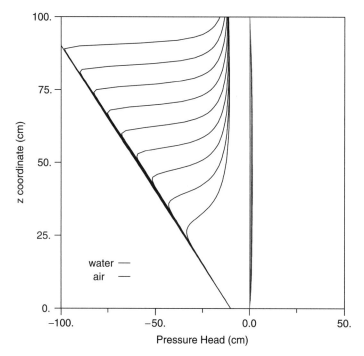

FIGURE 11.6. Plot of both water and air pressure heads as a function of depth, for different values of time. The water pressure head changes over a range of about 100 cm of water, while the air pressure head changes by only a small fraction of that amount (from [15]).

Figure 11.7 shows the evolution of this system in a somewhat different way. The dynamics of water infiltration are represented by shading to indicate moisture content, which is plotted on depth and time axes. Superimposed on this representation of moisture content are space–time traces of air particles, with particles initially placed at equal intervals in the vertical direction at time $t = 0$. The reversal from downward to upward movement of the air, after the water front passes, can be seen clearly in the figure.

This example calculation shows that air in the unsaturated zone is dynamic, and in some sense the unsaturated zone "breathes" as water infiltrates and then drains. Infiltration of water expels air, then subsequent drainage of the water draws air back into the soil. These dynamics can be important for vegetation health, and also for transport of volatile contaminants that partition into the air within the unsaturated zone.

To give a more complete picture of air dynamics, we present two additional cases. The first is the same one-dimensional problem that we just solved, but now we assume the bottom boundary is closed to flow. That means any air that escapes from the column must do so through the top boundary. In this case we see a larger pressure buildup in the air phase, as air from all parts of the column must eventually flow upward in order for water to be able to infiltrate—see Figure 11.8. Now buildup of air pressure becomes obviously larger than in the previous case. In these kinds of restricted flow environments, there may be cases when air pressure has to be included in the overall equations. Particle traces are again shown for this case in Figure 11.9. As expected, all air flow is upward behind the front, with essentially no movement ahead of the front. The slight downward

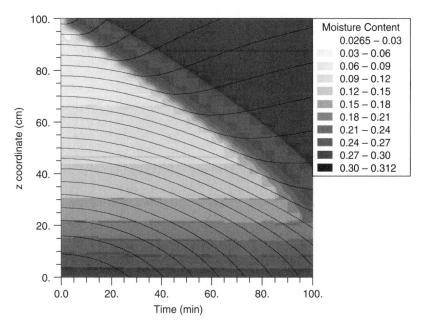

FIGURE 11.7. Water infiltration and air movement (from [15]).

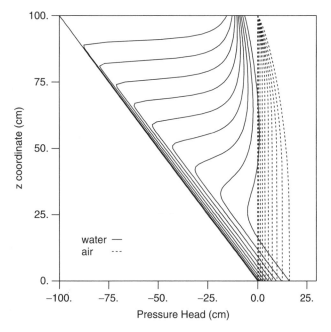

FIGURE 11.8. Plot of water and air pressure heads as a function of depth, for different times, for a system with impervious bottom boundary (from [15]).

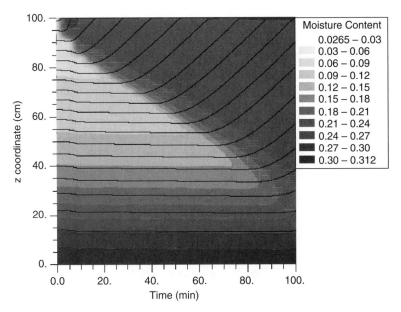

FIGURE 11.9. Plot showing water saturation as a function of depth and time (from [15]).

movement ahead of the front is caused by compressibility of the air, which is included in the model.

Finally, we show a multidimensional result, corresponding to infiltration from a strip of irrigated soil, where the water is applied along the top boundary between $x = 0$ and $x = 10$ cm. The simulation uses heterogeneous soil properties and calculates both air and water pressures and velocities. In this case the air pressure has very little buildup, and the flow directions, shown in Figure 11.10, for the air are more complex than in the one-dimensional cases but still consistent with the general observations from one dimension. In particular, we see the infiltrating water front pushing air out ahead of it, in this case causing movement both downward and laterally outward. Behind the front is a region of upward movement in the air. Again, we see a dynamic system with air being expelled from the unsaturated zone. After cessation of infiltration, drainage of the water into the saturated zone or uptake by root systems leads to pores changing from water to air occupancy, and the additional air to fill the spaces vacated by water comes in from the land surface.

11.3.4   Plant Roots and Evapotranspiration

The term *evapotranspiration* refers to loss of liquid water from the subsurface by a combination of *evaporation* and *transpiration*. Evaporation is the direct change of phase of water from liquid to gas, thereby representing a sink of liquid water. Transpiration refers to the uptake of liquid water from the soil into the plant root system and therefore is also a sink for liquid water in the root zone. Water taken up by the plant roots is then transported through the plant, nutrients dissolved in the water are extracted and utilized by the plant, and most of the water is eventually evaporated to the atmosphere via stomatal openings in the plant leaves. Evapotranspiration can return to the atmosphere most of

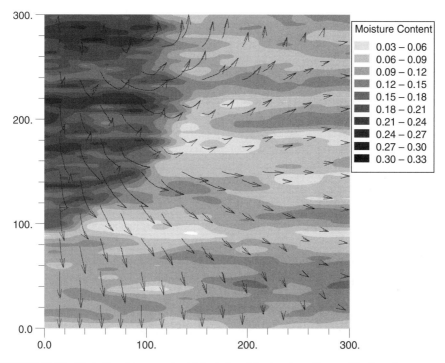

FIGURE 11.10. Plot of moisture content (shaded scale) at a given time, with air pathways super-imposed, for the case of infiltration into a heterogeneous two-dimensional system and air pathway traces (from [17]).

the rain that falls in arid and semiarid regions, and a significant fraction of rainfall in humid regions.

If Richards' assumption is appropriate, then we can describe water movement in the unsaturated zone with active plant roots using Richards' equation with a sink term on the right-hand side:

$$\varepsilon \frac{\partial S_w}{\partial t} + \nabla \cdot \mathbf{q}_w = Q^w. \tag{11.45}$$

This equation assumes compressibility effects are negligible. We already know that the volumetric flux term can be replaced using the Darcy equation for multiphase flow, so that we have an equation with two unknowns, S_w and ψ_w. The relationship between capillary pressure (in this case capillary pressure head, ψ_w) and saturation provides the second equation, and we have a system of two equations and two unknowns. The governing equation may be written with saturation or pressure head as the primary unknown, as shown in Eqs. (11.41a) to (11.41c). However, we now face the task of relating the sink term corresponding to evapotranspiration to the unknowns in the equation, namely, S_w and/or ψ_w. In fact, there are two relatively standard functional forms used to relate this function to either S_w or ψ_w. These two different functional forms are sometimes referred to as Type I and Type II functional representations (see Shani and Dudley [18]).

The simplest approach to root uptake is to relate uptake to saturation. A standard functional form is shown in Figure 11.11, where transpiration is plotted as a function of

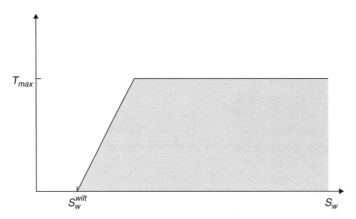

FIGURE 11.11. Typical functional relationship between transpiration and water saturation.

saturation. If saturation is low enough, no water can be taken up by the plant. As saturation increases, more and more water can be taken up by the plant, until it reaches a point where the demand for water is met. Beyond that value of saturation, the transpiration is not constrained by soil moisture, and therefore there is no dependence on S_w. The point below which no water is taken up by the plant is called the *wilting saturation*. The function depicted in Figure 11.11 serves to define the sink term on the right-hand side of Eq. (11.45). While this function is simple in form, there are some subtle points about how it is implemented over the spatially variable values of soil moisture within the root zone. These are discussed in some detail by, for example, Guswa et al. [19].

Evaporation may be represented by a simple saturation- and depth-dependent function included in the sink term Q^w, or an energy balance equation could be solved and phase changes of water from liquid to vapor can be tracked along with transport of water vapor in the air. For the simple case where evaporation is simply included as part of the sink term, a dependence like that shown in Figure 11.11 is sometimes used, with a dependence on depth that weights evaporation heavily close to the surface and then quickly reduces evaporation as depth increases.

While the expression for transpiration as a function of saturation is simple, and is used fairly frequently in hydrological studies, a more physically based approach uses the pressure head as the driving force for water flow into the plant root system. Let the plant be represented as having water within it at a specific potential, call it ψ_{plant}. Then the driving force for water to leave the soil and enter the plant is the difference between the water potential in the soil and the potential in the plant. The resistance to flow is taken to be a function of both the soil resistance (which is the inverse of the soil hydraulic conductivity) and the resistance to flow associated with the plant root. Therefore the flux to the plant roots may be written (e.g. see, [19])

$$T(\psi_w) = \frac{\psi_w - \psi_{\text{plant}}}{r_{\text{soil}} + r_{\text{root}}},$$

$$r_{\text{soil}} = \frac{C_{\text{soil}}}{K(S_w)\rho_{\text{root}}}, \tag{11.46}$$

$$r_{\text{root}} = \frac{C_{\text{root}}}{\rho_{\text{root}}},$$

where r_{soil} and r_{root} denote resistance to flow due to the soil and root, respectively, C_{soil} and C_{root} are constants associated with resistance to flow in the soil and roots, respectively, K is the unsaturated hydraulic conductivity of the soil, and ρ_{root} denotes a measure of the root density within the soil. Notice that this extraction function, $T(\psi_w)$, depends on spatial location within the soil column, since ψ_w and possibly the soil and plant characteristics (e.g., root density) may change with position. In addition, implementation of this functional form requires an overall constraint on the total, or integrated, amount of water that can be extracted from the soil, because it is the integrated (total) amount of water removed that serves to satisfy the demand of the plant. Notice that Eq. (11.46) treats the plant root system as a network of resistors distributed in space, with flux driven by potential differences. For details about this kind of treatment of uptake, see Shani and Dudley [18]. For our purposes, we simply note that functional forms of plant uptake exist in the literature, and therefore the right-hand side function in Q^w can be defined.

In larger-scale models that involve soil moisture, including large-scale atmospheric models, the root zone is often treated as a single effectively homogeneous block, within which a single value of soil moisture is assigned. This, in essence, corresponds to integration of the vertical dimension in a one-dimensional Richards's equation, resulting in a zero space-dimensional model that has only time dependence (see Section 4.5). If we integrate the one-dimensional Richards's equation, we find the following:

$$\int_0^{Z_R} \left(\varepsilon \frac{\partial S_w}{\partial t} + \frac{\partial q}{\partial z} \right) dz = \int_0^{Z_R} Q^w dz.$$

If we denote the spatially averaged saturation by $\overline{S}$, then the integrated equation maybe written

$$\varepsilon Z_R \frac{d\overline{S}}{dt} - q_{top} + q_{bot} = (ET)_{tot},$$

where we have taken z to be positive downward. This zero-dimensional equation is often written equivalently as

$$\varepsilon Z_R \frac{d\overline{S}}{dt} = I(\overline{S}, t) - L(\overline{S}) - T(\overline{S}) - E(\overline{S}),$$

where I denotes infiltration rate, L denotes leakage rate out the bottom of the root zone, T is total transpiration rate (over the entire root zone), and E is total evaporation rate. All of these terms have dimensions of volume per unit area per unit time, or $[LT^{-1}]$, and all are taken to be functions of average root-zone saturation. For this type of equation, with dependencies on saturation, the transpiration function shown in Figure 11.11 would typically be used, with a similar function used for evaporation. The advantage of this zero space-dimensional equation is that analytical expressions can be derived for soil moisture responses to infiltration events. Rodriguez-Iturbe and colleagues [20, 21] have presented a series of solutions to this equation, using random infiltration events to drive the dynamical system. These solutions are being developed for water-limited ecosystems to examine vegetative health under different climatic conditions and to try to understand the role of water in species competition and vegetative development.

11.4 SOLUTION METHODS: MULTIFLUID FLOW EQUATIONS

11.4.1 Simplified Solutions for Unsaturated Flow

Because the governing equation for unsaturated flow is nonlinear, it is generally impossible to derive analytical solutions. However, there are a number of simplifications, or special cases, that have been used to allow relatively simple solutions to be derived. We have already seen such a simplification, in the section discussing ecohydrology. There the one-dimensional Richards' equation was integrated in space to produce a zero space-dimensional, transient equation that governs the dynamics of the spatially averaged soil moisture, where the averaging was applied over the soil root zone. This kind of spatial averaging as a means to simplify equations is used in many ways—recall, for example, the idea of vertical integration of the saturated flow equation applied to essentially horizontal aquifers (Section 4.5). In the case of unsaturated flow, and averaging over the root zone, conditions under which this approach is appropriate are just now being defined in, for example, the recent work of Guswa et al. [19]. Overall, this approach is reasonable when the influence of individual wetting and drying fronts moving through the root zone do not have a significant impact on the evaporation, transpiration, and bottom leakage functions.

If we wish to resolve infiltrating fronts, then we cannot integrate away the vertical dimension. In that case, we have several options for equation simplification. One of the oldest approximations is due to Green and Ampt [22]. The *Green–Ampt approach* approximates the infiltrating water front as a sharp interface, with constant moisture content ahead of, and behind, the front, as shown schematically in Figure 11.12. Let the

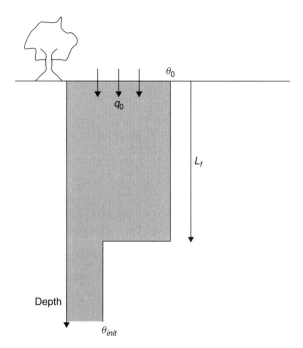

FIGURE 11.12. Schematic of saturation profile associated with Green–Ampt solution for infiltration.

moisture content behind the front be denoted by θ_0, while ahead of the front it is θ_{init}. Assume this jump in the moisture content across the infiltrating front corresponds to a fixed pressure head, call it ψ_f. If the pressure head at the land surface is assumed to be constant at ψ_0, and if the location of the wetting front is a distance $L_f(t)$ from the land surface, then the Darcy equation can be applied by assuming a constant pressure gradient behind the front to yield the flux at the land surface into the soil:

$$q_0 = K(\theta_0) \left(\frac{\psi_0 - \psi_f}{L_f} + 1 \right) = K(\theta_0) \left(\frac{\psi_0 - \psi_f + L_f}{L_f} \right).$$

The water added over a small time interval is $q_0 \, dt$. However, because of the constant nature of the water content behind the front, the volume added is also equal to $(\theta_0 - \theta_{\text{init}}) \, dL_f$. Therefore we have the following differential equation for the location of the infiltration front:

$$(\theta_0 - \theta_{\text{init}}) \, dL_f = q_0 \, dt$$

or

$$(\theta_0 - \theta_{\text{init}}) \frac{dL_f}{dt} = K_0 \left(\frac{\psi_0 - \psi_f + L_f}{L_f} \right),$$

where we use $K_0 \equiv K(\theta_0)$. This equation maybe solved analytically to yield

$$\frac{K_0 t}{(\theta_0 - \theta_{\text{init}})} = L_f - \psi_0 - \psi_f \, ln \left(1 + \frac{L_f}{\psi_0 - \psi_f} \right). \tag{11.47}$$

At sufficiently long times, the second term on the right-hand side of Eq. (11.47) becomes small relative to other terms, and the rate of movement of the infiltrating front approaches a constant, driven by gravity.

Notice that, in the case of horizontal infiltration, the gravity term does not appear in the equation, and the resulting equation for L_f is

$$(\theta_0 - \theta_{\text{init}}) \frac{dL_f}{dt} = K_0 \left(\frac{\psi_0 - \psi_f}{L_f} \right)$$

or

$$L_f \, dL_f = K_0 \left(\frac{\psi_0 - \psi_f}{\theta_0 - \theta_{\text{init}}} \right) dt.$$

This equation is easily solved to find

$$\frac{L_f^2}{2} = K_0 \left(\frac{\psi_0 - \psi_f}{\theta_0 - \theta_{\text{init}}} \right) t \equiv \widehat{D} t$$

or

$$L_f(t) = \sqrt{2 \widehat{D} t}.$$

This last equation indicates that the spatial distance covered by the wetting front increases as the square root of time. The relationship between a linear increase in space and a square-root increase in time, in which the relationship $x/\sqrt{t}$ is a constant, is characteristic of diffusion equations and will be used again shortly.

Note that in the Green–Ampt approximation, the key idea is that the moisture content takes on only two values, with an abrupt jump between those two values occurring at the pressure head value of ψ_f. If we use this idea to construct a relationship between (capillary) pressure head and moisture content, we have a curve that looks like that shown in Figure 11.13. This kind of curve might result from a soil with a single pore size for all pores; therefore the Green–Ampt approximation would be expected to apply to soils that have very uniform pore-size distributions. Also, because the function is a step function, and its derivative is a Dirac delta function, this approach is sometimes referred to as a *delta-function approximation* (e.g., see Philip [23]).

There are many other approaches to derive analytical expressions for infiltration dynamics in unsaturated soils. Some are based on the *Boltzmann transformation*, which takes advantage of the relationship we saw in the Green–Ampt analysis between distance and the square root of time. Others are based on series approximations. As an example of the Boltzmann transformation, consider the θ-based form of *Richards' equation*, with flow restricted to the horizontal plane so that we can ignore gravity. Then we have an equation of the form

$$\frac{\partial \theta}{\partial t} - \frac{\partial}{\partial x}\left(D(\theta)\frac{\partial \theta}{\partial x}\right) = 0, \quad x > 0, \quad t > 0. \tag{11.48}$$

This partial differential equation may be rewritten by introduction of a new variable, so that in terms of the new variable, the governing equation becomes an ordinary differential equation instead of a partial differential equation. Let us define the variable λ by

$$\lambda(\theta) = xt^{-1/2}.$$

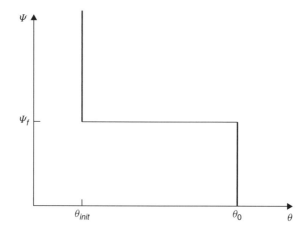

FIGURE 11.13. Plot of idealized relationship between capillary pressure head and moisture content, corresponding to the Green–Ampt formulation. The discontinuity in the curve at $\psi - \psi_f$ leads to the Dirac delta behavior.

If the chain rule of differentiation is then applied to the derivatives in Eq. (11.48), the following equation is obtained:

$$-\frac{\lambda}{2}\frac{d\theta}{d\lambda} = \frac{d}{d\lambda}\left(d(\theta)\frac{d\theta}{d\lambda}\right).$$

Solutions based on this equation have been derived by many researchers (see the discussion in standard books on unsaturated flow, like Hillel [24] and other references therein, for more information).

Finally, we mention the classic work of J. R. Philip, who has derived many analytical and quasianalytical solutions for vertical flow of water in unsaturated soils. Philip is generally credited with the first rigorous solution of Richards' equation for vertical infiltration [24]. His solution was in the form of a power series, given by

$$z(\theta, t) = f_1(\theta)t^{1/2} + f_2(\theta)t + f_3(\theta)t^{3/2} + \cdots + f_n(\theta)t^{n/2} + \cdots.$$

This is an equation that gives the vertical position (z) of a particular value of moisture content (θ) at any time t. It applies to a semi-infinite domain, with a constant initial moisture content, and a fixed moisture content at the land surface ($z = 0$) for all time greater than the initial time. The coefficients f_i are functions of the parameters of the soil (conductivity function and diffusivity function). Note that, at early time, the higher powers of t are small and the solution is dominated by the first term in the series. This implies that the distance traveled is proportional to the square root of time, which is behavior that corresponds to the case without gravity and fits with the Boltzmann transformation idea. At later times, it turns out that the rate of movement approaches a constant, which is again consistent with the observation from the simple Green–Ampt model. From these observations a number of practical results can be derived; the interested reader is referred to Hillel [24] for more details and additional references to related work.

11.4.2 Numerical Solution of Richards' Equation

While analytical solutions for Richards' equation can be derived for special cases, in general the equation is too complicated to solve analytically. Therefore numerical methods are usually employed to solve the equation. Because the equation is nonlinear (the coefficients in the equation are dependent on the unknown, e.g., relative permeability depends on moisture content), iterative methods are often used to calculate numerical solutions. In this section we give a brief overview of numerical approaches to solving Richards' equation (note that numerical solutions are discussed in detail in Chapter 7).

Let us begin with the so-called ψ-based form of the governing equation, written in one spatial dimension,

$$\left[C(\psi_w) + S_w S_s^w\right]\frac{\partial \psi_w}{\partial t} - \frac{\partial}{\partial z}\left[K_{\text{sat}}^w k_{rw}\left(\frac{\partial \psi_w}{\partial z} + 1\right)\right] = 0, \quad 0 < z < L, \quad t > 0.$$

$$(11.49)$$

One advantage of this form of the flow equation is that it applies directly to both saturated and unsaturated zones, so solving problems involving the water table is straightforward (in the saturated zone, $S_w = 1$, $k_{rw} = 1$, and $C = 0$). Of course, this equation must be supplemented with two boundary conditions (one at $z = 0$ and one at $z = L$), as well as an initial condition that provides values of ψ_w at all spatial locations, that is, $\psi_w(z, 0)$. If we discretize the domain in the spatial dimension z using finite-difference cells numbered by the integer counter i, and denote the location of the cell centers by z_i and the size of each cell by δz_i, then Eq. (11.49) may be written in discretized form as

$$
[C(\psi) + SS_s]_i \frac{\partial \psi_i}{\partial t} - \frac{1}{\delta z_i} \left[(K_{sat}k_r)_{i+1/2} \left(\frac{\psi_{i+1} - \psi_i}{\Delta z_i} + 1 \right) \right.
$$
$$
\left. - (K_{sat}k_r)_{i-1/2} \left(\frac{\psi_i - \psi_{i-1}}{\nabla z_i} + 1 \right) \right] = 0, \tag{11.50}
$$

where we use the difference operators defined by $\Delta z_i \equiv z_{i+1} - z_i$ and $\nabla z_i \equiv z_i - z_{i-1}$. Let the unsaturated hydraulic conductivity function be denoted by $K = K_{sat}k_r$. Then note that Eq. (11.50) represents a set of nonlinear ordinary differential equations in time, with the unknowns being the values of pressure head at each spatial node, $\{\psi_i\}_{i=1}^N$, where N is the number of spatial nodes. To solve the set of ordinary differential equations, we discretize in the time dimension by writing a simple approximation with the time derivative evaluated using successive discrete time levels, and the associated spatial derivatives evaluated within the time interval over which the time derivative is evaluated. Let the discrete time levels be denoted by superscript n. Then we begin with the known initial condition (at $n = 0$) and successively calculate the discretized spatial values of pressure head at each successive time level $n = 1, 2, \ldots$. Application of temporal discretization to the spatially discretized equation (Eq. (11.50)) produces the following discrete equation:

$$
[C(\psi) + SS_s]_i^{n+\nu_1} \frac{\psi_i^{n+1} - \psi_i^n}{\Delta t^n} - \frac{1}{\delta z_i} \left[K_{i+1/2}^{n+\nu_2} \left(\frac{\psi_{i+1}^{n+\gamma} - \psi_i^{n+\gamma}}{\Delta z_i} + 1 \right) \right.
$$
$$
\left. - K_{i-1/2}^{n+\nu_2} \left(\frac{\psi_i^{n+\gamma} - \psi_{i-1}^{n+\gamma}}{\nabla z_i} + 1 \right) \right] = 0. \tag{11.51}
$$

In Eq. (11.51), all terms evaluated at time level n are taken to be known, and values at time level $n + 1$ are to be solved for. Any evaluation between discrete time levels n and $n + 1$ would require interpolation based on the discrete values at times n and $n + 1$. The parameters ν_1, ν_2, and γ are used to indicate that the nonlinear coefficients and the spatial derivatives may be evaluated at any point in time between time levels n and $n + 1$, through choice of values $0 \le \nu_1 \le 1$, $0 \le \nu_2 \le 1$, and $0 \le \gamma \le 1$. Usually the parameters ν_1, ν_2, and γ are chosen so that $\nu_1 = \nu_2 = \gamma$, although that is not necessary. Choice of $\nu_1 = \nu_2 = \gamma = 0$ corresponds to an explicit calculation, because all terms in the equation are evaluated at the known time level n except the term ψ_i^{n+1} in the time-derivative approximation. Because each equation (the equation for any given spatial grid cell i) then has only one unknown, it can be solved independently of all other equations; hence it can be solved "explicitly." Any choice of $\gamma > 0$ leads to more than one unknown in each equation, so that the discrete equations become coupled and therefore the set of equations for the spatial values at time $n + 1$ involves solution of a matrix equation. If either $\nu_1 > 0$ or $\nu_2 > 0$, the unknowns in the equations, $\{\psi_i^{n+1}\}_{i=1}^N$, appear nonlinearly and the resulting set of equations is nonlinear. In that case, *iterative solutions techniques are required.*

While explicit methods offer simplicity of calculations, those methods usually have severe stability restrictions on allowable time-step size. Therefore implicit methods are often used to solve the discrete form of Richards' equation. Consider, for illustrative purposes, the choice $\nu_1 = \nu_2 = \gamma = 1$. Let us introduce a second integer superscript, m, to indicate iteration level. Then we may write the discrete version of Richards' equation using a simple iteration approach in which the nonlinear coefficients are evaluated at the new time level, but lagged in the iterations. If the first iteration level ($m = 0$) corresponds to the solution at the previous time level, so that $\psi_i^{n+1,0} = \psi_i^n$ is known, then we initiate the iterative calculation beginning with $m = 0$ and progress to larger integer values of m until the solution converges. For the simple iteration the equation takes the following form:

$$
[C(\psi) + SS_s]_i^{n+1,m} \frac{\psi_i^{n+1,m+1} - \psi_i^n}{\Delta t^n} - \frac{1}{\delta z_i} K_{i+1/2}^{n+1,m} \left(\frac{\psi_{i+1}^{n+1,m+1} - \psi_i^{n+1,m+1}}{\Delta z_i} + 1 \right)
$$

$$
- \frac{1}{\delta z_i} K_{i-1/2}^{n+1,m} \left(\frac{\psi_i^{n+1,m+1} - \psi_{i-1}^{n+1,m+1}}{\nabla z_i} + 1 \right) = 0. \tag{11.52}
$$

It turns out that when this equation is solved, the resulting solutions fail to conserve mass, meaning that when one uses the numerical solution to calculate the net flux entering the domain and compares that to the amount of fluid accumulated within the domain, the two do not match. For some problems, mass balance errors can be quite large, greater than 10%. This is an undesirable situation. The reason that mass is not conserved lies in the discretization of the time derivative. For simplicity, let us ignore compressibility and only focus on the time derivative of moisture content. The original balance equation has as a time derivative $\partial(\theta)/\partial t$, which we expand using the chain rule of differentiation because θ is a function of ψ. That gives $\partial \theta / \partial t = C \partial \psi / \partial t$. This expression is fine in the differential limit, but in an equation like Eq. (11.52), this chain rule is applied in a kind of discrete way. Because the coefficient C is highly nonlinear, the discrete expressions of the chain rule are not equivalent,

$$
\frac{\theta_i^{n+1} - \theta_i^n}{\Delta t} \neq C_i^{n+1} \frac{\psi_i^{n+1} - \psi_i^n}{\Delta t}.
$$

This leads to mass balance errors.

An example calculation can be used to show this problem clearly. It is taken from Celia et al. [25] and involves one-dimensional infiltration into an initially dry soil. The domain is 100 cm in length, the top corresponds to the soil surface and has a boundary condition of $\psi = -75$ cm, the bottom has a fixed pressure boundary condition of $\psi = -1000$ cm, and the entire spatial domain has an initial pressure condition of $\psi = -1000$ cm. Material properties are assigned using van Genuchten functional forms, where

$$
\theta(\psi) = \frac{\theta_s - \theta_r}{\left[1 + (\alpha |\psi|)^n \right]^m} + \theta_r,
$$

$$
K(\psi) = K_{\text{sat}} \frac{\left\{ 1 - (\alpha |\psi|)^{n-1} \left[1 + (\alpha |\psi|)^n \right]^{-m} \right\}^2}{\left[1 + (\alpha |\psi|)^n \right]^{m/2}}.
$$

Parameters used in the simulations correspond to measurements made at a field site in New Mexico and were $\alpha = 0.0335$ cm^{-1}, $\theta_s = 0.368$, $\theta_r = 0.102$, $n = 2$, $m = 0.5$, and

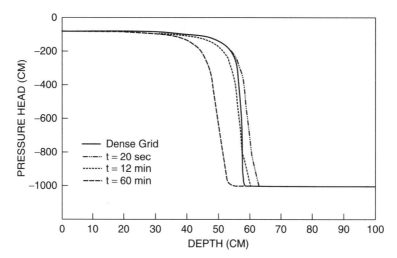

FIGURE 11.14. Finite-difference solutions based on Eq. (11.52), showing sensitivity to time-step size and loss of numerical mass balance (from [25]).

$K_{sat} = 0.00922$ cm/s. Figure 11.14 shows results using the approximation of Eq. (11.52), for different time-step sizes. The plot shows the location of the infiltration front as a function of depth, at a fixed time of 24 hours. The results indicate a strong dependence on time-step size, with the depth of the infiltration front moving progressively upward as Δt increases. This corresponds to an increasing mass balance error in the solution.

There are different options to alleviate this problem. The first is to decide to evaluate the coefficient C differently, so that the resulting product of $C^*(\psi_i^{n+1} - \psi_i^n)/\Delta t$ is required to equal the expression involving moisture content, $(\theta_i^{n+1} - \theta_i^n)/\Delta t$. See Abriola and Rathfelder [26] for more details on this approach. Another possibility is to use the so-called mixed form of Richards' equation, instead of the ψ-based form, and develop discretizations based on that equation. Following this idea, and applying the discretization and iteration methods outlined above for the ψ-based equation, one obtains

$$\frac{\theta_i^{n+1,m+1} - \theta_i^n}{\Delta t^n} - \frac{1}{\delta z_i} K_{i+1/2}^{n+1,m} \left(\frac{\psi_{i+1}^{n+1,m+1} - \psi_i^{n+1,m+1}}{\Delta z_i} + 1 \right)$$
$$+ \frac{1}{\delta z_i} K_{i-1/2}^{n+1,m} \left(\frac{\psi_i^{n+1,m+1} - \psi_{i-1}^{n+1,m+1}}{\nabla z_i} + 1 \right) = 0, \qquad (11.53)$$

where we have ignored the compressibility for simplicity. This equation now possesses the conservative property, meaning that mass is conserved, but we see that there are two primary unknowns at the unknown iteration level, $\theta_i^{n+1,m+1}$ and $\psi_i^{n+1,m+1}$. We need to eliminate one of these. One option is to use a truncated Taylor series to expand θ in terms of ψ, which takes the following form:

$$\theta_i^{n+1,m+1} = \theta_i^{n+1,m} + \left(\frac{d\theta}{d\psi} \right)_i^{n+1,m} \left(\psi_i^{n+1,m+1} - \psi_i^{n+1,m} \right) + \text{H.O.T.} \qquad (11.54)$$

where H.O.T. denotes *higher order terms*, in this case higher orders of the increment in pressure head $\psi_i^{n+1,m+1} - \psi_i^{n+1,m}$, beginning with the second-order term $\left(\psi_i^{n+1,m+1} - \psi_i^{n+1,m}\right)^2$. We ignore the higher-order terms and replace $\theta_i^{n+1,m+1}$ in Eq. (11.53) by the truncated right-hand side of Eq. (11.54), so that the only variables evaluated at the new iteration level are the nodal values of pressure head $\psi_i^{n+1,m+1}$:

$$
\frac{1}{\Delta t^n}\left[\theta_i^{n+1,m} + \left(\frac{d\theta}{d\psi}\right)_i^{n+1,m}\left(\psi_i^{n+1,m+1} - \psi_i^{n+1,m}\right) - \theta_i^n\right]
$$
$$
- \frac{1}{\delta z_i}K_{i+1/2}^{n+1,m}\left(\frac{\psi_{i+1}^{n+1,m+1} - \psi_i^{n+1,m+1}}{\Delta z_i} + 1\right)
$$
$$
+ \frac{1}{\delta z_i}K_{i-1/2}^{n+1,m}\left(\frac{\psi_i^{n+1,m+1} - \psi_{i-1}^{n+1,m+1}}{\nabla z_i} + 1\right) = 0 \qquad (11.55)
$$

Notice that, upon iteration, the time-derivative term recovers $\left(\theta_i^{n+1} - \theta_i^n\right)/\Delta t$, and therefore the converged solution conserves mass.

An example calculation using this algorithm is shown in Figure 11.15, taken from Celia et al. [25]. This is a solution to the same problem as was solved in Figure 11.14, except now the time-derivative approximation is treated in a way that conserves mass. Notice that the dependence on time-step size seen in Figure 11.14 is eliminated in Figure 11.15, and the solutions for different time-step sizes now all conserve mass in the system. The mismatch between the dense-grid solution and the other solutions is caused by the spatial discretization. More details about this procedure, as well as additional ideas about finite-element solutions and how they compare to finite-difference solutions, may be found in Celia et al. [25].

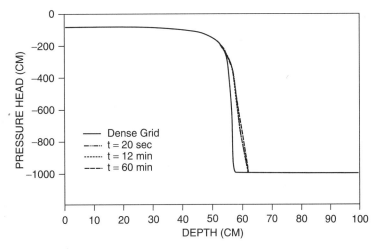

FIGURE 11.15. Finite-difference solutions based on Eq. (11.55), showing proper mass balance (from [25]).

11.4.3 Analytical Solutions for Two-Fluid Flow: The Buckley–Leverett Solution

The idea of the *Buckley–Leverett solution* derives from the classic work of Buckley and Leverett [27]. In this approach, the problem of two-fluid flow in a porous medium is treated as a kind of extended one-fluid problem, using a single *composite* fluid. The general idea is to assign an average pressure to the composite fluid and then write equations to solve for pressure and saturation in much the same way that we solve for pressure (or head) and concentration in groundwater contamination problems. Because the system with multiple-fluid phases is more complicated, the equations are nonlinear, even in simple cases of two-fluid flow in a horizontal column. The form of the resulting nonlinear equations is amenable to analytical solution for some simple cases, and this is the essence of the approach of Buckley and Leverett. Here we will treat a fairly simple case, to show the underlying concept and solution idea. In the process of rewriting the governing equations in a modified form, we introduce a new, derived material function called the *fractional flow function* f_w.

Let us begin with a case of two fluids, each assumed to be incompressible, with no source or sink terms. Then the governing equation for each fluid α is

$$\varepsilon \frac{\partial S_\alpha}{\partial t} + \nabla \cdot \mathbf{q}_\alpha = 0, \quad \alpha = w, nw. \tag{11.56}$$

If Eq. (11.56) is summed over the two fluid phases, the time derivative disappears (because $S_w + S_{nw} = 1$ and therefore $\partial (S_w + S_{nw}) / \partial t = 0$), and we have a simple statement that the total flux vector $\mathbf{q}_T \equiv \mathbf{q}_w + \mathbf{q}_{nw}$ must be divergence-free:

$$\nabla \cdot (\mathbf{q}_w + \mathbf{q}_{nw}) = \nabla \cdot \mathbf{q}_T = 0. \tag{11.57}$$

In the simple one-dimensional case, the equation reduces to

$$\frac{\partial q_w}{\partial x} + \frac{\partial q_{nw}}{\partial x} = \frac{\partial q_T}{\partial x} = 0. \tag{11.58}$$

This equation says that the total flux is constant in space, meaning that the total flux in one-dimensional problems with incompressible fluids can be a function of time only, $q_T = q_T(t)$. We will refer to Eq. (11.57) or (11.58) as the pressure equation, because substitution of the multifluid version of Darcy's law gives an equation with the fluid pressure as the unknown.

If we continue with the idea of a "composite" fluid, we can assign an average pressure to the fluid mixture, defined as some average of the individual fluid pressures. Consider a simple definition of average pressure as the arithmetic average of the two fluid pressures, such that

$$\overline{p} \equiv \tfrac{1}{2} (p_w + p_{nw}).$$

Then we may write the individual fluid pressures as

$$p_w = \overline{p} - \frac{P_c}{2},$$

$$p_{nw} = \overline{p} + \frac{P_c}{2}.$$

Substitution of these expressions into the Darcy equation yields the following expressions for the fluid fluxes as well as the total flux q_T:

$$\mathbf{q}_w = \frac{-\mathbf{k}k_{rw}}{\mu_w} \cdot \left[\mathbf{\nabla}\left(\overline{p} - \tfrac{1}{2}P_c\right) + \rho_w g \,\mathbf{\nabla}z\right],$$

$$\mathbf{q}_{nw} = \frac{-\mathbf{k}k_{rnw}}{\mu_{nw}} \left(\mathbf{\nabla}\left(\overline{p} + \tfrac{1}{2}P_c\right) + \rho_{nw} g \,\mathbf{\nabla}z\right),$$

$$\mathbf{q}_T = -\mathbf{k}\left(\frac{k_{rw}}{\mu_w} + \frac{k_{rnw}}{\mu_{nw}}\right)\cdot\mathbf{\nabla}\overline{p} - \mathbf{k}\left(\frac{k_{rw}}{\mu_w}\rho_w + \frac{k_{rnw}}{\mu_{nw}}\rho_{nw}\right)\cdot g\,\mathbf{\nabla}z$$
$$- \frac{\mathbf{k}}{2}\left(\frac{k_{rnw}}{\mu_{nw}} - \frac{k_{rw}}{\mu_w}\right)\cdot\mathbf{\nabla}P_c.$$

In the last equation, the first term on the right-hand side corresponds to the total flow due to gradients of the average pressure, the second term is the flow due to gravitational forces, and the third term corresponds to flow due to capillary pressure. Because the relative permeability functions, as well as the capillary pressure function, depend nonlinearly on the saturation, this equation is nonlinear in the pressure and saturation.

If we use Eq. (11.57) as one of our governing equations, then as the second equation we should choose one of the original fluid balance equations (Eq. (11.56)). Let us choose the equation for the wetting fluid. The idea of the Buckley–Leverett approach is to rewrite this equation by expanding the expression for the fluid volumetric flux $\mathbf{q}_w$ and relating this flux to the total flux $\mathbf{q}_T$, since we are focusing on the total flux in the pressure equation. At this point, we introduce the fractional flow function, which is based on relative permeabilities (Figure 11.16). We will comment on the physical meaning of this

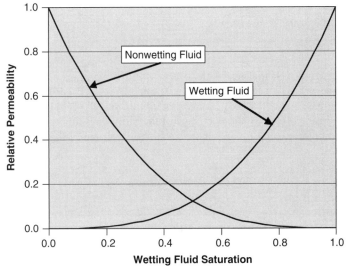

FIGURE 11.16. Relative permeability curves given by $k_{r\alpha}\left(S_\alpha\right) = \left(S_\alpha\right)^3$.

function shortly. For now, we simply give the definition as

$$f_w \equiv \frac{\dfrac{k_{rw}}{\mu_w}}{\dfrac{k_{rw}}{\mu_w} + \dfrac{k_{rnw}}{\mu_{nw}}}. \tag{11.59}$$

Similarly, we define the fractional flow function for the nonwetting fluid as

$$f_{nw} \equiv \frac{\dfrac{k_{rnw}}{\mu_{nw}}}{\dfrac{k_{rw}}{\mu_w} + \dfrac{k_{rnw}}{\mu_{nw}}}. \tag{11.60}$$

A typical form of this function is shown in Figure 11.17, where the fluids are assumed to have equal viscosities, and the relative permeabilities are taken as simple cubic functions of saturation, with $k_{r\alpha} = (S_\alpha)^3$ (Figure 11.16). Notice that the fractional flow curve is nonlinear and is concave upward for lower saturations but passes through an inflection point and then is concave downward for high saturations. Also, the fractional flow curves sum to one, so that $f_w + f_{nw} = 1$.

With this definition of the fractional flow function, we can write, using only algebraic manipulation, the following identity involving $\mathbf{q}_w$ and $\mathbf{q}_T$:

$$\mathbf{q}_w = f_w \left[\mathbf{q}_T + \frac{\mathbf{k} k_{rnw}}{\mu_{nw}} \cdot (\rho_{nw} - \rho_w) \, g \, \nabla z + \frac{\mathbf{k} k_{rnw}}{\mu_{nw}} \cdot \nabla P_c \right]. \tag{11.61}$$

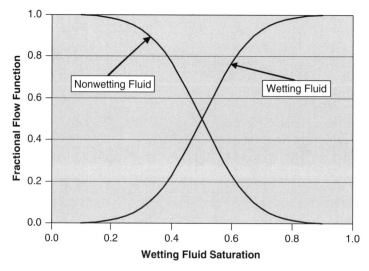

FIGURE 11.17. Fractional flow curves associated with the relative permeability curves of Figure 11.16, and assuming equal viscosities in the wetting and nonwetting fluids.

Substitution of this equation into the fluid balance equation yields the following equation, which we refer to as the *saturation equation*:

$$\varepsilon \frac{\partial S_w}{\partial t} + \nabla \cdot f_w \left[\mathbf{q}_T + \frac{k k_{rnw}}{\mu_{nw}} \cdot (\rho_{nw} - \rho_w) g \, \nabla z + \frac{k k_{rnw}}{\mu_{nw}} \cdot \nabla P_c \right] = 0. \quad (11.62)$$

This saturation equation, coupled with the pressure equation derived earlier, provide two governing equations, derived (without any approximation, just by equation manipulation) from the original fluid balance equations (Eq. (11.56)). To understand the utility of writing the equations in this way, we go to a simple example.

Consider a simple case of one-dimensional two-fluid flow in a horizontal column. In this case, $\nabla z = 0$, so there are no gravity terms that appear in the equations. Furthermore, let us assume that capillary pressure is small relative to the average pressure and can be ignored in the equations. We then obtain the following equations:

$$q_T = q_T(t), \quad (11.63)$$

$$\varepsilon \frac{\partial S_w}{\partial t} + \frac{\partial}{\partial x}(f_w q_T) = \varepsilon \frac{\partial S_w}{\partial t} + q_T \frac{\partial}{\partial x}(f_w) = 0. \quad (11.64)$$

The first equation simply says that the total flux is constant in space. It is derived from Eq. (11.57), which states that the spatial gradient of q_T is zero and therefore cannot be a function of space, but rather is a function only of time. We use that equation to remove the term q_T from the spatial derivative in the saturation equation. The second equation is a first-order, hyperbolic partial differential equation with the unknown being the wetting fluid saturation S_w. It turns out that much is known about first-order hyperbolic equations, even nonlinear cases like this one. Before providing the solution to this equation, it is instructive to analyze some basic characteristics of the equation. Let the derivative of the fractional flow function be expanded by use of the chain rule of differentiation, so that

$$\varepsilon \frac{\partial S_w}{\partial t} + q_T \frac{df_w}{dS_w} \frac{\partial S_w}{\partial x} = 0. \quad (11.65)$$

We can recognize this equation, by analogy to the simple equation for advective transport of a solute, as a kind of advection equation with the "velocity" given by the term $(q_T/\varepsilon)df_w/dS_w$. For comparison, the solute transport case for single-fluid flow corresponds to a linear fractional flow function, $f_w = S_w$. In that case the solute velocity is equal to the fluid velocity, which is simply q/ε. The nonlinear case makes the problem more difficult, because the "velocity" term is a function of saturation, with that functional dependence corresponding to the slope of the $f_w(S_w)$ curve.

Consider a case where the initial condition is given by $S_w = 0$, and the inflow boundary condition (say, at $x = 0$) is $S_w = 1$. In this case, a saturation front moves into the domain from $x = 0$. However, according to the values of df_w/dS_w, the initial step front does not move with speed q_t/ε, as would be the case for a solute in single-fluid flow. Rather, different parts of the curve move at different rates, due to the saturation dependence. We see that low saturations move more slowly than intermediate saturations, because of the varying slope of the fractional flow curve. But such a situation leads to a kind of "overrun" of the infiltrating front, where intermediate saturations overtake lower saturations and produce multivalued solutions for saturation, which is physically impossible. When this condition occurs, the physical result is the development of a shock front, in which an

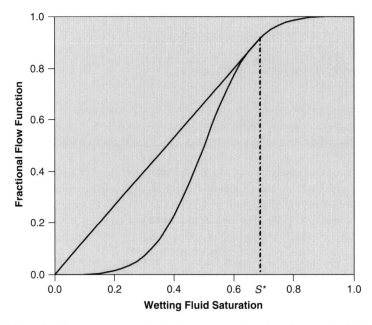

FIGURE 11.18. Tangent curve indicating the saturation limit for the shock front.

abrupt jump in saturation occurs, from a lower limit to an upper limit (Figure 11.18). That shock front then moves into the domain at a fixed velocity. Notice that this behavior does not occur in regions of high saturation, because there the slope of the fractional flow curve decreases as saturation increases (see Figure 11.17). Therefore in high-saturation regions we do not expect to have shock fronts.

It turns out that in the theory of hyperbolic partial differential equations, solutions for equations of the type given in Eq. (11.65) have been derived, and we can simply quote appropriate results here to obtain the solution to our saturation equation. For the initial condition of $S_w = 0$ and the boundary condition of $S_w = 1$, and assuming q_T to be a positive constant in time, then for the domain $x > 0$ we find a shock front defined by the saturation limits $S_w = 0$ and $S_w = S^*$, where S^* is given by the point on the $f_w(S_w)$ curve that is tangent to the straight line drawn through the origin (see Figure 11.18). The speed of the shock front is equal to the slope of this tangent line. All saturations above S^* are in the region where the slope of the fractional flow curve decreases as saturation increases, which forms a nonshock region that is referred to as a *rarefaction wave*. Therefore the solution to the saturation equation, Eq. (11.65), is composed of a shock front and a rarefaction wave. This solution was first presented by Buckley and Leverett [27], and is usually referred to as the Buckley–Leverett solution. This solution is shown schematically in Figure 11.19.

When capillary pressure is included in the analysis, the saturation equation has a capillary pressure term in it. That term may be viewed as a nonlinear diffusion-type term, because its form is

$$\nabla \cdot f_w \frac{\mathbf{k} k_{rnw}}{\mu_{nw}} \cdot \nabla P_c = \nabla \cdot f_w \frac{\mathbf{k} k_{rnw}}{\mu_{nw}} \frac{dS_w}{dP_c} \cdot \nabla S_w = \nabla \cdot \mathbf{D}(S_w) \cdot \nabla S_w.$$

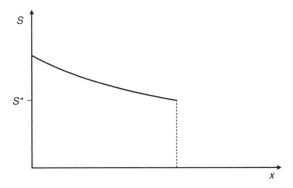

FIGURE 11.19. Typical form of the Buckley–Leverett solution for one-dimensional flow, showing both the shock front and the rarefaction wave.

The so-called capillary diffusion acts like a physical diffusion term, in that it smooths out the shock front. The diffusion function goes to zero at the limiting saturation values of 0 and 1. However, the nonlinear nature of the overall equation means that this diffusion does not act in the way that diffusion acts in the single-fluid solute transport case, where the width of the concentration front increases continuously in proportion to $\sqrt{t}$. In this nonlinear case, the nonlinear advection is always trying to create a shock front, and this tendency toward a shock is balanced by the diffusion, which is trying to spread out the sharp front. These two opposing actions come to a kind of balance, and a frontal structure is achieved that maintains its shape and is transported through the domain.

When fluids have different viscosities, like air and water, the fractional flow curves, as well as the capillary diffusion curve, are skewed toward one end of the saturation scale. For example, Figure 11.20 shows a case of $\mu_w > \mu_{nw}$. Now for infiltration of wetting fluid we would see almost no rarefaction wave and a front that infiltrates as a shock.

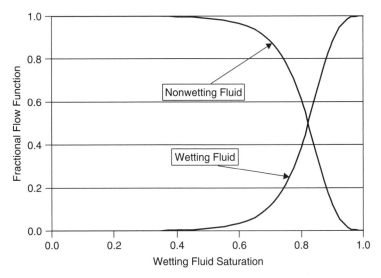

FIGURE 11.20. Fractional flow curves based on the relative permeability functions shown in Figure 11.16, but now with $\mu_{nw} < \mu_w$.

When capillary diffusion is added to the solution, the shock front is smoothed out to some extent, with more smoothing at higher saturations because the capillary diffusion curve $D(S_w)$ also shifts toward higher values of S_w. The result is an infiltration front that looks like the moisture fronts we have seen for unsaturated flow.

11.4.4   Numerical Solutions for Two-Fluid Systems

The IMPES Method   In this section, we outline a simple method for numerical solution of two-fluid problems known as the *implicit pressure explicit saturation*, or *IMPES method*. We also comment on how the IMPES algorithm can be extended to multicomponent fluids, leading us to an introduction of compositional simulations.

The IMPES algorithm solves one equation for pressure in the system. That equation is solved implicitly and is usually derived by summing the individual equations for the fluid phase mass balance. Then one of the fluid phase equations is used to solve for saturation, assuming the fluid fluxes are known from the pressure solution. The assumption of known fluxes allows the saturation equation to be solved explicitly.

The pressure equation is derived by summing the fluid phase equations, so that we arrive at

$$\sum_{\alpha=w,nw} \left[\frac{\partial}{\partial t} (\rho_\alpha \varepsilon S_\alpha) + \nabla \cdot (\rho_\alpha \mathbf{q}_\alpha) \right] = \sum_{\alpha=w,nw} \rho_\alpha Q^\alpha.$$

As we saw in the discussion of the Buckley–Leverett solution, when compressibility effects are negligible, this equation simplifies because the sum of the time-derivative terms is zero. In that case, we have

$$\sum_{\alpha=w,nw} \left[\varepsilon \frac{\partial S_\alpha}{\partial t} + \nabla \cdot \mathbf{q}_\alpha \right] = \phi \frac{\partial}{\partial t} \left(\sum_{\alpha=w,nw} S_\alpha \right) + \nabla \cdot \left(\sum_{\alpha=w,nw} \mathbf{q}_\alpha \right)$$

$$= \nabla \cdot (\mathbf{q}_w + \mathbf{q}_{nw}) = Q^w + Q^{nw}$$

$$= \nabla \cdot \mathbf{q}_T = Q^{\text{tot}}.$$

Note that in this development we have used the fact that the sum of the saturation of all phases is unity and that the sum of the volume fractions is the porosity ϕ. This equation tells us that the divergence of the "total flux" vector, $\mathbf{q}_T$, is equal to the total source or sink term, Q^{tot}.

We can substitute the Darcy expression for each fluid into the summed equation for mass balance. Substitution and subsequent collection of terms result in the following equation:

$$\nabla \cdot \left[-\mathbf{k} \left(\frac{k_{rw}}{\mu_w} + \frac{k_{rnw}}{\mu_{nw}} \right) \cdot \nabla \overline{p} + \mathbf{k} \left(\frac{k_{rw}}{\mu_w} \rho_w + \frac{k_{rnw}}{\mu_{nw}} \rho_{nw} \right) \cdot g \nabla z \right]$$

$$+ \nabla \cdot \frac{\mathbf{k}}{2} \left(\frac{k_{rnw}}{\mu_{nw}} - \frac{k_{rw}}{\mu_w} \right) \cdot \nabla P_c = Q^{\text{tot}}, \tag{11.66}$$

where the average pressure for the two fluids, $\overline{p}$, is taken to equal $(p_w + p_{nw})/2$. While other definitions of average pressure are possible, and some may be preferred (e.g. see Chavent and Jaffre [28]), for now let us accept this simple definition for average fluid pressure.

This equation may be rewritten to highlight the average pressure term by bringing the other terms to the right-hand side of the equation, so that

$$\nabla \cdot \mathbf{k} \left(\frac{k_{rw}}{\mu_w} + \frac{k_{rnw}}{\mu_{nw}} \right) \cdot \nabla \overline{p} = - \nabla \cdot \mathbf{k} \left(\frac{k_{rw}}{\mu_w} \rho_w + \frac{k_{rnw}}{\mu_{nw}} \rho_{nw} \right) \cdot g \nabla z$$

$$+ \nabla \cdot \frac{\mathbf{k}}{2} \left(\frac{k_{rnw}}{\mu_{nw}} - \frac{k_{rw}}{\mu_w} \right) \cdot \nabla P_c - Q^{\text{tot}}. \qquad (11.67)$$

As a second equation, we take the original mass balance equation for one of the fluids, say, the wetting fluid, and write it to highlight the changes in saturation with time:

$$\frac{\partial}{\partial t} (\rho_w \varepsilon S_w) = - \nabla \cdot (\rho_w \mathbf{q}_w) + \rho_w Q^w. \qquad (11.68)$$

Recall from Chapter 7 that an implicit time-marching calculation is one that involves discrete equations in which more than one unknown appears. Therefore in order to calculate numerical solutions at the next discrete time level, equations written for all of the spatial locations must be solved simultaneously. Such a time-marching algorithm is said to be implicit, because in any given equation the unknowns can only be solved implicitly in terms of other unknowns. Implicit methods usually involve approximations in which spatial derivative terms in the governing differential equation are evaluated at the new time level. Conversely, time-marching algorithms in which only one unknown appears in any equation do not require simultaneous solution for the unknowns, but instead the one unknown in an equation may be obtained explicitly. These kinds of time-marching algorithms are called explicit methods.

The IMPES method uses an implicit calculation to determine pressure, then uses an explicit calculation to determine new values of saturation. The algorithm proceeds along the following lines. First, discretize the spatial domain using finite-difference points or cells. For each discrete spatial location, write an approximation for the "pressure equation," Eq. (11.67). On a standard rectangular grid in one, two, or three dimensions, the result for diagonal permeability tensors would be three, five, or seven unknowns per discrete equation, respectively. However, the coefficients and the right-hand side terms in Eq. (11.67) are nonlinear functions of saturation, so their evaluation complicates the solution. In the IMPES algorithm, the saturation values from the previous time level are used to evaluate all coefficients and right-hand side terms in the equation. Therefore the only terms evaluated at the new time level are the average pressures at each finite-difference node. This set of linear algebraic equations for the discrete values of average pressure are solved by standard matrix methods, and the solution thereby provides numerical approximations for the pressure at the next time level.

The second step in the IMPES approach is to take the new values of pressure and insert them into the Darcy equation for fluid fluxes to calculate numerical values of fluid fluxes at the new time level. These are then used on the right-hand side of Eq. (11.68). With those values taken as knowns, the spatial-derivative terms in Eq. (11.68) are all known, and the only unknown in any finite-difference equation is the saturation at the finite-difference node at which the discrete equation is written. Therefore only one unknown appears in any discrete equation, so the calculation of saturation at the new time level is performed explicitly. Once the new saturation values are obtained for all nodes in the spatial domain, all coefficients that are nonlinear functions of saturation are updated to the new time level, including all relative permeability terms and the capillary pressure term. The time-step counter is then incremented by one, and the pressure and saturation calculations are repeated for the new time step. In this way, the calculation proceeds forward without iteration, with pressure calculated implicitly and saturation calculated explicitly. This, in outline form, is the basic idea of the IMPES method.

Compositional Simulations   There are many instances in which the fluids involved in the flow system are composed of multiple components, and our interest is in one or more of the component concentrations. In this case, knowledge about the overall fluid distribution may not be sufficient, and we may need to analyze the transport of individual components. When any mass exchange takes place involving transfer of a component between fluids, or between a fluid and the solid, we may use a systematic approach to writing and solving the transport equations of interest. The approach follows from our earlier discussion of mass transfer among fluids and phases, in which we observed that summation of component equations over fluid and solid phases was a reasonable approach. This avoids problems with very fast time scales associated with equilibrium partitioning. For numerical calculations, we can extend the IMPES concept to include multicomponent fluids and solids. The simplest approach is the following. First, sum the phase equations as in the standard IMPES approach, to derive an overall pressure equation. Again we refer to this as the pressure equation. Second, for each component of interest, write a summed mass balance equation, where summation is over all fluids and solids in the system. Summation of Eq. (11.25) results in an equation of the form

$$\sum_{\alpha=w,nw,s} \frac{\partial}{\partial t} \left(\rho_\alpha \varepsilon S_\alpha \omega_i^\alpha \right) + \nabla \cdot \left(\sum_{\alpha=w,nw,s} \mathbf{m}_i^\alpha \right) = \sum_{\alpha=w,nw,s} \rho_\alpha \varepsilon S_\alpha Q_i^\alpha,$$

where the summation is performed over all fluids and solids (in this case we assume only two fluids and one solid), and the mass flux vector $\mathbf{m}_i^\alpha$ includes advective and diffusive/dispersive fluxes (see Eq. (11.24)). Let this equation be discretized, so that it is applied to a grid cell. Then the first term may be seen as the time rate of change of the total amount of substance i within the grid cell, the second term is the net rate of accumulation of substance i within the cell due to advective, diffusive, and dispersive fluxes carrying the substance into or out of the cell through the boundaries of the cell, and the third term is the total mass added to or subtracted from the cell due to external sources or sinks. From this point of view, the unknowns in the system become the average pressure in each grid cell, and the total mass of substance i within each grid cell. However, in this case the update steps are a bit more complicated, because in order to know the compositions of the individual fluids, the total amount of substance i in a grid cell must be partitioned among the different fluids and perhaps the solid. This requires thermodynamic relationships that relate the relative amounts in each fluid or solid to the environmental variables of the system, such as system temperature and pressure, as well as other geochemical factors. The equations that describe this partitioning are nonlinear algebraic equations that can be difficult to solve. The overall process of partitioning the total amount of a substance into the different fluids is often called a flash calculation. Once this flash calculation is performed, all system properties can be updated, including fluid compositions, and from those fluid densities and viscosities. Once the compositions are known, the relative fluid phase amounts can also be determined, and thereby fluid saturation can be inferred.

Obviously multicomponent, multifluid systems can become quite complicated, and the details of the approach are beyond the scope of this book. But the overall idea should be clear from our simple discussion about the general approach. Note also that the noniterative extended IMPES method will eventually break down and give poor results as the system complexity increases, and coupled iterative methods would then be required.

11.5 SUMMARY

The focus of this chapter is the extension of earlier single-fluid flow and mass transport concepts to the case of multiple-fluid flow and transport. The chapter begins with a discussion of the two most commonly encountered examples of multiple-fluid flow, that is, unsaturated flow and nonaqueous phase liquid (NAPL) flow. Having defined these problems, we next developed the equations that govern the behavior of such systems, including the transport of multiple components in each fluid. Next, we addressed the area of unsaturated flow in greater detail, highlighting the role of Richards' equation, which is fundamental to air–water systems. In the final section we considered methods available to solve the multiphase flow equation, beginning with the relatively simple air–water system and expanding into the more complex multifluid flow case.

11.6 PROBLEMS

11.1. Consider a spill of oil at the land surface. The oil seeps into the ground and interacts with the resident fluids in the subsurface. How many fluid phases are involved in such a problem? Write the general governing equation for each of the phases, explaining what each term in the equation means.

11.2. Explain why the relative permeability coefficient is a function of fluid saturation. Then explain why the relative permeability coefficient always has values between 0 and 1.

11.3. The relative permeability function is almost always nonlinear. An example is Eq. (11.17), where the relative permeability is defined to be a power of the saturation, where the exponent is often chosen to be 3. Explain why the relative permeability function is nonlinear.

11.4. The capillary pressure as a function of saturation is both nonlinear and hysteretic. Explain why the functional relationship between P_c and S_w is hysteretic.

11.5. Consider a simple example of a *pore-scale network model*, which is composed of *pore bodies* and *pore throats*. As a simple example, consider a rectangular network in two dimensions, with the network having 25 pore bodies, each residing on a rectangular grid that is 5 by 5. Assume the pore bodies are connected to neighboring pore bodies by pore throats, so that the entire network forms a simple rectangular pattern. Assume all of the pore bodies are larger than all of the pore throats. Also assume that each pore body has a different size (chosen from some characteristic probability distribution for pore bodies), and that each pore throat also has a unique size, chosen from a characteristic probability distribution for pore throats. Assume that fluid–fluid interfaces follow the rule stated in Eq. (11.10), and illustrated in Figure 11.3. Next, assume the simple network you have created represents the porous-medium sample illustrated in Figure 11.4. Diagrammatically, with a nonwetting fluid reservoir on one end and a wetting fluid reservoir on the other (as illustrated in Figure 11.4), show how changes in capillary pressure (i.e., changes in the imposed reservoir pressures) lead to changes in fluid saturation within the porous medium. You should also observe that, given the size differences between pore bodies and pore throats, pore throats control drainage while pore bodies control imbibition. Use this observation to explain the hysteresis seen in the relationships between P_c and S_w.

11.6. Assume you wish to write a mathematical model of the unsaturated zone, which can also include zones of full saturation. Assume a one-dimensional description of the system is reasonable. Of the three equations: (11.42), (11.43), and (11.44), which is most appropriate? Which (if any) of the three is not appropriate? Explain your answers. Is any of Eq. (11.41a), (11.41b), or (11.41c) more appropriate than the others? Explain.

11.7. Consider Figures 11.7 and 11.9. Explain why the traces of air particle movements are consistent with the boundary conditions imposed at the top and bottom of the columns.

11.8. If the pressure in the air phase of an unsaturated soil remains essentially at atmospheric pressure, is it then reasonable to conclude that the air does not move? Explain.

11.9. List the major assumptions associated with the Green–Ampt solution. Is this solution more likely to apply to soils of uniform grain size or soils with very nonuniform grain size distributions?

11.10. Consider a two-fluid porous medium. For relative permeability functions that are equal to the saturation raised to the third power (Eq. (11.17) with $n = 3$), write the equations for the fractional flow functions. Show how the fractional flow functions depend on the viscosities of the two fluids.

BIBLIOGRAPHY

[1] J. C. Dooge, On the study of water, *Hydrol. Sci. J.* **28**(1): 23, 1983.

[2] F. A. L. Dullien, *Porous Media: Fluid Transport and Pore Structure*, 2nd ed., Academic Press, San Diego, CA, 1992.

[3] M. Sahimi, *Flow and Transport in Porous Media and Fractured Rocks*, VCH, Weinheim, 1995.

[4] N. R. Morrow (ed.), *Interfacial Phenomena in Petroleum Recovery*, Surfactant Science Series, Vol. 36, Marcel Dekker, New York, 1991.

[5] S. M. Hassanizadeh and W. G. Gray, Toward an improved description of the physics of two-phase flow, *Adv. Water Resour.* **16**: 53, 1993.

[6] R. J. Held and M. A. Celia, Modeling support of functional relationships between capillary pressure, saturation, interfacial areas, and common lines, *Adv. Water Resour.* **24**: 325, 2001.

[7] S. M. Hassanizadeh, M. A. Celia, and H. K. Dahle, Dynamic effects in the capillary pressure–saturation relationship and their impacts on unsaturated flow, *Vadose Zone Hydrol.* **1**: 38, 2002.

[8] R. H. Brooks and A. T. Corey, Properties of porous media affecting fluid flow, *ASAE J. Irr. Drainage Div.* **IR2**: 61, 1966.

[9] M. Th. van Genuchten, A closed-form equation for predicting the hydraulic conductivity of unsaturated soil, *Soil Sci. Soc. Am. J.* **44**: 892, 1980.

[10] N. T. Burdine, Relative permeability calculations from pore-size data, *Trans. AIME* **198**: 71, 1953.

[11] Y. Mualem, A new model for predicting the hydraulic conductivity of unsaturated porous media, *Water Resour. Res.* **12**: 513, 1976.

[12] J. C. Parker, R. J. Lenhard, and T. Kuppusamy, A parametric model for constitutive properties governing multiphase flow in porous media, *Water Resour. Res.* **23**: 618, 1987.

[13] P. C. Lichtner, Continuum formulation of multicomponent–multiphase reactive transport, in P. C. Lichtner, C. I. Steefel, and E. H. Oelkers (eds.), *Reactive Transport in Porous Media*, Reviews in Mineralogy Series, Vol. 34, p. 1, Mineralogical Society of America, Washington, DC, 1996.

[14] L. A. Richards, Capillary conduction of liquids through porous medium, *Physics* **1**: 318, 1931.

[15] M. A. Celia and P. Binning, A mass conservative solution for two-phase flow in porous media with application to unsaturated flow, *Water Resour. Res.* **28**(10): 2819, 1992.

[16] J. Touma and M. Vauclin, Experimental and numerical analysis of two-phase infiltration in a partially saturated soil, *Transport in Porous Media* **1**: 22, 1986.

[17] P. J. Binning, *Modeling Unsaturated Zone Flow and Transport in the Air and Water Phases*, Ph.D. Thesis, Department of Civil Engineering and Operations Research, Princeton University, 1994.

[18] U. Shani and M. Dudley, Modeling water uptake by roots under water and salt stress: soil-based and crop response root sink term, in Y. Waisel, A. Eshel, and U. Kafkafi (eds.), *Plant Roots: The Hidden Half*, 2nd ed., Marcel Dekker, New York, 1996, p. 635.

[19] A. J. Guswa, M. A. Celia, and I. Rodriguez-Iturbe, Models of soil moisture dynamics in ecohydrology: a comparative study, *Water Resour. Res.* **38**(9): 1166, 2002.

[20] I. Rodriguez-Iturbe, A. Porporato, F. Laio, and R. Ridolfi, Plants in water-controlled ecosystems: active role in hydrologic processes and response to water stress, I. Scope and general outline, *Adv. Water Resour.* **24**(7): 695, 2001.

[21] F. Laio, A. Porporato, L. Ridolfi, and I. Rodriguez-Iturbe, Plants in water-controlled ecosystems: active role in hydrologic processes and response to water stress, III. Vegetation water stress, *Adv. Water Resour.* **24**(7): 725, 2001.

[22] W. A. Green and G. A. Ampt, Studies on soil physics, 1. The flow of air and water through soils, *J. Agric. Sci.* **4**: 1024, 1911.

[23] J. R. Philip, Absorption and infiltration in two- and three-dimensional systems, in R. E. Rijtema and H. Wassink (eds.), *Water in the Unsaturated Zone*, IAHS/UNESCO Symposium, Wageningen, The Netherlands, Vol. 2, p. 503, 1966.

[24] D. Hillel, *Environmental Soil Physics*, Academic Press, San Diego, CA, 1998.

[25] M. A. Celia, E. T. Boul003as, and R. L. Zarba, A general mass conservative numerical solution for the unsaturated flow equation, *Water Resour. Res.* **26**(7): 1483, 1990.

[26] L. M. Abriola and K. Rathfelder, Mass balance errors in modelling two-phase immiscible flows: causes and remedies, *Adv. Water Resour.* **16**: 223, 1993.

[27] S. E. Buckley and M. C. Leverett, Mechanism of fluid displacement in sands, *Trans. AIME* **146**: 107, 1942.

[28] G. Chavent and J. Jaffre, *Mathematical Models and Finite Elements for Reservoir Simulation*, North Holland, Amsterdam, 1986.

[29] R. H. Brooks and A. T. Corey, *Hydraulic Properties of Porous Media, Hydrology Paper 3*, Colorado State University, Fort Collins, CO, 1964.

INDEX